T0137127

Advances in Intelligent Systems and Computing

Volume 771

Series editor

Janusz Kacprzyk, Polish Academy of Sciences, Warsaw, Poland
e-mail: kacprzyk@ibspan.waw.pl

The series "Advances in Intelligent Systems and Computing" contains publications on theory, applications, and design methods of Intelligent Systems and Intelligent Computing. Virtually all disciplines such as engineering, natural sciences, computer and information science, ICT, economics, business, e-commerce, environment, healthcare, life science are covered. The list of topics spans all the areas of modern intelligent systems and computing such as: computational intelligence, soft computing including neural networks, fuzzy systems, evolutionary computing and the fusion of these paradigms, social intelligence, ambient intelligence, computational neuroscience, artificial life, virtual worlds and society, cognitive science and systems, Perception and Vision, DNA and immune based systems, self-organizing and adaptive systems, e-Learning and teaching, human-centered and human-centric computing, recommender systems, intelligent control, robotics and mechatronics including human-machine teaming, knowledge-based paradigms, learning paradigms, machine ethics, intelligent data analysis, knowledge management, intelligent agents, intelligent decision making and support, intelligent network security, trust management, interactive entertainment, Web intelligence and multimedia.

The publications within "Advances in Intelligent Systems and Computing" are primarily proceedings of important conferences, symposia and congresses. They cover significant recent developments in the field, both of a foundational and applicable character. An important characteristic feature of the series is the short publication time and world-wide distribution. This permits a rapid and broad dissemination of research results.

More information about this series at http://www.springer.com/series/11156

Manuel Graña · José Manuel López-Guede
Oier Etxaniz · Álvaro Herrero
José Antonio Sáez · Héctor Quintián
Emilio Corchado
Editors

International Joint Conference SOCO'18-CISIS'18-ICEUTE'18

San Sebastián, Spain, June 6–8, 2018, Proceedings

Editors
Manuel Graña
Computational Intelligence Group
University of the Basque Country
Sarriena, Vizcaya
Spain

José Manuel López-Guede
Computational Intelligence Group
University of the Basque Country
Sarriena, Vizcaya
Spain

Oier Etxaniz
Computational Intelligence Group
University of the Basque Country
Sarriena, Vizcaya
Spain

Álvaro Herrero
Department of Civil Engineering
University of Burgos
Burgos, Burgos
Spain

José Antonio Sáez
University of Salamanca
Salamanca, Salamanca
Spain

Héctor Quintián
Department of Industrial Engineering
University of A Coruña
A Coruña, La Coruña
Spain

Emilio Corchado
University of Salamanca
Salamanca, Salamanca
Spain

ISSN 2194-5357 ISSN 2194-5365 (electronic)
Advances in Intelligent Systems and Computing
ISBN 978-3-319-94119-6 ISBN 978-3-319-94120-2 (eBook)
https://doi.org/10.1007/978-3-319-94120-2

Library of Congress Control Number: 2018946625

Printed on acid-free paper

This Springer imprint is published by the registered company Springer International Publishing AG part of Springer Nature
The registered company address is: Gewerbestrasse 11, 6330 Cham, Switzerland

Preface

This volume of Advances in Intelligent and Soft Computing contains accepted papers presented at SOCO 2018, CISIS 2018, and ICEUTE 2018; all conferences held in the beautiful and historic city of San Sebastian (Spain), in June 2018.

Soft computing represents a collection or set of computational techniques in machine learning, computer science, and some engineering disciplines which investigate, simulate, and analyze very complex issues and phenomena.

After a thorough peer review process, the 13th SOCO 2018 International Program Committee selected 41 papers which are published in these conference proceedings and represent an acceptance rate of 45%. In this relevant edition, a special emphasis was put on the organization of special sessions. Two special sessions were organized related to relevant topics as: Optimization, Modeling and Control by Soft Computing Techniques and Soft Computing Applications in the Field of Industrial and Environmental Enterprises.

The aim of the 11th CISIS 2018 conference is to offer a meeting opportunity for academic- and industry-related researchers belonging to the various, vast communities of computational intelligence, information security, and data mining. The need for intelligent, flexible behavior by large, complex systems, especially in mission-critical domains, is intended to be the catalyst and the aggregation stimulus for the overall event.

After a thorough peer review process, the CISIS 2018 International Program Committee selected 8 papers which are published in these conference proceedings achieving an acceptance rate of 40%.

In the case of 9th ICEUTE 2018, the International Program Committee selected 11 papers, which are published in these conference proceedings.

The selection of papers was extremely rigorous in order to maintain the high quality of the conference, and we would like to thank the members of the Program Committees for their hard work in the reviewing process. This is a crucial process to the creation of a high standard conference, and the SOCO, CISIS, and ICEUTE conferences would not exist without their help.

SOCO'18, CISIS'18, and ICEUTE'18 enjoyed outstanding keynote speeches by distinguished guest speakers: Prof. Hujun Yin—The University of Manchester (UK), Prof. Maya Dimitrova—St. Petersburg University (Russia), Prof. Iván Macía Oliver—Director of the eHealth and Biomedical Applications Area of Vicomtech (Spain).

SOCO'18 has teamed up with Cybernetics and Systems: An International Journal (Taylor and Francis), Expert Systems (Whiley) and the J. Applied Logics—IfCoLog Journal (College Publications) for a suite of special issue including selected papers from SOCO'18.

For this CISIS'18 special edition, as a follow-up of the conference, we anticipate further publication of selected papers in one special issue in the prestigious Logic Journal of the IGPL (Oxford Academic).

Particular thanks go as well to the conference main sponsors: Startup Ole, and University of Salamanca, University of Basque Country who jointly contributed in an active and constructive manner to the success of this initiative.

We would like to thank all the special session organizers, contributing authors, as well as the members of the Program Committees and the Local Organizing Committee for their hard and highly valuable work. Their work has helped to contribute to the success of the SOCO 2018, CISIS 2018, and ICEUTE 2018 events.

June 2018

Manuel Graña
José Manuel López-Guede
Oier Etxaniz
Álvaro Herrero
José Antonio Sáez
Héctor Quintián
Emilio Corchado

SOCO 2018

Organization

General Chairs

Manuel Graña Romay	University of Basque Country, Spain
Emilio Corchado	University of Salamanca, Spain

International Advisory Committee

Ashraf Saad	Armstrong Atlantic State University, USA
Amy Neustein	Linguistic Technology Systems, USA
Ajith Abraham	Machine Intelligence Research Labs—MIR Labs, Europe
Jon G. Hall	The Open University, UK
Paulo Novais	Universidade do Minho, Portugal
Amparo Alonso Betanzos	President Spanish Association for Artificial Intelligence (AEPIA), Spain
Michael Gabbay	Kings College London, UK
Aditya Ghose	University of Wollongong, Australia
Saeid Nahavandi	Deakin University, Australia
Henri Pierreval	LIMOS UMR CNRS 6158 IFMA, France

Program Committee Chairs

Emilio Corchado	University of Salamanca, Spain
Manuel Graña Romay	University of Basque Country, Spain
Álvaro Herrero	University of Burgos, Spain
Héctor Quintián	University of A Coruña, Spain

Program Committee

Emilio Corchado (Chair)	University of Salamanca, Spain
Manuel Graña Romay (Co-chair)	University of Basque Country, Spain
Adolfo R. De Soto	University of Leon, Spain
Alicia Troncoso	Universidad Pablo de Olavide, Spain
Andreea Vescan	Babes-Bolyai University, Romania
Andres Pinon	University of Coruna, Spain
Angel Arroyo	University of Burgos, Spain
Anna Bartkowiak	University of Wroclaw, Poland
Anna Burduk	Wrocław University of Technology, Poland
Anton Koval	Zhytomyr State Technological University, Ukraine
Antonio Bahamonde	University of Oviedo at Gijón, Spain
Bozena Skolud	Silesian University of Technology, Poland
Camelia Pintea	Technical University of Cluj-Napoca, Romania
Carlos Pereira	ISEC, Portugal
Carmen Benavides	University of León, Spain
Damian Krenczyk	Silesian University of Technology, Poland
Daniela Perdukova	Technical University of Kosice, Slovakia
Daniela Zaharie	West University of Timisoara, Romania
David Alvarez Leon	University of León, Spain
David Griol	University Carlos III of Madrid, Spain
Dilip Pratihar	Indian Institute of Technology, India
Dragan Simic	University of Novi Sad, Serbia
Eduardo Solteiro Pires	UTAD University, Portugal
Eleni Mangina	University College Dublin, Ireland
Eloy Irigoyen	UPV/EHU, Spain
Enrique De La Cal Marín	University of Oviedo, Spain
Enrique Dominguez	University of Malaga, Spain
Enrique Onieva	University of Deusto, Spain
Esteban García-Cuesta	Universidad Europea de Madrid, Spain
Esteban Jove	University of A Coruña, Spain
Eva Volna	University of Ostrava, Czech Republic
Fanny Klett	German Workforce ADL Partnership Laboratory, Germany
Fernando Sanchez Lasheras	Universidad de Oviedo, Spain
Florentino Fdez-Riverola	University of Vigo, Spain
Francisco Martínez-Álvarez	Universidad Pablo de Olavide, Spain
Francisco Moreno	University National of Colombia, Colombia
George Georgoulas	TEI of Epiruw, Greece
Georgios Ch. Sirakoulis	Democritus University of Thrace, Greece
Grzegorz Ćwikła	Silesian University of Technology, Poland

Héctor Quintián	University of A Coruña, Spain
Henri Pierreval	LIMOS-IFMA, France
Isaias Garcia	University of Leon, Spain
Ivan Ubero Martinez	University of Leon, Spain
Iwona Pisz	Opole University, Poland
Jaime A. Rincon	Universitat Politècnica de València, Spain
Javier Alfonso	University of Leon, Spain
Jesús D. Santos	University of Oviedo, Spain
Jiri Pospichal	University of Ss Cyril and Methodius, Slovakia
Jorge García-Gutiérrez	University of Seville, Spain
Jose Gamez	University of Castilla-La Mancha, Spain
José Valente de Oliveira	Universidade do Algarve, Portugal
Jose Alfredo Ferreira Costa	Federal University, UFRN, Brazil
José Antonio Sáez	University of Salamanca, Spain
Jose Luis Calvo-Rolle	University of A Coruña, Spain
José Luis Casteleiro-Roca	University of A Coruña, Spain
Jose M. Molina	Universidad Carlos III de Madrid, Spain
Jose Manuel Lopez-Guede	University of Basque Country, Spain
José Ramón Villar	University of Oviedo, Spain
Juan Gomez Romero	University of Granada, Spain
Juan Mendez	Universidad de La Laguna, Spain
Krzysztof Kalinowski	Silesian University of Technology, Poland
Leocadio G. Casado	University of Almeria, Spain
Lidia Sánchez González	Universidad de Leon, Spain
Luciano Alonso	University of Cantabria, Spain
Luís Nunes	Instituto Universitário de Lisboa (ISCTE-IUL), Portugal
Luis Alfonso Fernández Serantes	FH Joanneum University of Applied Sciences, Austria
Luis Paulo Reis	University of Minho, Portugal
M. Chadli	University of Picardie Jules Verne, France
Maciej Grzenda	Warsaw University of Technology, Poland
Manuel Castejon Limas	Universidad de Leon, Spain
Manuel Graña	University of Basque Country, Spain
Manuel Mejia-Lavalle	Cenidet, Mexico
Marcin Iwanowski	Warsaw University of Technology, Poland
Marcin Paprzycki	IBS PAN and WSM, Poland
Maria Luisa Sanchez	Universidad de Oviedo, Spain
María N. Moreno García	University of Salamanca, Spain
Marius Balas	Aurel Vlaicu University of Arad, Romania
Matilde Santos	Universidad Complutense de Madrid, Spain
Mehmet Emin Aydin	University of the West of England, UK
Michal Wozniak	Wroclaw University of Technology, Poland
Miguel Carriegos	RIASC, Spain
Mitiche Lahcene	University of Djelfa, Djelfa, Algeria

Oier Etxaniz	University of the Basque Country, Spain
Paul Eric Dossou	Icam, France
Paulo Moura Oliveira	UTAD University, Portugal
Paulo Novais	University of Minho, Portugal
Petr Dolezel	University of Pardubice, Czech Republic
Przemyslaw Korytkowski	West Pomeranian University of Technology in Szczecin, Poland
Reggie Davidrajuh	University of Stavanger, Norway
Richard Duro	University of A Coruña, Spain
Robert Burduk	Wroclaw University of Technology, Poland
Rosa Basagoiti	Mondragon University, Spain
Rui Sousa	University of Minho, Portugal
Sebastian Saniuk	University of Zielona Gora, Poland
Sebastián Ventura	University of Cordoba, Spain
Soco Conference	University of Salamanca, Spain
Stefano Pizzuti	ENEA, Italy
Sung-Bae Cho	Yonsei University, South Korea
Tzung-Pei Hong	National University of Kaohsiung, Taiwan
Urko Zurutuza	Mondragon University, Spain
Valeriu Manuel Ionescu	University of Pitesti, Romania
Vicente Matellan	University of Leon, Spain
Wei-Chiang Hong	Jiangsu Normal University, China
Wilfried Elmenreich	Alpen-Adria-Universität Klagenfurt, Austria
Zita Vale	GECAD/ISEP/IPP, Portugal

Special Sessions

Optimization, Modeling and Control by Soft Computing Techniques

Program Committee

Eloy Irigoyen (Organizer)	Universidad del País Vasco, Spain
Matilde Santos (Organizer)	Universidad Complutense de Madrid, Spain
José Luis Calvo-Rolle (Organizer)	University of A Coruña, Spain
Mikel Larrea (Organizer)	University of Basque Country, Spain
Luciano Alonso	University of Cantabria, Spain
Antonio Javier Barragán	Universidad de Huelva, Spain
José Luis Casteleiro-Roca	University of A Coruña, Spain
Oscar Castillo	Tijuana Institute of Technology, Mexico
José Luís Diez Ruano	Universitat Politècnica de València, Spain
Vicente Gomez-Garay	UPV/EHU, Spain
Albeto Herreros López	University of Valladolid, Spain

Esteban Jove	University of A Coruña, Spain
Anton Koval	Zhytomyr State Technological University, Ukraine
Luis Magdalena	Universidad Politécnica de Madrid, Spain
Javier Sanchis Saez	Universitat Politècnica de València, Spain
Maria Tomas-Rodriguez	The City University of London, UK

Soft Computing Applications in the Field of Industrial and Environmental Enterprises

Program Committee

Álvaro Herrero (Organizer)	University of Burgos, Spain
Alfredo Jiménez (Organizer)	KEDGE Business School, Spain
Angel Arroyo	University of Burgos, Spain
Jose Luis Calvo-Rolle	University of A Coruña, Spain
José Luis Casteleiro	University of A Coruña, Spain
Camelia Chira	Babes-Bolyai University, Romania
Sung-Bae Cho	Yonsei University, South Korea
Leticia Curiel	University of Burgos, Spain
David Griol	Universidad Carlos III de Madrid, Spain
Montserrat Jimenez	University Rey Juan Carlos I, Spain
Martin Macas	Czech Technical University in Prague, Czech Republic
Julio César Puche Regaliza	University of Burgos, Spain
Raquel Redondo	University of Burgos, Spain
Mercedes Rodriguez	University Rey Juan Carlos I, Spain
Dragan Simic	University of Novi Sad, Serbia
José Antonio Sáez	University of Salamanca, Spain

SOCO 2018 Organizing Committee

Emilio Corchado	University of Salamanca, Spain
Manuel Graña Romay	University of Basque Country, Spain
José Manuel López-Guede	University of Basque Country, Spain
Álvaro Herrero	University of Burgos, Spain
José Antonio Sáez	University of Salamanca, Spain
Héctor Quintián	University of A Coruña, Spain
Oier Etxaniz	University of Basque Country, Spain

CISIS 2018

Organization

General Chairs

Manuel Graña Romay	University of Basque Country, Spain
Emilio Corchado	University of Salamanca, Spain

International Advisory Committee

Ajith Abraham	Machine Intelligence Research Labs—MIR Labs, Europe
Michael Gabbay	Kings College London, UK
Antonio Bahamonde	University of Oviedo at Gijón

Program Committee Chairs

Emilio Corchado	University of Salamanca, Spain
Manuel Graña Romay	University of Basque Country, Spain
Álvaro Herrero	University of Burgos, Spain
Héctor Quintián	University of A Coruña, Spain

Program Committee

Emilio Corchado (Co-chair)	University of Salamanca, Spain
Manuel Graña Romay (Co-chair)	University of Basque Country, Spain
Adolfo R. De Soto	University of Leon, Spain
Agustin Martin Muñoz	CSIC. Spain
Alberto Peinado	University of Malaga, Spain
Amparo Fuster-Sabater	CSIC, Spain

Ana I. González-Tablas	University Carlos III de Madrid, Spain
Angel Arroyo	University of Burgos, Spain
Angel Martin Del Rey	University of Salamanca, Spain
Antonio Zamora Gómez	University of Alicante, Spain
Antonio J. Tomeu-Hardasmal	University of Cadiz, Spain
Araceli Queiruga-Dios	University of Salamanca, Spain
Carlos Pereira	ISEC, Portugal
Carmen Benavides	University of Leon, Spain
Cristina Alcaraz	University of Malaga, Spain
David Alvarez Leon	University of Leon, Spain
David Arroyo	CSIC, Spain
Debasis Giri	Haldia Institute of Technology, India
Eduardo Solteiro Pires	UTAD University, Portugal
Enrique Onieva	University of Deusto, Spain
Fernando Tricas	Universidad de Zaragoza, Spain
Gerardo Rodriguez Sanchez	University of Salamanca, Spain
Guillermo Morales-Luna	CINVESTAV-IPN, Mexico
Hector Alaiz Moreton	University of Leon, Spain
Héctor Quintián	University of A Coruña, Spain
Hugo Scolnik	University of Buenos Aires, Argentina
Isaias Garcia	University of Leon, Spain
Isaac Agudo	University of Malaga, Spain
Ivan Ubero Martinez	University of Leon, Spain
Javier Areitio	University of Deusto, Spain
Jesús Díaz-Verdejo	University of Granada, Spain
Joan Borrell	Universitat Autònoma de Barcelona, Spain
Jose A. Onieva	University of Malaga, Spain
José Antonio Sáez	University of Salamanca, Spain
Jose Luis Calvo-Rolle	University of A Coruña, Spain
Jose Luis Imana	Universidad Politécnica de Madrid, Spain
José Luis Casteleiro-Roca	University of A Coruña, Spain
Jose M. Molina	Universidad Carlos III de Madrid, Spain
Jose Manuel Lopez-Guede	University of Basque Country, Spain
Josep Ferrer	Universitat de les Illes Balears, Spain
Juan Jesús Barbarán	University of Granada, Spain
Juan Pedro Hecht	University of Buenos Aires, Argentina
Leocadio G. Casado	University of Almeria, Spain
Lidia Sánchez González	Universidad de Leon, Spain
Luis Hernandez Encinas	CSIC, Spain
Luis Enrique Sanchez Crespo	University of Castilla-la Mancha, Ecuador
Manuel Castejon Limas	University of Leon, Spain
Manuel Graña	University of Basque Country, Spain
Maria Fernandez Raga	Universidad de Leon, Spain
Michal Choras	ITTI Ltd, Poland
Oier Etxaniz	University of Basque Country, Spain

Paulo Moura Oliveira	UTAD University, Portugal
Paulo Novais	University of Minho, Portugal
Pilar Martinez	UPV/EHU, Spain
Pino Caballero-Gil	University of La Laguna, Spain
Rafael Alvarez	University of Alicante, Spain
Rafael Corchuelo	University of Seville, Spain
Ramón-Ángel Fernández-Díaz	Universidad de Leon, Spain
Raúl Durán	University of Alcalá, Spain
Robert Burduk	Wroclaw University of Technology, Poland
Roman Senkerik	TBU in Zlin, Czech Republic
Salvador Alcaraz	Miguel Hernandez University, Spain
Sorin Stratulat	Université de Lorraine, France
Urko Zurutuza	Mondragon University, Spain
Vicente Matellan	University of Leon, Spain
Wenjian Luo	University of Science and Technology of China, China
Zuzana Kominkova Oplatkova	Tomas Bata University in Zlin, Czech Republic

CISIS 2018 Organizing Committee

Emilio Corchado	University of Salamanca, Spain
Manuel Graña Romay	University of Basque Country, Spain
José Manuel López-Guede	University of Basque Country, Spain
Álvaro Herrero	University of Burgos, Spain
José Antonio Sáez	University of Salamanca, Spain
Héctor Quintián	University of A Coruña, Spain
Oier Etxaniz	University of Basque Country, Spain

ICEUTE 2018

Organization

General Chairs

Emilio Corchado	University of Salamanca, Spain
Manuel Graña Romay	University of Basque Country, Spain

Program Committee Chairs

Emilio Corchado	University of Salamanca, Spain
Manuel Graña Romay	University of Basque Country, Spain
José Manuel López-Guede	University of Basque Country, Spain
Álvaro Herrero	University of Burgos, Spain
Héctor Quintián	University of A Coruña, Spain

Program Committee

Emilio Corchado (Co-chair)	University of Salamanca, Spain
Manuel Graña Romay (Co-chair)	University of Basque Country, Spain
Igor Ansoategui	University of Basque Country, Spain
Oier Etxaniz	University of Basque Country, Spain
Unai Fernandez	University of Basque Country, Spain
Manuel Graña	University of Basque Country, Spain
Eloy Irigoyen	UPV/EHU, Spain
Vassilis Kaburlasos	TEI of Kavala, Greece
Jose Manuel Lopez-Guede	Basque Country University, Spain
Aitor Moreno Fdz. De Leceta	Ibermatica, Spain
J. David Nuñez-Gonzalez	University of Basque Country, Spain
Leire Ozaeta	UPV/EHU, Spain

Héctor Quintián	University of A Coruña, Spain
Jose Antonio Ramos-Hernanz	University of Basque Country, Spain
José Antonio Sáez	University of Salamanca, Spain
Andreea Vescan	Babes-Bolyai University, Cluj-Napoca, Romania

ICEUTE 2018 Organizing Committee

Emilio Corchado	University of Salamanca, Spain
Manuel Graña Romay	University of Basque Country, Spain
José Manuel López-Guede	University of Basque Country, Spain
Álvaro Herrero	University of Burgos, Spain
José Antonio Sáez	University of Salamanca, Spain
Héctor Quintián	University of A Coruña, Spain
Oier Etxaniz	University of Basque Country, Spain

Logos

Contents

Industry 4.0

Data Mining and Optimization

Soft Computing Methods in Manufacturing and Management Systems

Agents and Multi-agents Systems

An Investment Recommender Multi-agent System in Financial Technology

Elena Hernández[1(✉)], Inés Sittón[1], Sara Rodríguez[1(✉)], Ana B. Gil[1], and Roberto J. García[2]

[1] BISITE Digital Innovation Hub, University of Salamanca, Edificio Multiusos I+D+I, 37007 Salamanca, Spain
{elenahn, isittonc, srg, abg}@usal.es
[2] E. Politécnica Superior de Zamora, University of Salamanca, Edificio Politécnica, 236, Zamora, Spain
toles@usal.es

Abstract. In this article is presented a review of the state of the art on Financial Technology (Fintech) for the design of a novel recommender system. A social computing platform is proposed, based on Virtual Organizations (VOs), that allows to improve user experience in actions that is associated with the process of investment recommendation. The work presents agents functionalities and an algorithm that will improve the accuracy of the Recommender_agent which is in charge of the Case-based reasoning (CBR) system. The data that will be collected and will feed the CBR corresponds to user's characteristics, the asset classes, profitability, interest rate, history stock market information and financial news published in the media.

Keywords: Financial Technology · Virtual organization of agents
Recommender system · Hybrid A.I. algorithm · Investment decisions

1 Introduction

The emergence of Financial Technology (Fintech) is a result of the global economic-financial crisis that occurred in 2008. Companies known as Fintech distanced themselves from traditional banking in order to be able to offer the traditional services offered by banks, due to the cheapening of technology. In this way, small companies that grow in a technological environment have been able to use social networks to expand their market share [18]. Other aspects have also contributed to their expansion, such as the widespread use of smartphones, the bad reputation acquired by banks as well as the lack of transparency and the emergence of a new collaborative economy [2, 27].

In this paper, a research focuses on investment recommendation system for businesses is presented, in order to provide investment related suggestions. For this purpose, we identified different factors that could be extracted from the internet and from the information provided by the users. Perhaps, the biggest challenge is to gather relevant information to make through case-based reasoning (CBR), useful investment recommendations [3, 9, 21].

M. Graña et al. (Eds.): SOCO'18-CISIS'18-ICEUTE'18, AISC 771, pp. 3–10, 2019.
https://doi.org/10.1007/978-3-319-94120-2_1

The article is structured as follow: in Sect. 2 we analyzed the concept of Financial Technology, and the data oriented technology that this implies. We also describe Fintech's requirements, how it is being used to optimize business. The concept of Virtual Organizations is also described. VOs are our starting point in creating a recommendation system proposed in Sect. 3. Finally, conclusions and future work are presented in order to improve the recommender system in Sect. 4.

2 Financial Technology

Financial Technology (Fintech) can be considered as a consequence of the disruption of cloud computing, mobile devices, big data, cybersecurity and other Internet-related technologies, offering emerging business models that are more efficient, safer, innovative and more flexible than existing financial services [5, 11, 29]. IT Fintech companies present the following characteristics:

- Finance oriented
- Highly innovative companies
- New technologies are fundamental
- A challenging alternative to banking

Taking into account that it is necessary to handle large amounts of data, the starting point of many authors is data processing and its security [18]. Data-oriented techniques begin with the mining of operational data in the context of Big Data. This is the main technique for obtaining valuable information, it allows to analyze large volumes of data. In the field of banking, Big Data techniques have been considered an essential tool when dealing with financial data [23]. On the whole, the researchers developed on the handling of data in the field of Financial Technology are aimed at improving financial services or creating new ones. Obtaining datasets helps to distinguish processes, impacts and results and find solutions [15]. In this regard, some authors have used machine learning for large size datasets, in order to improve performance and guarantee privacy [34].

Due to the large amount of work, modern businesses use big data centers. For this reason, many businesses have been interested in the optimization of memory designs and energetic efficiency of the data centers [4, 14, 17, 27]. Researches intend to optimize computing performance with scalable and flexible systems. Data mining in a distributed environment is a tendency but at the same time it is a challenge (Lu *et al.* 2008). In [26] the authors propose a solution to the training problem in mining distributed data, a mechanism that could guarantee that different servers will process distributed data simultaneously, considering both the cost and efficiency. Yu *et al.*, 2015 added another variable: availability. Furthermore, they stressed the importance of integrating the Data Base Management System with storage, security and performance requirements. The result they obtained was that the input/output operation was 27 time faster than the traditional method [35].

In the field of management, the most commonly used approaches are optimization and machine learning at the time of making investment recommendations or when creating businesses strategies. Li and Hoi [24] applied machine learning as online decision support system. The study consisted in performing an online survey on investment portfolios. They presented selection as a sequential problem, obtaining five group categorization of solutions to the problem of online investment portfolio selection. However, they affirmed that precision continued being an unresolved problem. Wang in [32, 33] applied a different model to stock operations, it used fuzzy systems theory to transfer negotiation rules. Another approach to stock performance prediction was proposed by Hadavandi *et al.* in [20], they applied neural networks and integrated genetic fuzzy systems to predict performance on stock markets.

To sum up, many different data based techniques have been used in investment recommendation proposals: Machine learning, fuzzy logic algorithms, neural networks, etc. Nevertheless, it is necessary to include another approach to creating investment recommendations in the business sector. For this reason, the next section will overview the concept of Agent-based Virtual Organizations and the reasons for which they are a suitable recommendation model.

2.1 Virtual Organization of Agents

Agent technology is a branch of Distributed Artificial Intelligence (DAI). MAS (Multi-agent Systems) integrate different capabilities, such as: autonomy, reactivity, proactivity, learning, ubiquitous distributed communication and most importantly the intelligence of all their elements. These characteristics meet a large part of the requirements posed by Financial Technology, adapting to the needs of users in a ubiquitous, autonomous and dynamic manner [6, 16, 36].

In the field of computing, concretely in that of multi-agent systems, organizations are used to describe a group of agents who are coordinated through a series of behavioral patterns and roles aimed at achieving the system's objectives. A multi-agent system model has to be able to define organizations that can adapt dynamically to changes in the environment or to the specifics of the organization. Dynamic adaptation includes adapting to changes in the structure and behavior of the MAS, as well as the addition, deletion and substitution of components during system execution, without affecting its correct functioning [8, 12, 28].

Virtual Organizations (VOs) have a series of common characteristics:

- An organization of agents is made up of agents, roles, coordination and interaction rules.
- It pursues a common and global objective which is irrespective of the objectives of particular agents.
- Roles are assigned to the different agents. Thus, their tasks within the system are specialized in order to achieve the organization's global objectives.
- It divides the system in groups through departmentization, these groups are units of interaction between agents.
- It defines a series of limits for the agents belonging to the organization, their rules of interaction, its functionality and the services it offers.
- The input and output of agents of the organization determines their dynamics, therefore, their roles can change depending of the organization objective.

3 Investment Recommender System

Having described the characteristics of Agent-based Virtual Organizations, we propose a design based on VOs with new human-agent interaction modules which allow to improve user experience. This is associated with the process of investment recommendation [10, 13, 18, 20, 23, 26]. With this aim, the paper address the creation of a social computing platform which allows humans and machines to work collaboratively and transparently. Once functionalities are defined, this distributed design is going to facilitate subsequent development and allow for future modifications and extensions.

Our proposal is designed as a heterogeneous system in languages, applications and characteristics. Figure 1 illustrates the different elements of the platform together with the modules that make it up. Below, each of these modules is described, each of them will individually compose a Virtual Organization, with distinctive characteristics, rules and structures. This is described as follows:

- Identification (V.O.): User_agent is the interface that allows the user to Access recommendation functionalities. It is in charge of generating and updating a user profile.
- Information (V.O.): This organization is in charge of searching and processing information. In this case, we created different sub-organizations which are in charge of calculating the variables that are part of the recommendation system:
- User Risk Profile_Information_agent: This agent is responsible for collecting information on investment profiles, considering the level of risk that the investor is prepared to take (asset classes, profitability, interest rate, etc.).
- Share Price_Information_agent: It is in charge of obtaining the public process of shares.
- Financial news_Information_agent: This agent is responsible for obtaining financial news published in the media which list the transactions of businesses, both internal and external. Thy will be included in the Recommendation System in order to be able to extrapolate patterns and provide users with accurate recommendations.
- Recommendation (V.O.): It is responsible for making the different investment recommendations. The Recommender_agent is in charge of calculating factor and weight, and of managing the CBR System where the suggestions of the users are stored.
- Interface (V.O.): The Interface_agent is responsible for showing the user the investment recommendation when the access the system, this recommendation will be based on the user profile previously created by the User_agent for personalized investment recommendations.

3.1 Hybrid A.I. Algorithm for Investment Recommendation System

In [30] the authors refer to Machine learning algorithms like effective in fitting parameters automatically, avoiding over-fitting, and being capable of combining multiple inputs. Also mentioned that ranking investors' sentiment hence provides a natural way to select stocks based on the "portrayed performance" in news media.

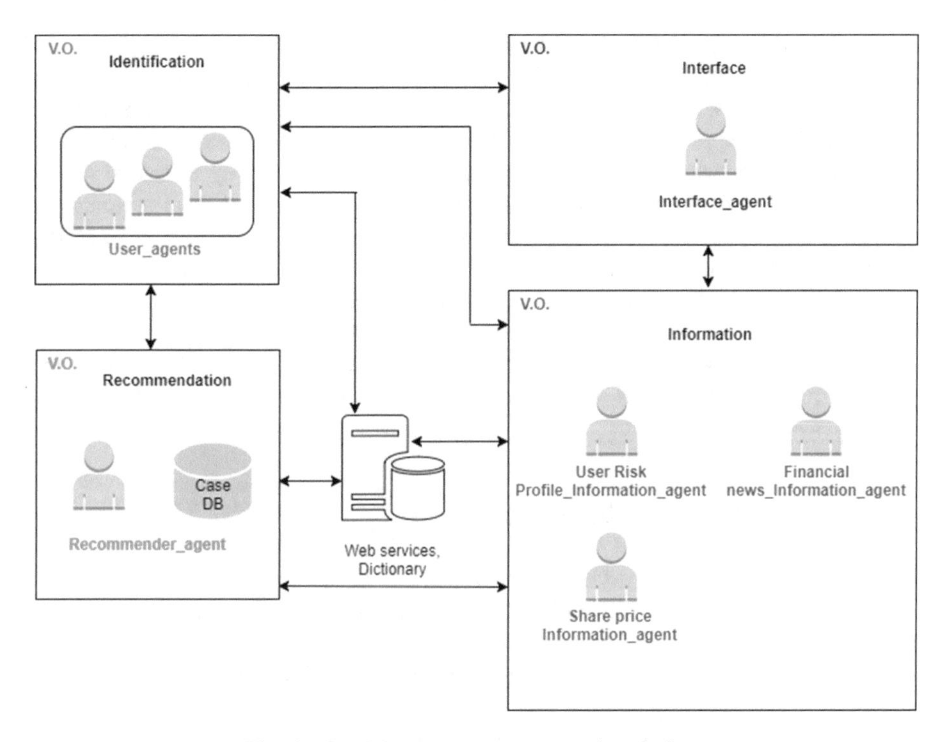

Fig. 1. Stock Investment recommender platform

Reviewing the literature, it is possible to find three kinds of machine learning algorithms for financial market prediction and trading strategies: price prediction, movement directions predictions and algorithms for rule-based optimization to determine optimal combinations. In our study, we will focus on those oriented to price prediction. Regression algorithms [10] and neural networks are able to perform approximations of the future performance of assets.

The concept of Support vector regression (SVR) is addressed in [25], the authors shown a typical regression problem to illustrate the SVR concept:

Consider a set of data $G = \{(x_i, q_i)\}_i^n$, *where* x_i *is a vector of the model inputs,* q_i *is actual value and represents the corresponding scalar output, and n is total number of data patterns. The objective of the regression analysis is to determine a function f(x), so as to predict accurately the desired (target) outputs (q). Thus, the typical regression function can be formulated as* $q_i = f(x_i) + \delta$, *where* δ *is the random error with distribution of* $N(0, \sigma^2)$. *The regression problem can be classified as linear and nonlinear regression problems. As the nonlinear regression problem is more difficult to deal with, SVR was mainly developed for tackling the nonlinear regression problem.* On the other hand, regarding neural networks, the definition given is that comprehensive system that considers numeric inputs, performs computations on these inputs, and creates outputs for one or more numeric values [1]. Neural networks improve the traditional statistical methods such as linear regressions, using function approximations, discriminant analysis, and logistic regression [22, 31].

The proposal is developed through a hybrid system: HBP-PSO algorithm (hybrid BP neural network combining adaptive Particles Swarm Optimization algorithm). BP networks with supervised learning rules is the most used in financial time series because can be capable of approximate any measurable function in a very precise manner [22]. However, the disadvantage of the BP neural network is its inability to search for the overall optimal value. That is why by using the PSO algorithm (intelligent optimization method to find the optimal value) together with the HBP an algorithm is obtained that maximizes the advantages of both [19]. Therefore, the algorithm will improve the accuracy of the Recommender_agent in charge of the CBR [7].

The data that will feed the CBR will be those collected by the Information V.O. and it corresponds to asset classes, profitability, interest rate, the public process of shares and financial news published in the media which list the transactions of businesses, both internal and external.

4 Conclusions and Future Work

This article overviewed the different techniques that have been implemented to create Fintech services. After a close study of the state of the art, our proposal consisted in adding Agent-based Virtual Organizations for investment recommendation. VOs were implemented with the aim of creating a light, well-structured, scalable and user-adapted system.

In a future work, the data that will be collected and will feed the CBR corresponds to user's characteristics, the asset classes, profitability, interest rate, history stock market information and financial news published in the media it is being considered to perform a sample analysis of the user's characteristics to identify the level of risk assumed. Once a significant sample has been obtained, the data collected from the IBEX35 history stock market could be used. In the case of the study, the hybrid algorithm HBP-PSO will be applied and the results obtained will be shown. Once the proposal has been tested in that market, and to give robustness to the work, the study could be replicated in other financial markets such as Dow Jones, NASDAQ, etc.

Acknowledgments. This work was supported by the Spanish Ministry, Ministerio de Economía y Competitividad and FEDER funds. Project: SURF, Intelligent System for integrated and sustainable management of urban fleets TIN2015-65515-C4-3-R.

References

1. Abdou, H.A., Pointon, J., El-Masry, A., Olugbode, M., Lister, R.J.: A variable impact neural network analysis of dividend policies and share prices of transportation and related companies. J. Int. Fin. Mark. Inst. Money **22**(4), 796–813 (2012)
2. Avital, M., Andersson, M., Nickerson, J., Sundararajan, A., Van Alstyne, M., Verhoeven, D.: The collaborative economy: a disruptive innovation or much ado about nothing? In: Proceedings of the 35th International Conference on Information Systems, ICIS, pp. 1–7. Association for Information Systems. AIS Electronic Library (AISeL) (2014)

3. Bach, K.: Knowledge engineering for distributed case-based reasoning systems. In: Synergies Between Knowledge Engineering and Software Engineering, pp. 129–147. Springer, Cham (2018)
4. Bajo, J., De la Prieta, F., Corchado, J.M., Rodríguez, S.: A low-level resource allocation in an agent-based Cloud Computing platform. Appl. Soft Comput. **48**, 716–728 (2016)
5. Casado-Vara, R., Corchado, J.M.: Blockchain for democratic voting: how blockchain could cast off voter fraud. orient. J. Comput. Sci. Technol. **11**(1). http://www.computerscijournal.org/?p=8042
6. Chamoso, P., Rivas, A., Rodríguez, S., Bajo, J.: Relationship recommender system in a business and employment-oriented social network. Inf. Sci. **433–434**, 204–220 (2017)
7. Corchado, J.M., Lees, B.: Adaptation of cases for case based forecasting with neural network support. In: Soft Computing in Case Based Reasoning, pp. 293–319. Springer, London (2001)
8. Corchado, J.M., Bajo, J., de Paz, Y., Tapia, D.: Intelligent environment for monitoring Alzheimer patients, agent technology for health care. Decis. Support Syst. **34**(2), 382–396 (2008). ISSN 0167-9236
9. Corchado, J.M., Laza, R.: Constructing deliberative agents with case-based reasoning technology. Int. J. Intell. Syst. **18**(12), 1227–1241 (2003)
10. De Paz, J.F., Bajo, J., González, A., Rodríguez, S., Corchado, J.M.: Combining case-based reasoning systems and support vector regression to evaluate the atmosphere–ocean interaction. Knowl. Inf. Syst. **30**(1), 155–177 (2012)
11. DeStefano, R.J., Tao, L., Gai, K.: Improving data governance in large organizations through ontology and linked data. In: 2016 IEEE 3rd International Conference on Cyber Security and Cloud Computing (CSCloud), pp. 279–284. IEEE, June 2016
12. Đurić, B.O.: Organisational metamodel for large-scale multi-agent systems: first steps towards modelling organisation dynamics. Adv. Distrib. Comput. Artif. Intell. J. **6**(3), 2017 (2017)
13. Elnagdy, S.A., Qiu, M., Gai, K.: Cyber incident classifications using ontology-based knowledge representation for cybersecurity insurance in financial industry. In: 2016 IEEE 3rd International Conference on Cyber Security and Cloud Computing (CSCloud), pp. 301–306. IEEE, June 2016
14. Gai, K., Du, Z., Qiu, M., Zhao, H.: Efficiency-aware workload optimizations of heterogeneous cloud computing for capacity planning in financial industry. In: 2015 IEEE 2nd International Conference on Cyber Security and Cloud Computing (CSCloud), pp. 1–6. IEEE, November 2015
15. Gai, K., Qiu, M., Sun, X., Zhao, H.: Security and privacy issues: a survey on FinTech. In: International Conference on Smart Computing and Communication, pp. 236–247. Springer, Cham, December 2016
16. García, E., Rodríguez, S., Martín, B., Zato, C., Pérez, B.: MISIA: middleware infrastructure to simulate intelligent agents. In: International Symposium on Distributed Computing and Artificial Intelligence, pp. 107–116. Springer, Heidelberg (2011)
17. Georgakoudis, G., Gillan, C.J., Sayed, A., Spence, I., Faloon, R., Nikolopoulos, D.S.: Methods and metrics for fair server assessment under real-time financial workloads. Concurrency Comput. Pract. Experience **28**(3), 916–928 (2016)
18. Gomber, P., Koch, J.A., Siering, M.: Digital Finance and FinTech: current research and future research directions. Bus. Econ. **87**, 537–580 (2017)
19. Han, F., Gu, T.Y., Ju, S.G.: An improved hybrid algorithm based on PSO and BP for feedforward neural networks. JDCTA: Int. J. Digit. Content Technol. Appl. **5**(2), 106–115 (2011)

20. Havandi, E., Shavandi, H., Ghanbari, A.: Integration of genetic fuzzy systems and artificial neural networks for stock price forecasting. Knowl. Based Syst. **23**(8), 800–808 (2010)
21. Hüllermeier, E., Minor, M. (eds.): Case-Based Reasoning Research and Development: Proceedings of the 22nd International Conference, ICCBR 2014, Cork, Ireland, September 29–October 1, 2014, vol. 8765. Springer (2015)
22. Kaastra, I., Boyd, M.: Designing a neural network for forecasting financial and economic time series. Neurocomputing **10**(3), 215–236 (1996)
23. Lazarova, D.: Fintech trends: the internet of things, January 2018. https://www.finleap.com/insights/fintech-trends-the-internet-of-things/
24. Li, B., Hoi, S.C.: Online portfolio selection: a survey. ACM Comput. Surv. (CSUR) **46**(3), 35 (2014)
25. Lu, C.J., Lee, T.S., Chiu, C.C.: Financial time series forecasting using independent component analysis and support vector regression. Decis. Support Syst. **47**(2), 115–125 (2009)
26. Lu, Y., Roychowdhury, V., Vandenberghe, L.: Distributed parallel support vector machines in strongly connected networks. IEEE Trans. Neural Netw. **19**(7), 1167–1178 (2008)
27. Owyang, J., Tran, C., Silva, C.: The Collaborative Economy. Altimeter, New York (2013)
28. Rodriguez, S., Julián, V., Bajo, J., Carrascosa, C., Botti, V., Corchado, J.M.: Agent-based virtual organization architecture. Eng. Appl. Artif. Intell. **24**(5), 895–910 (2011)
29. Sagraves, A., Connors, G.: Capturing the value of data in banking. Appl. Mark. Anal. **2**(4), 304–311 (2017)
30. Song, Q., Liu, A., Yang, S.Y.: Stock portfolio selection using learning-to-rank algorithms with news sentiment. Neurocomputing **264**, 20–28 (2017)
31. Tkáč, M., Verner, R.: Artificial neural networks in business: two decades of research. Appl. Soft Comput. **38**, 788–804 (2016)
32. Wang, L.X.: Dynamical models of stock prices based on technical trading rules part I: the models. IEEE Trans. Fuzzy Syst. **23**(4), 787–801 (2015)
33. Wang, L.X.: Dynamical models of stock prices based on technical trading rules—Part III: application to Hong Kong stocks. IEEE Trans. Fuzzy Syst. **23**(5), 1680–1697 (2015)
34. Xu, K., Yue, H., Guo, L., Guo, Y., Fang, Y.: Privacy-preserving machine learning algorithms for big data systems. In: 2015 IEEE 35th International Conference on Distributed Computing Systems (ICDCS), pp. 318–327. IEEE, June 2015
35. Yu, K., Gao, Y., Zhang, P., Qiu, M.: Design and architecture of dell acceleration appliances for database (DAAD): a practical approach with high availability guaranteed. In: 2015 IEEE 17th International Conference on High Performance Computing and Communications (HPCC), 2015 IEEE 7th International Symposium on Cyberspace Safety and Security (CSS), 2015 IEEE 12th International Conference on Embedded Software and Systems (ICESS), pp. 430–435. IEEE, August 2015
36. Zato, C., Villarrubia, G., Sánchez, A., Bajo, J., Corchado, J.M.: PANGEA: a new platform for developing virtual organizations of agents. Int. J. Artif. Intell. **11**(A13), 93–102 (2013)

Using Genetic Algorithms to Optimize the Location of Electric Vehicle Charging Stations

Jaume Jordán(✉), Javier Palanca, Elena del Val, Vicente Julian, and Vicente Botti

Universitat Politècnica de València, Camino de Vera s/n, Valencia, Spain
{jjordan,jpalanca,edelval,vinglada,vbotti}@dsic.upv.es

Abstract. The creation of a suitable charging infrastructure for electric vehicles (EV) is one of the main challenges to increase the adoption of this new vehicle technologies. In this article, we present a Multi-Agent System (MAS) that performs an analysis of a set of possible configurations for the location of EV charging stations in a city. To estimate the best configurations, the proposed MAS considers data from heterogeneous sources such as traffic, social networks, population, etc. Based on this information, the agents are able to analyze a large set of configurations using a genetic algorithm that optimizes the configurations taking into account a utility function.

Keywords: Multi-Agent Systems · Electric Vehicles
Charging stations · Genetic Algorithms

1 Introduction

Electric Vehicles (EV) are a key element that have become increasingly important in recent years due to the significant reduction in gas and noise emissions. This situation is also creating an emerging market around it [1], where governments are also focusing their attention and their next strategic moves.

The arrival of the electric vehicle comes with the need for new infrastructures to support the electric charging needs. These needs also have particular requirements such as the amount of electrical power needed at the electricity grid where the charging station is located. More and more electric cars are being made available to consumers every day, bringing great environmental benefits.

The lack of information is just one of the causes that may hinder the introduction of the use of the electric vehicle in cities. Also the technological uncertainty, since these technologies are not as well-known as the conventional vehicle technologies, the limitations on battery life and their charging times slow the adoption of EV [2]. The lack of charging infrastructures to meet the potential demand for electric vehicles is a point of special relevance [3]. This brings an effect known as *range anxiety*, which is the fear to have insufficient range to

M. Graña et al. (Eds.): SOCO'18-CISIS'18-ICEUTE'18, AISC 771, pp. 11–20, 2019.
https://doi.org/10.1007/978-3-319-94120-2_2

reach the destination [4]. However, since infrastructure development is expensive, it is necessary to direct investments towards the establishment of electric charging facilities in areas with maximum impact.

To help city policy-makers to allocate public resources efficiently to support the deployment of charging infrastructure, a systematic approach is needed to quantify the benefit of providing public charging opportunities, as well as to determine where to locate charging stations subject to limitations on the range of vehicle travel [5,6]. Shukla et al. [5] propose a mathematical programming for determining the best locations for establishing alternative transportation fuel stations while Nie and Ghamami [6] present conceptual optimization model to analyze travel by EVs along a long corridor. Other studies, as Wood et al. [7], analyze the distribution of electric stations in cities, focusing on the estimated number of charging stations needed to increase the utility of the vehicle. However, in the analyzed approaches other relevant factors for an effective charging infrastructure such as geography as well as to demographics are not considered.

This work addresses the problem of optimizing the placement of new infrastructures for EV charging. To do this, we design a system, made up of a set of agents, which gathers information from heterogeneous data sources of the city where the deployment of EV charging stations is intended. Finally, a search for the optimal solution is made by means of a genetic algorithm. This optimization checks the possible locations of the charging stations and tries to distribute throughout the city the required stations. In addition, it has to satisfy constraints such as the maximum number of piles per station, it must guarantee the supply of electricity for charging EVs in the city, and it has to consider urban information such as population and traffic per neighborhood to optimize the investment.

The article is structured in the following sections. Section 2 describes the main aspects of the proposed MAS and details the *Emplacement Optimizer Agent* which is the core of the system. Section 3 presents a case study for the city of Valencia including some experimental results. Finally, Sect. 4 gives some conclusions and future works.

2 Design of the Application

The application proposed in this work evaluates a set of points of interest throughout a city in order to optimize the best distribution of charging stations for EVs that minimizes the called *range anxiety* that is mainly determined by the infrastructure of the charging stations [8]. We have designed a genetic algorithm (GA), managed by a multi-agent system that tries to reduce the search problem by introducing information about the city and a heuristic that makes the search faster and prevents premature convergence to a less optimal solution.

The multi-agent system (MAS) was developed using the SPADE platform [9], which allowed us an easy prototyping and further development of the MAS by means of the instant messaging support provided by the XMPP[1] protocol in

[1] www.xmpp.org.

SPADE. All messages are sent in real-time between the agents described below using the presence notification mechanism that the platform provides.

2.1 Data Collecting and Processing

Before running the GA there is a set of agents of the application's MAS that collect information related with the city we are working on. This allows our application to be run and reused in other municipalities with not too much effort. The information collected by the application is:

- **Points of Interest:** The GA starts with a set of Points of Interest (PoI) that are candidates for hosting a charging station. These points are manually selected according to the city's urban development plan. Subsequently, the GA will be responsible for reducing this set, but the *PoI agent* is in charge of detecting and classifying PoIs candidates to be selected for the input of the algorithm. The agent also carries out a clustering process to eliminate PoIs that are too close and to define influence zones of the point. This clustering process joins points that are very close at the centroid of that set of points. This process reduces the initial number of PoIs, avoiding the computational expense of evaluating the placement of overlapping stations.
- **Population information:** The *Urban agent* gathers information about the population living in each area of the city. It queries census sections of open data portals and obtains the population in the different neighborhoods or blocks of the city under study.
- **Traffic information:** The amount of traffic that an area of a city has is obtained by the *Traffic agent*.
- **Popularity information:** The influx of an area determines its popularity. Based on the number of people that visits each PoI and how much time they spend in the area, we can estimate its popularity. The *Popularity agent* uses an exhaustive search on third party services (such as Google cards on the search engine) to locate this data.
- **Social Networks information:** Geo-tagged information of social networks can be used to estimate the activity that occurs in that area. The *Social Network agent* tracks some networks (Twitter and Instagram) to collect all the geolocated items in an area.

Once the data is collected, the *Data Processing Agent* aggregates all the information obtained by the previous agents and serves it through geo-queries that simplify obtaining the information around an individual Point of Interest.

2.2 Application's Data Flow

The application is run in six stages that include the selection of points of interest, transforming the points to a Voronoi diagram, the extraction of information about the city, the characterization with this data of the polygons representing PoIs, the execution of the GA and the visualization of the results (see Fig. 1).

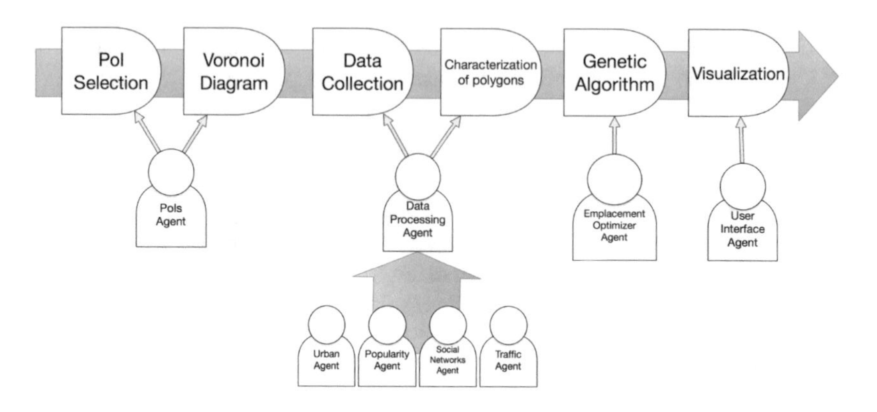

Fig. 1. Application's data flow

The output of the first stage is a set of PoIs P that are candidates for the location of a charge station s_i. The PoI agent usually uses as first set of PoIs the location of public parkings and garages, which have a large number of visitors per day. These public parkings are also good candidates because cars are usually parked long enough to make a charge and the electric infrastructure is suitable for EV charging stations [10].

At the second stage the PoI agent builds a Voronoi diagram, using the selected PoIs as centroids, to determine the area of influence of each point. This area allows us to better determine a full polygon, and not only a single point, to consider the location of a station. It also helps to calculate which information collected from the city at the next stage belongs to which polygon.

Third and fourth stages collect information from the city and aggregate it to the Voronoi polygons, which get characterized with population density, traffic intensity and social networks activity, among others. These stages are performed by the Data Processing Agent and the collecting agents described above.

The fifth stage is performed by the Emplacement Optimizer Agent, one of the most important agents of the application that is going to be described in depth in the next section. The Emplacement Optimizer Agent uses the characterized polygons and the constraints of the problem, such as the maximum number of stations to install in the city, to make a heuristic search using a GA.

The last stage of the application is performed by the User Interface Agent, which runs a website where the results of the application can be examined on a map, query the properties of each proposed station and run new executions.

2.3 The Emplacement Optimizer Agent

The *Emplacement Optimizer Agent* is responsible for determining the most appropriate locations for the emplacement of electric charging stations. In this subsection, we describe the formalization of the problem of finding the most suitable configuration for the location of the stations and the GA used.

Location of Electric Charging Stations Problem. The goal of the Emplacement Optimizer agent is to analyze a set of possible configurations and select the most appropriate according to a utility function and a set of predefined PoIs. We consider a set of possible locations (i.e., PoIs) $P = \{p_1, \ldots, p_n\}$ of the city to study. Each PoI p_i is characterized by a set of attributes $p_i = \{a_1, a_2, \ldots, a_n\}$. Specifically, we consider the following attributes:

- $a_{population}$: population in the area around p_i
- $a_{traffic}$: average traffic in the area
- a_{time}: average time spend by citizens in public places in the area
- a_{social}: geo-located social networking activity in the area
- $cost\_area$: cost depending on the area covered by the stations
- $cost\_per\_charger$: cost per each charging station

Besides the set of possible locations, the agent considers a set of charging stations $S = \{s_1, \ldots, s_n\}; 0 < s_i \leq max\_chargers\_per\_poi$, that could be deployed in the city. The number of chargers per station in a PoI ranges from 0 to a constant value $max\_chargers\_per\_poi$ (5 in our tests).

Considering the set of predefined PoIs P and the set of charging stations S, the agent is able to provide the most appropriate configuration C_i for the location of stations in PoIs. A configuration $C_i = \{\{p_1, s_1\}, \{p_2, s_2\}, \ldots, \{p_n, s_n\}\}$ consists of a set of pairs PoI-stations. Each configuration has associated a fitness value $V(C_i)$ according to its suitability.

Genetic Algorithm. Genetic Algorithms (GAs) [11] are adaptive methods that can be used to solve search and optimization problems gradually converging towards high-quality solutions through the application of a set of operators. Our GA is implemented using the *DEAP*[2] library of *Python*. We propose a GA that will create generations (i.e., sets of solutions), where each generation will inherit the properties of the previous generations (i.e., configuration of charging stations). Initially, the algorithm considers a random population of N individuals (up to 4000 in our tests through 200 generations). Each individual is a feasible solution C_i to the problem. Specifically, a chromosome is composed of a set of locations P and the number of stations per location S. An example of the chromosome encoding and functions of the GA is shown in Fig. 2.

To evaluate the suitability of each chromosome (i.e., solution), the agent uses a fitness function. This function evaluates the quality of the solution considering the suitability of placing charging stations in the selected points using the parameters defined above (i.e., characteristics of the points of interest). The fitness function considered is as follows:

$$V(C_i) = \sum_{\forall p_i \in C} ((\omega_p \cdot a_{population} + \omega_{tr} \cdot a_{traffic} + \omega_t \cdot a_{time} + \omega_s \cdot a_{social}) - (\omega_a \cdot cost\_area + \omega_c \cdot cost\_per\_charger \cdot |s_i|)), \quad (1)$$

[2] https://github.com/DEAP/deap.

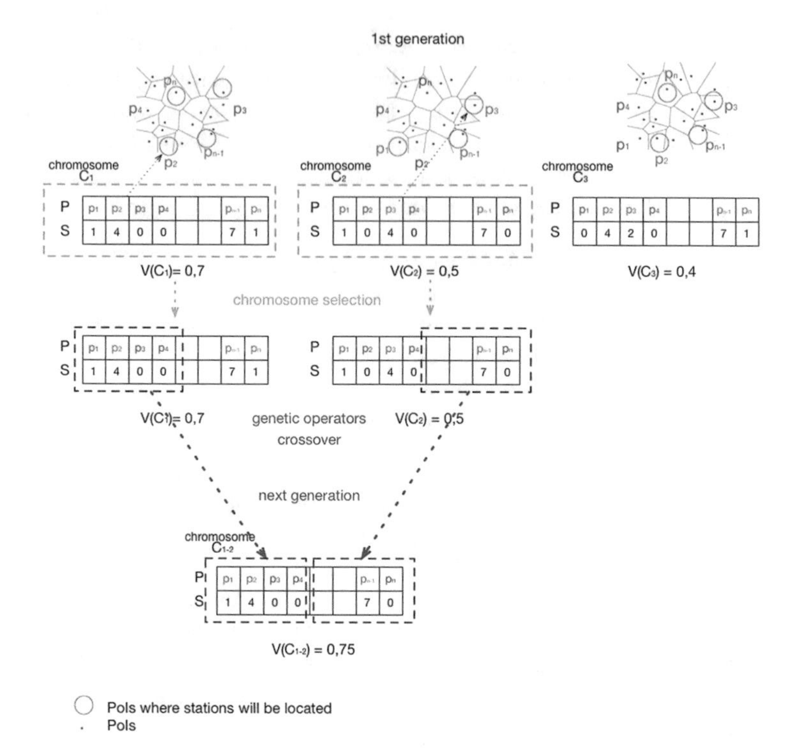

Fig. 2. The encoding of individuals (i.e., chromosomes) and genetic algorithm.

where $a_{population}$ refers to the population of the area covered by the charging stations located in p_i; $a_{traffic}$ refers to the traffic generated in the area of p_i; a_{time} refers to the average time citizens spend in public/commercial places in the area of p_i; a_{social} refers to the average social networks activity in the area of p_i; $cost\_area$ refers to the cost of locating stations in p_i that covers the demand of that area; and $cost\_per\_charger$ is a constant cost per each charger ($|s_i|$) located in p_i. The value of these parameters ranges in the interval [0,1]. Each parameter has associated a weight value ω established by the users of the system depending on the characteristics of the city (in our tests, these values are $\omega_p = 0.4$, $\omega_{tr} = 0.3$, $\omega_t = 0.2$, and $\omega_s = 0.1$).

In our implementation, we used the *tournament selection* method, which makes several random groups of individuals, called tournaments, and selects the best one of each group (in our tests, the size of groups is 3). The selected chromosomes are *combined* with others (crossover) and/or *mutated.*

The *crossover* operator selects two parents and a new individual child is created by combining the genes of the parents. We used the *cross uniform* technique, which swaps genes from parents taking a uniform number of genes from them. The probability of crossover in our tests is fixed to 0.5.

The *mutation* operator generates a new individual by mutating some of the genes of a selected individual (*flipbit* method) with a mutation probability (fixed to 0.05 in our case). This operator is used to maintain genetic diversity.

Once the operators have been applied to the population of a generation, the new individuals are inserted into the next generation. The best N individuals remain in the next generation and the others are removed. The process ends in the following situations: (i) when the number of generations exceeds a number established by the problem; (ii) when there is a certain number of generations where there is no individual in the next generation that improves the fitness value of the best individual in previous generations; (iii) when the algorithm exceeds a established time limit. At the end of the process, the *Emplacement Optimizer Agent* sends the obtained results to the *User Interface Agent.*

3 Example: City of Valencia

This section presents a case study in the city of Valencia using data from the open data portal supported by the city council[3]. The goal is to determine the most appropriated locations in Valencia where to place electric charging stations using the above presented system. Currently, there are 76 charging points in the province of Valencia, according to [12], and 24 of these are located in the city of Valencia. The Valencia City Council has carried out various initiatives aimed at improving infrastructure to facilitate the introduction of EV.

In the first phase, the MAS determines the P potential PoIs for the location of a charging station s_i taking into account data from the General Urban Development Plan. The points selected as potential EV charging stations were: shopping malls, workplaces, parkings, public thoroughfares, neighborhood communities, private garages, refueling stations, and vehicle fleet parkings. Considering these possible points, the system determines the area of influence of each of the PoIs creating a Voronoi diagram around the selected points.

After this, in a second phase, the MAS collects data about different aspects from the city of Valencia:

1. Information about the level of traffic in each street of the city. In particular, it is extracted from the Valencia City Council's open data portal. The average annual traffic is calculated for each street section on the map.
2. Information about the population that lives or works in each zone of the city. As in the previous case, this information is extracted from the Valencia City Council's open data portal. The amount of population is calculated for each block of buildings.
3. The average time spent in commercial and public spaces. This information is obtained from statistics published by the city council and extracted from Google cards on the search engine.

[3] http://gobiernoabierto.valencia.es/es/data/.

4. Information about the geo-located social activity from social networks. This information is obtained using uTool [13], which performs a real-time analysis on the activity of a city through the messages that users exchange inside a social network.

This data is collected and aggregated for each of the polygons around a PoI using the agents described in Sect. 2.1. The *Emplacement Optimizer Agent* receives all this data in order to determine possible solutions.

Finally, in the third phase, the *Emplacement Optimizer Agent* returns solutions through the proposed genetic algorithm. The best individual in the population is chosen based on the *fitness* value obtained by Eq. 1. Figure 3 shows an example of a solution given by the *User Interface Agent*, that is, the locations where each charging point would be located.

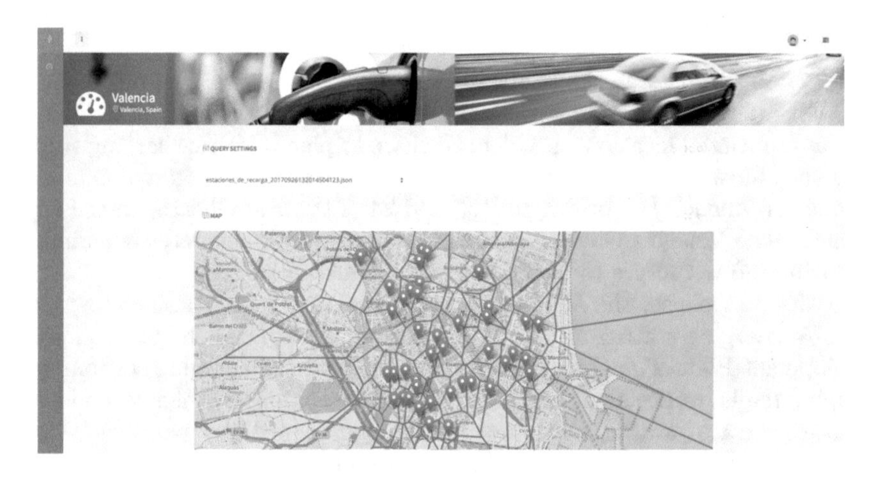

Fig. 3. Configuration of the location of the charging stations displayed by the *User Interface Agent.*

The proposed system has been tested with different data and configurations for the particular case study of Valencia in order to compare solutions of different quality. Figure 4 represents two example computed solutions. Figure 4(a) is a solution computed with an initial population of 250 with a fitness value of 0.563. The solution of Fig. 4(b) is computed with an initial population of 4000 that yields in a fitness of 0.639.

At a glance, both solutions are very similar. There is only a difference of 2 stations between the two proposals (solution of Fig. 4(a) has 40 charging stations, while Fig. 4(b) has 42). However, the quality of a solution is given by the disposition of the stations throughout the city. For instance, Fig. 4(a) places a charging station in the far south of the city because there is some activity there. However, this activity is not significant enough and it would be a waste of resources because that area does not need to be covered (in our scenario).

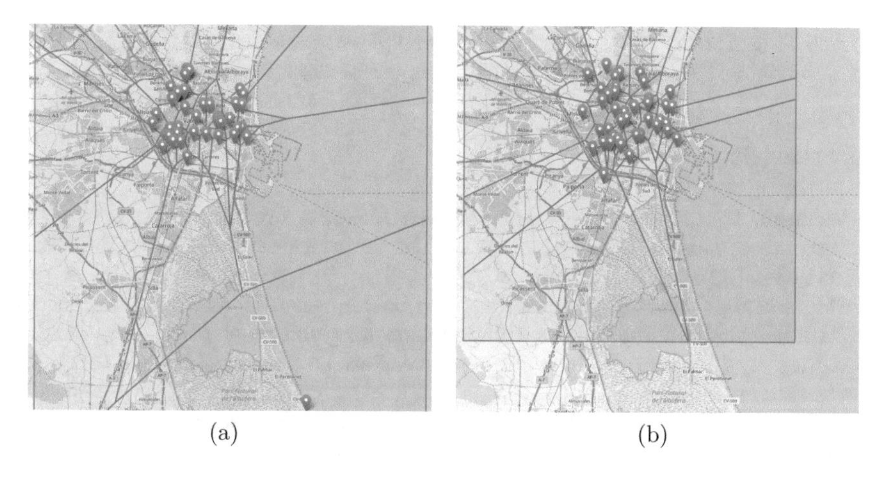

(a) (b)

Fig. 4. Computed solutions by the genetic algorithm for the city of Valencia

The solution of Fig. 4(b) places the charging stations more uniformly in the city, covering the full area where the main activity occurs. Concretely, there are several charging stations covering the center and north of the city which are not present in Fig. 4(a). Bearing this in mind, the solution proposed in Fig. 4(b) is more appropriate if we take into account the characteristics of the city.

4 Conclusions

One of the current challenges for increasing the use of electric vehicles at a general level and particularly in cities is the development of a reliable, accessible and comfortable charging infrastructure for the citizen. In this paper, a Multi-Agent System (MAS) has been proposed in order to facilitate the analysis of possible locating configurations of EV charging stations. The proposed MAS integrates information from heterogeneous data sources as a starting point to characterize the areas where charging stations could potentially be located. The core of the system is a genetic algorithm that takes such data as input and generates a proposal of possible locations taking into account several restrictions. The proposed system has been implemented in the city of Valencia where different experiments have been done in order to validate the implementation.

As future work, the proposed system can be extended allowing to define priorities among the different locations proposed. In this way, a phased installation schedule can be defined by the relevant authority. Moreover, the system can be extended for a more detailed analysis of the activity and/or mobility of citizen in the city. We also consider a deep analysis of the predicted energy consumption of the charging stations, especially in peak demand moments. Finally, we will also compare the solution proposed in this paper with the solution provided by a multi-objective approach exploring the Pareto frontier.

Acknowledgments. This work was partially supported by MINECO/FEDER TIN2015-65515-C4-1-R and MOVINDECI project of the Spanish government.

References

1. Wolfram, P., Lutsey, N.: Electric vehicles: Literature review of technology costs and carbon emissions. Technical report. The International Council on Clean Transportation (2016)
2. Skippon, S., Garwood, M.: Responses to battery electric vehicles: UK consumer attitudes and attributions of symbolic meaning following direct experience to reduce psychological distance. Transp. Res. Part D: Transp. Environ. **16**(7), 525–531 (2011)
3. Klabjan, D., Sweda, T.: The nascent industry of electric vehicles. In: Wiley Encyclopedia of Operations Research and Management Science (2011)
4. Dong, J., Liu, C., Lin, Z.: Charging infrastructure planning for promoting battery electric vehicles: an activity-based approach using multiday travel data. Transp. Res. Part C: Emerg. Technol. **38**, 44–55 (2014)
5. Shukla, A., Pekny, J., Venkatasubramanian, V.: An optimization framework for cost effective design of refueling station infrastructure for alternative fuel vehicles. Comput. Chem. Eng. **35**(8), 1431–1438 (2011)
6. Nie, Y.M., Ghamami, M.: A corridor-centric approach to planning electric vehicle charging infrastructure. Transp. Res. Part B: Methodological **57**, 172–190 (2013)
7. Wood, E., Neubauer, J.S., Burton, E.: Measuring the benefits of public chargers and improving infrastructure deployments using advanced simulation tools. Technical report, SAE Technical Paper (2015)
8. Needell, Z.A., McNerney, J., Chang, M.T., Trancik, J.E.: Potential for widespread electrification of personal vehicle travel in the United States. Nat. Energ. **1**, 16112 (2016)
9. Escriva, M., Palanca, J., Aranda, G., Garcia-Fornes, A., Julian, V., Botti, V.: A jabber-based multi-agent system platform. In: Proceedings of the Fifth International Joint Conference on Autonomous Agents and Multiagent Systems (AAMAS 2006), pp. 1282–1284. Association for Computing Machinery, Inc. (ACM Press) (2006)
10. Sedano Franco, J., Portal García, M., Hernández Arauzo, A., Villar Flecha, J.R., Puente Peinador, J., Varela Arias, J.R.: Sistema inteligente de recarga de vehículos eléctricos: diseño y operación. Dyna **88**(6), 644–651 (2013)
11. Goldberg, D.E.: Genetic Algorithms. Pearson Education, India (2006)
12. Electromaps: Electromaps: Puntos de recarga en Valencia (2017). https://www.electromaps.com/puntos-de-recarga/espana/valencia
13. del Val, E., Palanca, J., Rebollo, M.: U-tool: a urban-toolkit for enhancing city maps through citizens activity. In: Advances in Practical Applications of Scalable Multi-agent Systems. The PAAMS Collection, pp. 243–246. Springer (2016)

Case-Based Reasoning and Agent Based Job Offer Recommender System

Alfonso González-Briones[1], Alberto Rivas[1], Pablo Chamoso[1(✉)], Roberto Casado-Vara[1], and Juan Manuel Corchado[1,2,3]

[1] BISITE Digital Innovation Hub, University of Salamanca. Edificio Multiusos I+D+i, 37007 Salamanca, Spain
{alfonsogb,rivis,chamoso,rober,corchado}@usal.es
[2] Department of Electronics, Information and Communication, Faculty of Engineering, Osaka Institute of Technology, Osaka 535-8585, Japan
[3] Pusat Komputeran dan Informatik, Universiti Malaysia Kelantan, Karung Berkunci 36, Pengkaan Chepa, 16100 Kota Bharu, Kelantan, Malaysia

Abstract. The large amounts of information that social networks contain, makes it necessary for them to provide guides and aids that improve users' experience in the system. In addition to search and filtering tools, users should be presented with the content they wish to obtain before they take any action to find it. To be able to recommend content to users, it is necessary to analyse their profiles and determine what type of content they want to view. The present work is focused on an employability oriented social network for which a job offer recommender system is proposed, following the model of a multi-agent system. The recommendation system has a hybrid approach, consisting of a CBR system and an argumentation framework. The CBR system is capable of deciding, on the basis of a series of metrics and similar cases stored in the system, whether a job offer is likely to be recommended to a user. Besides, the argumentation framework extends the system with an argumentation CBR, through which old and similar cases can be obtained from the CBR system. Finally, based on the different solutions proposed by the agents and the experience gained from past cases, a process of discussion among agents is established. Here, a debate is held in which a final decision is reached, giving the best recommendation to the proposed problem.

Keywords: Agents · Machine learning · Recommender systems Social networks · User experience

1 Introduction

The role social networks play in people's daily lives is probably greater than much of our society can imagine. From their main function as tools that interconnect people and allow them to share content to more specific functions such as the collection of personal data, creation of customized advertising or even job search. These possibilities are offered to users through the data they generate on the

M. Graña et al. (Eds.): SOCO'18-CISIS'18-ICEUTE'18, AISC 771, pp. 21–33, 2019.
https://doi.org/10.1007/978-3-319-94120-2_3

Internet [1]. One of the most successful types of social networks today are job search-oriented social networks. The present work focuses on this type of social network and, more specifically, on a social network called beBee [2]. The main goal of beBee is to obtain the greatest possible affinity between job offers and the users searching for a job. Thus, the more users find employment through the social network, the greater its success rate.

When a user registers on the network, he/she completes a form answering a series of questions related to his/her profile, such as his/her level of education, the skills he/she possesses, his/her experience, the languages he/she knows, the location in which he/she resides and a current salary range. Likewise, companies can sign up on the social network and publish job offers by filling another form with fields for the level of studies required by the offer, necessary skills and desirable skills, previous experience, languages, location of the offer or the salary offered.

Therefore, there is a network of interconnected users that aim to apply to job offers published by different companies. Making use of the capabilities of Big Data Analytics [3] techniques, user profiles and job offers are analysed so that there is a recommendation system embedded in the social network.

The main objective of a recommendation system is to offer suggestions to users (potential candidates) that are suited to their preferences and can therefore be accepted by them. Nowadays, recommendation systems are used in multiple domains and are usually context-dependent. There are many works where on the basis of content, they accurately suggest music [4], e-commerce products such as in the case of Amazon [5,6], movies [7], people or new contacts through user profile information and even job offers [8].

These recommendation systems are classified according to how recommendations are made, as described in [9]. The categories described are the following:

- *Collaborative recommendations*: recommendations are made on the basis of similar cases that have been made by other users.
- *Content-based recommendations*: recommendations are made based on past cases of the user.
- *Hybrid approaches*: recommendations are made using a combination of content-based and collaborative methods.

This article proposes a hybrid recommendation system that pursues an architecture based on agent technology. It is similar to that developed in [10] but it employs a Multi-Agent System (MAS) with a recommendation system that is capable of learning over through a Case-Based Reasoning (CBR) model.

A MAS is composed of a series of agents that collaborate with each other to reach a common goal, that could not be reached if individual agents were used. However, in the case of the present work, the common objective is the evaluation and proposal of the best recommendation by considering different aspects.

The first part of the hybrid approach of the recommendation system, is the CBR system that uses content to suggest the best recommendations to users. The CBR system operates by storing all the previously solved problems together with their solution, as cases. These similar cases are then used in the search for a solution to a new problem that emerges in the system [11,12]. The accuracy of the solution provided by the CBR is evaluated through human interaction with the machine. The second part of the recommendation system is the argumentation framework in which a society of agents uses its knowledge to obtain multiple solutions to problems. Subsequently, these agents collaborate with each other by debating and arguing each of the proposed solutions in order to select a single and the most appropriate solution to be recommended to the user.

In the next section the proposed hybrid recommender system is described in more detail and each of its parts is explained separately. Hereafter, a special case study is conducted where the efficacy of the proposed system is verified. Finally, we outline the obtained results and the conclusions they led us to.

2 Hybrid Recommender System Proposal

As explained in the introduction section, the recommendation system proposed in this paper is focused on recommending job offers to registered social network users. It is a hybrid system and consists of: (i) a CBR system powered by a series of metrics that calculate the affinity between job offers and users; (ii) an argumentation framework that extends the functionality of the CBR, thus making it possible for agents to have a dialogue; individual solutions proposed by each agent are first debated and settled, before a single final suggestion is obtained.

2.1 Case-Based Recommender System

The proposed CBR system is defined by different metrics that calculate the affinity between job offers and users. These metrics are calculated for each of the parameters specified below:

- **Main Education** (me): calculates the value of affinity between the offer and the user according to the level of studies defined by the user in their profile and the level required by the offer.
- **Required Skills** (rs): calculates the value of affinity between the offer and the user according to the skills defined in the user's profile and in the description of the offer.
- **Desirable Skills** (ds): calculates the value of affinity between the offer and the user according to the skills defined in the user's profile and in the description of the offer.
- **Experience** (e): calculates the value of affinity between the offer and the user on the basis of the user's previous experience defined in their profile and the experience required by the offer.
- **Language** (l): calculates the value of affinity between the offer and the user on the basis of the languages defined by the user in their profile and those required in the description of the offer.
- **Salary Range** (sr): calculates the value of affinity between the offer and the user according to the salary range defined in the user profile and that stated in the description of the job offer.

- **Geographic Area** (ga): calculates the value of affinity between the offer and the user according to the geographic location of the offer and the one established by the user in his profile.

Each of the defined metrics has an initial weight in the system. In this way, a recommendation that includes skills, education and experience in a user-offer relationship (variables to which a company would give more weight), in comparison to another recommender that includes salary range and geographic area (variables to which a user would give more weight), which will suggest different results when recommending job offers to users.

Thus, from the values obtained by each metric and the weight of each of them, the final affinity $a_{o,u}$ of the offer-user relationship (o, u) is defined in (1). Moreover, each of these weights is fed back into the CBR system as a past case.

$$a_{o,u} = \{me', rs', ds', e', l', sr', ga'\} \tag{1}$$

Each element that provides a value of affinity is the result of weighing the value of the corresponding metric with the weight obtained from the historical values stored in previous cases.

On the basis of the affinity value ($a_{o,u}$), it can be determined whether an offer should or should not be recommended to the user. This depends on whether it exceeds the threshold (thr_u) that is established by each recommender role for each user, on the basis of past cases.

The scheme followed by the CBR proposed in the system is shown in Fig. 1, where we can see the case that is used when evaluating if a new job offer in the system should or should not be suggested to a particular user. Throughout the process all the different stages that make up the CBR system take place (retrieve, reuse, revise and retain).

First, text mining is performed to identify keywords both in the user profile and the job offer description. Subsequently, the cases that are similar to the extracted content are retrieved from the past cases that belonged to the same professional area as the offer published (retrieve stage).

In the reuse stage, the keywords that were retrieved using text mining are processed in order to match the user's abilities with those listed in the offer. If the result is not substantial, the recommendation is discarded by the system. Similarly, a filtering of the geographical area is performed to verify whether it coincides with that of the user, if not the offer is discarded. On the contrary, if both the skills and the geographic area of the offer coincide with those of the user, an extraction of offer-user affinities is conducted by obtaining the value of $a_{o,u}$.

In the revise stage the result that has been attained in the previous stage is used as input. This input is evaluated by the system to decide whether the suggestion should be recommended to the user or discarded. If recommended the user can choose to either accept it or reject it. If the user accepts it, he automatically gets applied for the offer and it is left to the company to decide whether it considers his application or rejects it. These two decisions are stored in order to evaluate the system.

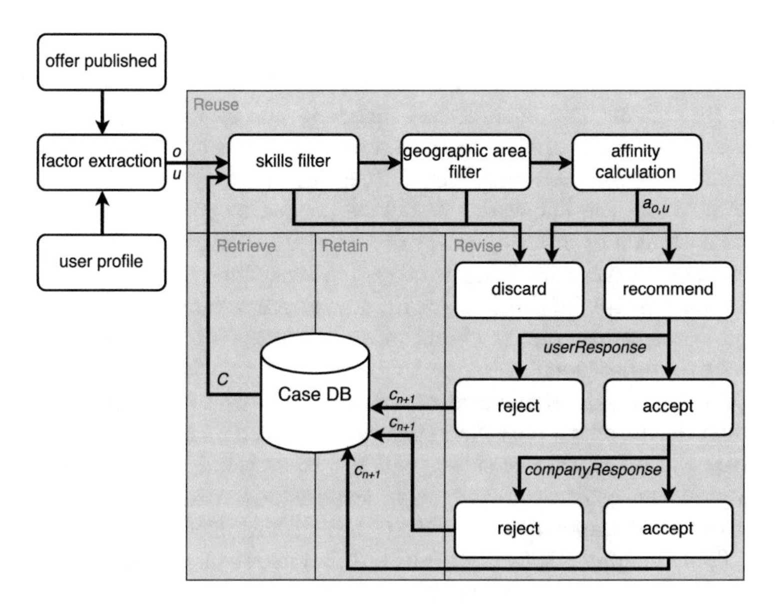

Fig. 1. State diagram.

Finally, each of the cases, defined as described in (2), must be stored in the case memory, where o is the variable that represents the factors extracted from the offer, u is the variable that represents the factors extracted from the profile of each user. The *userResponse* variable can take the value 0 or 1 depending on whether the offer has been rejected or accepted by the user. The variable *companyResponse* can also take the value 0 (if the company discards the user or if the user did not apply) or 1 (if the company considers the user).

$$C = \{o, u, a_{o,u}, userResponse, companyResponse\} \tag{2}$$

2.2 Argumentation Framework

The proposed system is entirely based on a MAS; it makes use of the context of an agent society. The concept of an agent society is defined as a set of agents that carry out different roles and follow a series of norms. In addition, there is a number of dependency relationships between agent roles. Furthermore, they are able to collaborate with each other and achieve the global goal of the whole agent society,something that could not be achieved individually, following a common language of communication [13,14].

However, the approach pursued in this work differs significantly from the classical definition of an agent society. In our system, each agent proposes its own solution to a given problem, on the basis of their knowledge. Subsequently, once agents propose multiple solutions to the problem, they generate arguments to refute the solutions of other agents and defend their own solution as the most optimal for the problem, until they finally reach an agreement on a single solution.

In this way the an argumentation framework is defined, where each of the agents proposes their own solution and defends it by attacking the solutions proposed by others. The agents that make up the system require knowledge resources in order to be able to provide solutions to the problems in the system, this knowledge is obtained with the CBR approach, a system that is proposed in Sect. 2.1. While the knowledge resources needed to generate debates and to refute the solutions of other agents, are stored in an argument-cases database, which contains all the past dialogues together with their final solution.

To give an example of the framework, an agent can represent a recommender who pursues a specific role by giving more importance to the user's skills and experience over other metrics. Hence, this recommender will generate a very large pool of potential candidates, however it is probable that many of them will not be interested in the offer, since their skills will be much more complete than those required by the offer and the salary will not correspond to their expectations. Similarly, another agent can represent a recommender with another type of role that gives greater importance to salary and geographical location over other metrics. This recommender will obtain a different set of possible candidates to the one generated by the previous recommender.

Therefore, to the prior definition of agent society we added the concepts that give value and preference to agents that follow the formal specification defined in [10]. Thus, an agent society is defined as follows:

Agent society. *Agent society in a certain time* t *is defined in* 3 *as a tuple* S_t:

$$S_t =< A_g, R_l, D, G, N, V, Roles, Dependency, Group, val, Valpref_Q > \qquad (3)$$

where:

- $A_g = \{a_{g1}, a_{g2}, ..., a_{gI}\}$ *is the set of* I *agent members* of S_t *in a certain time* t.
- $R_l = \{r_{l1}, r_{l2}, ..., r_{lJ}\}$ *is the set of* J *roles that have been defined in* S_t.
- $D = \{d_1, d_2, ..., d_K\}$ *is the set of* K *possible dependency relations in* S_t.
- $G = \{g_1, g_2, ..., g_L\}$ *is the set of groups that the agents of* S_t *form, where each* $g_L = \{a_1, a_2, ..., a_M\}$, $M \leq I$ *consist of a set of agents* $a_i \in A$ *of* S_t.
- N *is the defined set of norms that affect the roles that the agents play in* S_t.
- $V = \{v_1, v_2, ..., v_P\}$ *is the set of* P *values predefined in* S_t.
- $Roles : A_g \rightarrow 2^{R_l}$ *is a function that assigns an agent its roles in* S_t.
- *Dependency*: $<_D^{S_t} \subseteq R_l \times R_l$ *defines a reflexive, transitive and asymmetric partial order relation over roles.*
- $Group : A_g \rightarrow 2^G$ *is a function that assigns an agent its groups in* S_t.
- $val : A_g \rightarrow V$ *is a function that assigns an agent the set of values that it has.*
- $Valpref_Q \subseteq V \times V$, *where* $Q = A_g \vee Q = G$, *defines a irreflexive, transitive and asymmetric preference relation* $<_Q^{S_t}$ *over the values.*

Having described the argumentation framework, in the next section a case study is conducted with the job offer recommender system. In the case study the system follows the previously described model.

3 Case Study

This case study presents an implementation of the job offer recommender system for users on the social network, it follows the model proposed in Sect. 2. In this environment, the agents that make up the system propel the recommendation of offers to users on the social network. Each of these agents has its own personalized CBR system, through which it obtains the necessary knowledge resources to generate suitable recommendations.

In this manner, the system consists of an agent society S_t made up of a series of recommendation engines, which will act by taking on one of the three possible roles:

Economic recommender: its main objective is to favour the user economically. That is, the offers recommended to potential candidates give greater importance to the salary range stated by the user instead of strictly assessing the required skills. As a general rule, if a user is recommended a job with higher pay, he is more likely to accept the suggestion.

Safe recommender: its main objective is to find and offer the best offers to users according to the criteria provided in their profile. That is, they look for candidates that meet the requirements specified by the company. In this case, there may be overqualified users who will not be interested in the offers. Thus, it is not as likely for a user to accept the offer, but the companies will have a lower discard rate of the users who accepted the recommendation.

Manager: its main objective is to provide recommenders with new offers, so that they can be evaluated and recommended to the users.

As for dependency relationships between the different roles, they are based on the proposal made in [15]. The charity, authorisation and power relationships, which are applied to the described society S_t, are defined below:

- Economic recommender $<^{S_t}_{charity}$ Economic recommender; Economic recommender $<^{S_t}_{authorisation}$ Safe recommender; Economic recommender $<^{S_t}_{power}$ Manager
- Safe recommender $<^{S_t}_{Charity}$ Safe recommender; Safe recommender $<^{S_t}_{power}$ Manager
- Manager $<^{S_t}_{Charity}$ Manager

The manager has power over the safe and economic recommenders. Thus, the manager can impose his arguments about the correct solution on the safe and economic recommenders. Similarly, a safe recommender is superior to the economic recommender in making decisions, since its recommendation will always be more strict.

In addition, each of the agents has its own values that indicate their preferences (some may give more weight to skills, others to wages or languages etc.). The recommendation system is implemented in a MAS platform; the parts of this platform are described in the paragraphs below:

- **PANGEA:** PANGEA (Platform for Automatic coNstruction of orGanizations of intElligent Agents) is an agent platform for the development of open multiagent systems, the platform allows for the integral management of organizations and provides the end user with tools using the FIPA-ACL protocol [16]. Among the tools offered by PANGEA, we can find those that control communication through a series of norms, those that define virtual organizations and those for the secure control of persistent information.
- **Domain context CBR:** this module corresponds to the CBR system described in Sect. 2.1. It is initialized with a series of values that establish the level of importance of the metrics used, in this way the recommenders are able to consider different aspects depending on their role. In addition, it stores the values of each of the past cases, so that when a new case is presented, the most similar past cases are considered when generating a solution. Similarly, the case base is updated each time a new case is solved.
- **Argumentation CBR:** this is the other CBR system which stores all the past dialogues in the form of cases. They are used to generate new arguments, which are used to refute the proposals made by other agents. The cases stored in the CBR system are composed of the proposed problem and its solution, as outlined in Table 1.

Table 1. Argument-case example

Problem	Domain context	Premises ∪ $\{Evaluate\ offer\}$	
	Social context	Proponent	ID = R01
			Role = Economic recommender
			ValPref = $a_{o,u}(R01) < thr_u(R01)$
		Opponent	ID = R02
			Role = Safe recommender
			ValPref = $a_{o,u}(R02) > thr_u(R02)$
		Dependency relation = Authorisation	
Solution	Conclusion = "Discard offer"		
	Value = $a_{o,u}(R01) < thr_u(R01)$		
	Acceptability status = Accepted		
	Received attacks	Distinguishing premises = ∅	
		Counter examples = DC4	
Justification	Cases = ...		
	Dialogue graph = ...		

On the one hand, information regarding the problem is stored. A description of it with a series of premises, forms the domain context. In addition, the social context from which the problem proceeds, is stored. It is made up of the arguments held by the proponent and the opponent, both created depending on the role they follow, as well as their predefined values and the dependency relationship between them.

Next to the problem dealt with, an argument case stores the solution to the problem. That is; the conclusion, the resulting value, a level of acceptability of the solution and a series of records that count the number of attacks received when the case occurred.

Finally, the justification part of the argument-case is that which stores information on the cases used as knowledge resources, as well as a dialogue graph that represents the conversations maintained in the argument.

By means of this CBR system it is possible to determine if the proposed solution is likely or unlikely to be accepted as a final solution over the rest of the solutions provided by other agents.

- **Ontology parsers:** using ontologies is an advantage in that all agents can communicate more easily with each other, since a common language is used to define their characteristics and properties. In this way, the agents within the CBR domain context and the argumentation CBR are stored following the OWL 2 language [17]. All of these developed parsers provide an API through which the system's case data can persist.
- **Argumentation agent:** this agent is capable of initiating a conversation in a dialogue process in order to solve an incident. It contains a domain context CBR and an argumentation CBR, from which it can learn about both, the domain of the problem as well as the process of discussion. Thus, this agent updates the corresponding CBR, stores new cases and plays any of the roles defined in the system.
- **Commitment store**: it is a resource of the argumentation framework that stores all the information about the agents participating in the problem process. It has been implemented using the Information Agent provided by the PANGEA platform. Next, the workflow that is followed from the time a job offer arrives until a decision is made as to whether it should be recommended or not. This workflow is shown in Fig. 2, with each of the following steps establishing communication:

1. A company publishes a new job offer on the social network. In the proposed platform, recommender agents with different roles (economic and safe) are executed, in addition to the manager, who initiates the whole process. First, the Manager uses text mining techniques to extract the different factors from the job offer and the user profile. Subsequently, the Manager sends a request to recommender agents to start step 2.
2. Through its custom CBR, each recommender agent uses the existing metrics to evaluate the affinity ($a_{o,u}$) between the offer (o) and the profile of each user (u) that is found in the same professional sector as the offer. Hence, a decision is made on whether the offer should be recommended to the user or not. In addition, a comparison is made by extracting the past cases stored in the argumentation CBR.
3. Each agent stores its result in a common point within the commitment storage, waiting for other possible solutions from other agents to enter into debate.

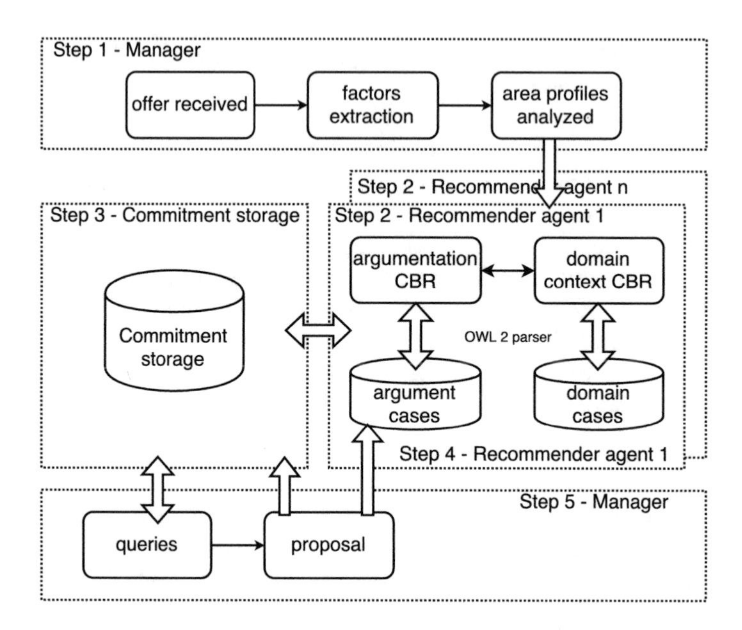

Fig. 2. Work-flow diagram.

4. In case there are other solutions for the same problem, each agent recovers its past cases from the argumentation CBR. The objective of these is to defend their own solution through the use of counter-examples that serve to refute the positions proposed by other agents.
5. Throughout this debate, the manager is continually making queries to commitment storage for counterexamples. After a certain timeout where no counterexample can be found, it ends the conversation and stays with the best positioned result in the common point as the final solution.
6. Finally, the case and its final solution are stored in the case base of each argumentation CBR. Moreover, the discussion is stored in commitment storage.

4 Results and Conclusions

This study presents a hybrid recommendation system based on agent technology. Through the use of intelligent techniques it is able to carry out analyses of user profiles and job offers found on the social network. The CBR system uses the contents extracted from the network to suggest the best recommendations to the users.

Apart of the agents that support the system's functionality, the hybrid system is also composed of a CBR system. It handles a series of metrics which calculate the affinity between job offers and users. Moreover, an argumentation framework is included which expands the functionality of CBR, in such way that the agents have the possibility of establishing a dialogue between them. When a query arises in the system, each agent first analyses the social network individually and each comes up with a solution, thus multiple solutions are proposed. Then a dialogue is established where each agent generates arguments in order to refute the solutions of other agents and defend their own solution as the most optimal for the problem, until finally an agreement is reached. Thus, dialogue between agents is an essential feature in the system that allows to find a unique and the most optimal solution.

The argumentation mechanism is capable of self-adapting to the needs of each individual recommendation (user-offer) in different aspects. Self-adaptation is achieved thanks to the following 3 actions: (i) extraction of parameters related to user profiles and job offer descriptions; (ii) the calculation of different thresholds used to determine the level of affinity between the user and the job offer; and (iii) the use of past cases. Argumentation provides precision and adapts well when determining the value of thresholds, given that the different aspects considered in the negotiation represent the two parties (users and companies) interested in the recommendation.

The safe recommender role has a negative influence on the percentage of recommendations that are accepted by the user, this is because it is restrictive with user skills and can therefore recommend certain offers to overqualified users. On the other hand, the economic recommender role would have a high recommendation acceptance rate among users, but consequently the rate of discards would also be high, since the user's qualifications may not exactly fulfil those required by the company. Therefore, negotiation between the different agents is necessary, creating a hybrid approach that balances the percentages in order to obtain adequate acceptance rates from both by the users and the companies, as can be seen in Table 2.

Table 2. Recommender results.

	User	Company
Accepted	5,192	1,129
Rejected	1,850	4,063
Ignored	7,853	5,192

More specifically, the system has been evaluated on a sample of active users searching for employment in the 15 days prior to the study. A total of 15,616 recommendations have been made. The majority of users have not interacted with the recommendation system, so unanswered recommendations are classified as "ignored". The acceptance rate of users who did interact with the recommendation system is 73.72%. Out of the 5,192 user applications made through the recommender system, the companies did not rule out 1.129, which is 21.74%, higher than the 18.23% the acceptance rate on the part of companies when the recommendation system was not used.

Therefore, we can assert that the recommendation system, functioning on the basis of the argumentation system, has improved the selection process of social network job applicants by 3.6 points, which results in more users joining the network and companies having greater interest in publishing their offers on it. Thus, user experience is improved from the point of view of all stakeholders.

Acknowledgments. This work was conducted within the framework of a project with Ref. RTC-2016-5642-6, financed by the Ministry of Economy, Industry and Competitiveness of Spain and the European Regional Development Fund (ERDF). The research of Alfonso González-Briones has been co-financed by the European Social Fund (Operational Programme 2014–2020 for Castilla y León, EDU/128/2015 BOCYL).

References

1. Serrat, O.: Social network analysis. In: Knowledge Solutions, pp. 39–43. Springer (2017)
2. beBee Affinity Social Network: bebee, successful personal branding, 20 July 2017. https://www.bebee.com/
3. Russom, P., et al.: Big data analytics. TDWI best practices report, fourth quarter, vol. 19, p. 40 (2011)
4. Song, Y., Dixon, S., Pearce, M.: A survey of music recommendation systems and future perspectives. In: 9th International Symposium on Computer Music Modeling and Retrieval, vol. 4 (2012)
5. Schafer, J.B., Konstan, J., Riedl, J.: Recommender systems in e-commerce. In: Proceedings of the 1st ACM Conference on Electronic Commerce, pp. 158–166. ACM (1999)
6. Linden, G., Smith, B., York, J.: Amazon.com recommendations: item-to-item collaborative filtering. IEEE Internet Comput. **7**(1), 76–80 (2003)
7. Miller, B.N., Albert, I., Lam, S.K., Konstan, J.A., Riedl, J.: MovieLens unplugged: experiences with an occasionally connected recommender system. In: Proceedings of the 8th International Conference on Intelligent User Interfaces, pp. 263–266. ACM (2003)
8. Bonhard, P., Sasse, M.A.: 'Knowing me, knowing you' – using profiles and social networking to improve recommender systems. BT Technol. J. **24**(3), 84–98 (2006)
9. Adomavicius, G., Tuzhilin, A.: Toward the next generation of recommender systems: a survey of the state-of-the-art and possible extensions. IEEE Trans. Knowl. Data Eng. **17**(6), 734–749 (2005)
10. Jordán, J., Heras, S., Valero, S., Julián, V.: An argumentation framework for supporting agreements in agent societies applied to customer support. In: Hybrid Artificial Intelligent Systems, pp. 396–403 (2011)
11. Aamodt, A., Plaza, E.: Case-based reasoning: foundational issues, methodological variations, and system approaches. AI Commun. **7**(1), 39–59 (1994)
12. Lorenzi, F., Ricci, F., Tostes, R., Brasil, R.: Case-based recommender systems: a unifying view. LNCS, vol. 3169, p. 89 (2005)
13. Dignum, M.: A model for organizational interaction: based on agents, founded in logic. SIKS (2004)
14. Artikis, A., Sergot, M., Pitt, J.: Specifying norm-governed computational societies. ACM Trans. Comput. Logic (TOCL) **10**(1), 1 (2009)
15. Dignum, F., Weigand, H.: Communication and deontic logic (1995)

16. Sánchez, A., Villarrubia, G., Zato, C., Rodríguez, S., Chamoso, P.: A gateway protocol based on FIPA-ACL for the new agent platform PANGEA. In: Trends in Practical Applications of Agents and Multiagent Systems, pp. 41–51. Springer (2013)
17. Group, W.O.W., et al.: OWL 2 web ontology language document overview (2009)

Soft Computing Applications

Retinal Blood Vessel Segmentation by Multi-channel Deep Convolutional Autoencoder

Andrés Ortiz[1(✉)], Javier Ramírez[2], Ricardo Cruz-Arándiga[1], María J. García-Tarifa[1], Francisco J. Martínez-Murcia[2], and Juan M. Górriz[2]

[1] Communications Engineering Department, University of Málaga, 29004 Málaga, Spain
{aortiz,rcruz,mj.gtarifa}@ic.uma.es

[2] Department of Signal Theory, Communications and Networking, University of Granada, 18060 Granada, Spain
{javierrp,fjesusmartinez,gorriz}@ugr.es

Abstract. The evaluation and diagnosis of retina pathologies are usually made by the analysis of different image modalities that allows to explore its structure. The most popular retina image method is the retinography, a technique to show the retina and other structures in the fundus of the eye. This paper deals with an important stage of the retina image processing for a diagnosis tool which aims to show the blood vessel structure. Our proposal is based on a deep convolutional neural network, that avoids any preprocessing stage such as gray scale conversion, histogram equalization, and other image transformations that determine the final result. Thus, we obtain the blood vessel segmentation directly from the original RGB color retinography image. The results obtained with our method are comparable to the state-of-the art methods but using a smaller network with less memory and computation requirements. Our approach has been assessed using the DRIVE database.

1 Introduction

Retina is a layered tissue that constitutes an essential part of the human vision system. It is a sensorial membrane that recover the inner part of the back of the eye. Different parts of the retina contain specialized cells called photoreceptors, which are responsible of the extraction of different features that eventually allows to distinguish not only objects but also object motion. Signals produced by photoreceptors are driven to the visual brain cortex for further and complex processing such as texture detection, object recognition, etc. As a consequence, the health of the retina directly affects to the quality of life of the population. This way, the precise diagnosis of retina pathologies allows an early treatment that maximizes its success probability as most retinal pathologies imply a degenerative process.

M. Graña et al. (Eds.): SOCO'18-CISIS'18-ICEUTE'18, AISC 771, pp. 37–46, 2019.
https://doi.org/10.1007/978-3-319-94120-2_4

The diagnosis of retinal pathologies is not always evident from retinal images, specially in the early stages of the disease. Thus, computer-aided image processing tools can help to clearly identify abnormal patterns linked to a disease [12]. Specifically, some of these diseases such as Diabetic Retinopathy (DR) can be discovered by abnormal patterns or damages in the retina blood vessel structure. Hence, the extraction of blood vessel from a retinography plays an important role in (1) the visual recognition task made by an ophthalmologist and (2) the construction of a computer-aided diagnosis tool for DR.

Some previous works have been focused in the segmentation of the retina blood vessels: [17] presents a classical segmentation approach which uses 2D Gabor wavelet and supervised classification by means of a Bayesian classifier with class-conditional Gaussian mixtures (GMM). This method uses only the green (G) channel from the RGB image to compute wavelet-based features and in a further step, a specific preprocessing algorithm to select the ROI pixels. [13] proposes the use of 2D Gabor filter bank as low-level oriented edge discriminators to extract features on a selected, balanced vessel/non vessel dataset for pattern learning. Again, a preprocessing stage using the green channel is used to enhance the image as well as a feature compression stage by Principal Component Analysis to filter out noisy components to improve the classification performance. In [15] a three-stage pipeline composed of a preprocessing stage that includes filtering and morphological transformations to enhance the image is used to extract the blood vessels by supervised classification by means of a GMM classifier. On the other hand, [2] is based on an ensemble of bagged and boosted decision trees and also includes a preprocessing stage and a feature extraction stage based on the orientation analysis of gradient vector field.

Novel proposals use the capabilities of deep learning to improve the segmentation performance. For instance, [7] uses a 10 layer, convolutional neural network as a classifier to generate the segmented images from the green channel in the original ones. Another deep learning approach using a large convolutional network is presented in [6], where the input data from the DRIVE [4] is preprocessed with global contrast normalization, zero-phase whitening, and augmented using different geometric transformations and gamma corrections. In addition, the *Unet* proposal [14] uses a convolutional autoencoder in a similar way that the work we present here, original images are preprocesed first converting them to gray scale and then performing a histogram equalization. In this paper we present a method based on a deep multichannel convolutional autoencoder to avoid the dependance to the preprocessing stage. Instead, some of the parameters related to the preprocessing aimed to enhance the image and to ease the extraction of patterns by convolution operators are computed by the network.

2 Methods

In this Section, the retinal image database and method used to extract the vessels are described.

2.1 Preprocessing

The main objective of this work consist on avoiding the use of predefined preprocessing stages, which may include gray scale conversion of the RGB original image, the use of only one color channel (typically the green channel as done in other works) or histogram equalization. The use of a mask for Field of View (FOV) selection is also avoided. On the contrary, each color channel is fed into the network in a multichannel approach. The only preprocessing we performed on the images consists on splitting each image into 48×48 overlapped patches with two objectives: (1) training the network using patches drastically diminishes the GPU memory requirements and (2) the training can be speeded up. Thus, each image is decomposed into 278166 2D (48×48) patches for each color channel.

2.2 Database

All the experiments in this work were carried out using the publicly-available DRIVE dataset [1,4,8], obtained from a diabetic retinopathy screening program in The Netherlands. These, Forty images composing the training and testing datasets are the result of the screening population consisted of 400 diabetic subjects between 25–90 years of age. 33 do not show any sign of diabetic retinopathy and 7 show signs of mild early diabetic retinopathy. Moreover, for training and testing images, a manual segmentation of the vasculature is available, which is used as gold standard in the learning process. These are generated by experienced ophthalmologists that were asked to mark all pixels for which they were for at least 70% certain that they were vessel.

2.3 Deep Convolutional Autoencoders

In general, an autoencoder can be seen as a symmetric neural network which is trained to reconstruct the original data. In an autoencoder, two parts can be differentiated. The first is called *encoder*, that maps the original image into a low-dimensional (namely, latent) space. The second part called *decoder* aims to reconstruct the original image from its low dimensional representation (i.e. in the latent space) with a minimum reconstruction error. Traditional autoencoders are composed of different linear layers with a diminishing number of neurons from the input layer of the encoder to the autoencoder's bottleneck. In this work we used a convolutional autoencoder, in which linear layers are replaced by convolutional layers to build two symmetric convolutional networks.

2.4 Convolutional Neural Networks

Convolutional Neural Networks (CNN) [5] are a type of deep learning networks originally designed for image recognition applications [3,11]. In fact, CNNs are one of the is one of the fields in which deep learning techniques have provided higher improvements in the classification performance with respect to previous statistical learning classifiers. CNNs work in a different way compared to these neural networks. They aim to extract the most representative features from the images, instead of using pre-computed ones by means of the convolution layers. The basic operation in CNNs is based on the convolution operator. Convolution allows to extract features from an image once defined a weight or kernel matrix $\mathcal{K}$. This weight matrix acts like a filter, which extracts specific local information when it is slided over the original image. Instead of using a fixed weights as in traditional convolution, CNNs learns the weight matrix that extracts the most discriminative information by a backpropagation-like method. In image applications, we use multidimensional arrays instead of time series. This way, x and $\mathcal{K}$ are tensors. For the sake of clarity, we refer to these as I and K, respectively. Moreover, the convolution operation is performed in more than one axis at a time.

In the case of two-dimensional convolutions, $\mathbf{V}_{i-1}$ is a tensor of size $H \times W \times C$ (height, width and number of channels), and $\mathbf{W}_i$ a tensor of size $P \times Q \times R \times K$, containing K filters of size $P \times Q \times R$, with $R_i = K_{i-1}$ and $R_1 = 1$ for the first layer. With this definitions, the k-th convolution term for the k-th filter is:

$$\mathbf{W}_{ik} * \mathbf{V}_{i-1} = \sum_{u=0}^{P-1} \sum_{v=0}^{Q-1} \mathbf{W}_{ik}(P-u, Q-v)\mathbf{V}_{i-1}(x+u, y+v) \tag{1}$$

In CNNs design, there are essentially three hyperparameters. The first is the filter size $P \times Q \times R$. The other two hyperparameters that control the output size of the activation layers are stride and zero-padding. Stride defines the step at which the convolution is computed, or in other words, how much overlapping there is between convolutions. That also defines the receptive field of a neuron in the convolution layer, which is the part of the image to which each convolution filter is connected to. Using a stride of 1, the convolution is performed at each voxel of the input. With a higher stride, there is less overlapping between receptive fields, and the output volumes will be smaller.

2.5 Segmentation Using Convolutional Autoencoders

Image segmentation consists in partitioning an image into different regions. A large variety of image segmentation approaches have been implemented for this purpose. Nevertheless, most of them use a priori knowledge about the pixel classification, which prevents figuring out other tissue classes different from the classes the system was trained for. These methods include classical statistical approaches, neural network based approaches and hybrid methods [9,10]. Recently, deep learning techniques have leveraged the performance of classification problems, especially in image applications. Thus, as segmentation is basically a classification problem, deep learning techniques can be applied. One method consist on the use of autoencoders trained in a supervised fashion. Autoencoders and, in particular, convolutional autoencoders learns to reconstruct the original data by unsupervised learning. This is achieved by modifying the network weights to match the input and output during training. However, the same neural structure can be used as a supervised classifier, which is useful for a number of applications such as image segmentation. In this case, while the original image is used as input in the encoder part, the decoder is trained to generate the segmented version of the image. In our case, we use a convolutional autoencoder in a similar way that [14]. As explained in the Introduction, most approaches use the green channel as it provides the more contrast between vessels and other part of the image. In other words, most part of the information is contained in the green channel.

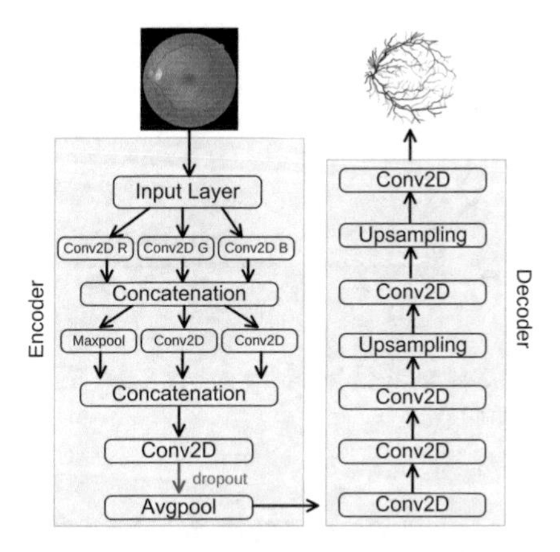

Fig. 1. Structure of the convolutional autoencoder used in this work

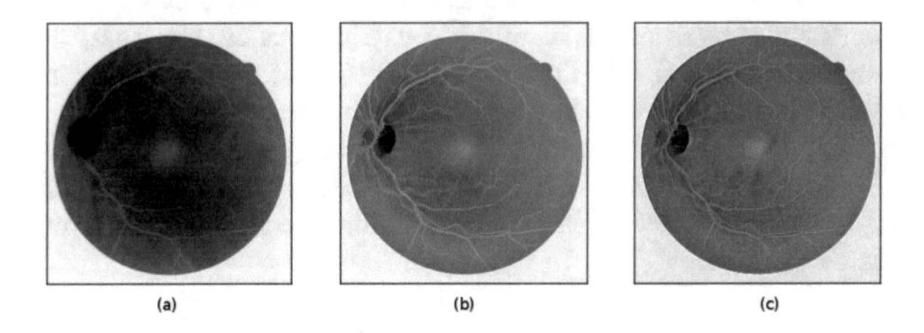

Fig. 2. Color channels for an example image. (a) Red, (b) Green, (c) Blue

The three color channels for an example image can be seen in Fig. 2. Most previous approaches also include other preprocessing in order to enhance the image, easing the learning process. In this work we used the structure shown in Fig. 1, where each channel is fed into a different convolutional layer. Moreover, the convolution results are concatenated and the split into three new layers, including maxpooling and convolutions with stride of 2 pixels. The use of this structure aims to process each channel separately The objectives of using a multichannel approach are (1) avoid the preprocessing stage, as it greatly determines the final prediction performance, (2) making the preprocessing as a part of the learning process, so that we can study how the neural network is mixing the color channels to obtain the best performance as well as the images generated by the network in the intermediate layers, which corresponds to the learnt features.

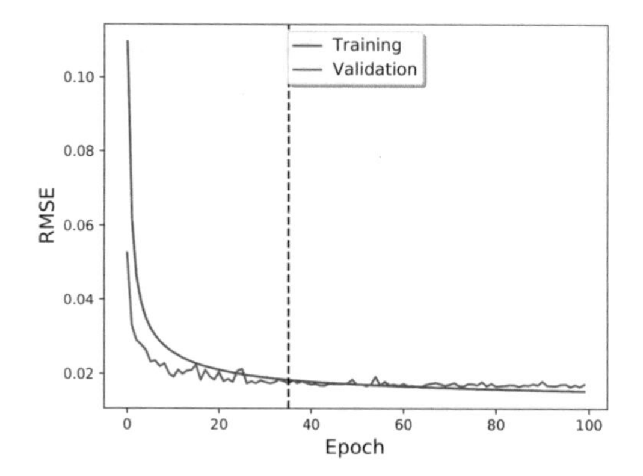

Fig. 3. Learning curve. 80% of available samples are used for training whereas remaining 20% are used for validation.

The network structure used in this work can be seen in Fig. 1. As shown in Table 2, although providing a lower performance, our approach uses 1.8 times less parameters than the Unet proposal, which implies less memory and computation requirements. In fact, the network in Fig. 1 learns to classify vessel pixels in just 25 epochs. Thus, autoencoder training is carried out in less than 3 h using a NVIDIA Geforce GTX-1080Ti GPU with 11GB of GDDR5X memory. The learning curve is shown in Fig. 3, where the evolution of the root mean squared reconstruction error (RMSE) is shown as the learning progresses. In this Figure, training is performed using 80% of available samples whereas validation is carried out with the remaining 20% of samples, obtaining a reconstruction error of 1.5% for validation samples.

2.6 Data Augmentation

The autoencoder in Fig. 1 is trained during 35 epochs along with data augmentation to increase the number of available samples. Data augmentation is performed by rotating all the images 180°. This way, a total amount of 40 images are available for training.

3 Results

This section shows the results obtained using the DRIVE training dataset to learn the model and the test dataset to assess the performance. It is worth noting that segmentation results show a membership probability map, in which black represents a probability of 1 and white represents a probability of 0. As an example, Fig. 6 show an original image (a), the segmentation obtained with our approach (b) and the manual segmentation (c).

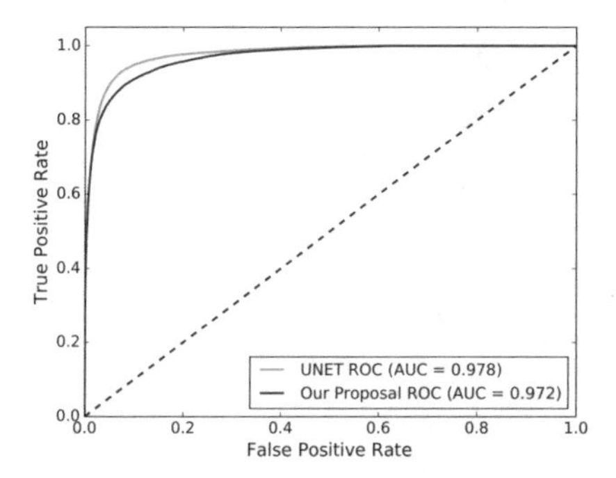

Fig. 4. ROC curves for the best performing method (unet [14]) and our approach

Table 1. Area under ROC Curve obtained for different segmentation methods.

Method	AUC
Soares et al. [17]	0.961
Osareh et al. [13]	0.965
Roychowdhury et al. [15]	0.970
Melinscak et al. [7]	0.975
Liskowski et al. [6]	0.979
Unet [14]	9.979
Our approach	0.972

Moreover, Fig. 5 show the segmentation result of the first four images in the DRIVE dataset, and Table 2 shows mean binary performance metrics for the test images. Additionally, 1 presents a performance comparison among different segmentation methods in terms of Area Under Roc Curve (AUC).

As explained above, the images in Fig. 5 show the probability map. These probabilities have been converted to binary values (vessel/non vessel) by computing the thresholding them using the mean ROC best cut-off point computed for the training images. On the other hand, Fig. 4 shows the ROC curves for the test images of the DRIVE database computed using the *Unet* method (orange line) and our approach (blue line). Both methods provide an AUC of 0.979 and 0.972, respectively.

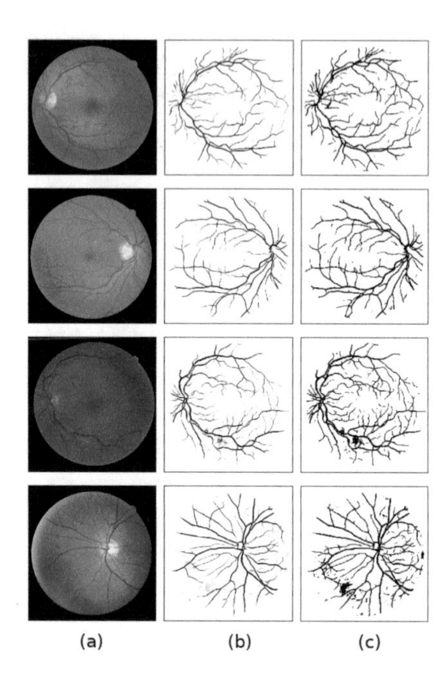

Fig. 5. Segmentation examples for the four first images in the DRIVE dataset. (a) original image, (b) probability map, (c) binarized image using the mean best cut-off point computed for the training set.

Table 2. Performance Comparison

Method	Performance metric			
	Accuracy	Sensitivity	Specificity	N^oparameters
Unet [14]	0.95	0.76	0.98	517,090
Our approach	0.96	0.73	0.92	277,681

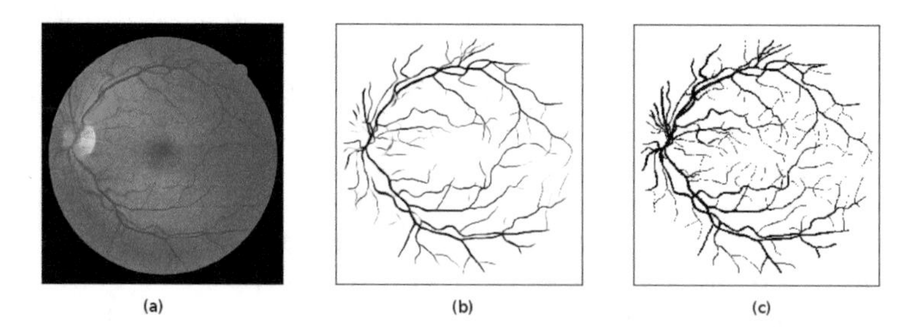

Fig. 6. Segmentation example. (a) original image, (b) Segmentation probability map obtained with our proposal and (c) ground truth

4 Conclusions and Future Work

In this paper, we present an alternative to previous approaches for blood vessel segmentation in retinal images. Our approach aims to avoid the preprocessing stages that greatly determines the final classification performance and, therefore, the segmentation accuracy. The proposed method is based on a deep convolutional autoencoder used as a supervised classifier, which is trained with the training set and its corresponding manual segmentations as ground truth. Unlike previous proposals, original images are neither preprocessed or converted to gray scale or equalized. Our approach aims to let the neural network to compute the optimal parameters to optimize the classification performance. Although the classification results do not outperform the *Unet* approach, they are good enough to continue the research in this direction to design a full Deep Learning approach without a priori determined preprocessing techniques. We plan to focus the research in the improvement of the first layers of the encoder. For instance, the optimization could be improved using some residual blocks at the encoder, in an attempt to improve the minimum RMSE, as well as to explore the saliency maps [16] to analyze the features the network is extracting and the relevant regions used from each color channel. On the other hand, we plan to improve the binarizing method using local techniques, in order to avoid pixel clusters in the image.

Acknowledgments. This work was partly supported by the MINECO/FEDER under TEC2015-64718-R and PSI2015-65848-R projects and the Consejeráa de Innovación, Ciencia y Empresa (Junta de Andalucía, Spain) under the Excellence Project P11-TIC-7103. We gratefully acknowledge the support of NVIDIA Corporation with the donation of the GPU used for this research.

References

1. Drive database: Digital Retinal Images for Vessel Extraction. https://www.isi.uu.nl/Research/Databases/DRIVE/
2. Fraz, M.M., et al.: An ensemble classification-based approach applied to retinal blood vessel segmentation. IEEE Trans. Biomed. Eng. **59**(9), 2538–2548 (2012)
3. Goodfellow, I., Bengio, Y., Courville, A.: Deep Learning. MIT Press, Cambridge (2016). http://www.deeplearningbook.org
4. Staal, J.J., Abramoff, M.D., Niemeijer, M., Viergever, M.A., van Ginneken, B.: Ridge based vessel segmentation in color images of the retina. IEEE Trans. Med. Imaging **23**, 501–509 (2004)
5. Krizhevsky, A., Sutskever, I., Hinton, G.E.: ImageNet classification with deep convolutional neural networks. Adv. Neural Inf. Process. Syst. **25**, 1097–1105 (2012)
6. Liskowski, P., et al.: Segmenting retinal blood vessels with deep neural networks. IEEE Trans. Med. Imaging **35**, 2369–2380 (2016)
7. Melinscak, M., et al.: Retinal vessel segmentation using deep neural networks. In: Proceedings of the 10th International Conference on Computer Vision Theory and Applications (VISIGRAPP 2015), pp. 577–582 (2015)

8. Niemeijer, M., Staal, J.J., Van Ginneken, B., Loog, M., Abramoff, M.D.: Comparative study of retinal vessel segmentation methods on a new publicly available database. In: Michael Fitzpatrick, J., Sonka, M. (eds.) SPIE Medical Imaging SPIE, vol. 5370, pp. 648–656 (2004)
9. Ortiz, A., Górriz, J.M., Ramírez, J., Salas-Gonzalez, D., Llamas-Elvira, J.M.: Two fully-unsupervised methods for MR brain image segmentation using SOM-based strategies. Appl. Soft Comput. **13**(5), 2668–2682 (2013)
10. Ortiz, A., Górriz, J.M., Ramírez, J., Salas-Gonzalez, D.: Improving MR brain image segmentation using self-organising maps and entropy-gradient clustering. Inf. Sci. **262**, 117–136 (2014)
11. Ortiz, A., Munilla, J., Górriz, J.M., Ramírez, J.: Ensembles of deep learning architectures for the early diagnosis of the Alzheimer's Disease. Int. J. Neural Syst. **26**, 07 (2016)
12. Ortiz, A., Górriz, J.M., Ramírez, J., Martínez-Murcia, F.J.: Automatic ROI selection in structural brain MRI using SOM 3D projection. PLoS ONE **9**(4), e93851 (2014)
13. Osareh, A., et al.: Automatic blood vessel segmentation in color images of retina. Iran. J. Sci. Technol. Trans. B Eng. **33**(B2), 191–206 (2009)
14. Ronneberger, O., Fischer, P., Brox, T.: U-Net: convolutional networks for biomedical image segmentation. arXiv:1505.04597v1 [cs.CV], 18 May 2015
15. Roychowdhury, S., et al.: Blood vessel segmentation of fundus images by major vessel extraction and subimage classification. IEEE J. Biomed. Health Inform. **19**(3), 1118–1128 (2015)
16. Simonyan, K., Vedaldi, A., Zisserman, A.: Deep Inside convolutional networks: visualising image classification models and saliency maps. CoRR, abs/1312.6034 (2013)
17. Soares, J.V., et al.: Retinal vessel segmentation using the 2-D Gabor wavelet and supervised classification. IEEE Trans. Med. Imaging **25**(9), 1214–1222 (2006)

Deep Convolutional Autoencoders vs PCA in a Highly-Unbalanced Parkinson's Disease Dataset: A DaTSCAN Study

Francisco Jesús Martinez-Murcia[1(✉)], Andres Ortiz[2], Juan Manuel Gorriz[1], Javier Ramirez[1], Diego Castillo-Barnes[1], Diego Salas-Gonzalez[1], and Fermin Segovia[1]

[1] Department of Signal Theory, Networking and Communications, University of Granada, 18071 Granada, Spain
fjesusmartinez@ugr.es

[2] Department of Communications Engineering, University of Málaga, 29071 Málaga, Spain

Abstract. The automated analysis of medical imaging, especially brain imaging, is a challenging high-dimensional task. Computer Aided Diagnosis (CAD) tools often require the images to be spatially normalized and then perform feature extraction to be able to avoid the small sample size problem. However, the spatial normalization often introduces artefacts, especially in functional imaging. Furthermore, variance-based decomposition techniques like PCA, which are extensively used in CAD tools, often perform poorly in highly-unbalanced dataset. To overcome these two problems, we propose a deep Convolutional Autoencoder (CAE) architecture that performs image decomposition -or encoding- in images that were not spatially normalized. A CAD system that used CAE for feature extraction and a Support Vector Machine Classifier (SVC) for classification was tested on a strongly imbalanced (5.69/1) Parkinson's Disease (PD) neuroimaging dataset from the Parkinson's Progression Markers Initiative (PPMI), achieving more than 93% accuracy in detecting PD with DaTSCAN imaging, and a area under the ROC curve higher than 0.96. This system paves the way for new deep learning decompositions that bypass the common spatial normalization step and are able to extract relevant information in highly-imbalanced datasets.

Keywords: Autoencoder · Convolutional neural networks
Deep learning · Parkinson's Disease · DaTSCAN

1 Introduction

Medical imaging, and especially neuroimaging, still poses a big challenge for automated systems that perform pattern recognition. Their high dimensional

M. Graña et al. (Eds.): SOCO'18-CISIS'18-ICEUTE'18, AISC 771, pp. 47–56, 2019.
https://doi.org/10.1007/978-3-319-94120-2_5

nature makes systems prone to the small sample size problem [3], that is, the loss of generalization ability of the models when the number of samples is much smaller than the number of features available. In this context, neuroimaging modalities range from hundred of thousands to millions of voxels, while cohorts are often in the range of tens or hundreds, with few notable exceptions such as the Parkinson's Progression Markers Initiative (PPMI) [8] or the Alzheimer's Disease Neuroimaging Initiative (ADNI) [30], which often contain a few thousand subjects.

Within the Computer Aided Diagnosis (CAD) paradigm, the small sample size problem is often addressed via feature extraction techniques. Many approaches have been applied to neuroimaging, among others decomposition techniques such as Independent Component Analysis (ICA) [2,13] or Partial Least Squares (PLS) [25], texture analysis [11,16] or morphological measures such as volume, area, etc [4,5]. Of these, the most popular is probably Principal Component Analysis (PCA), a technique that produces an internal, low-dimensional orthogonal representation of a dataset maximizing the variance explained by each component. The component scores have been extensively used to decompose brain images in the neuroimaging literature [7,10,12] that could help to understand the underlying patterns to a disease.

A common problem faced by CAD systems (and in particular, PCA) is spatial normalization. PCA needs that every image voxel represents the same anatomical position along all subjects. In high-resolution modalities such as Magnetic Resonance Imaging (MRI) this is easily achieved via common packages such as SPM [6], by estimating a series of parameters that minimize the differences between that image and a template in a common space. This is, however, more difficult to apply to nuclear imaging modalities (PET, SPECT), where patterns are more varying and noisier. Spatial normalization in these cases often creates artefacts that could lead to degraded performance. Furthermore, their dependence on dataset variance may make them unsuitable in the case of highly-imbalanced datasets [17], focusing in common patterns of the most prevalent class and obviating the most interesting patterns: the differences between classes.

Deep learning methodologies are a growing trend in neuroimaging, given their ability to extract relevant features for classification [19,31] without a need for spatial normalization. In some cases, complete architectures are created combining feature extraction with convolutional layers and classification, such as [15,29]. Autoencoders (or encoder-decoders) are a particular architecture aimed at obtaining an internal representation (encoding) of a given dataset that minimizes the reconstruction error. Their non-linear abstractions can adapt better to the underlying manifold structure of the data, making for a more effective representation in the latent space [20,21].

In this work, we propose a Convolutional Autoencoder (CAE) architecture for feature extraction in nuclear brain imaging of Parkinson's Disease (PD). The complete CAD system also includes classification using a Support Vector Machine classifier (SVC) with RBF kernel. These are introduced in Sect. 2. Then, in Sect. 3, we evaluate and analyse the proposed system and the internal representations, compared to a traditional CAD system using PCA. Finally, we draw some conclusions in Sect. 4.

2 Methodology

2.1 PPMI Dataset

Data used in the preparation of this article were obtained from the Parkinson's Progression Markers Initiative (PPMI) database (www.ppmi-info.org/data). For up-to-date information on the study, visit www.ppmi-info.org. The images in this database were imaged $4 + 0.5$ h after the injection of between 111 and 185 MBq of DaTSCAN. Raw projection data are acquired into a 128×128 matrix stepping each 3° for a total of 120 projection into two 20% symmetric photo-peak windows centred on 159 KeV and 122 KeV with a total scan duration of approximately $30 - 45$ min [8]. 1305 DaTSCAN images have been used, of which 195 belonged to control subjects (CTL) and 1110 to subjects affected by Parkinson's Disease (PD).

Image Preprocessing. The images have not been spatially normalized. However, since they are smooth images, only affected by some noise, we have downsampled them to a factor of 1.5. Their resulting shape of (48,52,48) was trimmed down to (48,48,48) by removing the first and last cuts of the coronal axis, which did not contain information about the brain, but some intensity related to the skull. Finally, the intensity of the images was divided by the average value of their 1% maximum intensity voxels, in a procedure commonly known as normalization to the maximum [14].

2.2 Principal Component Analysis

In this work, we apply the Principal Component Analysis (PCA) decomposition technique (which could be considered an encoding), which has been heavily used in image processing [18,26], medicine [12] or gene analysis [24] among others. PCA performs a decomposition of the dataset as a linear combination $\mathbf{s}_n$ of principal component loadings $\mathbf{W}$ (of size $p \times K$), which are estimated in order to maximize the variance explained:

$$\mathbf{x}_n = \mathbf{W}\mathbf{s}_n \tag{1}$$

The component scores $\mathbf{s}_n$ can be considered an encoded version of the original vector $\mathbf{x}_n$. Ideally, the length of $\mathbf{s}_n$ is infinite, so that all variance is captured. We usually estimate only L components of this decomposition, so that the reconstruction error ϵ is small:

$$\epsilon = \mathbf{x}_n - \hat{\mathbf{x}}_n, \quad \text{where} \quad \hat{\mathbf{x}}_n = \mathbf{W}\mathbf{s}_n^L \tag{2}$$

The most popular approach for PCA estimation is the Singular Value Decomposition (SVD) of the data matrix $\mathbf{X}$:

$$\mathbf{X} = \mathbf{U}\boldsymbol{\Sigma}\mathbf{V}^T \tag{3}$$

So that the component loadings are $\mathbf{W} = \mathbf{V}$, the eigenvalues $\boldsymbol{\Lambda} = \boldsymbol{\Sigma}^2$, and the transformed samples $\mathbf{S} = \mathbf{X}\mathbf{W} = \mathbf{U}\boldsymbol{\Sigma}$.

2.3 Convolutional Autoencoder

Autoencoders (also known as encoder-decoder architectures) are a particular type of neural network whose main aim is to learn an internal, low-dimensional representation of an image [1,22]. This is done by combining a encoder and a decoder neural network, in which the output layer of the encoder (also known as z-layer) is also the input layer of the decoder. This architecture is trained by minimizing the reconstruction error between the output and input data.

Traditional encoders and decoders are composed of Fully Connected (FC) layers, also known as dense layers. In those, all neurons in a layer are connected to all neurons in the following layers, in a similar way to a multilayer perceptron. Since convolutional architectures are gaining ground in image processing and analysis, convolutional layers can be used instead of dense layers, accounting for better features.

Convolutional neural networks use a wider variety of layers, but in this work we will specifically use convolutional and upsampling layers. Convolutional layers perform the convolution of the input $\mathbf{V}_{i-1}$ with a set of K filters $\mathbf{W}_i$. In a three-dimensional environment, the input, of size $H \times W \times D \times C$ (height, width, depth and number of channels) is convoluted with the filter stack, of size $P \times Q \times R \times S \times K$. The kth convolution term for the kth filter is:

$$\mathbf{W}_{ik} * \mathbf{V}_{i-1} = \sum_{u=0}^{P-1}\sum_{v=0}^{Q-1}\sum_{w=0}^{R-1}[\mathbf{W}_{ik}(P-u, Q-v, R-w)\cdot \mathbf{V}_{i-1}(x+u, y+v, z+w)] \tag{4}$$

The convolution operation is performed at a 'neuron', which convolutes the kth filter with a delimited region (a box in 3D) of usually the filter size (typically $P = Q = R$, with typical values of 3, 5 or 7). The centres of each neuron's area of influence can span over different intervals, in what is known as stride. Unless otherwise stated, stride is by default 1, but it can be two in case we want to downsample the images (see 'Conv_2'). After convolution, the activations of the K filters in layer i are stacked and passed to layer $(i + 1)$. Activations are performed using either the ReLU function by default or the sigmoid for Conv_7.

In this work we use the Convolutional Autoencoder (CAE) architecture displayed at Fig. 1. The encoder uses three convolutional layers and one dense layer to obtain an internal representation in the output layer, "Dense_2" . The decoder, for its part, uses one dense layer and four convolutional layers (three internal and one convolutional output layer) to transform back into the original space.

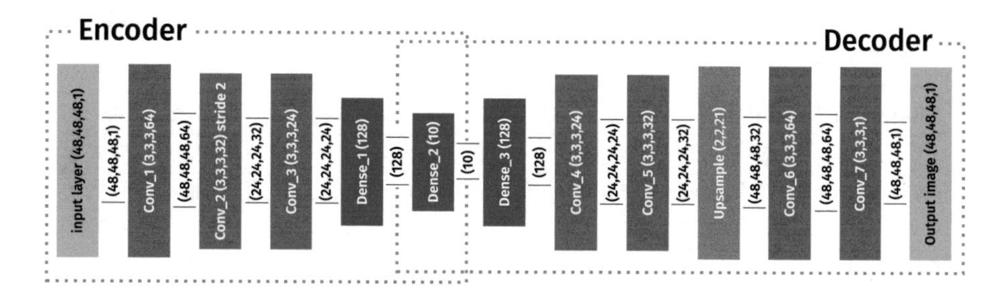

Fig. 1. Proposed encoder-decoder architecture

To obtain a feature reduction, we have chosen to make use of the stride property of convolutional layers at "Conv_2", since it can completely replace the pooling layer without loss in accuracy, according to a recent study [27]. An upsampling layer has been used in the decoder to counteract its effect. The CAE has been trained with the 'RMSprop' optimizer [28] using the mean squared error as loss function:

$$\mathcal{L} = \frac{1}{N}\sum_i(\mathbf{I}_i - \hat{\mathbf{I}}_i)^2 \tag{5}$$

where $\mathbf{I}_i$ is the ith image of the dataset and $\hat{\mathbf{I}}_i$ is its reconstructed version, obtained at the output of the CAE.

2.4 Evaluation

Two different CADs are proposed in this work, which include either PCA or the CAE in for feature extraction and a Support Vector Machine classifier (SVC) [13] with RBF kernel for classification. Additionally, class weights have been used in some cases to account for class prevalence. These two systems have been tested using 10-fold stratified cross-validation [9].

To evaluate the performance of each system we use different parameters that can be derived from the confusion matrix: the accuracy, sensitivity, specificity, the balanced accuracy (average of the sensitivity and specificity) and the F1-score (harmonic mean of sensitivity and recall) [23]. Additionally, the Receiver Operating Characteristic (ROC) curve has been used for a visual comparison, and the area under the ROC curve (AUC) is also provided.

3 Results and Discussion

The cross-validation results for the two systems when using or not class weighting are provided at Table 1. The CAE system was independently trained in each fold during 100 epochs. In both systems the internal representation had 10 features (10 components for PCA and 10 neurons in "Dense_2" for the CAE).

Table 1. Performance results of the two systems when using or not class weighting (W) to account for class prevalence.

Model	W	acc. [STD]	sens. [STD]	spec. [STD]	bal. Acc.	F1-Score	AUC
PCA	0	0.850 [0.005]	0.999 [0.003]	0.000 [0.500]	0.500	0.666	0.774
PCA	1	0.850 [0.005]	0.999 [0.003]	0.000 [0.500]	0.500	0.666	0.779
CAE	0	0.933 [0.016]	0.978 [0.017]	0.672 [0.172]	0.825	0.848	0.962
CAE	1	0.872 [0.060]	0.860 [0.070]	0.938 [0.080]	0.899	0.895	0.959

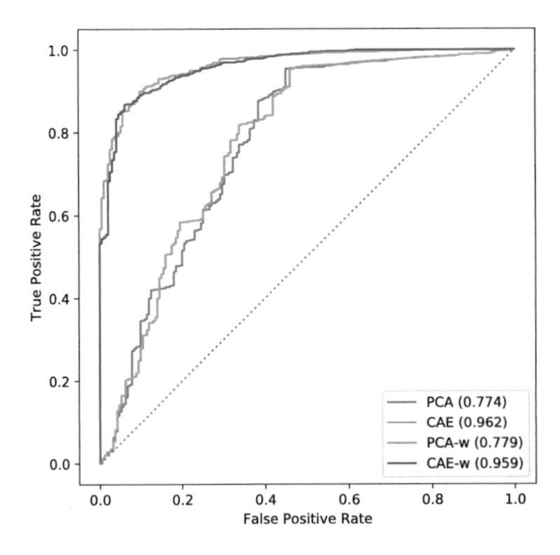

Fig. 2. ROC curves for the different feature extraction and classification algorithms.

The results show that the decomposition of the PCA system contained little or no information about the class at all (balanced accuracy 0.5, with specificity = 0), while the CAE decomposition correctly accounted for most part. While the plain accuracy was higher if not using class weights, the specificity was very low in that case. However, when using class weights in the SVC, the balanced accuracy and the F1-score increased up to almost 90%.

A more detailed analysis of the performance is provided by the ROC curve, shown at Fig. 2. There, the CAE is visibly superior in performance to the PCA-based CAD system. The ROC curve also confirms that the difference in performance between using or not class weights is related to the placement of the operation point, and depends on the balance between TPR and FPR, keeping similar AUC values.

The distribution of the projections of the images in the latent space are shown at Fig. 3. There we can observe that, while both decomposition achieved some degree of separability, the CAE produces a more regular description of the image space with two more condensed clusters, in contrast to PCA.

For its part, the CAE reconstruction (see Fig. 4) is softer than the original images, but the main features (relative intensities of the striata, the brain, skull

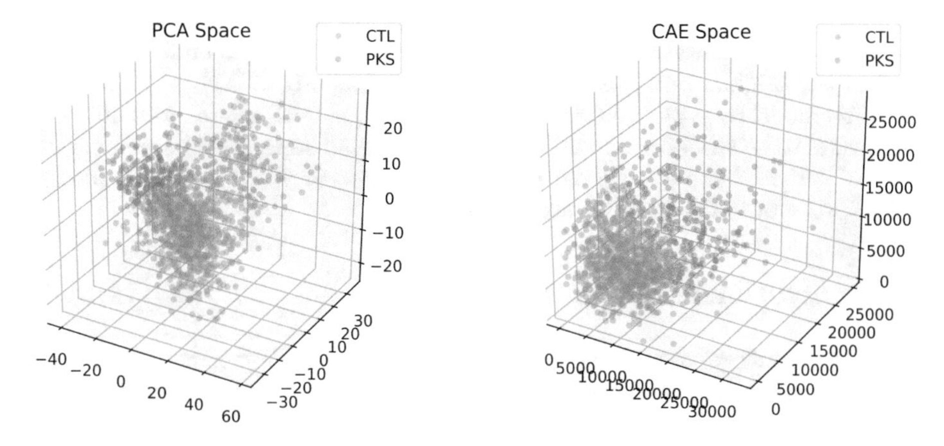

Fig. 3. Localization of the subjects in the latent space of PCA and the CAE. Visualization using the three components with higher variance.

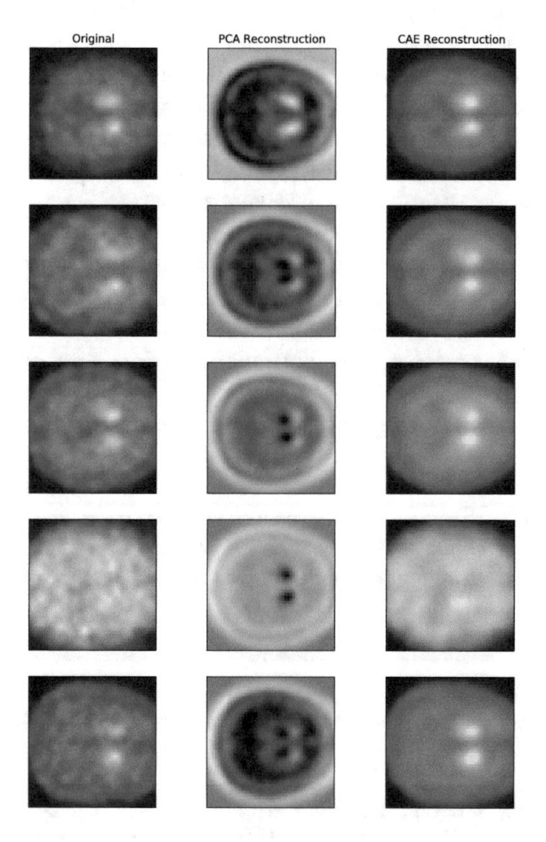

Fig. 4. Five examples comparing the original images (left column), and their reconstructions from the latent space using PCA (centre) and the CAE (right).

and background, and their asymmetry) are kept. We can therefore assume that the internal representation of the CAE is a better model of the underlying manifold of the original data, as it could be checked at Fig. 3. The advantages of the CAE model over PCA are then twofold: first, the CAE model does not need the images to be spatially normalized; second, the self-supervised decomposition of the CAE produces a more natural decomposition of the Parkinson's Disease (PD) spectrum in the latent space.

4 Conclusions and Future Work

In this work, we proposed a new methodology to overcome two common problems in neuroimaging studies: the need for spatial normalization and the performance of image decomposition methods in imbalanced datasets. We propose a Computer Aided Diagnosis (CAD) system based on a Convolutional Autoencoder (CAE) for feature extraction and a Support Vector Machine classifier (SVC) for classification.

When comparing the proposed methodology to more traditional image decomposition methods such as Principal Component Analysis (PCA), we proved that while the PCA devoted most components to model the heterogeneity within the most prevalent class (Parkinson's Disease, or PD), the CAE focused on more relevant features, accounting for a more natural decomposition of the disease spectrum in the latent space. The spatial heterogeneity also misled the PCA decomposition, causing a worse reconstruction of the images, whereas the CAE was able to identify the most relevant source of heterogeneity: the intensity ratio between the striata, the brain and the background.

This convolutional architecture paves the way for new CAD systems for nuclear imaging that do not need spatial normalization. In contrast to traditional methods such as PCA, convolutional autoencoders use images in their native space and producing a better model of the underlying structure of a certain disorder, which could be applied to many other modalities and disease, such as SPECT or Alzheimer's Disease.

Acknowledgements. This work was partly supported by the MINECO/FEDER under the TEC2015-64718-R project and the Consejera de Economía, Innovación, Ciencia y Empleo (Junta de Andalucía, Spain) under the Excellence Project P11-TIC- 7103.

References

1. Baldi, P.: Autoencoders, unsupervised learning, and deep architectures. In: Proceedings of ICML Workshop on Unsupervised and Transfer Learning, pp. 37–49 (2012)
2. De Martino, F., Gentile, F., Esposito, F., Balsi, M., Di Salle, F., Goebel, R., Formisano, E.: Classification of fMRI independent components using IC-fingerprints and support vector machine classifiers. Neuroimage **34**(1), 177–194 (2007)

3. Duin, R.P.W.: Classifiers in almost empty spaces. In: Proceedings 15th International Conference on Pattern Recognition, vol. 2, pp. 1–7. IEEE (2000)
4. Ecker, C., Marquand, A., Mourão-Miranda, J., Johnston, P., Daly, E.M., Brammer, M.J., Maltezos, S., Murphy, C.M., Robertson, D., Williams, S.C., Murphy, D.G.M.: Describing the brain in autism in five dimensions-magnetic resonance imaging-assisted diagnosis of autism spectrum disorder using a multiparameter classification approach. J. Neurosci. **30**(32), 10612–10623 (2010)
5. Fischl, B., Dale, A.M.: Measuring the thickness of the human cerebral cortex from magnetic resonance images. Proc. Nat. Acad. Sci. **97**(20), 11050–11055 (2000)
6. Friston, K., Ashburner, J., Kiebel, S., Nichols, T., Penny, W.: Statistical Parametric Mapping: The Analysis of Functional Brain Images. Academic Press, London (2007)
7. Hansen, L.K., Larsen, J., Nielsen, F.Å., Strother, S.C., Rostrup, E., Savoy, R., Lange, N., Sidtis, J., Svarer, C., Paulson, O.B.: Generalizable patterns in neuroimaging: how many principal components? NeuroImage **9**(5), 534–544 (1999)
8. Initiative, T.P.P.M.: PPMI. Imaging Technical Operations Manual, 2 edn., June 2010
9. Kohavi, R.: A study of cross-validation and bootstrap for accuracy estimation and model selection. In: Proceedings of International Joint Conference on AI, pp. 1137–1145 (1995). http://citeseer.ist.psu.edu/kohavi95study.html
10. Lila, E., Aston, J.A., Sangalli, L.M., et al.: Smooth principal component analysis over two-dimensional manifolds with an application to neuroimaging. Ann. Appl. Stat. **10**(4), 1854–1879 (2016)
11. Martínez-Murcia, F., Górriz, J., Ramírez, J., Moreno-Caballero, M., Gómez-Río, M., Parkinson's Progression Markers Initiative, et al.: Parametrization of textural patterns in 123I-ioflupane imaging for the automatic detection of parkinsonism. Med. Phys. **41**(1), 012502 (2014)
12. Martinez-Murcia, F.J., Górriz, J.M., Ramírez, J., Illán, I.A., Segovia, F., Castillo-Barnes, D., Salas-Gonzalez, D.: Functional brain imaging synthesis based on image decomposition and kernel modeling: application to neurodegenerative diseases. Front. Neuroinformatics **11**, 65 (2017)
13. Martínez-Murcia, F.J., Górriz, J., Ramírez, J., Puntonet, C.G., Illán, I.: Functional activity maps based on significance measures and independent component analysis. Comput. Methods Programs Biomed. **111**(1), 255–268 (2013)
14. Martínez-Murcia, F.J., Górriz, J.M., Ramírez, J., Illán, I.A., Puntonet, C.G.: Texture features based detection of Parkinson's Disease on DaTSCAN images. In: Natural and Artificial Computation in Engineering and Medical Applications, pp. 266–277. Springer, Heidelberg (2013)
15. Martinez-Murcia, F.J., Ortiz, A., Górriz, J.M., Ramírez, J., Segovia, F., Salas-Gonzalez, D., Castillo-Barnes, D., Illán, I.A.: A 3D convolutional neural network approach for the diagnosis of Parkinson's Disease. In: Natural and Artificial Computation for Biomedicine and Neuroscience, pp. 324–333. Springer International Publishing, Cham (2017)
16. Martinez-Torteya, A., Rodriguez-Rojas, J., Celaya-Padilla, J.M., Galván-Tejada, J.I., Treviño, V., Tamez-Peña, J.: Magnetization-prepared rapid acquisition with gradient echo magnetic resonance imaging signal and texture features for the prediction of mild cognitive impairment to Alzheimer's disease progression. J. Med. Imaging **1**(3), 031005 (2014)
17. Napierała, K., Stefanowski, J., Wilk, S.: Learning from imbalanced data in presence of noisy and borderline examples. In: International Conference on Rough Sets and Current Trends in Computing, pp. 158–167. Springer, Heidelberg (2010)

18. Nixon, M.: Feature Extraction & Image Processing. Academic Press, London (2008)
19. Ortiz, A., Górriz, J.M., Ramírez, J., Martinez-Murcia, F.J., Initiative, A.D.N., et al.: Automatic ROI selection in structural brain MRI using SOM 3D projection. PLOS ONE **9**(4), e93851 (2014)
20. Ortiz, A., Górriz, J.M., Ramírez, J., Martínez-Murcia, F.J.: LVQ-SVM based CAD tool applied to structural MRI for the diagnosis of the Alzheimer's disease. Pattern Recogn. Lett. **34**(14), 1725–1733 (2013)
21. Ortiz, A., Martínez-Murcia, F.J., García-Tarifa, M.J., Lozano, F., Górriz, J.M., Ramírez, J.: Automated diagnosis of parkinsonian syndromes by deep sparse filtering-based features. In: Innovation in Medicine and Healthcare 2016, pp. 249–258. Springer, Cham (2016)
22. Ortiz, A., Munilla, J., Martínez-Murcia, F.J., Górriz, J.M., Ramírez, J., for the Alzheimer's Disease Neuroimaging Initiative, et al.: Learning longitudinal MRI patterns by SICE and deep learning: assessing the alzheimers disease progression. In: Annual Conference on Medical Image Understanding and Analysis, pp. 413–424. Springer, Cham (2017)
23. Powers, D.M.: Evaluation: from precision, recall and f-measure to ROC, informedness, markedness and correlation (2011)
24. Quackenbush, J.: Computational analysis of microarray data. Nat. Rev. Genet. **2**(6), 418–427 (2001)
25. Segovia, F., Górriz, J.M., Ramírez, J., Chaves, R., Illán, I.Á.: Automatic differentiation between controls and Parkinson's Disease DaTSCAN images using a partial least squares scheme and the fisher discriminant ratio. In: KES, pp. 2241–2250 (2012)
26. Sonka, M., Hlavac, V., Boyle, R.: Image processing, analysis, and machine vision. Cengage Learning (2014)
27. Springenberg, J.T., Dosovitskiy, A., Brox, T., Riedmiller, M.: Striving for simplicity: The all convolutional net (2014). arXiv preprint: arXiv:1412.6806
28. Tieleman, T., Hinton, G.: Lecture 6.5-rmsprop: divide the gradient by a running average of its recent magnitude. COURSERA Neural Netw. Mach. Learn. **4**(2), 26–31 (2012)
29. Wang, X., Yang, W., Weinreb, J., Han, J., Li, Q., Kong, X., Yan, Y., Ke, Z., Luo, B., Liu, T., Wang, L.: Searching for prostate cancer by fully automated magnetic resonance imaging classification: deep learning versus non-deep learning. Sci. Rep. **7**(1), 15415 (2017)
30. Weiner, M.W., Veitch, D.P., Aisen, P.S., Beckett, L.A., Cairns, N.J., Green, R.C., Harvey, D., Jack, C.R., Jagust, W., Liu, E., Morris, J.C., Petersen, R.C., Saykin, A.J., Schmidt, M.E., Shaw, L., Siuciak, J.A., Soares, H., Toga, A.W., Trojanowski, J.Q.: The Alzheimer's disease neuroimaging initiative: a review of papers published since its inception. Alzheimer's Dement. J. Alzheimer's Assoc. **8**(Suppl. 1), S1–S68 (2012). http://www.ncbi.nlm.nih.gov/pubmed/22047634, PMID: 22047634
31. Xu, J., Xiang, L., Liu, Q., Gilmore, H., Wu, J., Tang, J., Madabhushi, A.: Stacked sparse autoencoder (SSAE) for nuclei detection on breast cancer histopathology images. IEEE Trans. Med. Imaging **35**(1), 119–130 (2016)

Steel Tube Cross Section Geometry Measurement by 3D Scanning

Álvaro Segura[1(✉)] and Alejandro García-Alonso[2]

[1] Vicomtech, Mikeletegi 57, 20009 San Sebastian, Spain
asegura@vicomtech.org

[2] Facultad de Informáitica, Universidad del País Vasco UPV-EHU,
Paseo Manuel Lardizabal 1, 20018 San Sebastian, Spain

Abstract. In this paper we describe a system aimed at obtaining several geometric measures from seamless steel tubes representative of their manufacturing quality. Traditionally, these measurements are taken manually using calipers and similar tools, which are error-prone and have low reproducibility. We introduce a system based on 3D scanning and computational geometry. Scanning adds surface noise and may leave small undetectable gaps depending on surface properties, and the tube itself may have small defects not representative of the overall tube shape. Also, processing high resolution scans containing millions of points can be too time consuming. Thus, special algorithms are needed to make the problem tractable and produce representative measures in a reasonable time.

1 Introduction

Tubos Reunidos Industrial (TRI) is a manufacturer of seamless steel tubes. The company produces steel tubes in a hot rolling process, with nominal diameters ranging from 26.7 mm to 180 mm, and up to 25 m long. Hot rolling incrementally shapes a hollow cylinder until it has the geometric properties requested. The plant produces batches of tubes which must meet tolerance requirements. Tubes are distinguished mainly by their cross section and several measurements are defined on it that quantify their shape and quality. These measurements are traditionally manually measured by operators using conventional tools such as calipers. Measurements are taken on small tube samples, a few centimetres long, that are cut from one of the tube ends, like those shown in Fig. 1.

Aspects of the geometrical quality of steel tubes to consider are highlighted in [1]. The industry still largely uses manual tools to perform measurements of produced tubes. In the standard procedure one or more small samples, a few centimetres long, are cut from the tube ends (which are normally discarded) and a set of measures are taken with the help of precision calipers. Some external surface defects are detected with the help of automated systems based on computer vision, such as the system described in [2].

M. Graña et al. (Eds.): SOCO'18-CISIS'18-ICEUTE'18, AISC 771, pp. 57–66, 2019.
https://doi.org/10.1007/978-3-319-94120-2_6

The measurements taken from tube samples are:

- **Outside diameter**: minimum and maximum values
- **Wall thickness**: minimum, maximum and average values
- **Eccentricity**: an indicator of hole decentering
- **Polygonality**: an indicator of non-circular inside shape
- **Ovality**: an indicator of out-of-roundness outside shape
- **Triangularity**: an indicator of out-of-roundness representative of the shape induced by the rolls placed 120° apart
- **Inside inscribed diameter**: maximum diameter of a circle inside the hole

Fig. 1. Tube samples for measurement

There are no universally agreed definitions of these measurements. In [3–5] we can find manufacturer supplied definitions of diameters, eccentricity and ovality.

One problem with the traditional manual procedure is its low accuracy and repeatability. For example, thickness values taken with a point-sample caliper, depend on the orientation of the caliper. If the point samples are not in the same cross section plane or are not aligned with the tube axis, the measured value will be too high. Values also vary when sampled points are not the same. Indirect measurements like poligonality are particularly sensitive to small variations in thickness measurements.

More advanced measurement systems exist. Alternatives include X-ray or radioactive isotope computed tomography (CT) [6] and Coordinate Measurement Machines (CMM). CT can used to obtain a rough approximation of the thickness profile and eccentricity, but does not provide the required accuracy. CMM are very accurate but impractical for use in a hot rolling line.

Initially, we sought an approach based on 2D computer vision. Images of the samples were taken with a conventional camera having its optical axis approximately aligned with the sample tube axis. These images were processed to find the outer and inner contours and thickness measurements were computed. The approach has several limitations. First, given the camera perspective, there is not a fixed relation between pixels and world units (e.g. millimeters) so measurements have a significant uncertainty (this could be solved with telecentric lenses adding cost and complexity). Then, cutting defects such as burrs modify the observed contours producing an unreal cross section and thickness profile. And finally, the obtained resolution was not high enough. The same would happen if the sample was put in a conventional 2D document scanner: the contour at the cut plane is irregular and not representative of the uncut tube.

This work presents a system for measurement of steel tubes based on 3D scanning and computational geometry. A 3D scanner produces a digital model containing in the order of one million points. Fully computing measurements on the whole sample model would be too time consuming. Therefore, some assumptions and simplifications are made in order to make the problem solvable with enough precision in a limited time.

2 Methodology

The final methodology developed is based on 3D surface scanning and computational geometry. Scanned parts are processed to locate the tube longitudinal axis and then sliced with planes perpendicular to the axis to obtain cross sections. These sections are processed to compute the desired measurements.

2.1 Acquisition

A 3D model of the tube sample is obtained by a structured light scanner with a 3-axis turntable. The device captures a set of depth images in different poses and then merges them to form triangle meshes. The scanned data may contain extra elements such as fixtures or parts of the turntable surface.

Structured light scanners work best with bright diffuse surfaces where their projected stripes are clearly visible. The newly formed tubes are dark and sometimes exhibit small shiny patches on their surface which leads to suboptimal scanning results. Dark surfaces require a higher exposure time in the scanner cameras and still may be undetected if they are too dark or shiny. For these reasons the quality of the scanned tube samples is variable and the generated meshes frequently contain holes. The cut plane surface, sometimes even manually polished, is often particularly shiny. But that is not a problem because the only geometrically relevant parts are the outside and inside tube surfaces.

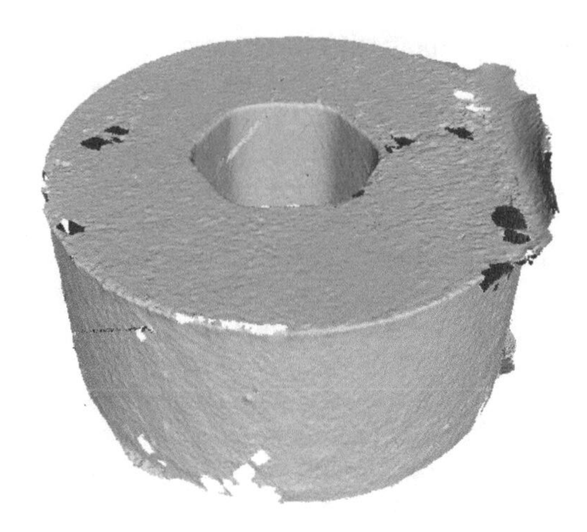

Fig. 2. Scanned tube sample 3D mesh

2.2 Part Location and Slicing

Locating the Tube Sample. The sample is manually placed on the scanner platform with its axis approximately vertical. It cannot be assumed to be at a specific location in the scanner coordinate system, nor perfectly oriented with the vertical axis. Thus, generating cross sections is not as simple as cutting the mesh with planes parallel with the XY plane.

Each sample is cut with a saw roughly perpendicularly to its axis, but the cut plane can be a few degrees off perpendicularity. Therefore, the virtual model cannot be assumed to have its longitudinal axis parallel to the vertical Z axis. Moreover, the sample is placed anywhere on the scanner turntable. Our system therefore must first locate the sample in the scanned data.

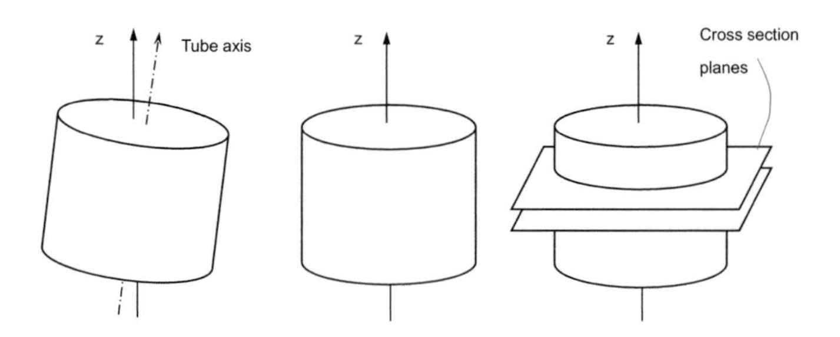

Fig. 3. Axis alignment and slicing

Location is done by finding a cylindrical surface in the scanned data using a RANSAC algorithm [7]. We consider the axis to be that of its outside approximately cylindrical surface. Since there may be more than one cylindrical surface in the acquired data (the outside and inside surfaces), we restrict the algorithm to cylinders having a diameter in an interval around the nominal diameter.

Once the tube axis is located, the whole 3D model is transformed to make this axis coincident with the Z axis. The affine transformation applied is explained in [8], in that case applied to cone axis alignment instead of cylinder axis.

Slicing and Average Section. With the tube longitudinal axis now aligned with the Z axis, cross sections parallel with the XY plane are generated. We compute a set of slices using parallel planes at several heights in Z. From the top of the model, we skip a short distance in order to avoid an area of frequent burr defects that would invalidate measurements. Then slices are computed at fixed Z intervals.

Measurements are not taken from a single cross section. If a section is cut at a position where a little protruding defect is found, the computed geometric parameters will be distorted. Measurement should produce overall geometric representative parameters irrespective of tiny rare point deviations. Additionally, the scanned mesh may have holes as explained above. Our solution is to compute a set of cross sections at several heights and to combine them into a representative section that we call the *average section*.

We model sections in polar coordinates. Slices are first computed from the triangle mesh at specific longitudinal positions along Z. This produces a set of polygonal, possibly open, contours which are then turned into a polar representation with uniform angular sampling. This is done by tracing rays from the center point to each angular direction α and finding their intersection with the contours, as shown in Fig. 4. Ideally, there are two intersections per direction ($r_1(\alpha)$ and $r_2(\alpha)$), one in the inside contour and another one in the outside contour, but that is not always the case. In the presence of acquisition errors, one or both intersections might be missing ($r_1(\alpha_2)$). And sometimes other scanned features (e.g. fixtures) add extra intersections. Rules are used to infer the most probable inside and outside intersections from the obtained intersections.

Rays are traced at uniform angular intervals using a predefined configurable resolution. This process is performed for each slice in the measurement length and a set of inside and outside polar profiles are obtained. The average section is built by obtaining the median of the radii for each angular direction. In contrast to the average, the median is less affected by outliers, spurious deviations due to small tube defects or scanning errors. If some slices contained holes at specific angles, by taking the median, the final section will have these holes filled with information from other slices.

2.3 Section Measurements

Once an average section representative of the tube is built, a set of geometric parameters are computed on it. The following sections describe the computation of the different parameters.

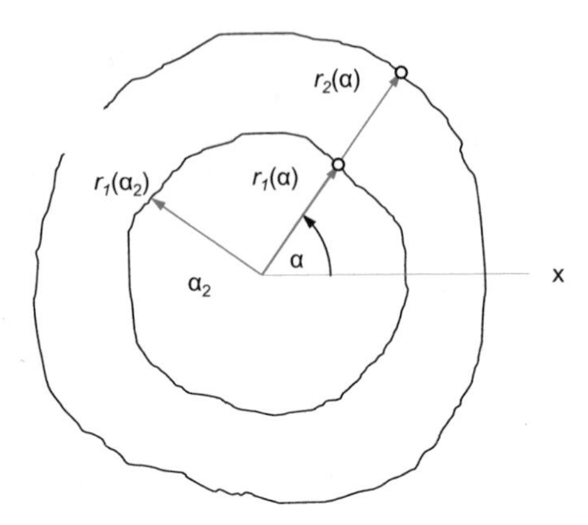

Fig. 4. Polar profiles by ray intersections

Eccentricity. Eccentricity is computed as the difference between the minumum and maximum wall thicknesses, normalized to the mean thickness. Thicknesses are obtained from the average section as the difference between the inside and outside radius at each angle.

Poligonality. There are several different definitions of polygonality, all of which compute a value from 12 thickness values taken at 30-degree intervals. Thicknesses have to be measured starting at the angle with the minimum thickness, which was identified in the previous step.

Ovality. Ovality is defined as the difference between the minimum and maximum diameters of the outside contour. In the conventional procedure a caliper is used to randomly sample the extents of the tube in different orientations to find maximum and minimum values. Emulating this procedure in the virtual model with an arbitrary polygonal shape would be too time consuming. A good approximation can be found by assuming the outside contour is roughly round. It can be slightly deformed and have approximately a two-fold or three-fold rotational symmetry, depending on the rolling process. But the deviation rarely exceeds 2% of the nominal diameter. Thus, the diameter at any angular direction can be approximated by the sum of the radii at opposite directions, as in Eq. 1.

$$d(\alpha) = r(\alpha) + r(\alpha + 180°) \tag{1}$$

The algorithm computes diameters at all angular directions and finds the maximum and minimum values.

Triangularity and Inside Inscribed Diameter. Triangularity is derived from the difference between the diameters circumscribed and inscribed in the outside contour. The circumscribed diameter is the diameter of the smallest circle that can fit the profile externally. That is approximately the diameter of the smallest circle that contains all contour points. And the inscribed diameter is the diameter of the largest circle that fits the profile internally. Or, equivalently, the diameter of the largest circle that keeps all profile points out.

To compute these diameters on the scanned mesh, the contour points are sampled in groups of three and a circle that contains them is defined. If the circle meets the relevant condition (it contains all contour points or leaves all out) it is considered a candidate. This is performed iteratively until the smallest candidate circle is found. To save computation time the section profile is first subsampled to reduce the number of vertices.

3 Results

A software system implementing the above described methodology has been developed and tested with tube samples of diameters up to 140 mm. The software requests the scanner to perform a scan and processes the produced data. Finally, the computed values are written and graphical representations of them are generated.

The following figures show the three graphs generated for a 34-millimeter nominal diameter tube. The tube average cross section is shown in Fig. 5. Figures 6 and 7 present a polar coordinate representation of the inner and outer profiles, and of the thickness profile, respectively. Processing has been done with 16 slices, 0.5 mm apart, 4 rays per degree and subsampling to 1 sample per degree when computing inscribed and circumscribed diameters.

To evaluate repeatability we performed 5 consecutive inspections of a 140-mm diameter tube to see how values differ between measurements. Table 1 lists the mean and standard deviation of the measured values in the 5 inspections. We can see a very low standard deviation that indicates high repeatability.

Table 1. Mean and standard deviation of resulting values of 5 inspections of the same 140 mm tube. The values are minumum and maximum diameter, eccentricity, poligonality, triangularity, ovality and inside inscribed diameter

	D_{min}	D_{max}	e	p	τ	o	D_{insc}
Mean	140.59	140.74	0.120	3.212	0.0495	0.151	121.10
Std.dev	0.0072	0.00228	0.000858	0.0235	0.00136	0.00547	0.00332

The accuracy of the thickness profile obtained by our method was compared with that obtained by a CMM. The mean absolute error was 0.02 mm with a maximum peak of 0.1 mm. The error was not symmetric: our thickness measurements tended to be slightly smaller than those given by the CMM. This can be due to the size of the CMM probe, which cannot go inside small pits. The deviation seen is considered acceptable.

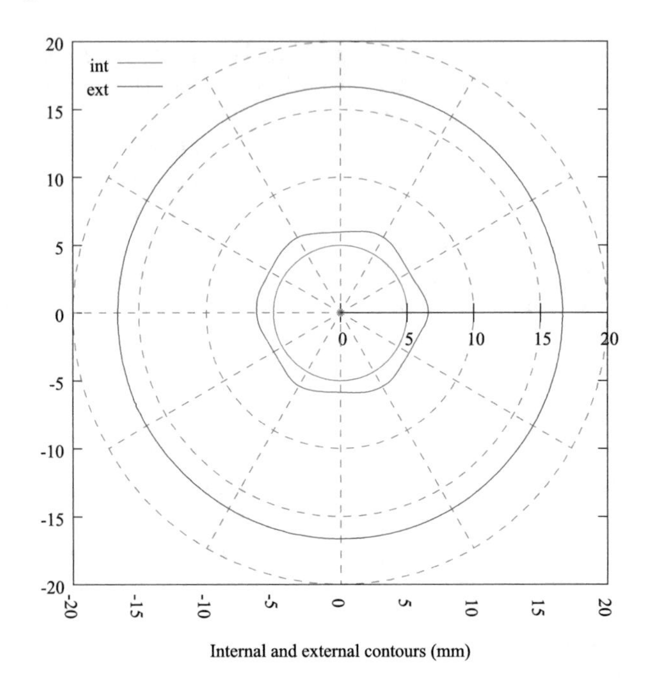

Fig. 5. Average cross section

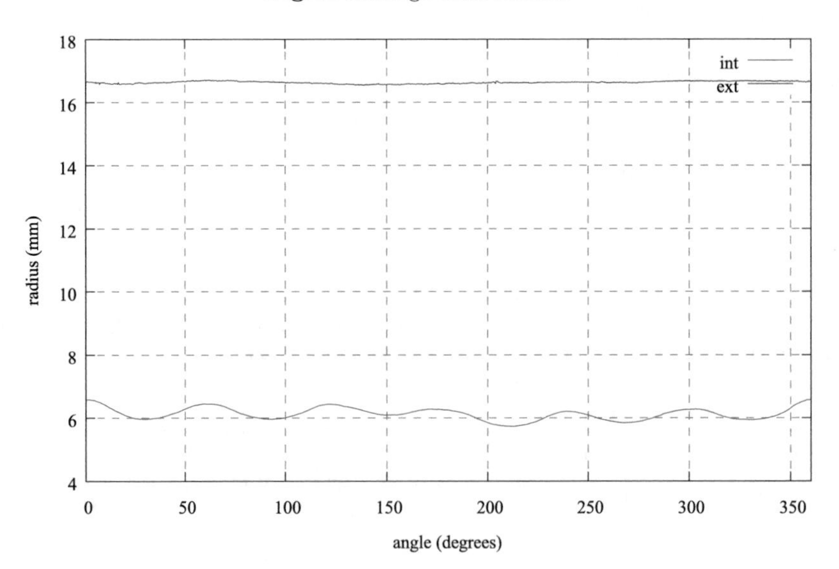

Fig. 6. External (red) and internal (blue) profiles

The total time for inspecting each sample includes scanning and processing, with the former being the most time-consuming phase. Scanning time, which encompasses multiple capture and merging depends on the tube sample size,

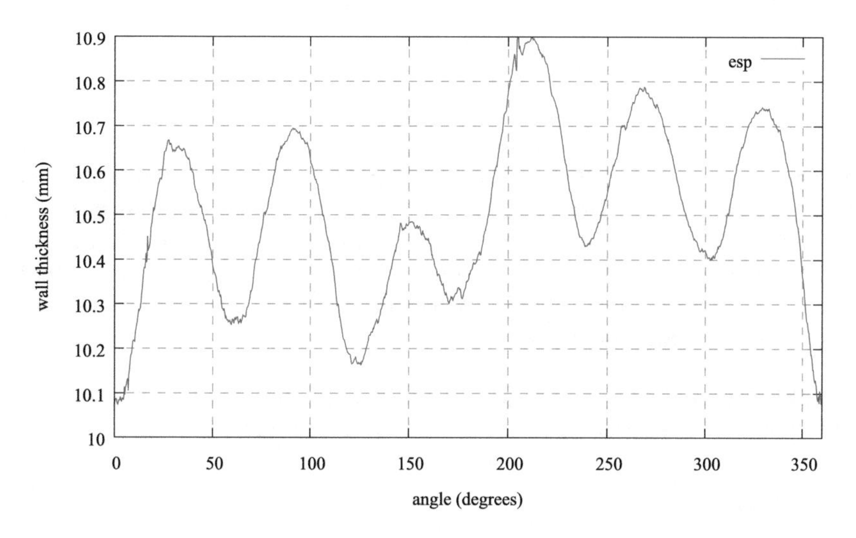

Fig. 7. Wall thickness profile

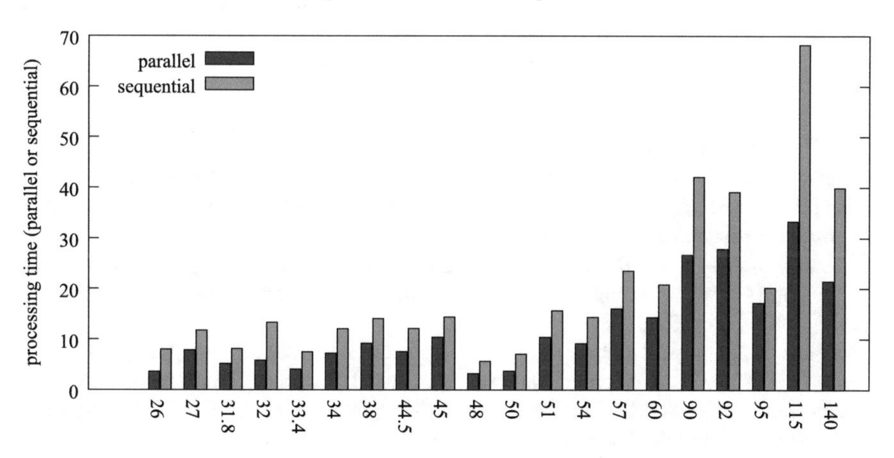

Fig. 8. Processing times: sequential vs parallel for tubes of different diameter

taking between 80 and 90 s. Processing time depends on part size and on processing parameters (such as number of slices, number of intersection rays per degree and amount of section subsampling).

We applied parallel computation where possible using the OpenMP interface [9]. Only some computations like processing the different slices support parallelization. Figure 8 shows the processing times for 24 tube samples of increasing diameter, using parallel and sequential computation. Computation time is reduced by 38% using parallelization on an 8-logical core processor.

4 Conclusions

We have presented a method for dimensional inspection of steel tubes based on 3D scanning suitable for deployment in a manufacturing plant. The method obtains results more accurate and repeatable than the traditional manual measurements and similar to results obtained by a CMM. These results highlight the benefit of 3D scanning with respect to manual measurements in metrology. Future work includes a comparison of repeatability with values obtained by manual methods, improvements in treatment of noisy or defective scans and the computation of additional measures.

Acknowledgments. The author acknowledges the support of Tubos Reunidos Industrial in this research, its supply of knowledge and physical samples.

References

1. Vashchenko, Y.I.: Rejection of tubes in rolling on installations with three-roll mills. Metallurgist **6**(10), 483–485 (1962)
2. Ayani, A., Lago, A., Cruz, A., Gutierrez, J.: Surface inspection of hot rolled seamless tubes. Metall. Plant Technol. Int. **31**(5), 76 (2008)
3. SteelTube: Sizes of tubes and pipes
4. FineTubes: Tube Definitions and Tolerances
5. EagleStainless: Do You Need Tube Ovality, Concentricity or Eccentricity
6. Peng, S.-J., Wu, Z.-F.: Research on the CT image reconstruction of steel tube section from few projections. NDT & E Int. **42**(5), 435–440 (2009)
7. Fischler, M.A., Bolles, R.C.: Random sample consensus: a paradigm for model fitting with applications to image analysis and automated cartography. Commun. ACM **24**(6), 381–395 (1981)
8. Sanchez, J., Segura, A., Barandiaran, I.: Fast and accurate mesh registration applied to in-line dimensional inspection processes. Int. J. Interact. Des. Manuf. (2017)
9. Dagum, L., Menon, R.: OpenMP: an industry standard API for shared-memory programming. IEEE Comput. Sci. Eng. **5**(1), 46–55 (2017)

How Blockchain Could Improve Fraud Detection in Power Distribution Grid

Roberto Casado-Vara[1(✉)], Javier Prieto[1], and Juan M. Corchado[1,2,3]

[1] BISITE Digital Innovation Hub, University of Salamanca. Edificio Multiusos I+D+i, 37007 Salamanca, Spain
{rober,javierp,corchado}@usal.es

[2] Department of Electronics, Information and Communication, Faculty of Engineering, Osaka Institute of Technology, Osaka 535-8585, Japan

[3] Pusat Komputeran dan Informatik, Universiti Malaysia Kelantan, Karung Berkunci 36, Pengkaan Chepa, 16100 Kota Bharu, Kelantan, Malaysia
https://bisite.usal.es/es

Abstract. Power utilities experience large losses of electricity in distribution from power plants to the end consumer. There are two types of losses: technical and non-technical. Among non-technical losses is a very prominent one: electrical fraud. In this paper we propose a new system to detect fraud. A blockchain is used to store the data collected by the WSN that monitors the power distribution grid. Using data stored in the blockchain, it is constructed directed directed acyclic graph (DAG) with non-technical losses and applied the clustering algorithm created to detect fraud. The main advantage of blockchain to our model is that every time the blockchain grows the stored data is more secure. Therefore, power utilities can perform an inspection in blockchain data stored.

Keywords: Blockchain · Power system · Fraud detection
Clustering · Directed acyclic graph

1 Introduction

Blockchain is currently gaining the interest of a wide variety of industries: from finance [35,40] and health-care [12,27,45], utilities and the government sector. The reason for this growing interest: With a blockchain, applications that worked only through a trusted intermediary. Now they can operate in a decentralized way, without the need for a verification system and achieve the same functionality with the same amount of reliability. This was not possible until the blockchain was created. With the implementation of blockchain trustless networks appeared. This is possible because in networks that use blockchain you can make transfers with no need to trust other users [5]. With the lack of intermediaries, transactions become faster between users. In addition, the use of cryptography in the blockchain ensures that the information is secure [24]. Blockchain is a large accounting ledger that records all transactions made by users. This makes

M. Graña et al. (Eds.): SOCO'18-CISIS'18-ICEUTE'18, AISC 771, pp. 67–76, 2019.
https://doi.org/10.1007/978-3-319-94120-2_7

researchers and developers on the Internet of Things (IoT) look for ways to link IoT with blockchain [25,46].

Power losses in electric power distribution systems is a reality for electric utilities. These losses can be classified as technical or non-technical losses. Energy losses classified as technical are due to the energy dissipated in the transfer of energy from the power plants to the final consumer [22]. On the other hand, non-technical losses are the result of subtracting technical losses from total losses (i.e., all other losses associated to the electric power distribution, such as energy theft, metering errors, errors in the billing process, etc.) [26,44]. To avoid these energy losses, electrical utilities invest a lot of money in detecting fraud, power losses and other measurement errors, among other things [18,23].

Methods to commit fraud include joining cables before the smart meter, tampering with the smart meter to stop or slow it down [7]. On the other hand, accidental malfunctioning of the smart meter can also lead to a net loss of energy for the company. The company actions and revenue after the detection will be very different, but both are issues that the company wants to detect and fix as early as possible. Advanced analysis of this data can help protect both consumer and utility interests by enabling fraud detection at both ends [6,16,43]. The use of machine learning has been a very common approach to energy fraud detection [10,29,42]. In this paper we present a new approach to fraud detection. To this end, we define a Wireless Sensor Network (WSN) model of users and miners (i.e., blockchain architecture) [5]. Then a blockchain analysis is performed using an algorithm created to detect energy losses. Finally, an analysis is made with a new cluster algorithm to identify most relevant electricity losses (REL).

2 Related Work

Blockschain is a distributed data structure that is shared and replicated among network members. It was introduced with Bitcoin [30] to resolve the double spending problem [11]. As a consequence of how the Bitcoin nodes (the so-called miners) mutually validate the agreed transactions, Bitcoin's Blockchain determines who owns what it [13]. How to build a Blockchain is through cryptography. Each block is identified by its own cryptographic hash. Each block refers to the hash of the previous block. This establishes a union between the blocks to form a Blockchain [3]. For this reason, users can interact with the Blockchain using a pair of public and private [32]. Miners in a Blockchain need to agree on the transactions and order in which they have occurred. Otherwise, individual copies of the Blockchain may differ in the production of a fork; miners then have a different view of how the transactions have occurred and it will not be possible to maintain a single blockchain until the fork is resolved [17,19].

Therefore, a consensus mechanism distributed across all networks in the block chain is needed to achieve this. Blockchain's approach to solve fork problem is that each node in blockchain can link the next block. Simply find the correct random number with SHA-256 [2,20] to have the number of zeros the block string waits for. Any node that can resolve this puzzle, has generated the so-called proof of work (pow) and comes to form next block in the blockchain [31].

Since these are one-way cryptographic functions, any other node can easily verify that the given response meets the requirement. Note that a fork can still occur in the network, when two competing nodes mine blocks almost simultaneously. These forks usually resolve automatically in the next block. On the other hand, a WSN is a network of sensors whose communications are transmitted by wireless signals. WSNs are used to collect and study a wide range of magnitudes, such as sound, humidity, temperature and many others. WSN is the preferred as the architecture of the sensor systems, but they are susceptible to the disadvantages caused by the limited lifetime of the operation. Unlike other sensor networks with physical support (wire), the use of WSN is limited by amount of energy it can store, its calculation capacities, memory, information flow and communication distance. WSN will be used to control and monitor a wide range of things [21]. Some works are achieving good results in fields such as energy efficiency using WSN [4], control of operations [9,15], optimal routing in WSN [41], and some other applications such as social good [34]. Sensor networks can also be used with other technologies such as multi-agent systems to manage data [38], for data mining [39], multi-agent localization [41] and WSN [43]. Clustering techniques will be used to find fraud cases. Among the machine learning algorithms that use clustering techniques, the most used technique to form clusters is the distance between the elements [14]. These algorithms are also being used for data mining [1,8].

Due of advantages offered by blockchain and WSN, applications are being created that combine both technologies. One of the first applications has been to authenticate and increase reliability in WSNs using blockchain [28]. Other applications are also available with the Internet of Things and the global commerce [33,36]. Other applications of blockchain with WSN is to increase the security of the transactions [37]. However, these applications have certain shortcomings because they are too general. In this paper we propose a blockchain application with WSN for monitoring a power distribution grid and with the new clustering algorithm that we created in this paper to locate the fraud in electrical consumption.

3 Methodology

In this section our methodology is presented. We propose to collect data from smart meters in order to detect electrical losses and possible associated frauds. To this end, a WSN is used. Data collected by the WSN is stored in a blockchain. Next, an algorithm checks the blockchain for energy losses. Once these energy losses are detected, machine learning algorithms are applied to classify these energy losses and determine if it is fraud or not.

WSN nodes are placed in every single element of power system grid shown in Fig. 1. Nodes of smart poles and smart meter are called users and can only transmit information. However, nodes placed in substations and street control center are called miners, and they have a double purpose: transmit data and operate as miners of the private blockchain that we defined. Users create new

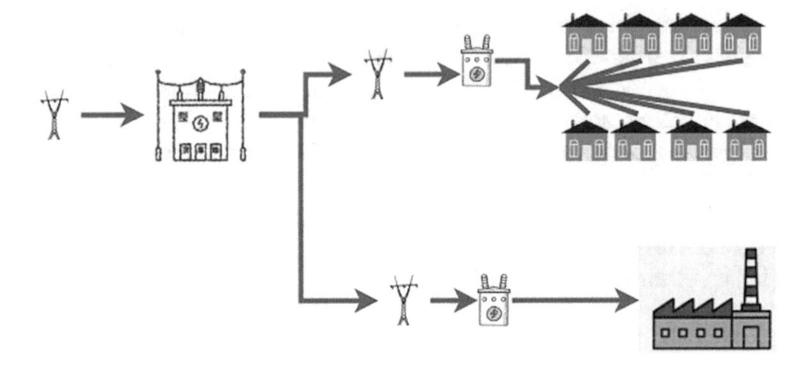

Fig. 1. Diagram of an electrical distribution network in a city. First, an electrical substation is displayed where the amount of electrical power is measured. In addition, before entering the homes or businesses there are also street control centers where electricity that reaches final consumers is measured. In the design a group of houses with a smart meter and a company with a smart meter can be shown.

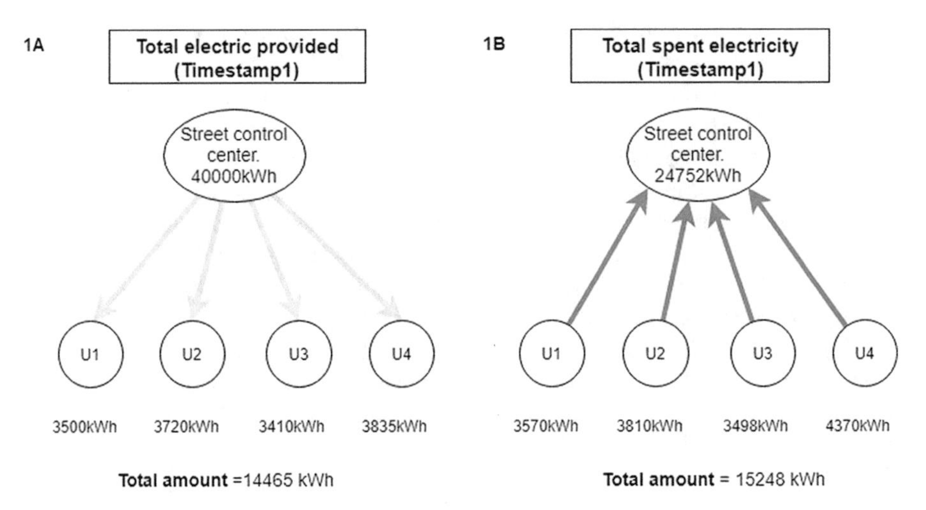

Fig. 2. Energy consumption graph is created with random data. 1(A) In this image you can notice that street control center registers 40000 kWh. Each user (i. e., smart meters) tracks the energy consumption of houses where it is located. Total amount of electricity used is 14465 kWh. 1(B) In this image is represented total amount of electricity that street control center registers that has been spent for each of graph vertices.

transactions transmitting all their data to miners. This data is incorporated into network of miners responsible for building the blockchain with this data. WSN nodes sent their ID, the amount of electricity passing through their power system grid element and a timestamp of when this measurement was made.

WSN nodes and data are encrypted with a cryptographic key provided by the electrical utility. This data is used to construct the private blockchain.

Once all data is in the blockchain, then amount of electricity circulating in each of the nodes of power distribution. In addition, the algorithm calculates difference between set of vertices of the graph. Value calculated by this algorithm will be the weight of the edges of the graph called electrical losses tree (ELT). The ELT formation process is shown in Fig. 2. Subtracting total power provided (TPP) of user from total power spent (TPS) that street control center records that user has spent is how you obtain the ELT. In Fig. 3 you can see how ELT is built using Fig. 2 data and Eq. 1.

$$V_i = \sum_i (TPS(V_i) - TPP(V_i)) \quad (1)$$

where V_i is the amount of losses power in vertice i, $TPS(V_i)$ is the total power spent by vertice i and $TPP(V_i)$ is the total power provided by vertice i.

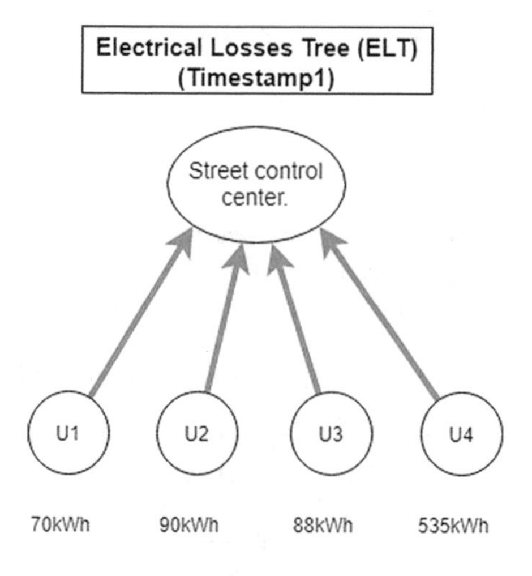

Fig. 3. ELT created by random data from Fig. 2 and Eq. 1. $V_1 = 70$, V_2=90, V_3=88 and V_4=535.

3.1 Cluster Algorithm and Simulations Test

Once ELT is created, our algorithm is applied to detect electrical losses. Graph vertices that are identified as relevant losses will be inspected by power company to see if that loss corresponds to electrical fraud or misclassified losses as non-technical. The algorithm builds a confidence interval with data users of each street control center. The confidence interval has a 1% error rate. Weights of ELT vertices outside that confidence interval are marked as vertices with relevant losses. We describe in deep our algorithm for our case study example.

First step the algorithm does is find out for vertice with the minimum weight. Once find it, calculate the distance from it to everyone else. Next step is to eliminate vertice with the maximum distance and build the confidence interval. Construction of this confidence interval has been using the Eq. 2.

$$NREL_{SCCID} = \left(EL_U \pm t_{(U-1,\frac{\alpha}{2})} \frac{S_U}{\sqrt{U}} \right) \tag{2}$$

where NREL = Not Relevant Electrical Losses, EL_U = *Electrical losses medium*, SCCID = Street Control Center ID, EL_{U_i} = Electrical Losses of user U_i, U = number of users, t = t-student, S_U = standard deviation of EL. Then, the algorithm calculate the max weight vertice (MWV) by:

$$MWV = V_{max\{|mean\{W(V_i)\}-W(V_j)|\}} \tag{3}$$

Once the MWV is calculated, we remove this node and $NREL_{SCCID}$ is calculated again. There are two different ways: the MWV is in the $NREL_{SCCID}$ interval, then the algorithm stop loops. In this case, $NREL_{SCCID}$ are relevant fraud risk losses $RFRL_{SCCID}$. In the other hand, MWV is not in the $NREL_{SCCID}$ interval, then the algorithm continues. Let's apply this algorithm to the example in Fig. 3. MWV is calculated:

$$MWV = V_{max\{|mean\{W(V_i)\}-W(V_j)|\}} = U_4 \tag{4}$$

After removing the vertice at maximum distance from the vertice with minimum weight (Eq. 4), $NREL_{SCCID}$ are now $RFRL_{SCCID}$ because the MWV is not in the interval. Remaining confidence interval is as follows:

$$RFRL_{SCCID} = (82.66 \pm 63.12) = (19.55; 145.78) \tag{5}$$

when all other vertices are in this range, those users are NREL. Finally, it is checked that the user of the vertice that was deleted is not in confidence interval. Otherwise, that user is a REL. In this case, all users are NREL and the algorithm stops. We add the following table with five simulations of our new algorithm to four nodes. In this simulations there are 4 ELT with 4 nodes. Then the algorithm is applied to find nodes which have risk of fraud.

4 Discussion

In this paper we present a new approach to improve fraud detection in power distribution grid. Novelty in the paper is a blockchain to authenticate each of power distribution grid elements. Blockchain is also used to account for electricity provided by electric company and that used by consumers. Furthermore, a new clustering algorithm is used to identify energy losses. Then, electric utilities can check the smart meters with electrical losses to determine if the non-technical losses are due to fraud or not. Results of simulations are shown in Table 1. In these results it can be seen how the algorithm detects nodes with risk of fraud.

Table 1. In this table four running simulations of our algorithm are shown. V_1, V_2, V_3 and V_4 are ETL vertices. When the algorithm finish its loops, several fraud risk node are noticed from 0 to all of them. Every loop is separated by 2 horizontal lines. "X" are the removes nodes by the algorithm.

V_1	V_2	V_3	V_4	loop count	MWV	$RFRL_{SCCID}$	Fraud risk node
70	90	88	535	1	U_4	(19.55;145.78)	U_4
349	221	481	105	1	U_3	(109.39;245.34)	U_3
349	221	X	105	2	U_1	(96.68;188.05)	U_1, U_3
233	81	183	73	1	U_1	(61.05;129.37)	U_1
X	81	183	73	2	U_3	(72.64;79.94)	U_1, U_3
91	242	148	227	1	U_2	(97.4;173.47)	U_2
91	X	148	227	2	U_4	(90.25;135.15)	U_2, U_4
103	83	141	7	1	U_4	(87.58;120.4)	U_4
103	83	141	X	1	U_4	(83.05;99.8)	U_3, U_4

This new method uses advantages of blockchain. We address that blockchain keeps all transactions in an immutable way. This allows you to keep track of all electrical transactions between utilities and consumers. On the other hand, a new clustering algorithm designed to locate REL and NREL users is introduced. Advantage of this new model compared to previous techniques for detecting fraud is that all data entering the blockchain will have a dual function, authenticate the nodes of power distribution grid and control amount of electricity flowing through the electrical network. We notice that blockchain can significantly reduce non-technical losses, as they are easier to track these non-technical losses and it is also possible to apply the appropriate measures to reduce them significantly.

Although with addition of blockchain the model achieves greater security. Our model is addressing certain disadvantages related to the use of blockchain. To keep a blockchain private, electric utilities require high computing power in their mining network. In addition, miners' Network are running 7 trans/sec and this is quite slow. Despite all these limitations, our model improves significantly security of power distribution grid communications, allowing electric utilities to have an immutable record of all transactions with users. In future work we will evaluate implementation of a case-based reasoning system (CBR) or a Multi-Agent system for optimization of our model. In addition, proposed algorithm will be improved to improve detection of electronic fraud cases.

Acknowledgments. This paper has been funded by the European Regional Development Fund (FEDER) within the framework of the Interreg program V-A Spain-Portugal 2014-2020 (PocTep) grant agreement No 0123_IOTEC_3_E (project IOTEC).

References

1. Agrawal, R., Gehrke, J., Gunopulos, D., Raghavan, P.: Automatic subspace clustering of high dimensional data for data mining applications. In: Proceedings of the ACM SIGMOD International Conference on Management of Data, pp. 94–105 (1998)
2. Announcing the Secure Hash Standard, 1 August 2002. http://csrc.nist.gov/publications/fips/fips180-2/fips180-2.pdf
3. Antonopoulos, A.M.: Mastering Bitcoin: Unlocking Digital Cryptocurrencies, 1st edn. O'Reilly Media Inc., Sebastopol (2014)
4. Biswas, S., Das, R., Chatterjee, P.: Energy-efficient connected target coverage in multi-hop wireless sensor networks. In: Industry Interactive Innovations in Science, Engineering and Technology, pp. 411–421. Springer, Singapore (2018)
5. Chamoso, P., Rivas, A., Martín-Limorti, J.J., Rodríguez, S.: A hash based image matching algorithm for social networks. In: De la Prieta, F., Vale, Z., Antunes, L., Pinto, T., Campbell, A.T., Julián, V., Neves, A.J.R., Moreno, M.N. (eds.) PAAMS 2017. AISC, vol. 619, pp. 183–190. Springer, Cham (2018). https://doi.org/10.1007/978-3-319-61578-3_18
6. Cody, C., Ford, V., Siraj, A.: Decision tree learning for fraud detection in consumer energy consumption. In: Proceedings of the IEEE 14th International Conference on Machine Learning and Applications, pp. 1175–1179, December 2015
7. Costa, Â., Novais, P., Corchado, J.M., Neves, J.: Increased performance and better patient attendance in an hospital with the use of smart agendas. Log. J. IGPL **20**(4), 689–698 (2012)
8. Corchado, J.M., Borrajo, M.L., Pellicer, M.A., Yáñez, J.C.: Neuro-symbolic System for Business Internal Control. In: Industrial Conference on Data Mining, pp. 1–10 (2004)
9. Tapia, D.I., Alonso, R.S., García, O., Corchado, J.M., Bajo, J.: Wireless sensor networks, real-time locating systems and multi-agent systems: the perfect team. In: 2013 16th International Conference on Information Fusion (FUSION), pp. 2177–2184 (2013)
10. Depuru, S.S.S.R., Wang, L., Devabhaktuni, V.: Support vector machine based data classification for detection of electricity theft. In: Power Systems Conference and Exposition (PSCE) 2011 IEEE/PES, pp. 1–8, 20–23 March 2011
11. Double-Spending—Bitcoin WiKi. https://en.bitcoin.it/wiki/Double-spending. Accessed 15 Mar 2016
12. Dwyer, G.: The economics of Bitcoin and similar private digital currencies. J. Financ. Stab. **17**, 81–91 (2015)
13. Eris Industries Documentation—Blockchains. https://docs.erisindustries.com/explainers/blockchains/. Accessed 15 Mar 2016
14. Fan, H., Mei, X., Prokhorov, D.V., Ling, H.: Multi-level Contextual RNNs with Attention Model for Scene Labeling. CoRR (2016). abs/1607.02537
15. Farooq, M.O., Kunz, T.: Operating systems for wireless sensor networks: a survey. Sensors **11**, 5900–5930 (2011)
16. Ford, V., Siraj, A., Eberle, W.: Smart grid energy fraud detection using artificial neural networks. In: IEEE Symposium on Computational Intelligence Applications in Smart Grid 2014, pp. 1–6, 9–12 December 2014
17. García-Valls, M.: Prototyping low-cost and flexible vehicle diagnostic systems. ADCAIJ: Adv. Distrib. Comput. Artif. Intell. J. Salamanca **5**(4), 93–103 (2016)

18. Coria, J.A.G., Castellanos-Garzón, J.A., Corchado, J.M.: Intelligent business processes composition based on multi-agent systems. Expert Syst. Appl. **41**(4), 1189–1205 (2014). Part 1
19. Greenspan, G.: Avoiding the Pointless Blockchain Project (2015). http://www.multichain.com/blog/2015/11/avoidingpointless-blockchain-project/
20. Hashcash-Bitcoin WiKi. https://en.bitcoin.it/wiki/Hashcash. Accessed 15 Mar 2016
21. Huang, C.F., Tseng, Y.C.: A survey of solutions to the coverage problems in wireless sensor networks. J. Int. Technol. **6**, 1–8 (2005)
22. Li, T., Sun, S., Corchado, J.M., Siyau, M.F.: Random finite set-based bayesian filters using magnitude-adaptive target birth intensity. In: FUSION 2014–17th International Conference on Information Fusion (2014). https://www.scopus.com/inward/record.uri?eid=2-s2.0-84910637788&partnerID=40&md5=bd8602d6146b014266cf07dc35a681e0
23. Li, T., Sun, S., Corchado, J.M., Siyau, M.F.: A particle dyeing approach for track continuity for the SMC-PHD filter. In: FUSION 2014–17th International Conference on Information Fusion (2014). https://www.scopus.com/inward/record.uri?eid=2-s2.0-84910637583&partnerID=40&md5=709eb4815eaf544ce01a2c21aa749d8f
24. Li, T., Sun, S., Bolić, M., Corchado, J.M.: Algorithm design for parallel implementation of the SMC-PHD filter. Sig. Process. **119**, 115–127 (2016). https://doi.org/10.1016/j.sigpro.2015.07.013
25. Lima, A.C.E.S., De Castro, L.N., Corchado, J.M.: A polarity analysis framework for Twitter messages. Appl. Math. Comput. **270**, 756–767 (2015). https://doi.org/10.1016/j.amc.2015.08.059
26. McLaughlin, S., Podkuiko, D., McDaniel, P.: Energy theft in the advanced metering infrastructure. In: Critical Information Infrastructures, Security, pp. 176–187 (2010)
27. Mettler, M.: Blockchain technology in healthcare: the revolution starts here. In: 2016 IEEE 18th International Conference on e-Health Networking, Applications and Services (Healthcom), Munich, pp. 1–3 (2016)
28. Moinet, A., Benoît, D., Jean-Luc, B.: Blockchain based trust & authentication for decentralized sensor networks. arXiv preprint arXiv:1706.01730 (2017)
29. Monedero, Í, Biscarri, F., León, C., Biscarri, J., Millán, R.: MIDAS: detection of non-technical losses in electrical consumption using neural networks and statistical techniques. In: Gavrilova, M.L., Gervasi, O., Kumar, V., Tan, C.J.K., Taniar, D., Laganá, A., Mun, Y., Choo, H. (eds.) ICCSA 2006. LNCS, vol. 3984, pp. 725–734. Springer, Heidelberg (2006). https://doi.org/10.1007/11751649_80
30. Nakamoto, S.: Bitcoin: A Peer-to-Peer Electronic Cash System (2008). https://bitcoin.org/bitcoin.pdf
31. Oprescu, F.: Method and apparatus for unique address assignment, node self-identification and topology mapping for a directed acyclic graph. U.S. Patent No. 5,394,556, 28, February 1995
32. Pinto, A., Costa, R.: Hash-chain-based authentication for IoT. ADCAIJ: Adv. Distrib. Comput. Artif. Intell. J. Salamanca **5**(4) 43–57 (2016)
33. Powell, A.P., Alex, M.W.: Constraint evaluation in directed acyclic graphs. U.S. Patent No. 9,892,529. 13, February 2018
34. Prieto Tejedor, J., Chamoso Santos, P., de la Prieta Pintado, F., Corchado Rodríguez, J.M.: A generalized framework for wireless localization in gerontechnology. In: 17th IEEE International Conference on Ubiquitous Wireless Broadband ICUWB'2017. IEEE, September 2017

35. Kelly, J., Williams, A.: Forty Big Banks Test Blockchain-Based Bond Trading System (2016). http://www.nytimes.com/reuters/2016/03/02/business/02reuters-bankingblockchain-bonds.html
36. Kupriyanovsky, Y., et al.: Smart container, smart port, BIM, Internet Things and blockchain in the digital system of world trade. Int. J. Open Inf. Technol. **6**(3), 49–94 (2018)
37. Kambourakis, G., Gomez Marmol, F., Wang, G.: Security and Privacy in Wireless and Mobile Networks, 18 (2018)
38. Rodríguez, S., De Paz, J.F., Villarrubia, G., Zato, C., Bajo, J.: Corchado, multi-agent information fusion system to manage data from a WSN in a residential home. Inf. Fusion **23**, 43–57 (2015)
39. Rodríguez, S., Zato, C., Corchado, J.M., Li, T.: Fusion system based on multi-agent systems to merge data from WSN. In: 2014 17th International Conference on Information Fusion (FUSION), pp. 1–8 (2014)
40. Satoshi, N.: Bitcoin: A Peer-to-Peer Electronic Cash System (2008). https://bitcoin.org/bitcoin.pdf
41. Sobral, J.V.V., Rodrigues, J.J.P.C., Saleem, K., de Paz, J.F., Corchado, J.M.: A composite routing metric for wireless sensor networks in AAL-IoT. In: Wireless and Mobile Networking Conference (WMNC), 2016 9th IFIP, pp. 168–173 (2016)
42. Spiric, J.V., Doi, M.B., Stankovi, S.S.: Fraud detection in registered electricity time series. Int. J. Electr. Power Energy Syst. **71**, 42–50 (2015)
43. Tapia, D.I., Fraile, J.A., Rodríguez, S., Alonso, R.S., Corchado, J.M.: Integrating hardware agents into an enhanced multi-agent architecture for ambient intelligence systems. Inf. Sci. **222**, 47–65 (2013)
44. Tang, Y., Ten, C.W., Brown, L.E.: Switching reconfiguration of fraud detection within an electrical distribution network. In: 2017 Resilience Week (RWS). Wilmington, DE, pp. 206–212 (2017)
45. Casado-Vara, R., Corchado, J.M., Blockchain for democratic voting: how blockchain could cast off voter fraud. Orient. J. Comp. Sci. Technol. **11**(1). http://www.computerscijournal.org/?p=8042
46. Redondo-Gonzalez, E., De Castro, L.N., Moreno-Sierra, J., Maestro De Las Casas, M.L., Vera-Gonzalez, V., Ferrari, D.G., Corchado, J.M.: Bladder carcinoma data with clinical risk factors and molecular markers: a cluster analysis. BioMed Res. Int. **2015**, 14 (2015)

Clustering and Classification

Enhancing Confusion Entropy as Measure for Evaluating Classifiers

Rosario Delgado[1] and J. David Núñez-González[2](✉)

[1] Department of Mathematics, Universitat Autònoma de Barcelona, Campus de la UAB, 08193 Cerdanyola del Vallès, Spain
delgado@mat.uab.cat

[2] Department of Mathematics, University of the Basque Country (UPV/EHU), Leioa, Spain
josedavid.nunez@ehu.eus

Abstract. Performance measures are used in Machine Learning to assess the behaviour of classifiers. Many measures have been defined on the literature. In this work we focus on Confusion Entropy (CEN), a measure based in Shannon's Entropy. We introduce a modification of this measure that overcomes its disadvantages in the binary case that disables it as a suitable measure to compare classifiers. We compare this modification with CEN and other measures, presenting analytical results in some particularly interesting cases, as well as some heuristic computational experimentation.

Keywords: Classifier · Performance measure
Confusion Entropy (CEN)

1 Introduction

Machine Learning is the subfield of Computer Science, as well as a branch of Artificial Intelligence, whose objective is to develop techniques that allow computers to learn. Machine learning tasks are typically grouped into two broad categories: Supervised and Unsupervised Learning. Classification falls in the category of Supervised Learning since it deals with some input variables (features or characteristics) and an output variable (the class), and uses an algorithm to infer the class (that is, to classify) a new case from its features.

Classifiers are compared by using different performance measures. In the binary case (in which the class variable has only two labels or classes) there are several classical measures that have been widely used: Accuracy, Sensitivity, Specificity and AUC (Area Under the Curve), only to mention the most commonly used. Not of all them allow a natural extension to the multi-class case (more than two labels), and in addition few measures have been specially designed for multi-class classification, which is a more complex scenario. Accuracy, by far the simplest and widespread performance measure in classification, extends seamlessly its definition in the binary case to multi-class classification.

M. Graña et al. (Eds.): SOCO'18-CISIS'18-ICEUTE'18, AISC 771, pp. 79–89, 2019.
https://doi.org/10.1007/978-3-319-94120-2_8

Another well known performance measure, formerly introduced in the binary case but that extends without problems is Matthiew's Correlation Coefficient (MCC) [6].

In this work, however, we focus on a different performance measure, named Confusion Entropy (CEN), which measures the uncertainty of a source of information and has been recently introduced by Wang et al. in [14] as a novel measure for evaluating classifiers based on the concept of Shannon's Entropy. CEN measures generated entropy from misclassified cases, considering not only how the cases of each fixed class have been misclassified into other classes, but also how the cases of the other classes have been misclassified as belonging to this class.

CEN is compared in [14] with Accuracy and other measures, showing a relative consistency with them: higher Accuracy tends to result in lower Confusion Entropy. This performance measure, which is more discriminating for evaluating classifiers than Accuracy, specially when the number or cases grows, has also been studied in [3], where the authors show the strong monotone relation between CEN and MCC. Both MCC and CEN improve over Accuracy. There are some works in the recent literature using Confusion Entropy. For example, in [2] the authors propose a novel splitting criterion based on CEN for learning decision trees with higher performance; experimental results on some data sets show that this criterion leads to trees with better CEN value without reducing accuracy. The authors of [7,8] use CEN, among other performance measures, to compare several common data mining methods used with highly imbalanced data sets where the class of interest is rare. Other works propose modifications of this measure, as [13], in which a Confusion Entropy measure based on a probabilistic confusion matrix is introduced, measuring if cases are classified into true classes and separated from others with high probabilities. A similar approach to that of [13] is followed in [1] to analyze the probability sensitivity of the Gaussian processes in a bankruptcy prediction context, by means of a probabilistic confusion entropy matrix based on the model estimated probabilities. In the context of horizontal collaboration, the system global entropy is introduced in [10] analogously to CEN (see also [11,12]), and it is used in the collaborative part of a clustering algorithm, which is iterative with the optimization process continuing as long as the system global entropy is not stable.

It is remarkable that CEN shows to have a weak point in the binary case that invalidates it as a suitable performance measure: this measure attributes high entropy even in situations of good accuracy and, what is more serious, it even gets values larger than one, unlike what happens in the multi-class case, in which CEN ranges between zero and one. Our aim is to introduce an enhanced CEN measure, that we denote by $\widetilde{\text{CEN}}$, and compare it with CEN, MCC and Accuracy. This new measure is highly correlated with CEN but presents two main differences regarding it: (a) definitions of probabilities involved in the construction of CEN have been modified in the sense of improve interpretability as real probabilities, (b) the disadvantages of CEN in the binary case are overcome.

2 Preliminaries and Notations

Given a multi-class classifier learned from a training dataset, with $N \geq 2$ classes labelled $\{1, 2, \ldots, N\}$, we apply it in order to classify data from a testing dataset, that is, to infer the class of the cases from their known features or characteristics. Since for the cases in the testing dataset we actually know the class, we obtain the $N \times N$ confusion matrix $C = (C_{i,j})_{i,j=1,\ldots,N}$, which collects the results issued by the classifier over the testing dataset in this way: $C_{i,j}$ is the number of cases in the testing dataset of class i that have been classified as belonging to class j. In this way, the diagonal elements of the confusion matrix C indicate how many cases in the testing dataset have been correctly classified while the rest of elements, out of diagonal, correspond to misclassification. We denote by S the sum of values of the matrix, that is, the total number of cases in the testing dataset, $S = \sum_{i=1}^{N} \sum_{j=1}^{N} C_{i,j}$.

In [14] the misclassification probability of classifying samples of class i as class j "subject to class j", denoted as $P_{i,j}^{j}$, is introduced as:

$$P_{i,j}^{j} = \frac{C_{i,j}}{\sum_{k=1}^{N} (C_{j,k} + C_{k,j})}, \quad i, j = 1, ..., N, \, i \neq j \,, \tag{1}$$

that is, $P_{i,j}^{j}$ is "almost" the relative frequency of samples or class i classified as j among all samples that are of class j or that have been classified as being of class j. But not exactly, since samples of class j that have been correctly classified, $C_{j,j}$, are counted twice in the denominator. Analogously, the misclassification probability of classifying samples of class i as class j "subject to class i", denoted as $P_{i,j}^{i}$, is defined in the same paper by:

$$P_{i,j}^{i} = \frac{C_{i,j}}{\sum_{k=1}^{N} (C_{i,k} + C_{k,i})}, \quad i, j = 1, ..., N, \, i \neq j \,. \tag{2}$$

Then, the Confusion Entropy associated to class j is defined by:

$$\mathrm{CEN}_j = - \sum_{k=1, k \neq j}^{N} \left(P_{j,k}^{j} \log_{2(N-1)}(P_{j,k}^{j}) + P_{k,j}^{j} \log_{2(N-1)}(P_{k,j}^{j}) \right) \tag{3}$$

with the convention $a \log_b(a) = 0$ if $a = 0$. Finally, the overall Confusion Entropy associated to the confusion matrix C is defined as a convex combination of the Confusion Entropy of the classes as follows:

$$\mathrm{CEN} = \sum_{j=1}^{N} P_j \, \mathrm{CEN}_j \,, \quad \text{where} \quad P_j = \frac{\sum_{k=1}^{N} (C_{j,k} + C_{k,j})}{2 \sum_{k,\ell=1}^{N} C_{k,\ell}} \,. \tag{4}$$

(note that the non-negative weights P_j sum 1). For $N > 2$ this measure ranges between 0 and 1, 0 is attained with perfect classification (the off-diagonal elements of matrix C being zero), while 1 under complete misclassification and balance in C, that is, if all diagonal elements in C are zero, and the off-diagonal elements take all the same value. In the binary case ($N = 2$), although CEN remains to be 0 with perfect classification, and is 1 under complete misclassification with balance, in intermediate scenarios we can also obtain CEN $= 1$ and even higher values, that is, in some cases CEN is out of range. See, for example, the confusion matrices in Table 1, which had also been considered in [3]. The lack of monotony when the situation monotonously goes from perfect classification to complete balanced misclassification, is a very inconvenient property of CEN in the binary case, and is our main motivation for introducing a modified version of it.

Note that CEN is an invariant measure; if we multiply all elements of the confusion matrix by a constant we obtain the same result. The same convenient property holds with Accuracy, MCC and the modified Confusion Entropy measure that we introduce in the next section.

3 The Modified Confusion Entropy $\widetilde{\text{CEN}}$

To alleviate its undesirable behaviour on binary class prediction, as well as to improve interpretation of involved terms as real probabilities, we propose $\widetilde{\text{CEN}}$ as an alternative to CEN measure.

3.1 Definition

Instead of (1), we propose to introduce the probability of classifying samples of class i as class j "subject to class j", as

$$\widetilde{P}_{i,j}^{j} = \frac{C_{i,j}}{\sum_{k=1}^{N}(C_{j,k}+C_{k,j}) - C_{j,j}}, \quad i,j = 1,...,N,\ i \neq j. \tag{5}$$

that is, we correct the fact that in (1) samples of class j that have been correctly classified are counted twice in the denominator. With this new definition, $\widetilde{P}_{i,j}^{j}$ really is the relative frequency of samples of class i classified as belonging to class j among all samples that are of class j or that have been classified as being of class j. Analogously, we correct definition (2) in the following sense:

$$\widetilde{P}_{i,j}^{i} = \frac{C_{i,j}}{\sum_{k=1}^{N}(C_{i,k}+C_{k,i}) - C_{i,i}}, \ , i,j = 1,...,N,\ i \neq j. \tag{6}$$

Then, $\widetilde{P}_{i,j}^{i}$ is really the relative frequency of samples or class i classified as j among all samples that are of class i or have been classified as being of class i. Next, we modify definition of the weights in (4) in the following way:

$$\widetilde{P}_{j} = \frac{\sum_{k=1}^{N}(C_{j,k}+C_{k,j}) - C_{j,j}}{2\sum_{k,\ell=1}^{N} C_{k,\ell} - \alpha \sum_{k=1}^{N} C_{k,k}}, \quad \text{where} \quad \alpha = \begin{cases} 1/2 & \text{if } N = 2 \\ 1 & \text{if } N > 2. \end{cases} \tag{7}$$

Then, we define the Confusion Entropy associated to class j as in (3) by

$$\widetilde{\text{CEN}}_j = -\sum_{k=1,k\neq j}^{N} \left(\widetilde{P}^j_{j,k} \log_{2(N-1)}(\widetilde{P}^j_{j,k}) + \widetilde{P}^j_{k,j} \log_{2(N-1)}(\widetilde{P}^j_{k,j})\right), \tag{8}$$

and the modified Confusion Entropy as in formula (4), that is,

$$\widetilde{\text{CEN}} = \sum_{j=1}^{N} \widetilde{P}_j \, \widetilde{\text{CEN}}_j. \tag{9}$$

Note that when $N > 2$, $\sum_{j=1}^{N} \widetilde{P}_j = 1$, so the overall Confusion Entropy is defined as a convex combination of the Confusion Entropy measures of the classes, while in the binary case ($N = 2$), it is just defined as a conical combination since although the weights $\widetilde{P}_j$ are non-negative, they do not necessarily sum up to 1 (indeed, their sum is 1 if and only if all the diagonal elements of the confusion matrix C are zero, that is, all samples have been misclassified).

We see from (4) and (9) that both measures are decomposable, which makes it easy to compute the effect on the behaviour of the classifier of a simple modification affecting only one class.

3.2 Some Basic Properties

Proposition 1. *In the symmetric and balanced case in which $C_{i,j} = F$ for all $i, j = 1, \ldots, N$, $i \neq j$ and $C_{i,i} = T$, with T, $F > 0$, we have that:*

$$If\ N > 2,\ \text{CEN} = \frac{2\,(\text{N}-1)}{\delta} \log_{2(\text{N}-1)}(\delta),\ \widetilde{\text{CEN}} = \frac{2\,(\text{N}-1)}{\widetilde{\delta}} \log_{2(\text{N}-1)}(\widetilde{\delta}), \tag{10}$$

$$If\ N = 2,\ \text{CEN} = \frac{1}{1+\gamma} \log_2(\delta),\ \widetilde{\text{CEN}} = \frac{1}{1+\frac{3}{4}\gamma} \log_2(\widetilde{\delta}), \tag{11}$$

where $\gamma = \frac{T}{F}$, $\delta = 2\,(N-1) + 2\,\gamma$ and $\widetilde{\delta} = 2\,(N-1) + \gamma$.

Note that CEN and $\widetilde{\text{CEN}}$ depend on the matrix values T and F only through the ratio γ, and that in (10) (case $N > 2$) they have the same expression except that CEN depends on δ while $\widetilde{\text{CEN}}$ does on $\widetilde{\delta} = \delta - \gamma$. Therefore,

$$\text{if } N > 2, \quad \widetilde{\text{CEN}}(2\,\gamma) = \text{CEN}(\gamma),$$

where in the notation we highlight the dependency on $\gamma = T/F$.
In case $T = 0$, $\widetilde{\text{CEN}} = \text{CEN} = 1$, while if $F = 0$, $\widetilde{\text{CEN}} = \text{CEN} = 0$.

Corollary 1. *In the symmetric and balanced case in which $C_{i,j} = F$ for all $i, j = 1, \ldots, N$, $i \neq j$ and $C_{i,i} = T$, with T, $F > 0$, we have that for any $N \geq 2$, $\widetilde{\text{CEN}}$ is monotonically decreasing as function of γ, with $\widetilde{\text{CEN}}(0) = 1$ and $\lim_{\gamma\to+\infty} \widetilde{\text{CEN}}(\gamma) = 0$. Moreover,*

$$
\begin{aligned}
&if\ N>2, \widetilde{\mathrm{CEN}}>\mathrm{CEN},\\
&if\ N=2, \gamma_0\in(5,\,6)\ exists\ such\ that\ \widetilde{\mathrm{CEN}}<\mathrm{CEN}\ if\ \ \gamma<\gamma_0,\\
&\qquad\qquad\qquad\qquad\qquad\qquad\qquad \widetilde{\mathrm{CEN}}=\mathrm{CEN}\ if\ \ \gamma=\gamma_0,\ and\\
&\qquad\qquad\qquad\qquad\qquad\qquad\qquad \widetilde{\mathrm{CEN}}>\mathrm{CEN}\ if\ \ \gamma>\gamma_0\,.
\end{aligned}
$$

Remark 1. Note that if $N = 2$, $\mathrm{CEN}(1) = 1$ while $\mathrm{CEN}(\gamma) > 1$ if $0 < \gamma < 1$, that is, CEN exhibits an undesirable behaviour, not showed by $\widetilde{\mathrm{CEN}}$.

Remark 2. Consider the particular case of the setting considered in Proposition 1 and Corollary 1, with $T = F$, that is, $\gamma = 1$. In other words, the confusion matrix is constant, say $\begin{pmatrix} 1\,1 \dots 1 \\ \vdots\ \vdots \dots \vdots \\ 1\,1 \dots 1 \end{pmatrix}$. Then, if $N > 2$, $\mathrm{CEN} = (1 - \frac{1}{N}) \log_{2(N-1)}(2N)$ and $\widetilde{\mathrm{CEN}} = (1 - \frac{1}{2N-1}) \log_{2(N-1)}(2N - 1)$, while if $N = 2$, $\mathrm{CEN} = 1$ and $\widetilde{\mathrm{CEN}} = \frac{4}{7}\log_2(3) < 1$. As a consequence, we can easily see that if $N > 2$, $\mathrm{CEN} < \widetilde{\mathrm{CEN}}$ and $\lim_{N\to\infty} \mathrm{CEN} = \lim_{N\to\infty} \widetilde{\mathrm{CEN}} = 1$. Moreover, $\widetilde{\mathrm{CEN}}$ is monotonically increasing, reaching the minimum when $N = 2$, while CEN reaches the minimum for $N = 3$, being a convex function of N.

4 Comparing $\widetilde{\mathrm{CEN}}$ with CEN and Other Measures

As we said earlier in the introduction, [3,14] give a comparison between CEN and other performance measures (Accuracy and MCC, among others). Definitions are as follows: $\mathrm{Accuracy} = \frac{\sum_{i=1}^{N} C_{i,i}}{S}$, and

$$
MCC = \frac{\sum\limits_{k,\ell,m=1}^{N} (C_{k,k}\, C_{\ell,m} - C_{m,k}\, C_{k,\ell})}{\sqrt{\sum\limits_{k=1}^{N} ((\sum\limits_{\ell=1}^{N} C_{k,\ell})\,(\sum\limits_{u,v=1,\,u\neq k}^{N} C_{u,v}))}\sqrt{\sum\limits_{k=1}^{N} ((\sum\limits_{\ell=1}^{N} C_{\ell,k})\,(\sum\limits_{u,v=1,\,u\neq k}^{N} C_{v,u}))}}.
$$

As MCC is in range $[-1, 1]$ while Accuracy, CEN and $\widetilde{\mathrm{CEN}}$ range in $[0, 1]$, we scale MCC as $\mathrm{MCC}^* = \frac{1-\mathrm{MCC}}{2} \in [0,\,1]$. Analogously, since Accuracy has an inverse relationship with both CEN and $\widetilde{\mathrm{CEN}}$, we compute ACC^* instead of Accuracy, where $\mathrm{ACC}^* = 1 - \mathrm{Accuracy}$.

Some particular pathological cases are studied in the multi-class setting, such as the Z_A family and the "dice rolling case", but before we consider the binary case.

4.1 The General Binary Case

Binary case ($N = 2$) can be fully studied. We will use the following notation for the confusion matrix:

$$
C = \begin{pmatrix} TP & FN \\ FP & TN \end{pmatrix} \tag{12}
$$

where TP is the true positive number of cases in class 1 that have been correctly classified, and the same for the true negative number of cases with class 2, TN. On the other hand, FP denote false positive cases and FN false negatives, taking class 1 as a reference.

Proposition 2. *If the confusion matrix* C *is given by* (12), *we have that*

$$\mathrm{CEN} = \frac{(FN+FP)\,\log_2\left(S^2-(TP-TN)^2\right)}{2\,S} - \frac{FN\,\log_2(FN)+FP\,\log_2(FP)}{S},$$

$$\widetilde{\mathrm{CEN}} = \frac{2\,(FN+FP)\,\log_2\left((S-TN)\,(S-TP)\right)}{3\,S+(FN+FP)} - \frac{4\left(FN\,\log_2(FN)+FP\,\log_2(FP)\right)}{3\,S+(FN+FP)},$$

where $S = TP+TN+FP+FN$.

Table 1 shows some examples of 2×2 confusion matrices, all of them with $S = 12$, that had been considered in [3]. We can observe the inconsistent behaviour of CEN regarding the other measures. It can easily seen that for 2×2 matrices of type $\begin{pmatrix} T & F \\ F & T \end{pmatrix}$, $\mathrm{ACC}^* = \mathrm{MCC}^*$.

Table 1. Examples in the binary case with $S = 12$.

	$\begin{pmatrix} 6 & 0 \\ 0 & 6 \end{pmatrix}$	$\begin{pmatrix} 5 & 1 \\ 1 & 5 \end{pmatrix}$	$\begin{pmatrix} 4 & 2 \\ 2 & 4 \end{pmatrix}$	$\begin{pmatrix} 3 & 3 \\ 3 & 3 \end{pmatrix}$	$\begin{pmatrix} 2 & 4 \\ 4 & 2 \end{pmatrix}$	$\begin{pmatrix} 1 & 5 \\ 5 & 1 \end{pmatrix}$	$\begin{pmatrix} 0 & 6 \\ 6 & 0 \end{pmatrix}$
$\mathrm{ACC}^* = \mathrm{MCC}^* =$	0.0000	0.1667	0.3333	0.5000	0.6667	0.8333	1.0000
$\mathrm{CEN} =$	0.0000	0.5975	0.8617	1.0000	1.0566	1.0525	1.0000
$\widetilde{\mathrm{CEN}} =$	0.0000	0.4678	0.6667	0.7925	0.8813	0.9479	1.0000

4.2 The Z_A family

As noted in [3], the behaviour of the Confusion Entropy is rather diverse from MCC and Accuracy for the pathological case of the family of confusion matrices $Z_A = (a_{i,j})_{i,j=1,\ldots,N}$, defined by $a_{i,j} = \begin{cases} A & \text{if } i=N,\ j=1 \\ 1 & \text{otherwise,} \end{cases}$ with $A \geq 1$.

Proposition 3

$$\textit{If } N>2,\ \mathrm{CEN(Z_A)} = \frac{1}{N^2+A-1}\Big((N-1)\,(N-2)\,\log_{2(N-1)}(2N) + (2N+A-3)\,\log_{2(N-1)}(2N+A-1) - A\,\log_{2(N-1)}(A)\Big),$$

$$\widetilde{\mathrm{CEN}}(Z_A) = \frac{2}{2\,(N^2+A-1)-N}\Big((N-1)\,(N-2)\log_{2(N-1)}(2N-1) + (2N+A-3)\log_{2(N-1)}(2N+A-2) - A\log_{2(N-1)}(A)\Big),$$

$$\textit{if } N=2,\ \mathrm{CEN(Z_A)} = \frac{1}{A+3}\Big((A+1)\,\log_2(A+3) - A\,\log_2(A)\Big),$$

$$\widetilde{\mathrm{CEN}}(Z_A) = \frac{2}{2\,A+5}\Big((A+1)\,\log_2(A+2) - A\,\log_2(A)\Big) < \mathrm{CEN}(Z_A).$$

Remark 3. Not from expressions in Proposition 3, but before being simplified, it can easily be seen that $\lim_{A\to+\infty} \mathrm{CEN}(\mathrm{Z_A}) = \lim_{\mathrm{A}\to+\infty} \widetilde{\mathrm{CEN}}(\mathrm{Z_A}) = 0$, that is, Confusion Entropy measures move towards the minimal entropy setting. Moreover, while $\widetilde{\mathrm{CEN}}(Z_A) < 1$, if $N = 2$ there exists $A_0 \in (1,2)$ such that $\mathrm{CEN}(\mathrm{Z_{A_0}}) = 1$ and $\mathrm{CEN}(\mathrm{Z_A}) > 1$ if $A \in (1,\, A_0)$.

Table 2 shows some examples of confusion matrices of the family Z_A. CEN and $\widetilde{\mathrm{CEN}}$ exhibit a very different behaviour comparing with Accuracy and MCC.

Table 2. Examples with different matrices Z_A in case $N = 4$.

	$\begin{pmatrix}1&1&1&1\\1&1&1&1\\1&1&1&1\\1&1&1&1\end{pmatrix}$	$\begin{pmatrix}1&1&1&1\\1&1&1&1\\1&1&1&1\\10&1&1&1\end{pmatrix}$	$\begin{pmatrix}1&1&1&1\\1&1&1&1\\1&1&1&1\\10^2&1&1&1\end{pmatrix}$	$\begin{pmatrix}1&1&1&1\\1&1&1&1\\1&1&1&1\\10^3&1&1&1\end{pmatrix}$	$\begin{pmatrix}1&1&1&1\\1&1&1&1\\1&1&1&1\\10^4&1&1&1\end{pmatrix}$
ACC* =	0.5000	0.5294	0.5556	0.5789	0.6000
MCC* =	0.5000	0.4722	0.4500	0.4318	0.4167
CEN =	1.0000	1.0104	1.0105	1.0040	0.9932
$\widetilde{\mathrm{CEN}}$ =	0.9057	0.9130	0.9102	0.9014	0.8889

4.3 The "Dice Rolling Case"

The "dice rolling case" is another pathological scenario corresponding to unbalanced classes, which has been considered in [3]. In this case the $N \times N$ confusion matrix has all entries equal to one but for the last row, whose entries are all equal to value $A \geq 1$.

Proposition 4

If $N > 2$,

$$\mathrm{CEN}(D_A) = \frac{N-1}{2N\,(N+A-1)}\Big(\,(2N+A-3)\,\log_{2(N-1)}(2N+A-1) + (A+1)\,\log_{2(N-1)}((N+1)A+N-1) - 2\,A\,\log_{2(N-1)}(A)\Big),$$

$$\widetilde{\mathrm{CEN}}(D_A) = \frac{1}{(2N-1)\,(A+N-1)}\Big((N^2+AN-(A+1))\log_{2(N-1)}(2N+A-2) + (N-1)\,(A+1)\log_{2(N-1)}(NA+N-1) - 2A\,(N-1)\log_{2(N-1)}(A)\Big),$$

if $N = 2$,

$$\mathrm{CEN}(\mathrm{D_A}) = \frac{1}{4(\mathrm{A}+1)}\Big((\mathrm{A}+1)\log_2(\mathrm{A}+3) + (\mathrm{A}+1)\log_2(3\mathrm{A}+1) - 2\mathrm{A}\log_2(\mathrm{A})\Big),$$

$$\widetilde{\mathrm{CEN}}(D_A) = \frac{2}{7(A+1)}\Big((A+1)\log_2(A+2) + (A+1)\log_2(2A+1) - 2A\log_2(A)\Big).$$

As a consequence, $\widetilde{\mathrm{CEN}}(D_A) < \mathrm{CEN}(\mathrm{D_A}) \leq 1$, *with* $\mathrm{CEN}(\mathrm{D_A}) = 1$ *if and only if* $A = 1$.

Remark 4. Not from expressions in Proposition 4, but before being simplified, it can easily be seen that $\lim_{A\to+\infty} \mathrm{CEN}(\mathrm{D_A}) = 0.4$ and $\lim_{A\to+\infty} \widetilde{\mathrm{CEN}}(D_A) = 0.29$.

See some examples of D_A confusion matrices with $N = 4$ in Table 3. Note the higher discriminant power of the Confusion Entropy measures, both CEN and $\widetilde{\mathrm{CEN}}$ with respect to Accuracy and MCC, that are invariant and do not depend on the value of A.

Table 3. Examples with different matrices D_A in case $N = 4$.

	$\begin{pmatrix} 1 & 1 & 1 & 1 \\ 1 & 1 & 1 & 1 \\ 1 & 1 & 1 & 1 \\ 10 & 10 & 10 & 10 \end{pmatrix}$	$\begin{pmatrix} 1 & 1 & 1 & 1 \\ 1 & 1 & 1 & 1 \\ 1 & 1 & 1 & 1 \\ 10^2 & 10^2 & 10^2 & 10^2 \end{pmatrix}$	$\begin{pmatrix} 1 & 1 & 1 & 1 \\ 1 & 1 & 1 & 1 \\ 1 & 1 & 1 & 1 \\ 10^3 & 10^3 & 10^3 & 10^3 \end{pmatrix}$	$\begin{pmatrix} 1 & 1 & 1 & 1 \\ 1 & 1 & 1 & 1 \\ 1 & 1 & 1 & 1 \\ 10^4 & 10^4 & 10^4 & 10^4 \end{pmatrix}$
$\mathrm{ACC}^* =$	0.7500	0.7500	0.7500	0.7500
$\mathrm{MCC}^* =$	0.5000	0.5000	0.5000	0.5000
$\mathrm{CEN} =$	0.6459	0.4021	0.3464	0.3381
$\widetilde{\mathrm{CEN}} =$	0.679	0.4053	0.3424	0.3333

4.4 Empirical Experimentation

Finally, we take profit of some experimentation on UCI imbalanced datasets [4] with common classifiers, performed in [9], where we can find some confusion matrices that can be used to compute the performance measures. More specifically, from [9] we borrow confusion matrices in tables 3.6–3.9 there, and show them in Table 4 below, corresponding to the medical dataset Cardioctocography [5]. Two classifiers have been used: Classification Tree (CT) and Naïve Bayes (NB), both without feature selection (baseline), and also with feature selection, that intuitively must produce better results comparing with baseline. We can see from Table 4 that all the performance measures behave in accordance with intuition except in case (b), being $\widetilde{\mathrm{CEN}}$ the only one showing improvement with respect to the baseline case (a).

Table 4. Cardioctocography dataset. (a) CT baseline; (b) CT with feature selection; (c) NB baseline; (d) NB with feature selection.

	$\begin{pmatrix} 519 & 12 & 10 \\ 31 & 408 & 0 \\ 9 & 0 & 74 \end{pmatrix}$ (a)	$\begin{pmatrix} 514 & 12 & 15 \\ 34 & 405 & 0 \\ 3 & 0 & 80 \end{pmatrix}$ (b)	$\begin{pmatrix} 476 & 37 & 28 \\ 24 & 408 & 7 \\ 6 & 2 & 75 \end{pmatrix}$ (c)	$\begin{pmatrix} 490 & 28 & 23 \\ 11 & 427 & 1 \\ 1 & 0 & 82 \end{pmatrix}$ (d)
$\mathrm{ACC}^* =$	0.05833	0.06021	0.09784	0.06021
$\mathrm{MCC}^* =$	0.05155	0.05281	0.08421	0.05148
$\mathrm{CEN} =$	0.15736	0.15755	0.23890	0.15510
$\widetilde{\mathrm{CEN}} =$	0.24711	0.24540	0.35983	0.24127

5 Conclusion

We introduce a modified Confusion Entropy performance measure for multi-class classification, $\widetilde{\text{CEN}}$, proving some of its properties. We compare $\widetilde{\text{CEN}}$ with CEN, MCC and Accuracy, showing that in the binary case, $\widetilde{\text{CEN}}$ has a consistent behaviour with Accuracy and MCC, overcoming the lack of reliability of CEN. Nevertheless, in two multi-class pathological settings we show that $\widetilde{\text{CEN}}$ is consistent with CEN, but that they are inconsistent with both Accuracy and MCC. The reason is that while nor Accuracy nor MCC can distinguish among different misclassification distribution, the Confusion Entropy measures have an high level of discrimination.

Acknowledgements. This work have been supported by Ministerio de Economía y Competitividad, Gobierno de España, project ref. MTM2015 67802-P.

References

1. Antunes, F., Ribeiro, B., Pereira, F.: Probabilistic modeling and visualization for bankruptcy prediction. Appl. Soft Comput. **60**, 831–843 (2017)
2. Jin, H., Wang, X.-N., Gao, F., Li, J., Wei, J.-M.: Learning Decision Trees using Confusion Entropy. Proceedings of the 2013 International Conference on Machine Learning and Cybernetics, Tianjin, 14–17 July (2013)
3. Jurman, G., Riccadonna, S., Furlanello, C.: A comparison of MCC and CEN error measures in multi-class prediction. Plos One **7**(8), 1–8 (2012)
4. Lichman, M.: UCI Machine Learning Repository. University of California, Irvine, School of Information and Computer Sciences (2013). https://archive.ics.uci.edu/ml/index.php
5. Marques de S., J.-P., Bernardes, J., Ayres de Campos, D.: UCI Machine Learning Repository: Cardiotocography Data Set (2010)
6. Matthews, B.: Comparison of the predicted and observed secondary structure of T4 phage lysozyme. Biochimica et biophysica acta. Vol 405, Num 2, 442–451 (1975)
7. Roumani, Y.-F., May, J.-H., Strum, D.-P.: Classifying highly imbalanced ICU data. Health Care Manag. Sci. **16**, 119–128 (2013)
8. Roumani, Y.-F., Rouman, Y., Nwankpa, J.-K., Tanniru, M.: Classifying readmissions to a cardiac intensive care unit. Ann. Oper. Res. **263**(1–2), 429–451 (2018)
9. Sherman, I.-B.: On the Role of Genetic Algorithms in the Pattern Recognition Task of Classification. Master's Thesis, University of Tennessee, 2017. http://trace.tennessee.edu/utk_gradthes/4780
10. Sublime, J., Grozavu, N., Cabanes, G., Bennani, Y., Cornuéjols, A.: From Horizontal to Vertical Collaborative Clustering using Generative Topographic Maps. International Journal of Hybrid Intelligent Systems, vol. 12(4), 245–256 (2015). https://doi.org/10.3233/HIS-160219
11. Sublime, J., Matei, B., Murena, P.-A.: Analysis of the influence of diversity in collaborative and multi-view clustering. In: 2017 International Joint Conference on Neural Networks (IJCNN), Anchorage, AK, 4126–4133 (2017). https://doi.org/10.1109/IJCNN.2017.7966377
12. Sublime, J., Matei, B., Cabanes, G., Grozavu, N., Bennani, Y., Cornuéjols, A.: Entropy based probabilistic collaborative clustering. Pattern Recogn. **72**, 144–157 (2017)

13. Wang, X.-N., Wei, J.-M., Jin, H., Yu, G., Zhang, H.-W.: Probabilistic Confusion Entropy for Evaluating Classifiers. Entropy **15**, 4969–4992 (2013)
14. Wei, J.-M., Yuan, X.-Y., Hu, Q.-H., Wang, S.-Q.: A novel measure for evaluating classifiers. Expert Syst. Appl. **37**, 3799–3809 (2010)

The Right to Honour on Social Networks: Detection and Classifications of Users

Rebeca Cordero-Gutiérrez[1,2], Pablo Chamoso[1(✉)], Alfonso González Briones[1], Alberto Rivas[1], Roberto Casado-Vara[1], and Juan Manuel Corchado[1]

[1] BISITE Research Group, University of Salamanca, Edificio I+D+I, 37007 Salamanca, Spain
rcorderogu@upsa.es, {chamoso,alfonsogb,rivis,rober,corchado}@usal.es
[2] Faculty of Computer Science, Pontifical University of Salamanca, Salamanca, Spain

Abstract. It is clear that social networks have come to stay. In recent years they have become the media par excellence. The users who participate in them help to spread information quickly and easily so that everyone can benefit. Users are becoming more inclined to voice their opinions, networks are willing to listen and technology has an enormous outreach. A priori, something that seems a great advantage can become a big problem when the news spread violates an individual's right to honour. This paper proposes a tool that detects and collects information from users who publish or disseminate offensive information to an individual. It establishes parameters that determine the level of damage certain individuals can make on social media and makes a ranking that is based on their characteristics and publications. This proposal is an example of the infinite possibilities that automatic data collection and processing provide us with. Without these technologies it would have been impossible to protect the rights of individuals on social networks, due to the large number of users.

Keywords: Social networks · Right to honour · Text analytic

1 Introduction

Social networks have revolutionized our lives; they have changed the way we share and obtain information, the way in which we behave and how we communicate. Today, much of the information we need in our daily lives is just a click away. Social profiles are tools which have given users the freedom and the power to give advice, recommendations, opinions or value judgements. Messages, data and publications deluge these virtual platforms every second, as a result new information systems must be able to extract large amounts of data. [1,2].

In the thicket of today's social networks, Twitter is undoubtedly the microblogging network par excellence. Monthly, it has 330 million active users [3] and generates 7,706 tweets per second, a total of 665,798,400 tweets per

M. Graña et al. (Eds.): SOCO'18-CISIS'18-ICEUTE'18, AISC 771, pp. 90–99, 2019.
https://doi.org/10.1007/978-3-319-94120-2_9

day [4]. These data only confirm Twitter's dominant position in the field of online and worldwide dissemination of information. Information is generated by various agents, from anonymous or known individuals to companies, products and services. Not so long ago users did not have access to such a vast amount of information, nor did they have the power to create and share content with the help of a single click [5]. Users are well aware of their new position, the ability to give their opinions and judgements publicly on anything they wish and at any time. These opinions may simply regard the quality of a product or a service, or a bad experience with a company, however there may also be comments that harm and offend other users.

Relationships between individuals can be represented thanks to Graph Theory [6]. Social networks approach a composition of nodes (individuals or institutions) united by lines that express the relationships between them [7]. One of the most useful applications of this theory is social network analysis [8]. The relationships created in these environments generate the creation of groups with common interests, in which individuals are influenced by their peers, a fact that has been addressed by the Theory of social influence [9] that has a great parallelism with the Theory of Social Networks [10].

Some social network users become true gurus and have great influence over others. Their influence may be of such extent that their followers share the information they post without considering whether or not this infromation is correct or true. These social network gurus may be public figures, they may have a large number of followers or have a lot of knowledge of these platforms. Information becomes viral in these online environments, especially on Twitter where some users' opinions on different topics may have strong influence over the public. When comments that express negative feelings or harmful infromation about an indiviual are spread they cause irreparable damage to that individual's right to honour.

The following section describes the legal situation regarding the right to honour in social networks, with a special focus on the situation in Spain, where this research was conducted. Section 3 describes how the tool was used to retreieve messages and identify users. Section 4 presents a case study and its results. Finally, Sect. 5 outlines the conclusions drawn from the work.

2 The Right to Honour on Social Networks

Although it is an extremely topical issue, the line between the right to honour and freedom of expression has still not been clearly established [11].

In the present work, we will not discuss this dilemma. Rather, we will present a tool designed to help identify messages on social networks which violate a person's right to honour. In addition, we will consider a possible way of classifying the most harmful users according to their characteristics and the nature of their messages, as well as the impact they have caused.

As already discussed in the previous section, social networks provide users with many advantages in terms of infromation access and diffusion and communication [12], but they also become a perfect means of harming the honour, intimacy and image of individuals. Many messages are diluted in the large information flow and in many cases, individuals post their messages anonymously (hidden behind a fictional character created on their profile).

In Spain, the rights to honour, privacy and personal image are recognised in Article 18.1 CE12, and this provision has been developed by Organic Law 1/1982, on the 5 of May, Civil Protection of the Right to Honour, Personal and Family Privacy and the Personal Image (known as LOPDH in Spain). Article 7 of the LOPDH lists the actions that are considered to be illegitimate interferences, among them and in the matter at hand, the imputation of facts or the manifestation of value judgements through actions or expressions that in any way damage the dignity of another person, undermining their fame or violating their own estimation [13].

In addition, the Constitutional Court [14] stated that "honour is a personality right that fits under social projection rights, it is the respect paid to an individual who abides to the moral code of the society, it is determined by the person's view of themselves as well as their reputation, that is, their social standing in the opinion of others".

Before the advent of social networks, mass media could not violate the right to honour as easily and therefore such incidents did not occur often, but nowadays the spread of a harmful message on a platform like Twitter is uncontrollable in the majority of cases, as pointed out by various authors [11], because this message can be linked or shared.

The great novelty of the present work is that it allows the unequivocal and simple location of all those individuals who have been able to damage the honor of another through social networks, even if it is a very high number of individuals who cause the damage. In addition, it proposes an indicator to quantify in which cases this damage has been greatest.

Considering the fact that on a network like Twitter damage can be caused not only by viral messages but also by a single message, it is necessary to gather all the information automatically, including the user or users who published or shared it. This must be done with a low expenditure of monetary and temporal resources.

3 Automatic Location of Users on Twitter and Collection of Relevant Information

The amount of tweets that can be generated for a single event is so large that gathering information about all of them can be a complicated task, the complexity of analysing the gathered information adds further to the problem. In the case of large-scale events, manual data collection can be a very costly and and time-consuming task, as mentioned in the previous section.

The tool for locating individuals and obtaining information was developed on Twitter due to the advatangeous features of this platform. As the vast majority of profiles on this network are public, the collection of data from these profiles does not entail any legal issues (as opposed to Facebook). Moreover, this is one of the most dynamic and active platforms because of the speed of information dissemination.

The developed tool allows to track specific words (keywords) on Twitter and collect information chronologically from the messages that contain these keywords and individuals who publish.

The program allows to select different types of data representations such as geographic activity and sentiment maps, representation of virtual communities or clouds of terms (some of these representations can be seen in Figs. 1 and 2).

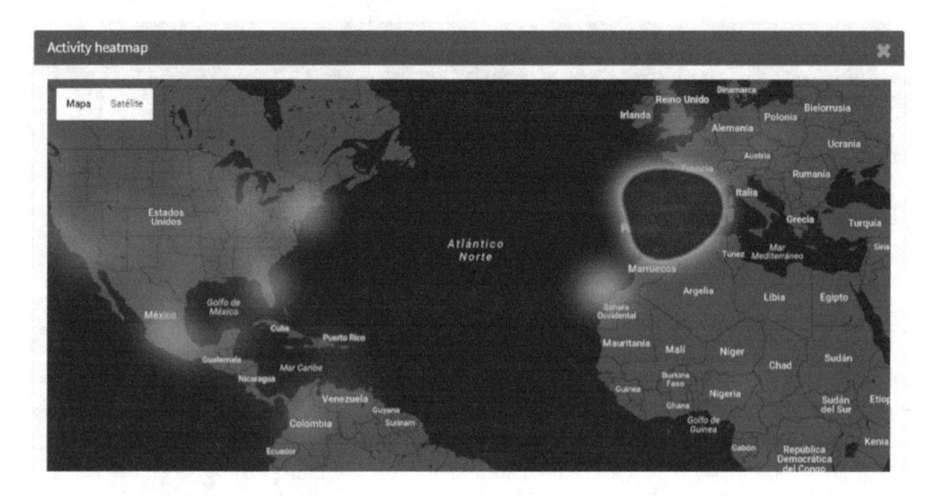

Fig. 1. Heat map generated by the software tool. The figure shows the heat map when the software tool searched the word "Blesa". The red spots show negative comments and the green ones positive.

3.1 Software Tool Characteristics

The community analysis performed by the tool is based on the Louvain method [15]. Louvain's algorithm is based on the optimization of modularity between communities (measurement of the density of links between members of a community and between communities, in an interval of -1 to 1). Its calculation begins by identifying individual members who, because of their relationships with each other, are grouped into communities. Later, he relates these communities with each other. The process is repeated iteratively.

This method is weighed against other factors such as the degree of relationship between users, the degree of influence of the user's home (mathematical approximation of the Klout algorithm), the sentiment of the comments (for which Brooke *et al.*'s proposal was modified [16]), allowing us to determine whether a tweet contains a positive, negative or neutral feeling.

In addition to the information mentioned above, the tool obtains the entire tweet and extracts information from it on the number of retweets, likes, mentions, hashtags and links. In addition, it unambiguously identifies the user who

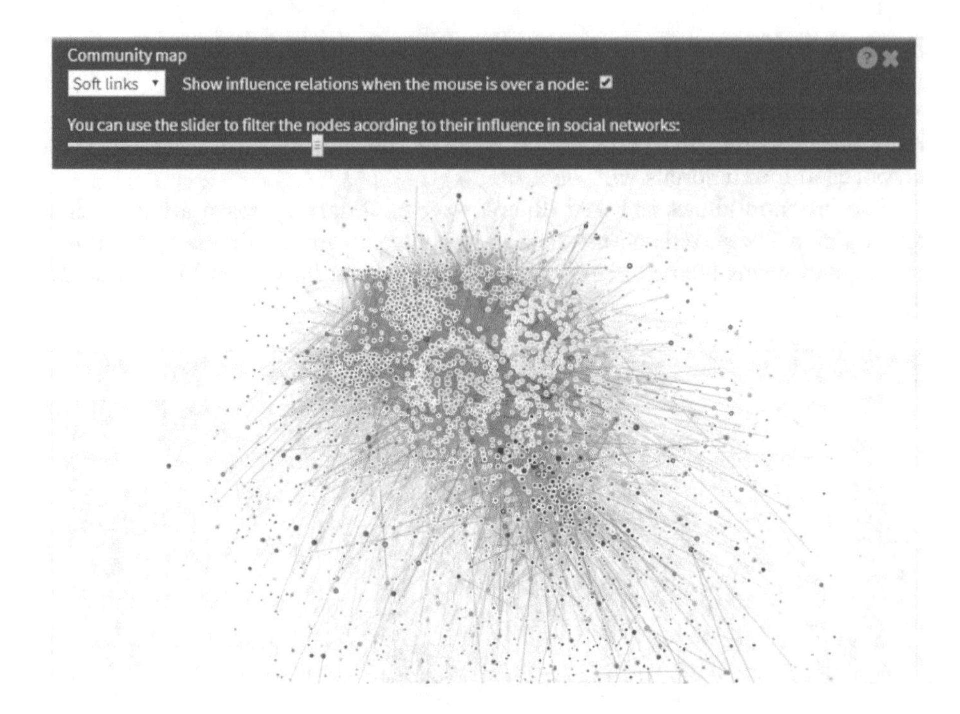

Fig. 2. User community generated by the software tool. The figure shows the user community generated when the software tool searched the word "Blesa". The circles represent individuals and the stripes represent the connections between them. The different colors indicate the communities formed around the keyword.

published it and collects data on their activity (number of followers, followings, favourite tweets, identifier, it even determines if geolocation is available). An example of this can be seen in Fig. 3 where we can select from the generated user community, an individual and with a simple click, access their information. Remember that this data can only be obtained from the users who chose to create public profiles.

Fig. 3. Twitter user (example of unequivocal location).

The tool has already been used in various fields, such as business [17]. However, its possibilities in the field of security are very high, some applications can be the identification of users who may incur in various criminal acts, from violations of the right to honour, glorification of terrorism to the collection of information for the prevention of certain crimes.

The next section presents a case study conducted on Twitter, in which the robustness of the developed tool was tested. It focused on a specific case where the right to honour of a deceased person was damaged.

4 Case Study: Death of Ex-banker Miguel Blesa

As explained in previous sections, the purpose of our work is to show how the use of data mining tools can help identify and localize individuals who violated the right to honour of another person. Specifically, in a case where The amount of information generated for this event would have been difficult to collect manually. Thus, the presented tool facilitates the task of gathering and analysing information which help identify harmful messages and their authors, enabling authorities to take legal action.

The case at hand is just one example of many other cases of damage to the right to honour on social networks, which unfortunately occur with more frequency.

Miguel Blesa was a well-known Spanish banker, chairman of the board of directors of a large Spanish savings bank, Caja Madrid (which was closely associated with the creation of Bankia). In early 2017, he was sentenced to six years' imprisonment for a continuing misappropriation offence in the case of "black cards". For this reason, in recent months his person has been severely criticised.

On July 19 he was found dead on a farm in the province of Córdoba (Spain) with a gunshot to the chest. The main hypothesis was that he had committed a suicide. When this news came out there was a stir on social networks, especially Twitter. Upon this event, many of the tweets insluted Miguel Blesa and even rejoiced at or mocked his death.

To demonstrate that the proposed tool is very useful for this purpose, we have collected information about the tweets in Spanish containing the keyword "Blesa". In order to choose the search word, an internet search was carried out for individuals who could be suffering damage to the right to honour. This search coincides with Blesa's death and is a perfect example to show the power of the tool presented and also the validity of the model presented to evaluate the damage to the right to honour of the individual. Information from the 24 h after the news of his death was collected, during that period a total of 39,439 tweets were generated by 5,636 different users.

This information has been downloaded, cleaned and filtered in order to be able to carry out an assessment of the damage that each individual has caused. Thus, certain variables have been taken into account that, due to the characteristics of Twitter, make one message more harmful than another, either because of its reach, the words it has used or a combination of factors. The variables extracted from the tool that have been used for the proposal are as follows:

- Influence of the individual on the network (i): as we have mentioned in previous sections, certain individuals have a greater influence than others, in many cases modifying user behaviour in certain contexts. Thus, a user who is considered influential in a community has a greater outreach to other users, increasing the possibilities of retweet and diffusion of this message to a wider spectrum, which increases the damage done to the individual.
- Number of followers (f): these are the individuals to whom the tweets reach, so that the greater this number is, the more people will be able to read and therefore, they will have the possibility of sharing the harmful tweet in the form of retweet. Thus, a greater number of followers makes it easier to popularize the content of a message and spread information [18].
- Sentiment (s): Several works have shown that the more powerful the sentiment is in a tweet, the more viral it is on this microblogging network [19]. Therefore, messages with a more harmful sentiment would be more likely get diffused to other individuals.
- Number of message retweets (r): The retweet diffuses a message written by another individual to reach a larger network of individuals, therefore a harmful tweet would be read by an increasing spectrum of individuals, thus favouring its propagation and impact.
- Message likes (fv): giving a like to a tweet gives a leading position to other tweets. Liking a harmful tweet could be interpreted as a way of being proud of something harmful that has been said against another user.
- Additional elements in the tweet such as hashtags (h), mentions (m) or links (l): the messages that contain these elements extend the information already contained in the tweet as in the case of links which name other individuals, as in the case of mentions or allow easy tracking of a topic such as hashtags [20]. These elements make it easier to find and diffuse the message because they act as heuristic clues, as stated by the model of systematic heuristic processing.

With all of them, and taking into account the dispersion of the variables, a proposal (described in Eq. (1)) has been made for an indicator that makes it possible to rank which individual has been most harmful with their message. In order to do this, the variables social influence, followers, sentiment and number of retweets have been recoded, since these variables are not restricted by Twitter or the data extraction tool, i.e. an individual can have a substantial amount of followers that if introduced directly into our proposal can blur the results. Thus, the variables indicated have been calculated taking into account the deciles of the original variable, so that each individual can obtain a maximum of 10 points for each variable, thus eliminating the effect size. The new variables, therefore, are an index of the original ones: influence index (i_index), number of followers index (f_index), sentiment index (s_index) and number of retweets index (r_index). In the case of the variables number of favourites, mentions, hashtags and links, we have chosen to discriminate only if these elements were present or not (fv0,1, m0,1, h0,1, l0,1), because even though an individual wants to mention a very large number of users could not by character restriction. Therefore, this variable is limited. In addition, a correction factor has been added, taking into account whether the individual has written more than one message (repetition of the harmful action, rp) uttering insults, humiliations or mocking the deceased. Of course, this proposal is subject to corrections that a lawyer can make. This correction factor was calculated by searching if a user wrote more than one message about the person they want to harm. But they can be a good starting point in cases where the number of individuals who have been involved in infringing the right to honour is very high. The large amount of data offered by the tool and its great flexibility allows the user to filter the information in order to obtain what he or she needs at any given time to take the legal action he or she deems appropriate.

$$h = i\_index + f\_index + s\_index + r\_index + fv0,1 + m0,1 + h0,1 + l0,1 + rp \quad (1)$$

Thus, after making all the pertinent calculations, we can obtain a ranking of individuals, which is presented below anonymously in which we observe that, of a total maximum of 45 points, an individual has obtained a h (harm) value of more than 27 points (Table 1).

Table 1. Fragment of the ranking of the most harmful users in the Blesa Case in the analysed sample.

User id	Date	Name	Value
6031152	2017-07-20 – 3:05:33 pm	G[...]	27.0861914007
954441708	2017-07-21 – 7:19:07 am	Me[...]	25.0275740176
894920761	2017-07-20 – 5:56:01 pm	AN[...]	24.9920055374
1333566120	2017-07-20 – 4:16:15 pm	Mar[...]	23.9421845192
14954153	2017-07-20 – 12:37:52 pm	M.A	23.9290575676
780756926	2017-07-21 – 9:27:41apm	Car[...]	23.5304099221
2285269720	2017-07-20 – 12:10:20 pm	Al[...]	21.5982901913
967073047	2017-07-20 – 11:37:49 am	ANT[...]	21.0054429472
816326328	2017-07-20 – 12:46:02 pm	inb[...]	21.0016858738
6015212	2017-07-21 – 0:55:06 am	Ga[...]	20.7421914007
780756926	2017-07-20 – 3:15:35 pm	Ca[...]	20.5076099221
1643105568	2017-07-20 – 1:35:35 pm	Vic[...]	20.3808726389
970822632	2017-07-20 – 4:03:35 pm	ga[...]	20.2499000000
1634105568	2017-07-20 – 2:14:41 pm	Vic[...]	19.6308726389
591713765	2017-07-20 – 9:26:50 pm	Tu[...]	19.6079662828
595728565	2017-07-20 – 9:19:53 pm	Ale[...]	19.4342758968

5 Conclusions

Throughout this work we have seen how social networks revolutionize the daily lives of millions of individuals. The way in which we communicate is changing and with it many other aspects of our lives. Users feel increasingly free to express their opinions, and social networks are the perfect megaphone that allows these reflections to reach a large number of recipients quickly. Although this fact is a major step forward for our society, it also entails certain disadvantages and dangers which have been described in this work.

It is difficult to estimate the degree of damage that harmful messages on a social network can cause to an individual's honour. This tool has been proposed in order to help deal with such cases and to ensure that those who cause such damage do not get away without legal consequences. This work proposed the use of the PIAR tool in the collection and analysis of a large amount of information generated on Twitter, for the detection of messages that violate the right to honour of an individual. The case study looked at a particular case in which Twitter users damaged the right to honour of a deceased person.

The classification allows us to propose a ranking of individuals who, due to their characteristics, could have caused more damage than others, so that a human agent can decide whether or not to take legal action.

The presented tool facilitates the task of collecting information and identifying individuals and unambiguously locating them, as well as the relationships of individuals in their network. These weaknesses have already been described in the studies of other authors.

Despite clear examples of violations of the right to honour, the border between honour and freedom of expression remains unclear. More defined regulations are needed in this respect. New technologies have advanced to facilitate the collection and analysis of information, but the regulation of many aspects in this regard is obsolete.

Acknowledgments. This work was conducted under the framework of the project with Ref. RTC-2016-5642-6, financed by the Ministry of Economy, Industry and Competitiveness of Spain and the European Regional Development Fund (ERDF).

The research of Alfonso González-Briones has been co-financed by the European Social Fund (Operational Programme 2014–2020 for Castilla y León, EDU/128/2015 BOCYL).

References

1. López Sánchez, D., Revuelta Herrero, J., Prieta Pintado, F.d.l., Dang, C.: Analysis and visualization of social user communities (2016)
2. Sánchez, D.L., Revuelta, J., De la Prieta, F., Gil-González, A.B., Dang, C.: Twitter user clustering based on their preferences and the Louvain algorithm. In: Trends in Practical Applications of Scalable Multi-Agent Systems, the PAAMS Collection, pp. 349–356. Springer (2016)
3. Omnicore Agency: (2018). https://www.omnicoreagency.com/twitter-statistics/. 30th March 2018
4. Internet live stats: (2018). http://www.internetlivestats.com/one-second/. 19th February 2018

5. Steyn, P., Ewing, M.T., Van Heerden, G., Pitt, L.F., Windisch, L.: From whence it came: understanding source effects in consumer-generated advertising. Int. J. Advert. **30**(1), 133–160 (2011)
6. Euler, L.: Solutio problematis ad geometriam situs pertinensis. Comm. Acad. Sci. Imper. Petropol. **8**, 128–140 (1736)
7. Mena Díaz, N.: Redes sociales y gestión de la información: un enfoque desde la teoría de grafos. Ciencias de la Información **43**(1), 29–37 (2012)
8. Orta, C., Pardo, F., Salas, J.: Análisis de redes de influencia en twitter (2012)
9. Kelman, H.C.: Compliance, identification, and internalization three processes of attitude change. J. Conflict Resolut. **2**(1), 51–60 (1958)
10. Colina, C.L.: La teoría de redes sociales. Papers: Revista de Sociologia (48), 103–126 (1996)
11. Vallilengua, L.G.: Los derechos al honor, a la intimidad ya la propia imagen en las redes sociales: la difusión no consentida de imágenes. Revista Electrónica del Departamento de Derecho de la Universidad de La Rioja, REDUR (14), 161–190 (2016)
12. Chamoso, P., Rivas, A., Rodríguez, S., Bajo, J.: Relationship recommender system in a business and employment-oriented social network. Inf. Sci. (2018)
13. Atienza Navarro, M.L., Carrión Olmos, S., Cobiella, C., Cobiella, C., Chaparro Matamoros, P., Chaves Pedrón, C., de la Maza Gazmuri, Í., García Ortega, J., López Orellana, M., de Leonardo, G., et al.: La Ley orgánica 1/1982, de 5 de mayo, de protección civil del derecho al honor, a la intimidad personal y familiar ya la propia imagen. Editorial de la Universidad del Rosario (2011)
14. Constitutional Court: (2008). https://supremo.vlex.es/vid/derecho-honor-personalidad-as-42930503. 30th March 2018
15. De Meo, P., Ferrara, E., Fiumara, G., Provetti, A.: Generalized Louvain method for community detection in large networks. In: 2011 11th International Conference on Intelligent Systems Design and Applications (ISDA), pp. 88–93. IEEE (2011)
16. Brooke, J., Tofiloski, M., Taboada, M.: Cross-linguistic sentiment analysis: from English to Spanish. In: Proceedings of the International Conference RANLP-2009, pp. 50–54 (2009)
17. Lahuerta-Otero, E., Cordero-Gutiérrez, R.: Looking for the perfect tweet. The use of data mining techniques to find influencers on Twitter. Comput. Hum. Behav. **64**, 575–583 (2016)
18. Zhang, L., Peng, T.Q., Zhang, Y.P., Wang, X.H., Zhu, J.J.: Content or context: which matters more in information processing on microblogging sites. Comput. Hum. Behav. **31**, 242–249 (2014)
19. Zhang, L., Peng, T.Q.: Breadth, depth, and speed: diffusion of advertising messages on microblogging sites. Internet Res. **25**(3), 453–470 (2015)
20. Lee, K.: What analyzing 1 million tweets taught us. The Next Web (2015)

Classification Improvement for Parkinson's Disease Diagnosis Using the Gradient Magnitude in DaTSCAN SPECT Images

Diego Castillo-Barnes(✉), Fermin Segovia, Francisco J. Martinez-Murcia, Diego Salas-Gonzalez, Javier Ramírez, and Juan M. Górriz

Department of Signal Theory, Networking and Communications, University of Granada, Periodista Daniel Saucedo Aranda, 18071 Granada, Spain
diegoc@ugr.es
http://sipba.ugr.es/

Abstract. In this work, we propose a novel imaging preprocessing step based on the use of the gradient magnitude for medical DaTSCAN SPECT images. As Parkinson's Disease (PD) is characterized by a marked reduction of intensity at *striatum* area, measuring intensities in this region is considered as a good marker for this neurological disorder. To extend this idea, we have been studying how quick these values decrease. A simple way to do this was using the gradient of each image. Applying Machine Learning algorithms, we have classified the gradient images and obtained an accuracy improvement of almost **2%**. These results prove that the gradient magnitude is even a better marker for PD diagnosis and opens the door to new future investigations about this pathology.

Keywords: Parkinson's Disease · Gradient · SVM Classification · Machine Learning · α-Stable distributions Parkinson's Progression Markers Initiative (PPMI) · SPECT DaTSCAN · Neuroimaging

1 Introduction

Parkinson's Disease (PD) is a neurological disorder characterized by symptoms such as tremor, rigidity, bradykinesia, cognitive alterations, lack of emotion expressiveness and autonomy problems [1]. Although origins or triggers that makes appear the PD are still unknown, several studies point that is related to the destruction of pigmented neurons in the substantia nigra [2].

A marked reduction in dopaminergic neurons is the most significative feature of PD. In this sense, multiple radio-ligands have been used in order to study Parkinsonism. One of them is the Iodine-123-fluoropropyl-carbomethoxy-3-β-(4-iodophenyltropane) [3,4] also known as FP-CIT, 123I-Ioflupane or DaTSCAN.

M. Graña et al. (Eds.): SOCO'18-CISIS'18-ICEUTE'18, AISC 771, pp. 100–109, 2019.
https://doi.org/10.1007/978-3-319-94120-2_10

This kind of SPECT images give us a quantitative measure of the spatial distribution of transporters in the brain.

All this information can be used to differentiate Healthy Control subjects (HC) from subjects with PD [5]. Although several Computer-Aided-Diagnosis (CAD) systems have been developed for automatic diagnosis of PD [6–8], to date, no studies have focused on how quick intensity is spatially lost in the brain.

In short, unlike it happens with HC subjects, PD patients are expected to present their *striatum* area with poor intensity levels. Measuring the difference between striatum area of a subject and its borders we could determine if that person is probably suffering PD or not.

This paper is organized as follows: Sects. 2.1 and 2.2 presents the image database used in this work and statistics about subjects. Sections 2.3 and 2.4 consist of preprocessing steps carried out in order to normalize the full dataset. Sections 2.5 and 2.6 describe the gradient algorithms and their use to this proposal. The last two sections from Materials and Methods, Sects. 2.7 and 2.8, summarize the methodology for classification and validation. Finally, the experimental results and discussion are given in Sect. 3 and the conclusions are drawn in Sect. 4.

2 Materials and Methods

2.1 PPMI Dataset

Data used in the preparation of this article were obtained from the Parkinsons Progression Markers Initiative (PPMI) database (www.ppmi-info.org/data). For up-to-date information on the study, visit www.ppmi-info.org. PPMI - a public-private partnership - is funded by the Michael J. Fox Foundation for Parkinsons Research and funding partners, including all partners listed on fundingpartners section of the website. Informed consents to clinical testing and neuroimaging prior to participation of the PPMI cohort were obtained, approved by the institutional review boards (IRB) of all participating institutions.

2.2 Demographics

For this work, we retrospectively selected the baseline (BL) results of 719 participants in the PPMI cohort study including HC subjects, patients diagnosed with PD and patients with PD whose scans have not evidence of dopaminergic deficit (SWEDD) [9]. Demographic information is included in Table 1.

Table 1. Demographics

Subjects	Number	Mean age (Std)	Sex (Male)
HC	194	53.04 (2.27)	66%
PD	447	53.13 (2.32)	64%
SWEDD	78	53.28 (2.26)	63%

2.3 Preprocessing Step - Spatial Normalization

All DaTSCAN images have been spatially registered using the software tool `SPM12` available from www.fil.ion.ucl.ac.uk/spm/software/spm12/. It was checked that matching between voxels and anatomical structures was unaltered. After being co-registered and averaged, each cerebral image was reoriented into a standard image grid. Obtained images had a dimension of $79 \times 95 \times 78$ voxels and a voxel size of $2.0 \times 2.0 \times 2.0$ mm.

2.4 Preprocessing Step - Intensity Normalization Based on the Use of α-Stable Distributions

Figure 1 shows the variability of the histogram for the subjects of the database. As this figure reveals, variability between intensity values requires a preprocessing step capable of homogenizing the input data. Since variability does not only depend on a multiplicative factor (some histograms are clearly shifted in the x-axis), it is necessary to use an intensity normalization method different from the traditional specific/nonspecific Binding Ratio (BR) intensity normalization model. In order that, we have followed the method proposed in [10]. This model is based on the use of *alpha*-Stable distributions. The procedure is able to homogenize better the information than other approaches like the BR or an equivalent model based on Gaussian distributions [11].

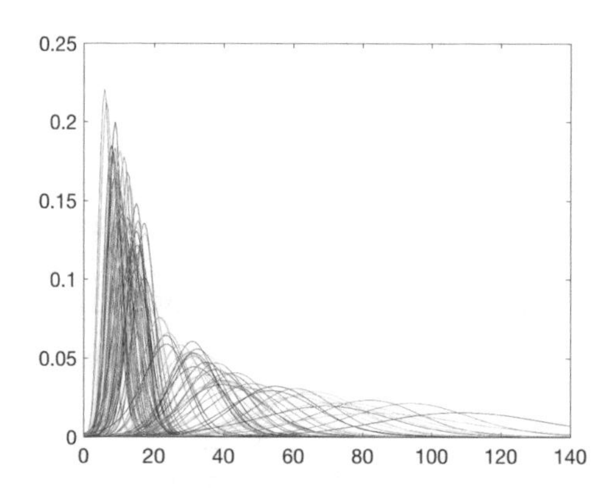

Fig. 1. Intensity distribution of the first 100 DaTSCAN images before applying the α-Stable intensity normalization.

To perform the α-Stable normalization, we have to apply a linear transformation as shown in (1):

$$Y = aX + b \tag{1}$$

with a and b as described in (2):

$$a = \frac{\gamma^*}{\gamma} \qquad b = \mu^* - \frac{\gamma^*}{\gamma}\mu \tag{2}$$

where γ^* and μ^* represent the mean of γ (dispersion) and μ (location) parameters, respectively, that are computed for the whole database.

Procedure is carried out as follows:

- A mask is applied to source images in order to consider only voxels in the brain outside the *striatum* region as explained in [12].
- Once histogram from the selected voxels is obtained, an α-Stable distribution has to be fitted.
- Compute γ^* and δ^* parameters as mean of all γ and δ parameters.
- Obtain a and b parameters following (2).
- Apply the linear transformation (1).

Once this normalization step ends, we can verify our dataset has been homogenized as depicted in Fig. 2:

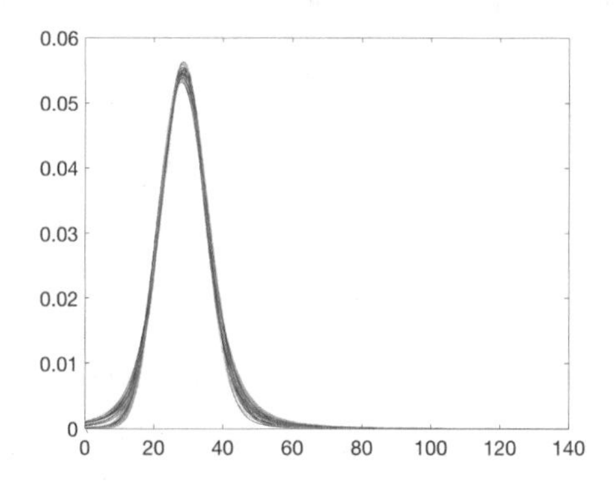

Fig. 2. Intensity distribution of the first 100 DaTSCAN images after applying the α-Stable intensity normalization.

2.5 Gradient

Once dataset is normalized, next step is the computation of gradient. We could define a volume gradient as the directional change of intensity in the volume. Mathematically, the gradient of a 3-variable function $f(x, y, z)$, is a vector whose components are given by the directional derivatives of x, y and z as shown in (3):

$$\nabla f(x,y,z) = \frac{\partial f}{\partial x}\mathbf{i} + \frac{\partial f}{\partial y}\mathbf{j} + \frac{\partial f}{\partial z}\mathbf{k} \tag{3}$$

At each point of the volume, gradient vector points in the direction of largest possible intensity increase (maximum slope), and the length of the gradient vector corresponds to the rate of change in that direction. Moreover, as derivative of a 3D image can be approximated by finite differences, we can rewrite (3) in spherical coordinates as follows in (4) and, consequently, in terms of magnitude r and directions θ and φ as shown in (5):

$$\nabla f(r,\theta,\varphi) = \frac{\partial f}{\partial r}\mathbf{e}_r + \frac{1}{r}\frac{\partial f}{\partial \theta}\mathbf{e}_\theta + \frac{1}{r\sin\theta}\frac{\partial f}{\partial \varphi}\mathbf{e}_\varphi \tag{4}$$

$$r = \sqrt{x^2 + y^2 + z^2} \qquad \theta = \arctan \frac{\sqrt{x^2 + y^2}}{z} \qquad \varphi = \arctan \frac{x}{y} \tag{5}$$

In practice, gradient is approached by using the Sobel-Feldman Operator Algorithm (SOA) [13]. Given an image, the Sobel-Feldman Operator convolves the image with a small, separable and integer-valued filter in the horizontal and vertical directions to calculate approximations of the derivatives. Algorithm procedure is summarized as follows for 2D images:

- Declare matrices M_x and M_y as defined in (6).
- Convolve source image I with M_x and M_y matrices as expressed in (6) to obtain two new matrices: G_x and G_y. These matrices represent two images which at each point contain the horizontal and vertical derivative approximations respectively.
- Merge results using expressions in (8). G represents the gradient magnitude and Θ its direction.

$$M_x = \begin{bmatrix} -1 & 0 & 1 \\ -2 & 0 & 2 \\ -1 & 0 & 1 \end{bmatrix} \qquad M_y = \begin{bmatrix} 1 & 2 & 1 \\ -2 & 0 & 2 \\ -1 & 0 & 1 \end{bmatrix} \tag{6}$$

$$G_x = M_x * I \qquad G_y = M_y * I \tag{7}$$

$$G = \sqrt{G_x^2 + G_y^2} \qquad \Theta = \arctan\left(\frac{G_y}{G_x}\right) \tag{8}$$

A generalization of this procedure for 3D images is included in [14] and an example of its application is depicted in Fig. 3.

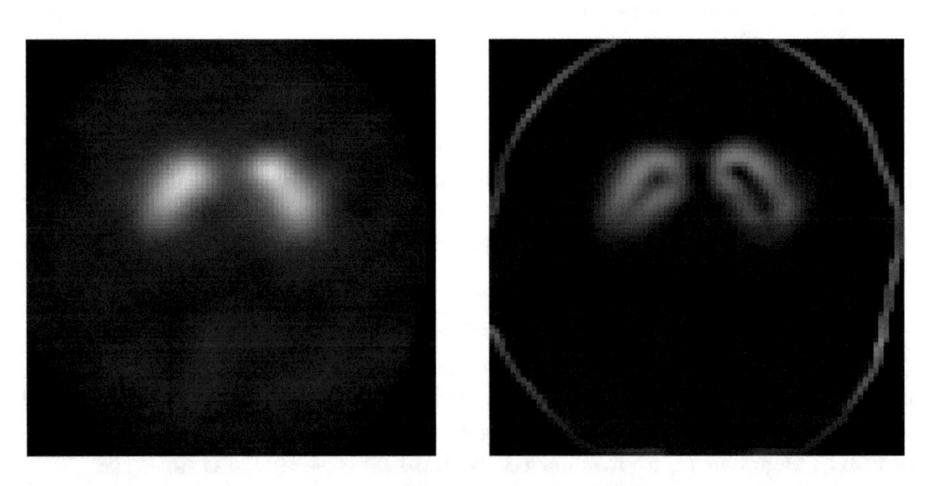

Fig. 3. Original DaTSCAN SPECT image from a subject (left) and its gradient magnitude transformation (right).

2.6 Feature Extraction

In order not to compare directly all volumes (each image has a total of 585, 390 intensity values), we selected a set of voxels whose posterior classification could give us the most precise results with a lower computing cost. As a way to determine this interclass separation, we made use of the Mann-Whitney-Wilcoxon U-Test for two reasons: It is more robust to outliers in the data than t-Test and it does not assume Gaussian data [15].

Once we observed the interclass separation map of all original images (as shown in Fig. 4), we determined that our set of voxels should be at least the twice of the *striatum* total volume (44,126 voxels). This approach will result in consider not only the inner *striatum* region but also its borders.

2.7 Classification

The final classification was carried out using a linear SVM (*Support Vector Machine*) classifier with feature regularization. As input data was not heterogeneous, Kernel functions or similarity matrices were not considered necessary as in a multi-modal analysis [16,17]. In this case, a simple two-class (binary) classifier is considered as sufficient to separate HC subjects from patients with PD through the gradient images dataset. The SVM problem can be defined as follows: consider (x_i, y_i), $i = 1, \ldots, N$ where $x \in \Re^P$ and $y_i \in -1, 1$, we have to find the maximum margin hyperplane, by solving the optimization problem that minimizes (9):

$$\frac{1}{2\,\|\mathbf{w}\|^2} \tag{9}$$

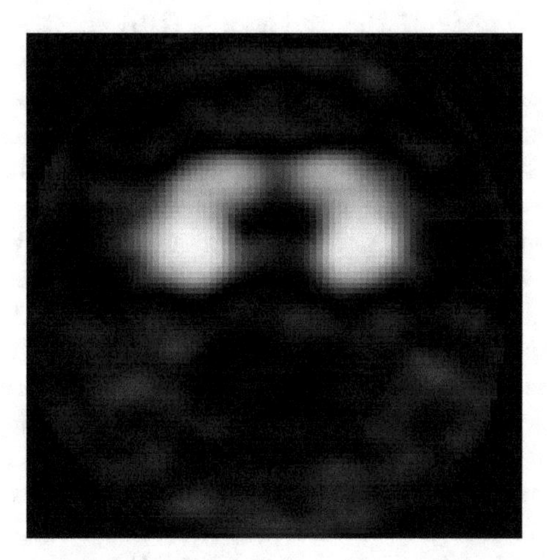

Fig. 4. Interclass separation of original images in terms of Mann-Whitney-Wilcoxon U-Test.

subject to (10):

$$y_i(\mathbf{w}^T\mathbf{x}_i - b) \leq 1 \tag{10}$$

This problem can be solved using standard quadratic programming techniques and programs.

2.8 Validation

The main idea of this section is to split dataset into two groups: a training data group, which we use to train the prediction model, and a test data group, that is then used to measure the classifier's performance through the cross-validation strategy selected. In this work, validation results are analyzed considering: correct rate or accuracy (Acc), sensitivity or true positive rate (Sens), specificity or true negative rate (Spec) and precision (Prec) as follows in (11):

$$\begin{aligned} \text{Accuracy} &= \frac{T_P+T_N}{T_P+T_N+F_N+F_P} \qquad & \text{Sensitivity} &= \frac{T_P}{T_P+F_N} \\ \text{Specificity} &= \frac{T_N}{T_N+F_P} \qquad & \text{Precision} &= \frac{T_P}{T_P+F_P} \end{aligned} \tag{11}$$

where T_P is the number of PD patients correctly classified (true positives), T_N is the number of HC subjects correctly classified (true negatives), F_P is the number of HC subjects classified as PD (false positives) and F_N is the number of PD patients classified as HC (false negatives). Statistics results have been estimated by means of a leave-one-out cross-validation strategy [18].

3 Results and Discussion

The proposed methodology has been tested using 719 different SPECT images (194 HC, 78 SWEDD and 447 PD) in baseline (BL) as cited in Table 1.

Once data sources have been properly pre-processed (spatial normalization, intensity normalization and gradient map), the Mann-Whitney-Wilcoxon U-Test is performed to determine voxels with the highest interclass separation. The map of higher interclass separation obtained is used for both the original dataset and the gradient dataset classification.

The classification and validation is carried out using a SVM classifier with a leave-one-out cross-validation strategy. In a first experiment, we have compared HC subjects with PD+SWEDD patients. Inclusion of SWEDD labelled subjects is because they are clinically considered as PD diagnosed patients. In a posterior experiment also performed, we made a classification considering only HC versus PD labelled subjects. Classification results have been included in Table 2.

In this proposed scenario, the novel methodology based on classification of gradient images outperforms the original method in all cases considered. Maximum classification results are reached classifying HC versus PD patients using the gradient dataset (96.6%). Note that as SWEDD subjects are patients whose scans have not evidence of dopaminergic deficit, their intensity distribution is quite similar to HC subjects and this results in a less accurate classification.

Table 2. Classification results.

Experiment	Source	Accuracy	Sensitivity	Specificity	Precision
HC vs. PD+SWEDD	Original	81.50%	88.00%	63.92%	86.84%
HC vs. PD+SWEDD	Gradient	83.03%	88.76%	67.53%	88.09%
HC vs. PD	Original	95.32%	96.42%	92.78%	96.85%
HC vs. PD	Gradient	96.57%	97.32%	94.85%	97.75%

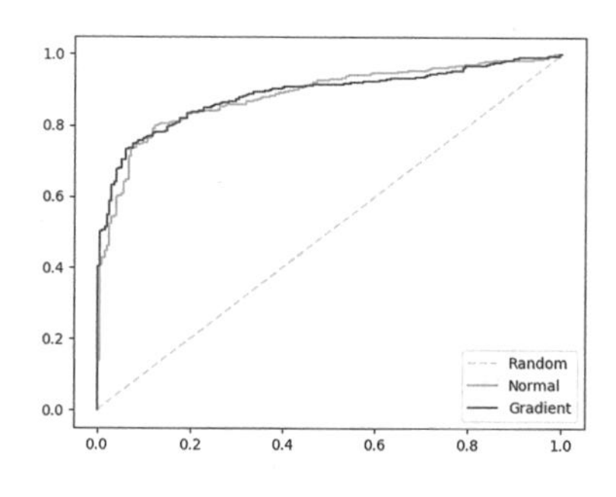

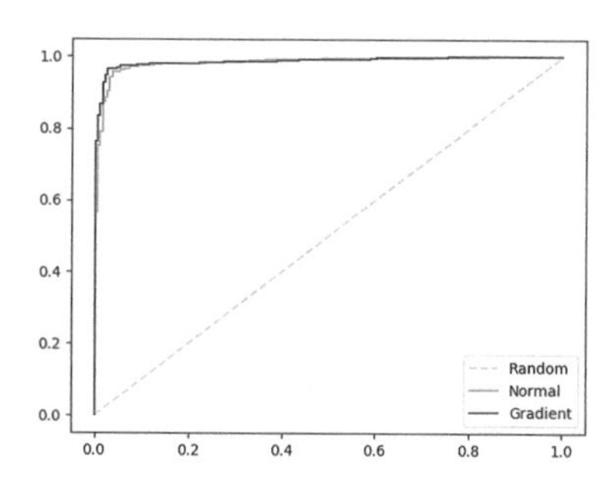

Fig. 5. ROC curves for HC vs. PD+SWEDD (left) and HC vs. PD (right).

Table 3. Area Under the Curve (AUC) obtained for each ROC curve.

Experiment	Source	AUC
HC vs. PD+SWEDD	Original	0.8833
HC vs. PD+SWEDD	Gradient	0.8871
HC vs. PD	Original	0.9840
HC vs. PD	Gradient	0.9874

To highlight the difference between considering HC vs. PD+SWEDD or HC vs. PD, a further comparison was performed by means of the Receiver Operating Characteristic (ROC) curves [19]. As can be checked in Fig. 5, both experiments are more reliable when considering the gradient transformation. Area Under the Curve (AUC) parameters for each ROC curve are reflected in Table 3.

4 Conclusion

In this work, an alternative transformation in the preprocessing step of DaTSCAN SPECT images has been proposed. This method consists of the fact that intensity differences between striatum voxels and non-striatum voxels should be profuse in healthy conditions, i. e. indicative of disease in case of slightly differences.

Once normalized all images, spatially and through an intensity normalization step based on the use of *alpha*-Stable distributions, a gradient transformation has been performed. This transformation has proven that differences in contiguous voxels result in a more accurately classification of HC subjects versus patients with PD. This idea opens the door to new lines of work about PD and how cerebral structures and tissues behave in the progression of this neurodegenerative pathology.

Acknowledgements. This work was supported by the MINECO/FEDER under the TEC2015-64718-R project and the Ministry of Economy, Innovation, Science and Employment of the Junta de Andalucía under the Excellence Project P11-TIC-7103.

References

1. Pohl, A.: Impaired emotional mirroring in Parkinsons disease - a study on brain activation during processing of facial expressions. Front. Neurol. **8**, 682 (2017)
2. Kordower, J.H.: Disease duration and the integrity of the nigrostriatal system in Parkinsons disease. Brain **136**(8), 2419–2431 (2013)
3. Neumeyer, J.L.: [123I]-2β-carbomethoxy-3β-(4-iodophenyl)tropane: high-affinity SPECT (single photon emission computed tomography) radiotracer of monoamine reuptake sites in brain. J. Med. Chem. **34**(10), 3144–3146 (1991)
4. Sixel-Döring, F.: The role of 123I-FP-CIT-SPECT in the differential diagnosis of Parkinson and tremor syndromes: a critical assessment of 125 cases. J. Neurol. **258**(12), 2147–2154 (2011)
5. Marek, K.L.: [123I]β-CIT SPECT imaging assessment of the rate of Parkinson's disease progression. Neurology **57**(11), 2089–2094 (2001)
6. Illán, I.A.: Automatic assistance to Parkinsons disease diagnosis in DaTSCAN SPECT imaging. Med. Phys. **39**(10), 5971–5980 (2012)
7. Segovia, F.: Improved Parkinsonism diagnosis using a partial least squares based approach. Med. Phys. **39**, 4395–4403 (2012)
8. Martínez-Murcia, F.J.: Parametrization of textural patterns in 123I-ioflupane imaging for the automatic detection of Parkinsonism. Med. Phys. **41**(1), 012502 (2014)
9. Wyman-Chick, K.A.: Cognition in patients with a clinical diagnosis of Parkinson disease and scans without evidence of dopaminergic deficit (SWEDD): 2-year follow-up. Cognit. Behav. Neurol. **29**(4), 190–196 (2016)
10. Salas-González, D.: Finite mixture of α-stable distributions. Digit. Signal Process. **19**(2), 250–264 (2009)
11. Salas-González, D.: Linear intensity normalization of FP-CIT SPECT brain images using the α-stable distribution. NeuroImage **65**(C), 449–455 (2013)
12. Brahim, A.: Intensity normalization of DaTSCAN SPECT imaging using a model-based clustering approach. Appl. Soft Comput. **37**(C), 234–244 (2015)
13. Sobel, I.: History and Definition of the Sobel Operator (2014)
14. Engel, K.: Real-time Volume Graphics, 2nd edn, pp. 109–113. A. K. Peters Ltd., Natick (2006)
15. Martínez-Murcia, F.J.: Computer aided diagnosis tool for Alzehimer's disease based on Mann-Whitney-Wilcoxon U-Test. Expert Syst. Appl. **39**(10), 9676–9685 (2012)
16. Tong, T.: Multi-modal classification of Alzheimers disease using nonlinear graph fusion. Pattern Recognit. **63**(C), 171–181 (2017)

17. Li, Q.: Multi-modal discriminative dictionary learning for Alzheimers disease and mild cognitive impairment. Comput. Methods Programs Biomed. **150**(C), 1–8 (2017)
18. Kohavi, R.: A study of cross-validation and bootstrap for accuracy estimation and model selection. In: Proceedings of the 14th International Joint Conference on Artificial Intelligence (IJCAI 1995), vol. 2, pp. 1137–1145 (1995)
19. Zweig, M.H.: Receiver-operating characteristic (ROC) plots: a fundamental evaluation tool in clinical medicine. Clin. Chem. **39**(4), 561–577 (1993)

Experimental Study on Modified Radial-Based Oversampling

Barbara Bobowska(✉) and Michał Woźniak

Faculty of Electronics, Department of Systems and Computer Networks,
Wrocław University of Science and Technology,
Wybrzeże Wyspiańskiego 27, 50-370 Wrocław, Poland
{barbara.bobowska,michal.wozniak}@pwr.edu.pl

Abstract. Although, imbalanced data analysis gained significant attention in the past years, it still remains an underdeveloped area of research posing many difficulties due to the difference in the number of objects in the examined classes, rendering traditional, accuracy driven machine learning methods useless. With many modern real-life applications being examples of imbalanced data classification i.e. fraud detection, medical diagnosis, oil-spills detection in satellite images or network anomaly detection, it is crucial to develop new algorithms suitable to use in such situations. One of the approaches to deal with the disproportion between the instances of objects in classes are either over- or undersampling techniques. In this paper, we propose a modification of an existing RBO algorithm. Due to the additional constraint the modified algorithm eliminates instances which may be problematic to classify. Additionally, a recursion mechanism was added in order to make the search of synthetic points more robust. The results obtained from computer experiments carried out on the benchmark datasets prove that the presented algorithm is applicable.

1 Introduction

While many existing machine learning algorithms achieve satisfying results for classification tasks without class imbalance, most cannot be applied to data sets with a significant disproportion between the instances of different classes due to the underlying rules under which they operate. The majority of methods are accuracy driven i.e. the objective is to maximize the classification accuracy [5]. Given that the traditional algorithms were developed on the assumption, that the number of instances of objects in classes are comparable and thus treated as equally important. As a result the classifier is biased towards the majority class. In order to deal with the issues specific to imbalanced data analysis researchers have been working on developing new methods which can be divided into three categories:

M. Graña et al. (Eds.): SOCO'18-CISIS'18-ICEUTE'18, AISC 771, pp. 110–119, 2019.
https://doi.org/10.1007/978-3-319-94120-2_11

1. Preprocessing methods - such as under- and oversampling, focus on altering the datasets before said data is applied to a given classifier [6]. The aim of both is to lessen the disproportion in the number of the instances between the classes, while maintaining invariable class distribution. Randomized methods, while having low computational cost, may be dangerous, as oversampling the minority class can result in adding instances that aren't beneficial and undersampling some instances may disrupt certain subregions of the majority class. Thus mostly utilized are the methods that add artificial or delete instances in a guided manner.
2. Algorithm-level methods - aim at altering the learning step of the classifier in a way that would make it immune to the skewed class distribution. Here, the popular techniques are either one-class classifiers of cost-sensitive methods [8,11]. For the former one class is designated as the so-called target while the remaining instances are considered to be outliers. The latter introduces the concept of costs. For each misclassification of either the instance of the majority or minority class the classifier is penalized by different cost. The erroneous recognition of the object from the minority class will be more costly thus reducing the bias towards the majority class.
3. Hybrid methods - a combination of the aforementioned methods and in most cases ensemble learning [12].

In this paper we propose a novel oversampling method based on the Radial-Based oversampling (RBO) technique introduced in [9]. The RBO algorithm considers all instances belonging in the minority class as suitable when choosing the initial value for the object used to initiate the method, thus taking into considerations all regions in which at least one minority class instance is present. That leads to creating synthetic samples hard to classify. In the modification a criterion potentially mitigating such situation was applied. To ensure that the considered point is indeed the best in the area a recursion was added. The paper is organized as follows: the second sections explains the concept of radial functions and their usage in estimating the potential of a given instance, as well as the mutual class distribution and the basic principles of the proposed algorithm. In section three, the results of the experimental study are explored. Finally, section four offers the conclusions and discusses future works.

The main contributions of this work are as follows:

- Presentation of the modification of the Radial Based Oversampling algorithm [9]
- Experimental evaluation of the presented approach on the basis of diverse benchmark datasets and a detailed comparison with the original version and state-of-art methods.

2 Radial-Based Oversampling Technique

As described in Sect. 1 the simplest approach to oversampling is to randomly copy existing examples from the minority class, until the balance between the distribution of the classes is achieved. However, using existing instances causes the minority class distribution to be focused in the area where those instances were present. Because of that learning on a set altered this way is prone to overfitting. To eliminate the drawback of this approach a Synthetic Minority Oversampling Technique (SMOTE) [3] algorithm was created. SMOTE and all techniques derived from it [2,4,10] are neighborhood-based and instead of duplicating existing instances they create new synthetic ones. Those synthetically generated samples are created by inserting a new object in between the given minority instance and one of its nearest neighbors of the same class. In other words, SMOTE algorithm finds regions in which synthesized samples can be added. However, this approach may inadequate, in a situation where no regions between nearest neighbors are fit to create new objects. It is often observable, that the instances of the minority class form clusters, that can be sparsely spread throughout the object space, often bounded by a majority objects cluster of much bigger size (Fig. 1), making it impossible to find appropriate neighborhood, and thus trespassing the majority cluster when generating new instances (Fig. 2). Additionally, the decision on the parameter k, corresponding to the size of the neighborhood is also not a trivial matter. Due to the limitations of the SMOTE algorithm, the value of the parameter has to be constant for the entirety of the object space, however, since the sizes of the minority clusters vary, it is impossible to choose one suitable k value.

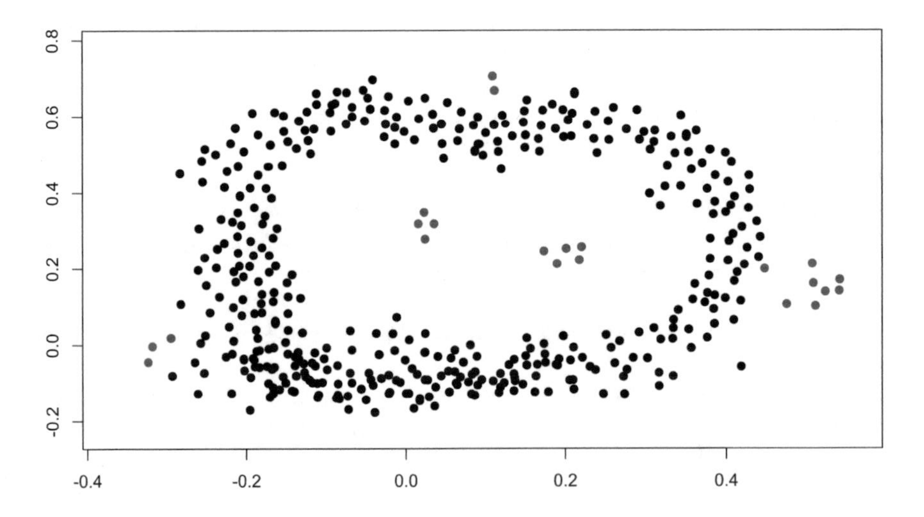

Fig. 1. A situation, where due to the lack of appropriate regions between neighbors, SMOTE algorithm may have difficulties.

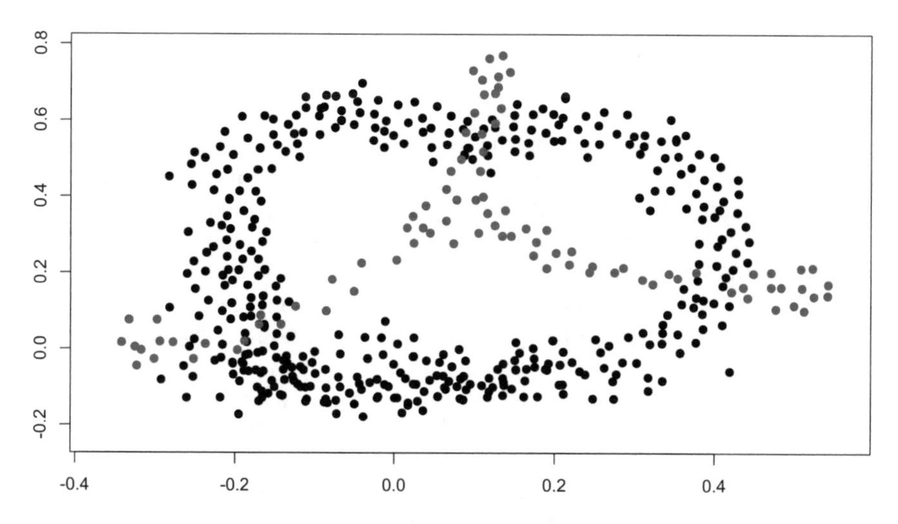

Fig. 2. An undesired outcome of applying SMOTE algorithm - synthetic samples trespassed the majority cluster.

To mitigate aforementioned issues, we propose an alternative approach for approximating the regions of interest. Instead of generating binary regions between the minority objects, we propose using a real-valued potential surface, with potential at each point in space representing our preference towards that point belonging to either minority or majority class. To calculate that potential, we assign a Gaussian radial basis function (RBF) to every object in our training dataset, with the polarity dependent on its class. Given a set of majority objects K, a set of minority objects κ, and a parameter γ representing a spread of a single RBF, we define the potential in a point x as

$$\Phi(x, K, \kappa, \gamma) = \sum_{i=1}^{|K|} e^{-(\frac{\|K_i - x\|_1}{\gamma})^2} - \sum_{j=1}^{|\kappa|} e^{-(\frac{\|\kappa_j - x\|_1}{\gamma})^2}. \quad (1)$$

where K_1 stands for the ith object from majority class and $kappa_j$ denotes jth object form minority class respectively.

Therefore a high potential at a given point will indicate that the point most likely belongs to the majority class. Consequently low potential at a given point will be taken as an indication towards belonging to the minority class. If a high potential is observed for an instance of a minority class said point is assumed to be present in an area dominated by majority class examples making it tough to classify.

3 RBO_modified Algorithm

The proposed algorithm uses the potential to guide the process of artificial creation of new samples of the minority class. Its goal is to find regions that are suitable i.e. where the potential indicates proximity of instances form the minority class, for the injection of generated samples. Firstly, a random point form the minority class is chosen and its potential is calculated. If the value indicates the point is surrounded by mostly majority class instances, such point is considered to be hard to classify and is omitted, while the algorithm tries for another point and repeats the procedure. If the point satisfies the condition the algorithm tries then to find a candidate to create a synthetic sample, by comparing the values of the potentials in the initially chosen point and a proposed one. The new candidate is calculated by translating the initial point by a given step size in a randomly chosen direction. It is ensured that all directions are possible to be selected. If the proposed one indeed is better the algorithm adds it to the set of artificial instances. If not the algorithm calls the procedure again. The methods was presented in Algorithms 1 and 2.

Algorithm 1. Modified Radial-Based Oversampling

1: **Input:** collections of majority objects K and minority objects κ
2: **Parameters:** spread of radial basis function γ, optimization *step size*, number of *iterations* per synthetic sample, probability of early stopping p
3: **Output:** collection of synthetic minority objects S
4:
5: **function** RBO(K, κ, γ, *step_size*, *iterations*):
6: initialize empty collection S
7: **while** $|\kappa| + |S| < |K|$ **do**
8:     *point* $\leftarrow$ randomly chosen object from κ
9:     **if** $0.5 * \sum_{i=1}^{|K|} e^{-(\frac{\|K_i - x\|_1}{\gamma})^2} > \sum_{j=1}^{|\kappa|} e^{-(\frac{\|\kappa_j - x\|_1}{\gamma})^2}$ **then**
10:         **break**
11:     **else**
12:         *original_point* $\leftarrow$ *point*
13:     **end if**
14:     *point* $\leftarrow$ OPTRBO(*point*,K, κ, γ, *step_size*, *iterations*)
15:     **if** *point* $\neq$ *original_point* **then**
16:         add *point* to S
17:     **end if**
18: **end while**
19: **return** S

Algorithm 2. Optimization

1: **Input:** collections of majority objects K and minority objects κ
2: **Parameters:** *point*, spread of radial basis function γ, optimization *step size*, number of *iterations* per synthetic sample
3: **Output:** *syntetic_point*
4:
5: **function** OPTRBO(*point*, K, κ, γ, *step_size*, *iterations*):
6: *original_value* $\leftarrow$ *point*
7: **for** $i \leftarrow 1$ **to** *iterations* **do**
8:     *direction* $\leftarrow$ randomly chosen standard basis vector in a d-dimensional Euclidean space
9:     *translated* $\leftarrow$ *point* + *direction* $\times$ *step_size*
10:     **if** $|\Phi(translated, K, \kappa, \gamma)| < |\Phi(point, K, \kappa, \gamma)|$ **then**
11:         *point* $\leftarrow$ *translated*
12:     **end if**
13: **end for**
14: **if** *point* $\neq$ *original_value* **then**
15:     **return** OPTRBO(*point*,K, κ, γ, *step_size*, *iterations*)
16: **else**
17:     **return** *point*
18: **end if**

4 Experimental Study

The evaluation of the results obtained by the proposed algorithm was conducted on the basis of computer experiments. In this section the set-up as well as the obtained results will be presented.

As a measure of comparison three state-of-art algorithms were used. Namely SMOTE [3], ADASYN [7] and the RBO (Radial Based Oversampling) technique [9], which served as a basis for the proposed RBO_modified method. A case where no oversampling was performed was also included. The data sets obtained from these methods were then used to train two classifiers: Naive Bayes (NB) and k-nearest neighbors (k-NN).

For the RBO and RBO_modified methods following parameters were used: coefficient, corresponding to the spread of radial basis function, was set to 0.01. Step size was set to 0.001. The number of iterations per synthetic sample was set to 500. The probability of stopping the optimization early used only in RBO was set to 0.02. The number of recursive calls used in RBO_modified was set to 50. The parameter k used in SMOTE and ADASYN was set to 5. The experiments were run on 7 databases, presented in Table 1, from the KEEL repository [1]. The processed data were normalized to range from 0 to 1. To partition the datasets 5-folds stratified cross-validation was used.

The results in Tables 2, 3, and 4 present the obtained values of accuracy, precision and F-measure for the tested methods, on two selected classifiers (NB and k-NN). The overall performance of the proposed algorithm is closest to the algorithm from which it was derived i.e. RBO. It is plausible to state that the

Table 1. Details of the datasets used in the experiments

No	Name	IR	Samples	Features
1	ecoli1	3.36	336	7
2	ecoli2	5.46	336	7
3	ecoli3	8.6	336	7
4	ecoli4	15.8	336	7
5	glass-0-1-6_vs_2	10.29	192	9
6	glass2	11.59	214	9
7	yeast-0-5-6-7-9_vs_4	9.35	528	8

Table 2. Accuracy achieved for chosen datasets for k-NN/NB

Dataset	None	SMOTE	ADASYN	RBO	RBO_mod
1	**0.9**/0.62	0.88/0.66	0.87/0.6	0.88/**0.81**	0.88/**0.81**
2	**0.96**/0.46	0.91/0.61	0.87/0.39	0.91/0.85	0.91/0.88
3	0.92/0.76	0.86/0.8	0.86/0.75	0.86/0.86	0.87/**0.88**
4	**0.98**/0.87	0.97/0.9	0.97/0.88	0.96/0.94	0.97/**0.95**
5	**0.9**/0.45	0.77/0.45	0.76/0.46	0.79/0.4	0.85/**0.53**
6	**0.92**/0.48	0.76/0.51	0.76/0.51	0.8/0.45	0.85/**0.59**
7	**0.92**/0.2	0.82/0.23	0.81/0.21	0.83/0.68	0.85/**0.9**

modifications in the proposed method proved to be successful in improving the accuracy of the algorithm. As observed in Table 2 the modified RBO algorithm obtained satisfactory results in accuracy, preforming best out of all the oversampling techniques, especially for the NB classifier. As shown in Table 3, the proposed method works well for datasets with high imbalance ratio. The results presented in Table 4 demonstrate that the algorithm produces outcomes comparable to the state-of-art methods for k-NN classifier. However it is crucial to note that for the NB classifier the results in some instances are of poorer quality when compared to other tested oversampling techniques. The reason behind such occurrence may be, that the initial condition, ensuring that certain regions, recognized as hard to classify, where omitted when applying the oversampling procedure. Thus further study on the relationship between the administration the rule constraining the insertion of new synthetic objects in regions deemed as hard to classify needs to be conducted. What is more the proposed algorithm obtained best results out of all the oversampling techniques on a data set with the highest number of instances, which may indicate that for datasets of a more considerable volume the performance of the RBO_modified algorithm may further improve.

Table 3. Precision achieved for chosen datasets for k-NN/NB

Dataset	None	SMOTE	ADASYN	RBO	RBO_mod
1	**0.83**/0.44	0.69/**0.46**	0.66/0.43	0.72/0.45	0.7/0.31
2	**0.86**/0.24	0.66/0.32	0.57/0.21	0.64/**0.6**	0.64/0.48
3	**0.61**/0.32	0.43/**0.39**	0.42/0.32	0.42/0.2	0.43/0.23
4	**0.97**/0.44	0.72/0.45	0.68/0.38	0.64/**0.51**	0.71/0.47
5	0/**0.12**	**0.23**/0.11	0.22/0.11	0.2/**0.12**	0.21/**0.12**
6	0/0.11	**0.18**/0.11	**0.18**/0.11	**0.18**/**0.12**	0.17/**0.12**
7	**0.64**/0.12	0.3/0.12	0.3/0.12	0.32/0.19	0.39/**0.44**

Table 4. F-measure achieved for chosen datasets for k-NN/NB

Dataset	None	SMOTE	ADASYN	RBO	RBO_mod
1	**0.77**/0.56	0.75/**0.59**	0.75/0.56	**0.77**/0.38	0.75/0.37
2	**0.87**/0.38	0.77/**0.46**	0.7/0.34	0.75/0.45	0.75/0.41
3	**0.58**/0.47	0.56/**0.51**	0.56/0.47	0.56/0.27	0.55/0.29
4	**0.83**/0.56	0.8/**0.58**	0.77/0.53	0.74/0.47	0.77/0.28
5	0/0.2	**0.34**/0.19	0.33/0.19	0.28/**0.21**	0.25/0.2
6	0/0.19	**0.27**/0.19	**0.27**/0.2	0.25/**0.21**	0.19/**0.21**
7	0.4/0.19	0.42/0.2	0.42/0.2	0.43/0.3	**0.47**/**0.34**

5 Conclusion and Future Work

In this paper, we proposed a modification to the novel approach to imbalanced data oversampling. The algorithm is resilient towards the potential pitfalls of the neighborhood-based methods described in the article by utilizing radial basis functions to estimate the potential of a class in a point, a concept that describes the preference towards belonging to a certain class. The obtained results indicate that while the proposed algorithm can achieve satisfactory results in comparison with the existing state-of-art methods it may outperform them on datasets of higher volume. The modifications implemented in the technique addressed the potential drawbacks of the original RBO method, such as the oversampling the regions the a proper classification is considered to be difficult, by adding a criterion ensuring that the observed point is not in close proximity to many instances of the majority class. However, the criterion needs further analysis, since not considering too many instances may cause the deterioration of the technique or possible malfunction due to the fact that the number of instances of minority and majority cannot be balanced. Moreover a situation, where an object is surrounded by a considerable amount of instances from both classed, needs further examining. Additionally, the modified method ensures that all directions are possible to be generated when choosing a successor to the observed instance, thus making the search more robust. Unfortunately, the technique is time-consuming and requires a lot of computational resources, due to the fact that in each step the potential is calculated in relation to all existing points as well as the recursion mechanism. The number of times it is possible for a method to call itself must be bounded, as not to exhaust the memory of the stack and for the execution time to be feasible. It's important to notice that from a certain moment increasing the number of recursion steps is not in any way beneficial for the quality of the method. Similarly, since the importance of the instances surrounding an observed point is digressive with the distance from it, it is reasonable to consider only those instances from a certain area in proximity of the object.

Acknowledgements. This work is supported by the Polish National Science Center under the Grant no. UMO-2015/19/B/ST6/01597 as well the statutory funds of the Department of Systems and Computer Networks, Faculty of Electronics, Wrocław University of Science and Technology.

References

1. Alcalá-Fdez, J., Fernández, A., Luengo, J., Derrac, J., García, S., Sánchez, L., Herrera, F.: Keel data-mining software tool: data set repository, integration of algorithms and experimental analysis framework. J. Multiple Valued Log. Soft Comput. **17**, 255–287 (2011)
2. Bunkhumpornpat, C., Sinapiromsaran, K., Lursinsap, C.: Safe-level-SMOTE: safe-level-synthetic minority over-sampling technique for handling the class imbalanced problem. In: Pacific-Asia Conference on Knowledge Discovery and Data Mining, pp. 475–482. Springer (2009)
3. Chawla, N.V., Bowyer, K.W., Hall, L.O., Kegelmeyer, W.P.: SMOTE: synthetic minority over-sampling technique. J. Artif. Intell. Res. **16**, 321–357 (2002)
4. Chawla, N.V., Lazarevic, A., Hall, L.O., Bowyer, K.W.: SMOTEBoost: improving prediction of the minority class in boosting. In: European Conference on Principles of Data Mining and Knowledge Discovery, pp. 107–119. Springer (2003)
5. Ganganwar, V.: An overview of classification algorithms for imbalanced datasets. Int. J. Emerg. Technol. Adv. Eng. **2**(4), 42–47 (2012)
6. García, S., Herrera, F.: Evolutionary undersampling for classification with imbalanced datasets: proposals and taxonomy. Evol. Comput. **17**(3), 275–306 (2009)
7. He, H., Bai, Y., Garcia, E.A., Li, S.: ADASYN: adaptive synthetic sampling approach for imbalanced learning. In: IEEE International Joint Conference on Neural Networks, IJCNN 2008. IEEE World Congress on Computational Intelligence, pp. 1322–1328. IEEE (2008)
8. Hempstalk, K., Frank, E., Witten, I.H.: One-class classification by combining density and class probability estimation. In: Joint European Conference on Machine Learning and Knowledge Discovery in Databases, pp. 505–519. Springer (2008)
9. Koziarski, M., Krawczyk, B., Woźniak, M.: Radial-based approach to imbalanced data oversampling. In: International Conference on Hybrid Artificial Intelligence Systems, pp. 318–327. Springer (2017)

10. Ramentol, E., Caballero, Y., Bello, R., Herrera, F.: SMOTE-RSB*: a hybrid preprocessing approach based on oversampling and undersampling for high imbalanced data-sets using smote and rough sets theory. Knowl. Inf. Syst. **33**(2), 245–265 (2012)
11. Sun, Y., Kamel, M.S., Wong, A.K., Wang, Y.: Cost-sensitive boosting for classification of imbalanced data. Pattern Recogn. **40**(12), 3358–3378 (2007)
12. Woźniak, M., Graña, M., Corchado, E.: A survey of multiple classifier systems as hybrid systems. Inf. Fusion **16**, 3–17 (2014)

Deep Learning

Deep Learning for Big Data Time Series Forecasting Applied to Solar Power

J. F. Torres[1], A. Troncoso[1(✉)], I. Koprinska[2], Z. Wang[2], and F. Martínez-Álvarez[1]

[1] Division of Computer Science, Universidad Pablo de Olavide, 41013 Seville, Spain
jftormal@alu.upo.es, {atrolor,fmaralv}@upo.es
[2] School of Information Technologies, University of Sydney, Sydney, Australia
{irena.koprinska,zheng.wang}@sydney.edu.au

Abstract. Accurate solar energy prediction is required for the integration of solar power into the electricity grid, to ensure reliable electricity supply, while reducing pollution. In this paper we propose a new approach based on deep learning for the task of solar photovoltaic power forecasting for the next day. We firstly evaluate the performance of the proposed algorithm using Australian solar photovoltaic data for two years. Next, we compare its performance with two other advanced methods for forecasting recently published in the literature. In particular, a forecasting algorithm based on similarity of sequences of patterns and a neural network as a reference method for solar power forecasting. Finally, the suitability of all methods to deal with big data time series is analyzed by means of a scalability study, showing the deep learning promising results for accurate solar power forecasting.

Keywords: Deep learning · Big data · Solar power
Time series forecasting

1 Introduction

Solar energy is a very promising renewable energy source that is still underused. However, in recent years there has been a considerable increase world while in the production and use of solar power. This is due to the lower cost of solar panels and also the bigger number of large-scale solar plants which have been especially efficient. In many countries the cost of electricity produced by solar energy is now comparable to that of using conventional energy sources. This competitive cost, coupled with the fact that solar is a clean and abundant energy source, has led to a huge growth in solar capacity. This trend is expected to continue - by 2020, the global solar capacity is projected to reach 700 GW, an increase of about 140 times compared to 2005 [6]. In Australia it is expected that by 2050 30% of the electricity supply will come from solar energy [1].

Solar energy suffers a great variability since it depends on meteorological conditions such as solar radiation, cloud cover, rainfall and temperature. This dependency creates uncertainty about how much power will be generated, which is

M. Graña et al. (Eds.): SOCO'18-CISIS'18-ICEUTE'18, AISC 771, pp. 123–133, 2019.
https://doi.org/10.1007/978-3-319-94120-2_12

important to ensure reliable electricity supply, and makes the integration of solar power into electricity markets more difficult. Hence, the ability to predict the generated solar power is a task of utmost importance and relevance for both energy managers and electricity traders, in order to minimize the aforementioned uncertainty when this kind of renewable energy is used.

Historical photovoltaic power data with high frequency is easily available, and therefore, advanced computing technologies and machine learning approaches for big data can be used to analyze very large time series. Deep learning is an emerging branch of machine learning that extends artificial neural networks to deal with big data. One of the main drawbacks that classical artificial neural networks exhibit is that, with many layers, their training typically becomes too complex [9]. In this sense, deep learning involves the use of a set of learning algorithms to train artificial neural networks with a large number of hidden layers.

In this paper we propose a new approach based on deep learning to forecast big solar power time series data. We firstly compare the performance of the proposed algorithm with two other advanced methods for forecasting published in [18]. In particular, Pattern Sequence-based Forecasting (PSF [10]) based on similarity of patterns and a Neural Network (NN) as a reference method for solar power forecasting. In addition, we also conduct a scalability study in order to evaluate the suitability of all methods to deal with big data time series.

The rest of the paper is structured as follows. Section 2 reviews of the existing literature related to time series forecasting of solar data. Section 3 introduces the proposed methodology to forecast big data time series. Section 4 presents the experimental results corresponding to the prediction of solar energy. Finally, Sect. 5 summarizes the main results and provides final conclusions.

2 Related Work

In this section, we review the recently published approaches related to photovoltaic (PV) power forecasting.

The methods for time series forecasting can be divided into two groups: classical statistical and data mining techniques [10]. With regard to the statistical methods, autoregressive integrated moving average and exponential smoothing have been the most popular for predicting PV time series [5,12]. Concerning to data mining techniques, neural networks, Support Vector Machine (SVM) or k nearest neighbors techniques have been recently applied to PV solar data. For instance, a NN optimized by means of a genetic algorithm is proposed in [3] to obtain a forecasting for the intra-hour power of a PV plant. In [17], the training data is split into clusters based on the weather characteristics. Next, the solar power output for the previous day and the cluster label are used to compute the forecasting for the next day. In [13] a SVM is used as prediction algorithm to obtain interval forecasts, which are more suitable for the highly variable nature of the solar data. A forecasting method based on the weather and power data for the previous days and the weather forecast for the next day is proposed to one-day-ahead prediction in [16].

In the last years, several studies in time series forecasting have focused on ensembles that combine the predictions of several forecasting models as they have shown to be very competitive and more accurate than single forecasting models [2,8,11], including for PV power forecasting [18].

Currently, deep learning techniques are being explored in many applications due to their excellent results [7]. Moreover, deep learning models have been shown to be effective for energy demand forecasting in the area if big data. In [14] a novel method based on deep learning was proposed to predict big data times series using electricity consumption in Spain, with a ten minute frequency sampling rate, from 2007 to 2016. In [4] a deep learning model is used for disaggregated household energy demand forecasting. In this case, a Graphics Processing Unit (GPU) architecture is proposed in order to accelerate time series learning. In [15] a hybrid method based on wavelet transforms and deep convolutional neural networks is proposed for PV power forecasting. With wavelet transforms the original data are decomposed into several frequency series, and the deep convolutional neural network is used to extract the features in PV power data for each series. Later, a probabilistic model is applied to forecast each series separately.

After a wide literature review, to the best of our knowledge, there are no previous studies that have addressed the problem of forecasting big solar data by using deep learning techniques. This work tries to fill this gap by proposing and evaluating an algorithm for forecasting big PV solar data.

3 Methodology

This section presents the methodology proposed to forecast time series in solar PV data context.

The main goal of this work is to predict future values, expressed as $[x_1,...,x_h]$, where h means the number of values to predict. To predict these h values, the process is based on a historical value window called w. In this way, the problem can be formulate as:

$$[x_{t+1}, x_{t+2}, \ldots, x_{t+h}] = f(x_t, x_{t-1}, \ldots, x_{t-w-1}) \tag{1}$$

where f refers to the model to be found in the training phase by the algorithm to forecast the next h values.

In order to use in-memory data, we use the Apache Spark cluster-computing. For the deep learning implementation, we chose the H2O package written in R. This framework provides a simple syntax for parallel and distributed programming. However, H2O does not support the multi-step forecasting. To avoid this problem, a possible solution consist of splitting the problem in h forecasting sub-problems. Therefore, it is necessary to compute a model for each sub-problem. This new formulation can be expressed as:

$$\begin{aligned} x_{t+1} &= f_1(x_t, x_{t-1}, ..., x_{t-w-1}) \\ x_{t+2} &= f_2(x_t, x_{t-1}, ..., x_{t-w-1}) \\ &\ldots \\ x_{t+h} &= f_h(x_t, x_{t-1}, ..., x_{t-w-1}) \end{aligned} \tag{2}$$

From this formulation, we can see that each of the h values from the prediction horizon is predicted separately, thus removing the error propagation due to previously predicted samples being used to predict the next one. Nevertheless, the computational cost of this methodology is higher than building just one model to predict all h values from the prediction horizon. The deep learning architecture used for solving each subproblem is presented in Fig. 1.

It is well known that the values of hyper-parameters of the deep learning algorithm may influence the results. To find a good combination of hyper-parameters, we employed the grid search method of H20. The grid-search was used separately for each sub-problem to obtain the best parameter setting.

The parameters used in the grid search are described in Sect. 4. A flow diagram of the proposed methodology is depicted in Fig. 2.

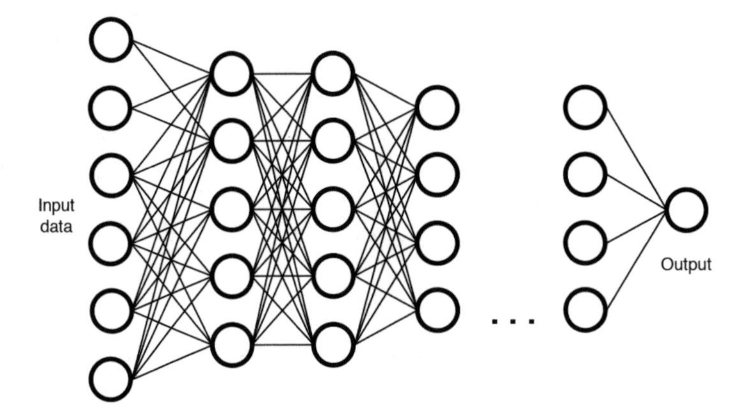

Fig. 1. Illustration of the DL architecture.

As can be seen in Fig. 2, the original dataset (in column vector format) is transformed depending on the data history (w) and the prediction horizon (h), where each column of this prediction horizon corresponds to the class of each subproblem. To compute each model, the dataset is divided into training, validation and test sets. First, the training and validation sets are used for the grid search. The grid search computes a model for each combination of hyper-parameters, for each sub-problems. These models are evaluated on the validation set and the best one is chosen to predict the test set.

4 Results

This section summarizes the results obtained after applying the methodology proposed in Sect. 3 for forecasting PV solar time series. This methodology has been compared to the methodology and implementation described by Zheng et al. in [18]. We firstly describe the dataset and experimental setup, and then discuss the results.

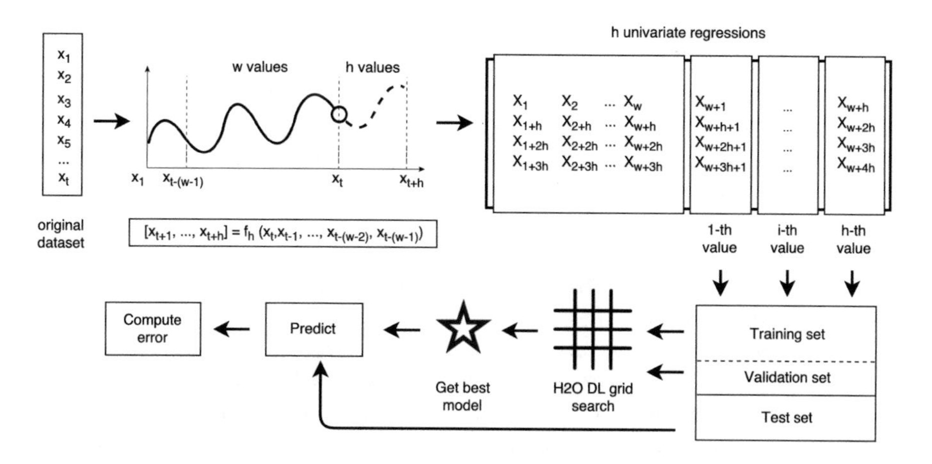

Fig. 2. Illustration of the proposed methodology.

4.1 Dataset Description

The time series considered in this study is related to PV power, collected from a rooftop PV plant located at the University of Queensland, Australia. The dataset is composed of two years, exactly from January 1st, 2015 to December 31st, 2016 in 30 min intervals between each measure. Due to the context of the study, only the daylight period have been considered, selecting the data between 7 a.m. and 5 p.m. As a result, the dataset is composed of 14620 samples.

The dataset has been pre-processed in order to adapt it to the chosen historical data window and prediction horizon. Concretely, a historical window of one day has been considerated to forecast the full next day. These values correspond to 20 past samples as historical window and also 20 future samples as prediction horizon. Thus, the final dataset considered in this research has 730 rows and 40 columns, resulting in a total of 29200 measures. Furthermore, the data has been normalized to [0,1].

4.2 Experimental Setup

The experimentation carried out consisted in comparing the results obtained by the proposed methodology and the results described by the authors in [18], which discusses the results of a traditional neural network (NN) and the pattern sequence forecasting (PSF) algorithm, applied to the same datset. In particular, we compare the accuracy and scalability of the methods. All experiments have been run in an Intel Core i7-5820K at 3.3 GHz with 15 Mb of cache, 6 cores with 12 threads and 16 GB of RAM memory, working under Ubuntu 16.04 operating system.

To evaluate the accuracy of the models, we use the Root Mean Squared Error (RMSE) and the Mean Absolute Error (MAE).

4.3 Analysis of Results

4.3.1 Parameter Selection

As stated before, we applied the grid search strategy available in H2O to find optimal parameters for each sub-problem. Many of the grid search parameters can be customised. In this experiment, we used the following settings:

- The dataset has been split into training and test sets, corresponding to 2015 for the training set and 2016 for the test set. The training set has also been split into 70% for training and 30% for validation.
- The number of hidden layers for applying the deep learning ranges from 1 to 5 and the number of neurons for each layer from 10 to 40.
- The initial weight distribution was set to uniform distribution.
- As activation function, the hyperbolic tangent function (tanh) has been chosen.
- The distribution function has been set to Gaussian distribution.

After training a model for each combination of the above described parameters for each sub-problem, it has been tested on the validation set and the best model has been obtained. Table 1 shows the parameters of the best model obtained for each sub-problem using the above-mentioned grid search: number of hidden layers and neurons per layer, and also the errors on the training and validation set.

We can see that the best network configuration varied and most often (for 40% of the sub-problems) included 3 hidden layers, with number of neurons in these 3 layers between 17 and 32. The training and validation errors followed the same pattern, they increased till step 13–14 from the prediction horizon (sub-problem 13–14), and then decreased. As expected the error on the validation set was higher than the error on the training set.

4.3.2 Accuracy

For the optimal configuration of the network for each subproblem, a new run was launched to predict the test set. The results are shown in Table 2 and compared with the PSF and NN algorithms from [18]. The PSF algorithm first applies clustering to the training set, adding a class tag. Next, the prediction of a new data is based on the similarity of tag sequences previous to the point to be predicted. The NN model is a multi-layer NN with one hidden layer, trained with the Levenberg-Marquardt version of the backpropagation algorithm.

It can be seen that the deep learning algorithm slightly improves the PSF and NN results, but not enough to decide to use it instead of PSF or NN.

To study these errors in more detail, the best and worst predicted day have been obtained. These results are depicted in Fig. 3 which presents the evolution of actual and forecasted solar data for the NN, PSF and DL algorithms. The best and worst days are: Apr. 7 2016 and Jun. 19 2016 for NN, Apr. 3 2016 and Jun. 19 2016 for PSF, and finally, Sept. 11 2016 and Jun. 18 2016 for DL.

Table 1. Best models for each sub-problem

Sub-problem	Hidden layers	Neurons per layer	RMSE training	MAE training	RMSE validation	MAE validation
1	5	39	58.01	40.94	128.31	109.13
2	1	13	86.83	62.32	145.66	120.24
3	3	27	90.96	69.57	158.33	132.08
4	1	37	120.60	90.32	174.85	140.98
5	2	30	128.22	98.39	184.39	147.77
6	2	11	146.58	116.47	189.90	162.55
7	4	14	161.54	128.87	208.80	179.44
8	3	23	167.14	134.46	212.02	170.35
9	2	39	168.24	135.11	217.07	177.33
10	3	32	161.17	130.43	219.82	180.26
11	2	31	166.59	134.74	218.45	181.73
12	5	32	158.69	131.25	211.29	174.76
13	4	37	165.03	138.96	202.01	168.33
14	3	17	165.03	138.59	213.21	184.85
15	5	14	155.30	127.95	196.42	167.23
16	1	39	132.54	107.20	184.21	155.12
17	5	38	117.94	92.98	152.45	130.06
18	4	34	86.55	65.72	122.07	100.04
19	4	40	74.16	53.33	96.01	79.49
20	3	28	63.70	48.37	57.09	45.80

Table 2. Accuracy of the NN, PSF and DL algorithms.

	NN	PSF	DL
RMSE	154.16	149.52	148.98
MAE	116.64	119.17	114.76

Table 3 summarizes the MAE and RMSE for the above stated days. For the best day, overall PSF is the best performing method and NN is the worst, while for the worst day NN is the best and PSF is the worst.

4.3.3 Scalability

Finally, a scalability comparison -in terms of runtime- between these methods have been accomplished. To conduct this, the optimal values described in Table 1 have been set. Furthermore, the time series length has been multiplied by 2, 4, 8, 16, 32 and 64, respectively. The results obtained are shown in Table 4.

Table 3. Best and worst day for NN, PSF and DL.

	Best day		Worst day	
	MAE	RMSE	MAE	RMSE
NN	58.87	106.88	191.52	221.58
PSF	31.72	36.15	252.77	279.12
DL	31.66	41.91	206.33	233.00

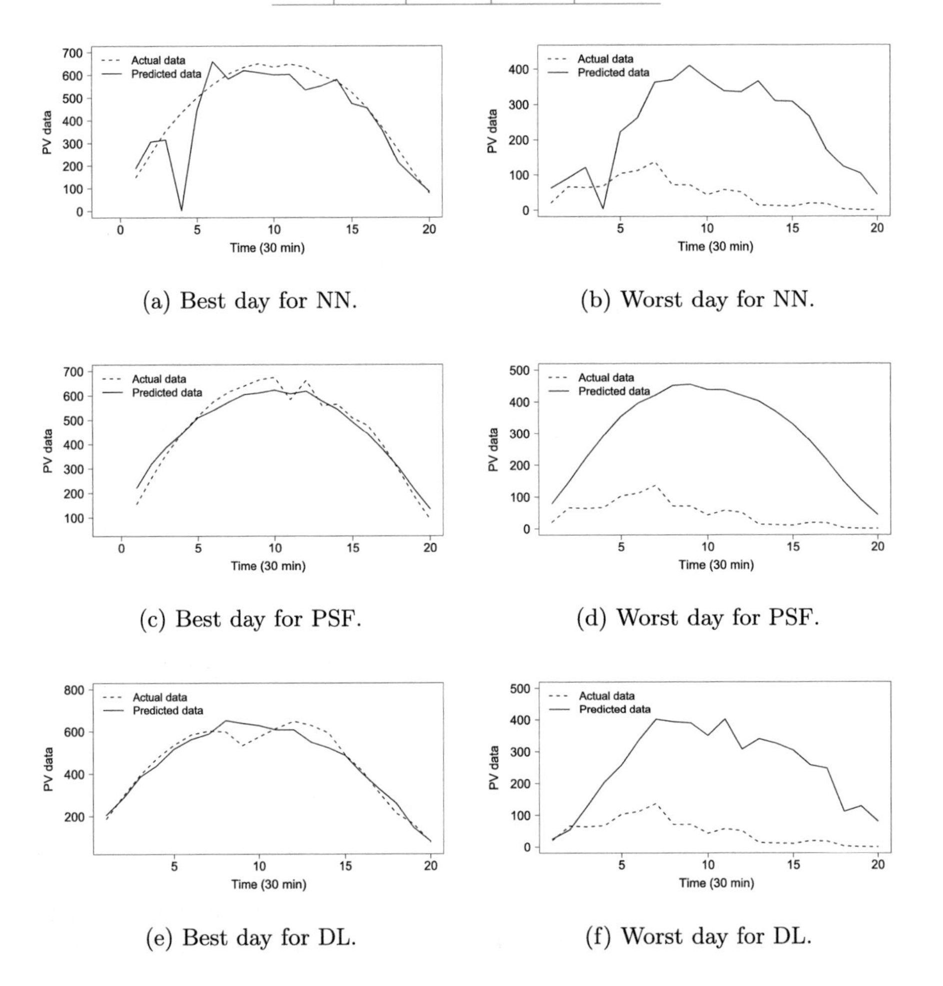

(a) Best day for NN.

(b) Worst day for NN.

(c) Best day for PSF.

(d) Worst day for PSF.

(e) Best day for DL.

(f) Worst day for DL.

Fig. 3. Best and worst day for NN, PSF and DL algorithms.

As it can be seen in Table 4, for short time series the NN and PSF algorithm are faster than DL. However, as the size of the data set increases with a factor of 32 or bigger, the deep learning method is much faster than the other algorithms.

Table 4. Computing times (in seconds) for different time series lengths.

	x1	x2	x4	x8	x16	x32	x64
NN	0.8020	1.8885	5.4975	24.7970	114.1169	378.0876	2098.0432
PSF	2.4858	14.6286	9.6493	28.9169	101.3701	365.4012	1345.8199
DL	23.0470	23.0480	23.0540	23.0400	22.9600	43.1210	63.2050

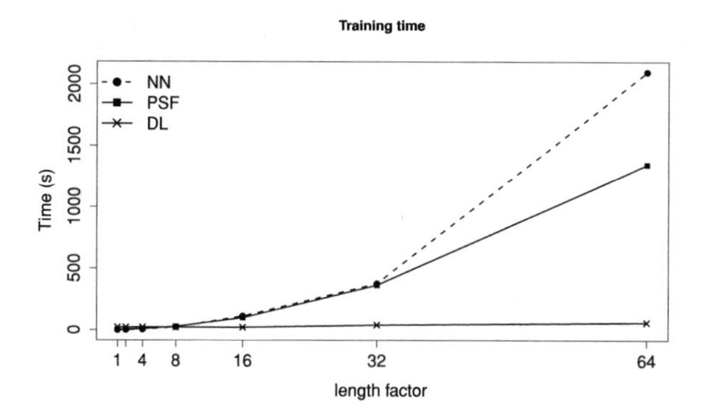

Fig. 4. Scalability of NN, PSF and DL algorithms.

This is because the H2O framework supports distributed and parallel computing, while the Matlab implementations of NN and PSF were single-thread.

Figure 4 graphically summarizes the results collected from Table 4. We can see that the proposed deep learning model is scalable as its training time increase in a linear way while the training time of the other two methods increases exponentially.

5 Conclusions

In this paper we introduced a novel approach for predicting the electricity power generated by solar photovoltaic systems. Our approach has three main novel features. Firstly, it uses deep learning which hasn't been investigated previously for solar power forecasting. The deep neural network has been implemented using the H2O R package. Second, it has been specifically developed to handle big time series data, and, hence, has been implemented using an Apache Spark cluster-computing framework. And, third, it uses a novel multi-step methodology which decomposes the forecasting problem into several sub-problems, allowing arbitrary prediction horizons. The performance of our approach has been evaluated on real Australian data and compared with two well-established algorithms, PSF and NN, showing competitive results. Finally, a scalability analysis has also been conducted demonstrating that the proposed deep learning approach is particularly suitable for big solar data, given its linear time increase behavior, contrary to PSF and NN which show an exponential time increase.

Acknowledgments. The authors would like to thank the Spanish Ministry of Economy and Competitiveness and Junta de Andalucía for the support under projects TIN2014-55894-C2-R and P12-TIC-1728, respectively.

References

1. Climate Commission. The critical decade: Australia's future - solar energy (2013)
2. Cerqueira, V., Torgo, L., Pinto, F., Soares, C.: Arbitrated ensemble for time series forecasting. In: Proceedings of the European Conference on Machine Learning and Principles of Knowledge Discovery in Databases, pp. 478–494 (2017)
3. Chu, Y., Urquhart, B., Gohari, S.M.I., Pedro, H.T.C., Kleissl, J., Coimbra, C.F.M.: Short-term reforecasting of power output from a 48 mwe solar pv plant. Solar. Energ. **112**, 68–77 (2015)
4. Coelho, I.M., Coelho, V.N., Luz, E.J.S., Ochi, L.S., Guimarães, F.G., Rios, E.: A GPU deep learning metaheuristic based model for time series forecasting. Appl. Energ. **201**, 412–418 (2017)
5. Dong, Z., Yang, D., Reindl, T., Walsh, W.M.: A novel hybrid approach based on self-organizing maps, support vector regression and particle swarm optimization to forecast solar irradiance. Energy **82**, 570–577 (2015)
6. Solar Power Europe. Global market outlook for solar power/2016–2020 (2016)
7. Kamilaris, A., Prenafeta-Boldú, F.X.: Deep learning in agriculture: a survey. Comput. Electron. Agric. **147**, 70–90 (2018)
8. Koprinska, I., Rana, M., Troncoso, A., Martínez-Álvarez, F.: Combining pattern sequence similarity with neural networks forforecasting electricity demand time series. In: Proceedings of the International Joint Conference on Neural Networks, pp. 1–8 (2013)
9. Livingstone, D.J., Manallack, D.T., Tetko, I.V.: Data modelling with neural networks: advantages and limitations. J. Comput. Aided Mol. Des. **11**, 135–142 (1997)
10. Martínez-Álvarez, F., Troncoso, A., Riquelme, J.C., Aguilar, J.S.: Energy time series forecasting based on pattern sequence similarity. IEEE Trans. Knowl. Data Eng. **23**, 1230–1243 (2011)
11. Oliveira, M., Torgo, L.: Ensembles for time series forecasting. In: Proceedings of the Sixth Asian Conference on Machine Learning, pp. 360–370 (2015)
12. Pedro, H.T.C., Coimbra, C.F.M.: Assessment of forecasting techniques for solar power production with no exogenous inputs. Solar Energ. **86**, 2017–2028 (2012)
13. Rana, M., Koprinska, I., Agelidis, V.G.: 2D-interval forecasts for solar power production. Solar Energ. **122**, 191–203 (2015)
14. Torres, J.F., Fernández, A.M., Troncoso, A., Martínez-Álvarez, F.: Deep learning-based approach for time series forecasting with application to electricity load, pp. 203–212. In: Biomedical Applications Based on Natural and Artificial, Computing (2017)
15. Wang, H., Yi, H., Peng, J., Wang, G., Liu, Y., Jiang, H., Liu, W.: Deterministic and probabilistic forecasting of photovoltaic power based on deep convolutional neural network. Energ. Convers. Manag. **153**, 409–422 (2017)
16. Wang, Z., Koprinska, I.: Solar power prediction with data source weighted nearest neighbors. In: Proceedings of the International Joint Conference on Neural Networks, pp. 1411–1418 (2017)

17. Wang, Z., Koprinska, I., Rana, M.: Solar power prediction using weather type pair patterns. In: Proceedings of the International Joint Conference on Neural Networks, pp. 4259–4266 (2017)
18. Wang, Z., Koprinska, I., Rana, M.: Solar power forecasting using pattern sequences. In: Artificial Neural Networks and Machine Learning (ICANN), pp. 486–494 (2017)

Deep Dense Convolutional Networks for Repayment Prediction in Peer-to-Peer Lending

Ji-Yoon Kim and Sung-Bae Cho(✉)

Department of Computer Science, Yonsei University, Seoul, Republic of Korea
{jiyoon_kim, sbcho}@yonsei.ac.kr

Abstract. In peer-to-peer (P2P) lending, it is important to predict default of borrowers because the lenders would suffer financial loss if the borrower fails to pay money. The huge lending transaction data generated online helps to predict repayment of the borrowers, but there are limitations in extracting features based on the complex information. Convolutional neural networks (CNN) can automatically extract useful features from large P2P lending data. However, as deep CNN becomes more complex and deeper, the information about input vanishes and overfitting occurs. In this paper, we propose a deep dense convolutional networks (DenseNet) for default prediction in P2P social lending to automatically extract features and improve the performance. DenseNet ensures the flow of loan information through dense connectivity and automatically extracts discriminative features with convolution and pooling operations. We capture the complex features of lending data and reuse loan information to predict the repayment of the borrower. Experimental results show that the proposed method automatically extracts useful features from Lending Club data, avoids overfitting, and is effective in default prediction. In comparison with deep CNN and other machine learning methods, the proposed method has achieved the highest performance with 79.6%. We demonstrate the usefulness of the proposed method as the 5-fold cross-validation to evaluate the performance.

Keywords: Deep learning · Dense Convolutional Networks · Big data Fintech · P2P social lending

1 Introduction

Peer-to-Peer (P2P) lending is the practice of lending money to businesses or individuals through online services that directly match borrowers and lenders without a financial intermediary such as a bank [1]. It is rapidly developing around the world. P2P lending platforms attract large number of users and generate huge loan transaction data.

The lenders are able to look for potential borrowers and select them [2]. If the borrowers do not pay or pay only part during the repayment period, the lenders become the financial loss [3]. The lenders may suffer from the financial risk of default, and to reduce this, it is important to predict the repayment of the borrowers by distinguishing the characteristics of the borrowers.

M. Graña et al. (Eds.): SOCO'18-CISIS'18-ICEUTE'18, AISC 771, pp. 134–144, 2019.
https://doi.org/10.1007/978-3-319-94120-2_13

Since P2P lending transactions are processed online, a large amount of online data is generated and it can be used effectively to support credit risk management [4]. Related studies extract features from data for predicting the repayment or assessing the credit risk. Statistical methods [5] and manual methods [2], which are mainly used for feature extraction in social lending, limit to extracting features because the amount of data is huge and various.

The recent study proposed an architecture of deep convolutional neural networks (CNN) with large amounts of financial data to predict the success of bank telemarketing [6]. Leveraging deep CNN, one of the deep learning methods that can provide prediction models for large and complex data, local features are automatically extracted by using hierarchical features and relationships between the attributes of financial data. They solved the problem of feature extraction by automatically extracting features from big data.

Deep CNN can automatically extract useful features as the layer is deeper. By stacking the convolutional layer and the pooling layer several times, the basic features are extracted from the lower layer and the complex ones are derived from the higher layer [7]. However, in plain CNN, as the model is complex or deeper layers are stacked, overfitting occurs and the information about input vanishes [8].

Figure 1 shows the cross-entropy and accuracy for the repayment prediction model based on deep CNN using P2P lending data provided by the Lending Club. The more training the network goes, the more accurate the training set is, and the less the cross-entropy is, while the test set shows the opposite tendency. It implies that the CNN model is overfitting.

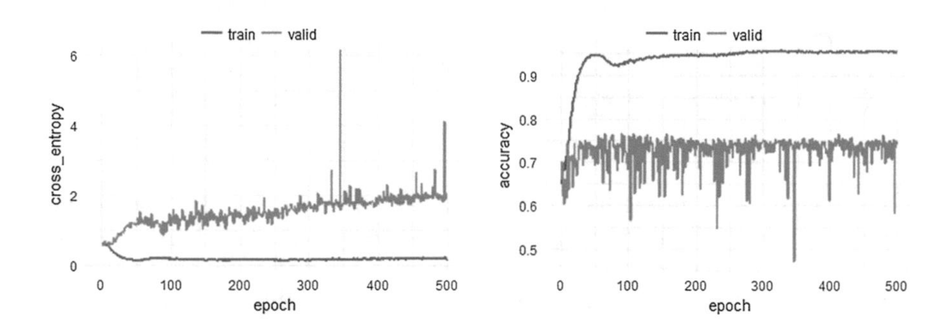

Fig. 1. Cross-entropy and accuracy per epoch for CNN model

Recently, various networks have been proposed to solve the problem that vanishes information about input and gradient. Dense convolutional networks (DenseNet) [9] has a structure directly connected to feed-forward from a layer to a subsequent layer based on existing CNN. This dense connectivity ensures the information flow and reduces the overfitting problem.

In this paper, we propose a DenseNet structure for default prediction in social lending. Social lending data has characteristics of borrower and information of loan product. Dense block, which is a component of DenseNet, extracts discriminative features of borrowers through a convolutional layer. In the transition layer, the pooling layer merges similar features into one. As dense connectivity, it maintains the information of the lending data, reuses the features, and obtains basic and powerful representations through various layers. DenseNet classifies the loan status of the borrower by learning patterns while maintaining and extracting distinctive features.

2 Related Work

As P2P lending transaction increases, it is becoming more and more important to predict the repayment [10]. As shown in Table 1, in recent years, there has been a growing study of credit assessment and default on the borrowers in social lending.

Table 1. Related studies in P2P social lending

Year	Author	Dataset	#Data	#Attribute	Method
2017	Kim and Cho [11]	Lending club	332,844	17	Decision tree
2017	Lin *et al.* [5]	Yooli	48,784	10	Logistic regression
2017	Jiang *et al.* [12]	Eloan	39,538	32	Logistic regression, naïve Bayes, support vector machine, random forest
2017	Fu [13]	Lending club	1,320,000	13	Combination of random forest and neural network
2017	Zhang *et al.* [14]	Paipai	193,614	21	Logistic regression
2016	Serrano-Cinca and Gutiérrez-Nieto [15]	Lending club	40,907	26	Linear regression, decision tree

In the studies of using small data and few attributes, they mainly extracted features based on statistical methods or manual ones and then proposed the model for predicting the repayment or assessing credit risk as a machine learning method. For example, Jiang et al. [12] proposed a default prediction model combined with soft information related to the text description. Seven features were extracted from the text using latent Dirichlet allocation to provide subjective and qualitative information for support decision making in addition to objective quantitative information. The performance is compared by using four types of machine learning methods depending on the number of features. It is shown that the classification performance may be degraded if invalid features are included. These studies are difficult to compare the performance because they derive unique features [7].

On the other hand, in the studies of using a large amount of data, they mainly selected the features through preprocessing using data from Lending Club which provides large data. To utilize the big data, they experimented with newly labeled class. Fu [13] predicted the default of the borrower by labeling "fully paid" as "good," "charged off" and "default" as "bad" for 1,320 K data. Kim and Cho [11] used two semi-supervised learning methods to label unlabeled data. Even if there is huge data, it is necessary to extract the feature, but it is difficult to extract distinctive features in big data.

3 The Proposed Method

3.1 Dense Convolutional Network

Figure 2 shows the overall architecture for repayment prediction in social lending using DenseNet. We train DenseNet to obtain powerful features from P2P social lending data. As information passes through multiple layers, it learns feature space that captures the characteristics of the borrower and the inherent characteristics of the loan product, and continues to train the classifier to model these characteristics. The social lending data are projected into the learned representation space. We then use the softmax classifier to predict the repayment of the borrowers.

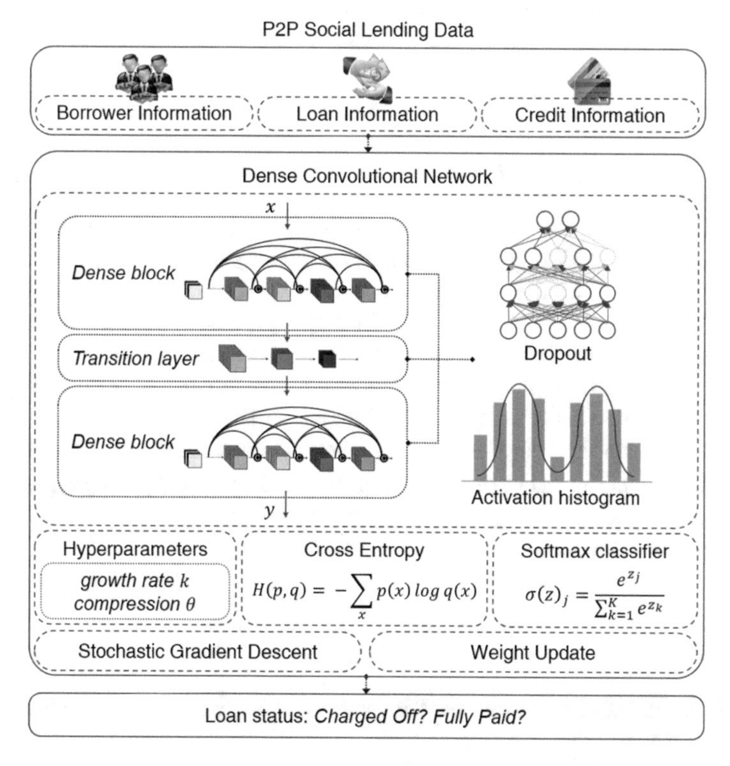

Fig. 2. The overall architecture of the proposed method

DenseNet directly connects each layer to every other layer using dense connectivity [9]. The extracted features through layers are reused leveraging it. From the P2P lending data, the output passed through the lth layer is defined as x_l, and $H_l(\cdot)$ is defined as a composite function consisting of an activation function and a convolution layer. Each feature-map up to the $(l-1)^{th}$ layer is concatenated and used as the input of the lth layer and expressed as in Eq. (1). The use of concatenation operation can maintain information of lending data.

$$x_l = H_l([x_0, x_1, \cdots, x_{l-1}]) \quad (1)$$

The concatenation operation can be allowed to feature-maps of the same size, so it is limited when a pooling layer uses. The down-sampling layer, an essential part of CNN, changes the size of the feature-maps. We divide the network into dense blocks and transition layers to facilitate pooling layer. The dense block consists of a directly connected convolutional layer, and the transition layer consists of convolution and pooling layers. Through the dense block, discriminative features are extracted from the characteristics of the borrower or inherent property of the loan product in the Lending Club data, and the similar features are merged into one through the pooling layer of the transition layer.

Because DenseNet combines feature-maps with concatenation, the number of feature-maps is excessively increased. We use the hyper-parameter k as growth rate to adjust the size of the feature-maps. When H_l generates k feature-maps, the number of input feature-maps in the lth layer is given by (2).

$$l^{th} layer = k_0 + k \times (l-1) \quad (2)$$

where k_0 is the number of feature-maps in the input layer of the first dense block. Growth rate can be used to control the amount of newly generated information. k does not need to be large because each layer can always reuse the original lending data.

To make the model compact, we use the hyper-parameter θ as the compression to reduce the number of feature-maps in the transition layer. When the number of feature-maps passed through the dense block is n, the number of feature-maps passed through the transition layer is θn, where θ ranges in [0, 1].

The feature-maps generated by repeating several dense blocks and transition layers from the social lending data are arranged in the feature vector $p^l = [p_1, p_2, \cdots, p_l]$ as one dimension through the global average pooling layer. The last layer, softmax classifier, predicts the loan status d of borrowers.

$$P(d|f) = \underset{d \in D}{\operatorname{argmax}} \frac{\exp(f^{L-1} w^L + b^L)}{\sum_{k=1}^{N_D} \exp(f^{L-1} w_k)} \quad (3)$$

where L is the index of the last layer, b is the bias, w is the connected weight, and N_D is the number of classes for loan status.

When lending data is presented to the network, it is propagated forward through the network. The weights of DenseNet are updated using backpropagation algorithm based on the stochastic gradient descent that minimizes the categorical cross-entropy in the mini-batches of the social lending data.

$$\theta = \theta - \eta \sum\nolimits_{i=1}^{n} \nabla_{\theta} J(\theta)/n \tag{4}$$

where θ is the parameters of DenseNet, η is the learning rate, and $J(\theta)$ is the gradient of the objective function. Forward propagation and backpropagation are repeated until the stopping region is satisfied.

Dropout is a technique to avoid overfitting by randomly dropping out a node in the network [16]. It is performed independently for each node of the lending data and each training data. If the dropout probability is too large or too small, overfitting or underfitting can occur, which affects to the performance. We set the dropout at a probability of 0.25 after passing through the convolutional layer at the dense block and transition layer.

3.2 Architecture

Table 2 shows the architecture of DenseNet. Our architecture consists of two dense blocks and one transition layer. The lending data form the 64 feature-maps through the initial convolution operation. Each dense block has the same number of layers and a 1 × 1 convolution reduces the number of parameters. All lending data through 3 × 3 convolution is set so that the size of feature-maps is not changed using zero-padding. In the transition layer, the size of feature-maps is reduced as 1 × 1 convolution and the average pooling is used. After the last dense block, global average pooling is performed and attached to the next softmax classifier.

Table 2. The proposed DenseNet architecture

Layers	Output size	# Parameters	Configuration
Convolutional layer	(70, 64)	256	1 × 3 conv, stride 1
Pooling layer	(71, 64)	–	1 × 2 max pool, stride 1
Dense block	(71, 448)	196,608	$\begin{pmatrix} 1 \times 1\,conv \\ 1 \times 3\,conv \end{pmatrix} \times 3$
Transition layer	(70, 224)	100,352	1 × 1 conv, stride 1 1 × 2 avg pool, stride 1
Dense block	(70, 608)	196,608	$\begin{pmatrix} 1 \times 1\,conv \\ 1 \times 3\,conv \end{pmatrix} \times 3$
Classification layer	(608)	–	Global avg pool
	(2)	1,218	Fully-connected, softmax

DenseNet has various architectures depending on the combination of hyper-parameters. It affects to feature extraction in lending data, learning time and performance. Designing the architecture of DenseNet requires understanding of lending data. The Lending Club data has a 1 × 72 size. We use a small window size in the convolutional and pooling layers to minimize the loss of information for each attribute. We also use nonlinear activation functions such as rectified linear unit (ReLU) [17] to extract the complex relationships between properties. The growth rate k is changed from 32 to 128 and the compression θ is changed from 0.25 to 0.75.

4 Experiments

4.1 Lending Club Dataset

In this paper, we use lending transaction data provided by Lending Club, a representative P2P lending platform in the United States. We use 855,502 data in the 2015–2016 year, and the data consist of 110 attributes such as information of borrowers, credit information, loan product information, and loan status, which are predictor variables. Attributes that cannot be used for prediction such as borrower's ID, URL, description of loans, duplicate attributes, attributes that have missing values with more than 80%, and attributes that are filled after starting to repay are removed. After the preprocessing, 143,823 data with 63 attributes are used.

DenseNet has an input format in the range [0, 1]. We prepossess that the categorical variables are created as dummy variables to represent binary variables, and continuous variables are normalized as follows:

$$X' = \frac{x - x_{min}}{x_{max} - x_{min}} \tag{5}$$

4.2 Result and Analysis

Accuracy. We train with different depth L, growth rate k, and compression θ. The main results are shown in Table 3. DenseNet-BC with k = 128 and two dense blocks achieved the highest performance, and overall performance is higher than the existing CNN model. θ do not have a significant effect on the performance. The results show steady performance when $k = 128$.

Table 3. Comparison of performance for CNN and DenseNet

Model	# Block	k	θ	Accuracy	F1-score
CNN	–	–	–	75.9%	85.4%
DenseNet-BC	2	64	0.25	79.3%	87.4%
DenseNet-BC	2	128	0.25	79.2%	**88.0%**
DenseNet-BC	2	128	0.5	**79.6%**	87.9%
DenseNet-BC	3	128	0.5	79.2%	**88.0%**
DenseNet-BC	3	128	0.75	79.0%	**88.0%**

We save a model achieved the highest performance by tuning many hyper-parameters to 500 epochs in the validation set. To verify the effectiveness of the proposed model, 5-fold cross-validation is performed. Figure 3 shows the comparison of the performance with other methods on 5-fold cross-validation.

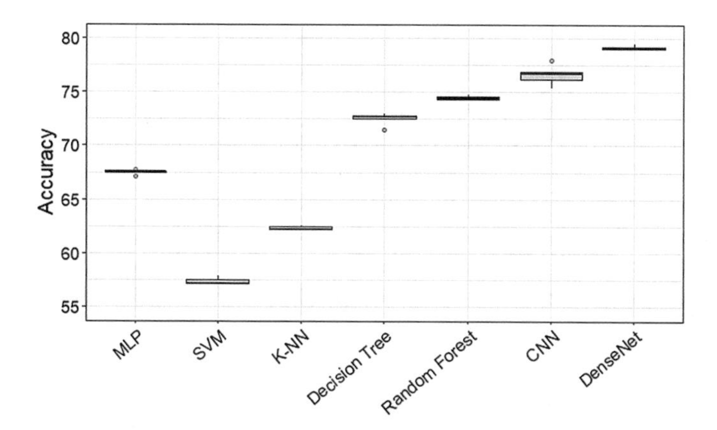

Fig. 3. Comparison of performance in 5-fold cross validation

DenseNet has achieved the highest performance compared to other machine learning methods, followed by CNN, random forest, and decision tree. We set other methods as deep CNN with four layers, k-NN with k = 3, multi-layer perceptron with 15 hidden layers, a decision tree with a depth of 25, and a random forest with a depth of 30.

Overfitting. We compare the cross-entropy and accuracy of the validation set for the two models to confirm the overfitting problem raised in Sect. 1. As CNN trains, the loss increases, while the loss of DenseNet decreases, and the accuracy is steadily increasing. Figure 4 shows the cross-entropy and accuracy per epoch for CNN and DenseNet.

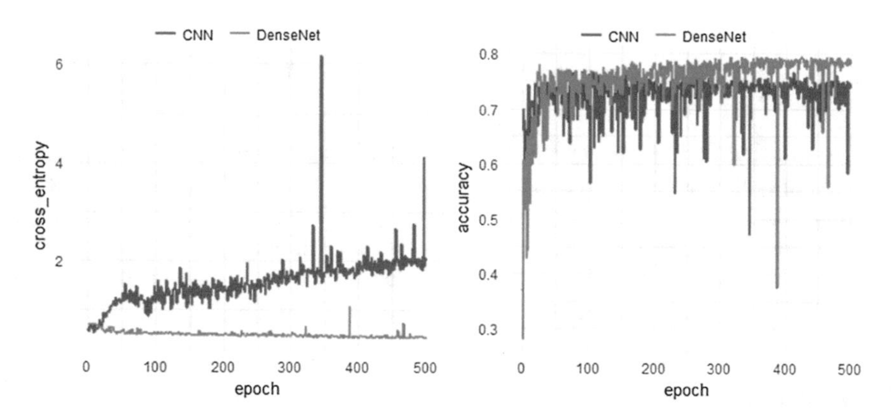

Fig. 4. Comparison of cross-entropy and accuracy per epoch

Analysis of Misclassification Cases. Table 4 shows the confusion matrix of DenseNet. Our model tends not to classify "Charged Off." In fact, it seems to be due to class imbalance problems because there are fewer non-repaid borrowers in social lending.

Table 4. Confusion matrix

Predicted\Actual	Fully paid	Charged off
Fully paid	32,033 (TP)	7,169 (FP)
Charged off	1,657 (FN)	2,286 (TN)

We analyze the four cases (TP, FP, FN, TN) based on the properties that we mentioned as important variables in the related studies [18]. Figure 5-(a) shows the boxplot of samples TP and TN, and (b) shows samples FN and FP where color means the true class.

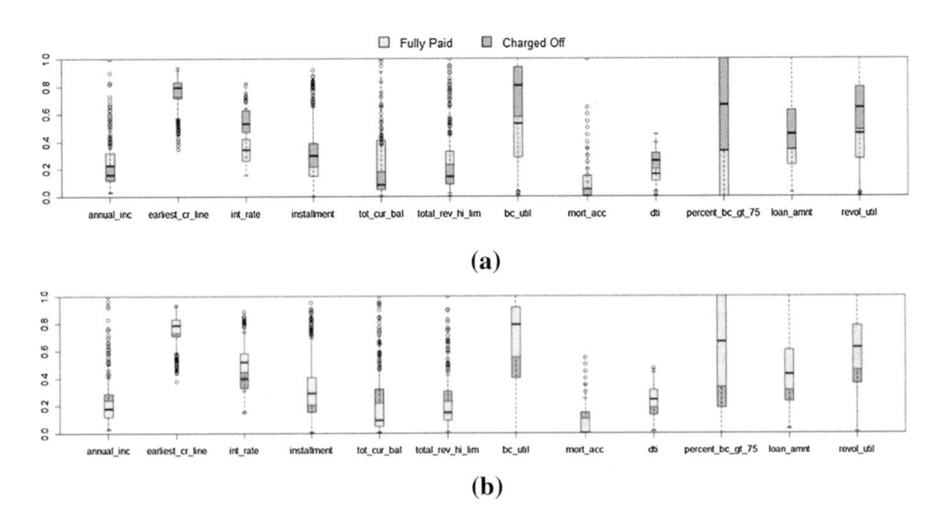

Fig. 5. Comparison of boxplots from classified samples

The distribution of variables that are significant in repayment prediction tends to be opposite to each other in (a) and (b). In particular, compared with (a), the distributions of "fully paid" and "charged off" tend to overlap each other in (b). For example, in actual cases, "charged off" was a high-interest rate in loans, "fully paid" was a low-interest rate, but in the misclassified data, "charged off" was a low-interest rate and "fully paid" was a high-interest rate.

5 Conclusions

In this paper, we have proposed an architecture of DenseNet for predicting the repayment in social lending. DenseNet maintains the flow of lending information with dense connectivity and provides a method to automatically extract relevant and powerful features from lending data. In experiments using Lending Club data, we showed that the proposed DenseNet model achieved the high accuracy and reduced overfitting compared to existing CNN. The analysis of misclassification cases based on confusion matrix confirmed the characteristics of them. Our model is very effective in predicting the repayment and helps the lenders to select borrowers who can pay for P2P lending.

Although the proposed method has proved useful in predicting the repayment in social lending, some additional work is needed to improve the current study. We can divide future work into short- and long-term studies. In the short term, we need to improve the class imbalance problem raised in the experiment. To solve this problem, we provided f1-score, but we can also adjust the training set by oversampling or undersampling. In the long term, we need to develop an automatic deep-learning system for predicting the repayment in P2P lending. It is aimed at developing a repayment prediction system that introduces techniques to find the optimal architecture automatically.

Acknowledgement. This research was supported by the MSIT (Ministry of Science, ICT), Korea, under the ITRC (Information Technology Research Center) support program (IITP-2018-2015-0-00369) supervised by the IITP (Institute for Information & communications Technology Promotion).

References

1. Zhao, H., Ge, Y., Liu, Q., Wang, G., Chen, E., Zhang, H.: P2P lending survey: platforms, recent advances and prospects. ACM Trans. Intell. Syst. Technol. **6**(8), 72–101 (2017)
2. Malekipirbazari, M., Aksakalli, V.: Risk assessment in social lending via random forests. Expert Syst. Appl. **42**(10), 4621–4631 (2015)
3. Xu, J., Chen, D., Chau, M.: Identifying features for detecting fraudulent loan requests on P2P platforms. In: IEEE Conference on Intelligence and Security Informatics, pp. 79–84 (2016)
4. Yan, J., Yu, W., Zhao, J.L.: How signaling and search costs affect information asymmetry in P2P lending: the economics of big data. Financ. Innov. **1**(1), 19 (2015)
5. Lin, X., Li, X., Zheng, Z.: Evaluating borrower's default risk in peer-to-peer lending: evidence from a lending platform in China. Appl. Econ. **49**(35), 3538–3545 (2017)
6. Kim, K.-H., Lee, C.-S., Jo, S.-M., Cho, S.-B.: Predicting the success of bank telemarketing using deep convolutional neural network. In: IEEE Conference of Soft Computing and Pattern Recognition, pp. 314–317 (2015)
7. Ronao, C.A., Cho, S.-B.: Human activity recognition with smartphone sensors using deep learning neural networks. Expert Syst. Appl. **59**, 235–244 (2016)
8. Huang, G., Sun, Y., Liu, Z., Sedra, D., Weinberger, K.Q.: Deep networks with stochastic depth. In: European Conference on Computer Vision, pp. 646–661 (2016)
9. Huang, G., Liu, Z., Matten, L., Weingerger, K.-Q.: Densely connected convolutional networks. In: IEEE Conference on Computer Vision and Pattern Recognition, pp. 4700–4708 (2017)
10. Milne, A., Parboteeah, P.: The business models and economics of peer-to-peer lending. European Credit Research Institute, no. 17, pp. 1–31 (2016)
11. Kim, A., Cho, S.-B.: Dempster-Shafer fusion of semi-supervised learning methods for predicting defaults in social lending. In: International Conference on Neural Information Processing, pp. 854–862 (2017)

12. Jiang, C., Wang, Z., Wang, R., Ding, Y.: Loan default prediction by combining soft information extracted from descriptive text in online peer-to-peer lending. Ann. Oper. Res. 1–19 (2017). https://doi.org/10.1007/s10479-017-2668-z
13. Fu, Y.: Combination of random forests and neural networks in social lending. J. Financ. Risk Manag. **6**(4), 418–426 (2017)
14. Zhang, Y., Li, H., Hai, M., Li, J., Li, A.: Determinants of loan funded successful in online P2P lending. Procedia Comput. Sci. **122**, 896–901 (2017)
15. Serrano-Cinca, C., Gutiérrez-Nieto, B.: The use of profit scoring as an alternative to credit scoring systems in peer-to-peer (P2P) lending. Decis. Support Syst. **89**, 113–122 (2016)
16. Srivastava, N., Hinton, G., Krizhevsky, A., Sutskever, I., Salakhutdinov, R.: Dropout: a simple way to prevent neural networks from overfitting. J. Mach. Learn. Res. **15**(1), 1929–1958 (2014)
17. Nair, V., Hinton, G.E.: Rectified linear units improve restricted Boltzmann machines. In: International Conference on Machine Learning (2010)
18. Emekter, R., Tu, Y., Jirasakuldech, B., Lu, M.: Evaluating credit risk and loan performance in online peer-to-peer (P2P) lending. Appl. Econ. **47**(1), 54–70 (2015)

A Categorical Clustering of Publishers for Mobile Performance Marketing

Susana Silva[1], Paulo Cortez[1](✉), Rui Mendes[2], Pedro José Pereira[1], Luís Miguel Matos[1], and Luís Garcia[3]

[1] ALGORITMI Centre, Department of Information Systems, University of Minho, 4804-533 Guimarães, Portugal
a64871@alunos.uminho.pt, pcortez@dsi.uminho.pt

[2] ALGORITMI Centre, Department of Informatics, University of Minho, 4710-057 Braga, Portugal
rcm@di.uminho.pt

[3] OLAmobile, Spinpark, 4805-017 Guimarães, Portugal
luis.garcia@olamobile.pt

Abstract. Mobile marketing is an expanding industry due to the growth of mobile devices (e.g., tablets, smartphones). In this paper, we explore a categorical approach to cluster publishers of a mobile performance market, in which payouts are only issued when there is a conversion (e.g., a sale). As a case study, we analyze recent and real-world data from a global mobile marketing company. Several experiments were held, considering a first internal evaluation stage, using training data, clustering quality metrics and computational effort. In the second stage, the best method, COBWEB algorithm, was analyzed using an external evaluation based on business metrics, computed over test data, and that allowed an identification of interesting clusters.

Keywords: Categorical clustering · Mobile marketing · Big data

1 Introduction

The mobile performance marketing industry is currently experiencing a great evolution due to the growth of mobile device usage (e.g., tablets, smartphones). In this industry, compensation only occurs when an ad performs well (e.g., purchase). There are several Demand-Side Platforms (DSP) that match users to ads. In this market, user traffic is generated by publishers, that have a popular free website or app (e.g, news, game), capable of attracting a vast audience, and that is financed by dynamic link ads. Each time a user clicks on an ad, the DSP redirects the user to a marketing campaign from an advertiser. When there is a conversion (e.g., purchase), the DSP returns a percentage of the advertiser revenue to the publishers [10]. From the publisher's point of view, the Return On Investment (ROI) is highly relevant, since there are several costs for maintaining

M. Graña et al. (Eds.): SOCO'18-CISIS'18-ICEUTE'18, AISC 771, pp. 145–154, 2019.
https://doi.org/10.1007/978-3-319-94120-2_14

and improving their service (e.g., producing content, rent digital space, technological costs). Thus, a key decision factor is the expected revenue when deciding about joining or leaving a particular DSP. Often, this involves the analysis of several performance marketing metrics adopted by the industry, including [6,19]: Click Through Rate (CTR), the rate between the number of advertisement clicks and the number of impressions; the Conversion Rate (CVR), the percentage of ad clicks (redirects) that originated a conversion sale (e.g., purchase); and client Lifetime Value (LTV), the amount of revenue generated per sale.

Currently, most DSPs only provide global metrics, computed using all publishers. In this work, we explore a clustering approach to automatically group publishers into similar profiles, such that more informative and realistic revenue metrics could be provided. As a case study, we work with recent data from OLAmobile, a global mobile marketing company. This DSP generates big data with its volume and velocity properties (e.g., millions of redirects and thousands of sales per hour).

Although clustering is a popular data mining approach in many real-world applications [2], its usage in mobile performance marketing is scarce, since studies are mostly focused on user CTR or CVR response prediction [6,20]. The current clustering approaches on advertising data include, mostly, understanding user behaviour. In particular, organizing user's search terms by subject, using clustering, and then relating it with user's intention to purchase and CTR [14]. In [15] the authors use k-means on user data, ending up with 10 user profiles, and mapping each profile to the types of advertisement that profile is more likely to interact with. In a different context, clustering was also applied to model users' behaviour in e-commerce web-pages, with the purpose of optimizing these websites and providing personalized recommendations to users, using click-stream data [18]. More recently, a study on click-stream data used clustering and biclustering to group users [11]. Furthermore, the authors stated that click-stream data clustering is a recent and challenging problem for many applications, with the main difficulty being related to grouping categorical data sequences, since most traditional clustering algorithms are not applicable to categorical data. Moreover, the problem complexity increases when handling big data, which is the case of mobile performance marketing.

In the mobile marketing performance industry, there is a lack of studies concerning clustering approaches related to publishers. This is probably due to the clustering complexity associated with the amount of data produced by this industry. There are vast amounts of records, most variable attributes are categorical and several of these attributes have a large cardinality (e.g., with dozens or hundreds of levels). This study fulfills this gap by addressing a categorical data clustering approach that is capable of grouping publishers in different clusters based on their similarities. In particular, we first compare five distinct clustering algorithms using two internal (training) metrics and also execution time. The comparison is executed using one week of data and a rolling window evaluation. The best clustering method is then selected and evaluated in terms of an external (test) metric that includes several business goals, showing its potential value in this domain.

This document is organized as follows: Sect. 2 discusses data collection, clustering methods and the evaluation procedure; Sect. 3 describes the experiments performed and analyzes the results obtained; finally, Sect. 4 draws the main conclusions and discusses future work.

2 Materials and Methods

2.1 Data Collection

For the data collection process, the organization under study, OLAmobile, provided a web-service enabling the connection to their data center, where data is divided in two streams: redirects and sales. The first one is related to data generated by user clicks on advertisements, where each click originates a record, while the latter is from redirects that originated a purchase, with each record containing also the information about the corresponding redirect. It is important to notice the great discrepancy existing between the number of redirects and the number of sales, since OLAmobile deals with millions of hourly records concerning redirects and only approximately 1% culminate in a purchase. Moreover, due to computational limitations associated with our server for this project, it was impossible to retrieve all redirect records from OLAmobile, thus, our ratio between collected redirects and sales was much higher (around 36%) than the real one. In Sect. 2.3, we adjust the CVR metric to realistic values (around 1%) by adopting an α correction factor.

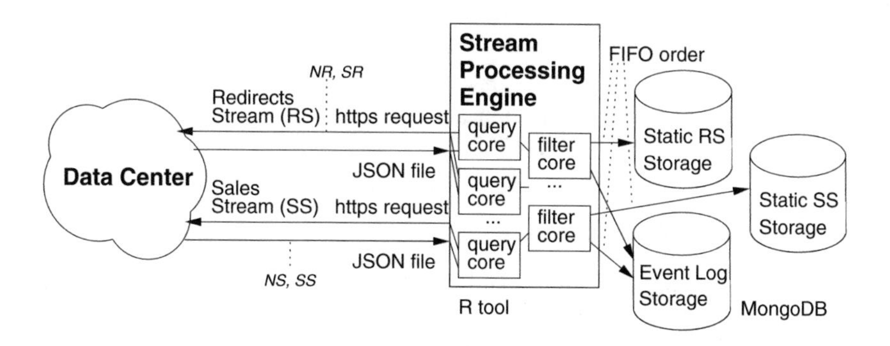

Fig. 1. Representation of the data collection process

Figure 1 represents the data collection process that was carried out and it consists in a multi-core system constantly gathering the most recent records concerning both redirects and sales (NR and NS denote the amount of redirects and sales events requested, while SR and SS represents the sleep time between two consecutive requests). The first core was constantly collecting records regarding redirects, the second core was collecting sales while a third core was filtering data from both redirects and sales. Our filtering process consisted in selecting best mode events and clients from Europe. The best mode is related with product campaigns that have obtained a minimal performance in a testing mode and it corresponds to most DSP traffic. After filtering, data was stored in two different MongoDB databases (one for redirects and another for sales) with a predefined size, using a FIFO (First In First Out) writing process, and, once the database was full, the data was stored in two static files. This process was performed during 7 days, from 31th October to 6th November 2017, ending up with 48,900 records with 17,600 concerning sales and nearly 31,300 regarding redirects.

After collecting and storing data, we joined both redirect and sales events into a single file. Table 1 shows a summary of the analyzed data attributes. In particular, the first six clustering attributes were identified by OLAmobile as interesting to associate with distinct publisher profiles and were used to estimate internal (int.) metrics, while the last two attributes were used to compute business metrics in the external (ext.) evaluation.

Table 1. Description of analyzed data attributes

Feature	Use	Description
Vertical	int.	Ad type selected by the publisher, with 4 levels (e.g., VOD - video on demand, mainstream)
Country	int.	Code of country, with 45 levels (e.g., Russia, Spain)
Idos	int.	Operating system, with 10 levels (e.g., Android, iOS)
Account	int.	Account type, with 11 levels (e.g., network, advertiser, app)
Hardware	int.	Type of device, with 3 levels (smartphone, tablet, smart TV)
Network	int.	Type of network access (mobile, WiFi)
Target	ext.	If there is a conversion (no sale, sale)
Revenue	ext.	Publisher revenue if sale (in EUR)

2.2 Clustering Methods

There is an increasing need for clustering algorithms to support categorical data, since most of the traditional clustering algorithms were built on the assumption that they would work only with numerical data, while most real-world problems deal with other types of data, such as categorical, temporal or structural [1]. Several challenges are associated with clustering of categorical data, namely, the lack of ability to understand how similar two different classes are. Thus, clustering algorithms for numerical data cannot be directly applied to this type of problems [2].

Based on [1], there are several clustering methods that support categorical data, namely: K-modes, COOLCAT, ROCK, CLOPE and COBWEB. The K-modes algorithm results from k-means and aims to fill its inability to deal with non-numerical data, preserving its efficiency, by using dissimilarity measures for categorical objects, replacing the mean from the k-means algorithm for modes and using a frequency based method for modes detection [1]. Its main disadvantage is that it may produce locally optimal solutions that strongly depend on the initial modes and the objects' order in the dataset.

The COOLCAT algorithm is based on the same principles of k-modes but uses the clusters' centers instead of modes. This algorithm performs in an incremental way and it is capable of grouping new data without processing the whole dataset, which makes it suitable for clustering data streams, as well as categorical Big Data [17]. A main disadvantage is the fact that the data processing order influences clustering quality and, to increase the quality, an offline data sample must be reprocessed [7].

Previous clustering algorithms belong to partitioning methods, while the next ones are hierarchical. ROCK is a clustering agglomeration technique that uses the number of common neighbours between two objects as a connection measure, aiming to group objects with higher number of connections, which represents higher similarity between them [3].

The COBWEB algorithm produces a cluster dendrogram, which is a classification tree, that characterizes each cluster with a probabilistic description [16]. This algorithm uses an heuristic evaluation metric, termed category utility, to guide the tree's construction, and it has the ability of automatically adjusting the number of classes, without the need for the user's intervention.

Finally, the CLOPE algorithm was developed from the heuristic method of increasing the ratio between height and width of clusters histograms. It is considered to be fast and scalable in transactional databases and other data sources with a high number of dimensions [1,17].

2.3 Evaluation

The experimental design includes two stages. In the first stage, using only clustering attributes and training data, we use internal objective metrics to select the clustering method and number of clusters. To get a more robust and realistic assessment, we adopt three clustering iterations in this stage, under a realistic rolling window evaluation [12]. The first four days are used in the first iteration. Then, the data is slided by one day, leading to the second iteration, and so on. Each day contains around 7,000 redirect samples, translating into a training window with around 28,000 samples. The internal metrics are computed using average rolling window values of two popular clustering metrics (silhouette and Dunn index) [2] and the computational effort. The silhouette varies between -1 (poor clustering) to 1 (excellent) [1,5]. The Dunn index represents the ratio between the lower distance among observations on different clusters and the higher distance among objects within the same cluster. Its values range from 0 to 1, and higher values are better [5]. To compute the silhouette and Dunn index metrics we used the Gower distance measure, which is suitable for mixed (categorical and numeric) attributes and that ranges from 0 (identical) to 1 (most dissimilar) [9]. To measure the similarity between two clustering results, we used the rand index, which presents the highest value of 1 when both clusterings are equal [4].

In the second stage, we apply a time order holdout split to the best clustering algorithm of the previous stage. The first four days of data are used to training and the last three days of data for testing. We compute silhouette and Dunn index metrics for the test data and also external business metrics, which allow the selection of interesting clusters. These clusters are then described in terms of their main characteristics in order to get feedback from human experts. The business metrics include: N - the total number of redirects associated with a cluster of publishers; CVR - the conversion rate for the cluster; and LTV - the average publisher revenue for the cluster when there is a conversion. Since our collected sampled data includes a ratio of sales that is much higher than the real market, we adjusted the conversion rate to its realistic version: $CVR = S/N \times \alpha$, where S is the number of cluster sales and $\alpha = 1/35.99$ is a correction factor, such that the global CVR value, over all data, is near 1%.

3 Results

We conducted all experiments using the R tool [13] and a Linux server with an Intel Xeon 1.70 GHz processor with 56 cores and 64 GB of RAM. The packages used for cluster algorithms were **klaR** for k-modes, **coolcat** for COOLCAT, **cba** for ROCK and **RWeka** for both COBWEB and CLOPE. Furthermore, the **fpc** package was used for computing the clustering metrics.

The first experiments aimed to find a good number of clusters (K) and we tested four different possibilities: $K \in \{10, 20, 30, 40\}$. We only used the k-modes algorithm for this initial experiment, since it is easy to set K for this method and also it requires a reasonable computational effort. Table 2 presents the average clustering rolling window metric results. After analyzing Table 2, we decided to choose $K = 20$, since it accomplished good results with an acceptable computational cost and it is within the granularity level desired by OLAmobile.

Table 2. Results from tests using K-modes testing different numbers of clusters (best values in **bold**)

Metric	Number of clusters			
	10	20	30	40
Silhouette	**0.18**	0.17	**0.18**	0.17
Dunn index	0.17	0.18	0.18	**0.20**
Execution time (s)	**4579.31**	8133.19	12710.18	20048.49

After setting K, we compared the clustering methods by executing the first experimental design phase. The COOLCAT and ROCK algorithms did not provide results in useful time. We executed these algorithms and waited during 3 days and, besides the time they took executing, they have shown to be unable to deal with this quantity of data, often returning processing errors. As such, both

Table 3. Average first stage clustering results (best values in **bold**)

Metric	K-modes	COBWEB	CLOPE
Silhouette	0.17	**0.30** (0.27)	−0.10
Dunn index	0.18	**0.20** (0.20)	0.06
Clusters	20	19.67	19.33
Execution time (s)	8133.19	6093.06	**2432.38**

Table 4. Rand-index results for cluster similarity comparison

Algorithms	Rand-Index
COBWEB and CLOPE	0.64
K-Modes and CLOPE	0.63
K-Modes and COBWEB	0.90

of these algorithms were excluded from the second phase. Thus, in Table 3 we report only the K-modes, COBWEB and CLOBE average rolling window results. Also, the similarity between the cluster methods (in terms of the rand-index) is presented in Table 4. Regarding the remaining algorithms, it is important to notice that only K-modes receives the number of clusters (K) as a parameter. Thus, we have adjusted the COBWEB (cutoff) and CLOPE (repulsion) parameters to indirectly control the number of clusters, such that on average $K \approx 20$. Although it was the faster algorithm, CLOPE presented low values for the silhouette and Dunn index metrics. The second best performing algorithm, COBWEB revealed a high similarity with K-Modes (rand-index of 0.9) and was selected for the second phase, since it achieved better internal metric values under a lower computational cost.

In the second phase, COBWEB provided 25 different clusters when trained with data from the first four days. The respective silhouette and Dunn index test set results (computed using the last three days) are shown in brackets in Table 3. The average rolling window training and holdout test clustering metrics are similar (e.g., silhouette of 0.30 and 0.27), showing that the clusters have stable quality values through time.

The left of Fig. 2 presents the external evaluation results in terms of CVR (x-axis) versus LTV (y-axis). In the graph, each cluster label is set within a circle that is proportional to the cluster size (N). To simplify the analysis, the graph includes only the nine biggest clusters (such that $N > 300$). Using a multi-objective analysis to the business metrics, we selected four interesting clusters: 1 – with 4,179 redirects, highest CVR (1.4%) and LTV (2.1) values; 15 – largest cluster with 4,518 redirects, low LTV and CVR slightly above average (1.07%); 21 – second largest cluster with 4,419 redirects, good LTV (1.0) and low CVR (0.6%); and 25 – small cluster with 355 redirects, low CVR and LTV values. The right of Fig. 2 shows the main differences between a high (1) and bad (25)

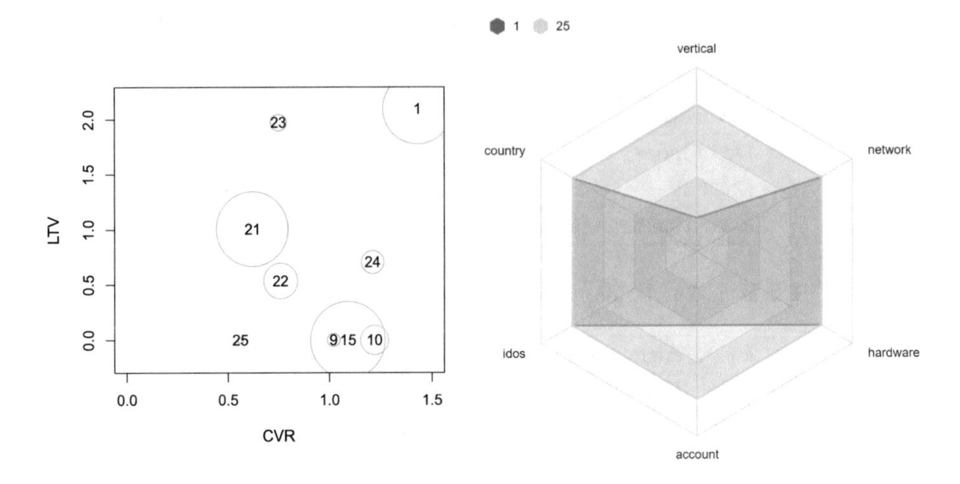

Fig. 2. Business metrics for the obtained clusters (left) and characterization of clusters 1 and 25 (right)

performing clusters, in terms of the mode values for the six clustering attributes. The graph shows differences between the clusters in terms of the vertical and account attributes. The full mode values for the four selected clusters are shown in Table 5, which highlights the main cluster differences.

Table 5. Characterization of the selected clusters (mode values)

Cluster	Vertical	Country	Idos	Account	Hardware	Network
1	VOD	Russia	Android	Advertiser	Smartphone	Mobile
15	Mainstream	Russia	Android	Advertiser	Smartphone	Mobile
21	Mainstream	Russia	Android	Advertiser	Smartphone	WiFi
25	Mainstream	Russia	Android	Network	Smartphone	Mobile

These results were shown to OLAmobile experts and the feedback was positive. In particular, the selected clusters are aligned with their business knowledge and were considered interesting, presenting a good potential for supporting publisher's decisions in terms of adhering to a DSP market or renting digital spaces.

4 Conclusions

Mobile performance marketing is a growing industry due to current ubiquitous usage of personal devices (e.g., smartphones, tablets). This industry includes several players: publishers, which attract users but need to be funded by ads; Demand-Side Platforms (DSP), which act as brokers, matching users to ads and managing payouts when there is a conversion (e.g., sell); and advertisers, which resort to DSPs for promoting their products and increase sales.

In this paper, we adopt a novel categorical clustering approach for this industry, which groups publishers into similar segments, in order to provide them with more informative expected revenue metrics. As a case study, we worked with data from OLAmobile, a global mobile performance marketing company that receives millions of ad clicks per hour. We collected a sample dataset with 48,900 records related with European ad clicks during one week. Then, we conducted a first set of experiments in order to compare five categorical clustering methods using a first internal evaluation that used objective clustering quality metrics. The best performing algorithm (COBWEB) was then analyzed under a second evaluation phase, based on business metrics, namely the Conversion rate (CVR) and Lifetime value (LTV). Such analysis allowed the particular identification of four interesting clusters, including a large cluster with high CVR and LTV values, and a smaller cluster with low CVR and LTV values. These clusters were characterized in terms of their mode values (considering six clustering attributes), resulting in a positive feedback from domain experts in terms of their validity and usefulness for publishers. In the future, we wish to collect more data (e.g., worldwide) and also study data scalability issues. For instance, by adapting the categorical clustering algorithms to the MapReduce framework [8].

Acknowledgements. This article is a result of the project NORTE-01-0247-FEDER-017497, supported by Norte Portugal Regional Operational Programme (NORTE 2020), under the PORTUGAL 2020 Partnership Agreement, through the European Regional Development Fund (ERDF). This work was also supported by COMPETE: POCI-01-0145-FEDER-007043 and FCT Fundação para a Ciência e Tecnologia within the Project Scope: UID/CEC/00319/2013.

References

1. Agarwal, D., Long, B., Xin, D.: LASER: a scalable response prediction platform for online advertising. In: Proceedings of the 7th ACM International Conference on Web Search and Data Mining, WSDM 2014, pp. 173–182 (2014)
2. Aggarwal, C.C., Reddy, C.K.: Data Clustering: Algorithms and Applications. CRC Press, Boca Raton (2013)
3. Alamuri, M., Surampudi, B.R., Negi, A.: A survey of distance/similarity measures for categorical data. In: Proceedings of the International Joint Conference on Neural Networks, pp. 1907–1914 (2014)
4. Amini, A., Wah, T.Y., Saboohi, H.: On density-based data streams clustering algorithms: a survey (2014)
5. Brock, G., Pihur, V., Datta, S.S., Datta, S.S.: clValid: an R package for cluster validation. J. Stat. Softw. **25**, 1–28 (2008)
6. Du, M., State, R., Brorsson, M., Avenesov, T.: Behavior profiling for mobile advertising. In: Proceedings of the 3rd IEEE/ACM International Conference on Big Data Computing, Applications and Technologies - BDCAT 2016, pp. 302–307 (2016)
7. Gan, G., Ma, C., Wu, J.: Data Clustering: Theory, Algorithms, and Applications. ASA-SIAM Series on Statistics and Applied Probability. Society for Industrial and Applied Mathematics, Philadelphia (2007)

8. Garcia, K.D., Naldi, M.C.: Multiple parallel mapreduce k-means clustering with validation and selection. In: 2014 Brazilian Conference on Intelligent Systems (BRACIS), pp. 432–437. IEEE (2014)
9. Gower, J.C.: A general coefficient of similarity and some of its properties. Biometrics **27**, 857–871 (1971)
10. Hu, Y., Shin, J., Tang, Z.: Pricing of online advertising: cost-per-click-through vs. cost-per-action. In: Proceedings of the Annual Hawaii International Conference on System Sciences (2010)
11. Melnykov, V.: Model-based biclustering of clickstream data. Comput. Stat. Data Anal. **93**, 31–45 (2016)
12. Oliveira, N., Cortez, P., Areal, N.: The impact of microblogging data for stock market prediction: using twitter to predict returns, volatility, trading volume and survey sentiment indices. Expert Syst. Appl. **73**, 125–144 (2017)
13. R Core Team: R: A Language and Environment for Statistical Computing. R Foundation for Statistical Computing, Vienna, Austria (2016)
14. Regelson, M., Fain, D.: Predicting click-through rate using keyword clusters. In: Proceedings of the Second Workshop on Sponsored Search Auctions, vol. 9623, pp. 1–6 (2006)
15. Reps, J., Aickelin, U., Garibaldi, J., Damski, C.: Personalising mobile advertising based on users' installed apps. In: IEEE International Conference on Data Mining Workshops, ICDMW, pp. 338–345 (2015)
16. Sharma, N., Bajpai, A., Litoriya, R.: Comparison the various clustering algorithms of weka tools. Int. J. Emerg. Technol. Adv. Eng. **2**(5), 73–80 (2012)
17. Sora, M., Roy, S., Singh, S.I.: FLoMSqueezer: an effective approach for clustering categorical data stream. Int. J. Comput. Sci. Issues **8**(6), 284–291 (2011)
18. Su, Q., Chen, L.: A method for discovering clusters of e-commerce interest patterns using click-stream data. Electron. Commer. Res. Appl. **14**(1), 1–13 (2015)
19. Yuan, S., Abidin, A.Z., Sloan, M., Wang, J.: Internet advertising: an interplay among advertisers, online publishers, ad exchanges and web users. CoRR, abs/1206.1754 (2012)
20. Zhang, W., Du, T., Wang, J.: Deep learning over multi-field categorical data. In: European Conference on Information Retrieval, pp. 45–57. Springer (2016)

Spatial Models of Wireless Network Efficiency Prediction by Turning Bands Co-simulation Method

Anna Kamińska-Chuchmała(✉)

Faculty of Computer Science and Management,
Wrocław University of Science and Technology,
Wybrzeże Wyspiańskiego 27, 50-370 Wrocław, Poland
anna.kaminska-chuchmala@pwr.edu.pl

Abstract. Modernly, main Internet traffic is by the means of wireless network, due to the high mobility requirements and surging number of portable electronic devices users. Rapid growth of demand for uninterrupted access to vast amounts of information that are available in Internet forces network operators to focus on network reliability and to anticipate potential events such as network overload, that could pose threat for sustained delivery of data. In this paper Author investigated efficiency of WiFi open network in building located at main campus of Wrocław University of Science and Technology (WUST). The database analyzed in this paper consist data from two monthly periods, namely May of 2014 and 2015. The idea of research was to create spatial model prediction of WiFi network efficiency. Models of prediction contains two important parameters of WiFi network: number of users and load channel utilization. Spatial (3D) predictions for two database were made with using geostatistical co-simultaion method Turning Bands. Obtained results were compared with each other and conclusions with future research directions to WiFi network efficiency predictions were drawn.

Keywords: WiFi · Wireless network efficiency · Spatial prediction
Turning bands · Geostatistical co-simulation method

1 Introduction

Increased request for access to the Internet by mobile devices have influence on the operation of the whole network. It means that more Internet traffic is generated by users from wireless network. In 2016, wired devices accounted for the majority of IP traffic at 51% and it is confirmed by Cisco Visual Networking Index (Cisco VNI) document [1] about global IP traffic forecast. It states also that and traffic from wireless and mobile devices will account for more than 63% of total IP traffic by 2021 and wired devices will account only 37% of IP traffic. On basis of this situation Author decided to conduct research to make

M. Graña et al. (Eds.): SOCO'18-CISIS'18-ICEUTE'18, AISC 771, pp. 155–164, 2019.
https://doi.org/10.1007/978-3-319-94120-2_15

3D prediction of wireless network utilization and their efficiency dedicated for network operators and administrators.

In the paper, next section contains related works. After that, simulation methods are described and then preliminary and structural analysis are presented. Subsequently, section with models of spatial prediction and results are discussed. The end of the paper provides conclusions and plan of future research.

2 Related Work

In this section brief review of research concerning prediction of wireless network efficiency is given.

In [2] Author proposed Wi-Fi performance estimation model based on Machine Learning method, which allows service providers to predict Wi-Fi saturated throughput from easy measurement on the access point. A SVM-based classification model was proposed to work as a prediction function. Moreover, proposed Wi-Fi throughput estimation model consists of saturated throughput as a function of device RSSI and noise floor, resource contention. Different SVM kernel functions conducted to evaluate the proposed model and results show that classification accuracy can be up to 0.88.

Another estimation of throughput in wireless networks is presented in [3]. Research are based on active measurements and statistical learning tools. Authors presented a methodology where the system is trained during short periods with application flows and probe packets bursts. As a result a continuous non intrusive methodology is obtained and it allows to determine the maximum throughput of a wireless connection only knowing some characteristics of the network. Authors also use Support Vector Machines (SVM) for regression whereas results They obtained by simulations.

Analysis of Quality of Service (QoS) performance metrics for networked systems (networked robots) using an analytic model based on Markov chain theory are proposed in [4]. The Markov model describes the IEEE 802.11 Distributed Coordination Function (DCF) for periodic round trip traffic of Wireless Networked Control System (WNCS). The goal of Authors is to adjust Bianchi Markov model to represent the behavior of periodic round trip traffic of a control loop under unsaturated conditions.

Paper [5] contains machine learning techniques to learn implicit performance models, from a limited number of measurements. These models do not require to know the internal mechanics of interfering Wi-Fi links. Authors improved accuracy by at least 49% compared to measurement-seeded models based on SINR. Learned models were shown as a new algorithm that uses such a model as an oracle to jointly allocate spectrum and transmit power. This algorithm is utility-optimal, distributed, and it produces efficient allocations that significantly improve performance and fairness.

Concluding, on basis of actual literature there is wide spectrum of research wireless network performance, but no one till now using geostatistical methods to spatial prediction efficiency of wireless networks. Thus, in this paper 3D prediction models with using Turning Bands simulation method are conducted.

3 Turning Bands Simulation Method

The Turning Bands method (TBM) is one of the earliest geostatistical simulation methods designed by Matheron [6,7] and developed by Lantuejoul [8]. TBM is a stereologic tool used for reduction of multidimensional simulation to a one-dimensional one. The first step of the geostatistical simulation is modeling process variables and then simulate these variables using grid file. The Turning Bands method consists in the reduction of a Gaussian random function of covariance C to the simulation of an independent stochastic process of covariance C_θ. Let $(\theta_n,\ n \in \mathbf{N})$ be a sequence of directions $\mathbf{S}_d^+$, and let $(X_n,\ n \in \mathbf{N})$ be a sequence of independent stochastic processes of covariance C_{θ_n}. Random function:

$$Y^{(n)}(x) = \frac{1}{\sqrt{n}} \sum_{k=1}^{n} X_k(< x, \theta_k >), \quad x \in \mathbb{R}^d \tag{1}$$

assumes covariance equal to:

$$C^{(n)}(h) = \frac{1}{n} \sum_{k=1}^{n} C_{\theta_k}(< h, \theta_k >). \tag{2}$$

Turning Bands Algorithm Method

1. *Transform input data using Gaussian anamorphosis.*
   In the transformation of Gaussian type variable y into new variable z of arbitrary distribution the anmorphosis function is used. Empirical Gaussian anamorphosis is represented by infinite series of Hermite polynomials:

   $$\varphi(y) = \sum_{k=0}^{\infty} \frac{\varphi_k}{k!} H_k(y). \tag{3}$$

   Finally the limited anamorphosis is used in the random diagram of Gaussian curve to obtain a random graph of the original random variable.
2. *Choose a set of directions $\theta_1, ..., \theta_n$ such that $\frac{1}{n}\sum_{k=1}^{n} \delta_{\theta_k} \approx \varpi$.*
   Simulation of covariance C_3 is obtained by summing, the projection of the simulation for a given number of lines of covariance C_1. Particular lines are named *turning bands.*
3. *Generate independent standard stochastic processes $X_1, ..., X_n$ with covariance functions $C_{\theta_1}, ..., C_{\theta_n}$.*
   The standard approach is to calculate the spectral measure χ of C, then derive the spectral measure χ_θ of C_θ and take its Fourier transform.
   Considerable simplifications occur when C is isotropic, i.e. it can be written as

   $$C(h) = C_d(|h|) \tag{4}$$

   for some scalar function C_d defined on $\mathbf{R}^+$.
   In this case, all of the C_θ are equal to a single covariance function, say C_1. Matheron gives the relationship between C_1 and C_d:

   $$C_d(r) = 2\frac{(d-1)\omega_{d-1}}{d\omega_d} \int_0^1 (1-t^2)^{\frac{d-3}{2}} C_1(tr)dt \tag{5}$$

where: ω_d stands, as usual, for the d-volume of the unit ball in $\mathbb{R}^d$. If $d = 3$, this formula reduces to:

$$C_3(r) = \int_0^1 C_1(tr)dt \tag{6}$$

or equivalently:

$$C_1(r) = \frac{d}{dr}(rC_3(r)) \tag{7}$$

Knowing variogram and variance *C(0)* for a stationary random function Z is tantamount to knowing its covariance. Thus to determine a covariance function for each of the variables, a variogram function is calculated, it is defined for any vector h as:

$$\gamma(h) = \frac{1}{2}E((Z(x+h) - Z(x))^2). \tag{8}$$

4. *Compute* $Y^{(n)}(x) = \frac{1}{\sqrt{n}} \sum_{k=1}^{n} X_k(< x, \theta_k >)$ *for any* $x \in D$.
   There is a wide variety of simulation methods for stochastic processes with a given covariance function C_1. The most common methods are:
   - spectral,
   - dilution,
   - migration.
5. *Make kriged estimate* $y^*(x) = \sum_c \lambda_c(x)y(c)$ *for each* $x \in D$.
   Let Y be a stationary Gaussian random function $\in \mathbb{R}^d$ with average value 0, variance equal to 1, and covariance function C. The goal is to conduct simulation Y satisfying the conditions: $Y(c) = y(c)$ for each c in finite subset $C \in \mathbb{R}^d$.
   Simple kriging estimation (linear regression) $Y(x)$ to $Y(c)$ is a linear combination:

$$Y^*(x) = \sum_{c \in C} \lambda_c(x)Y(c), \tag{9}$$

which minimizes the mean square error $E(Y^*(x) - Y(x))^2$. Coefficients λ_c are the solutions of the system of linear equations:

$$\sum_{c' \in C} \lambda_{c'} C(c, c') = C(c, x) \quad \forall c \in C. \tag{10}$$

The mean square error, the kriging variance, is then:

$$E(Y^*(x) - Y(x))^2 = 1 - \sum_{c \in C} \lambda_c C(c, x). \tag{11}$$

For the calculations of the system of linear equations in kriging, it is crucial to know an estimated neighborhood point.
There are two types of neighborhood:
- unique,
- moving.

A characteristic feature of an unique neighborhood is a single calculation of the inverse kriging matrix. It remains the same for all considered points. Thus the type of an unique neighborhood can be computed much more quickly than moving neighborhood. In case of moving neighborhood if the location of a point x is known, the closest points in a sphere or ellipsoid are selected to be correlated with the considered point x. The selected points are subject to a sequence algorithm which takes into account such criteria as rotation, search ellipsoid (it defines the maximum distance along the main axes U, V, W after rotation), anisotropy, minimum number of points considered in the range search, etc.

6. *Simulate a Gaussian random function with mean 0 and covariance C in domain $\boldsymbol{D}$ on condition points.*
   Values obtained as a result of calculations: $(z(x), x \in D)$ and $(z(c), c \in C)$.
7. *Make kriged estimate.*
   Kriging estimation:
$$z^*(x) = \sum_c \lambda_c(x) z(c), \tag{12}$$
   for each $x \in D$.
8. *Obtain the result $(y^*(x) + z(x) - z^*(x), x \in D)$.*
   Finally, as a result of a conditional simulation, a random function is obtained:
$$W(x) = y^*(x) + z(x) - z^*(x), \tag{13}$$
   where $x \in D$.
9. *Perform a Gaussian back transformation to return to the original data.*

In this paper Turning Bands co-simulation method is used. Simulation of a co-model is called co-simulation. It means that two variables are taken into account during conduction of prediction modelling.

4 Description of Database and WiFi Network

Wireless network named PWR-WiFi is an open university network located in main campus at Wrocław University of Science and Technology (WUST) in Poland. The data collected to this research are obtained from eleven Access Points (APs) given in five-storey building named B4 (Fig. 1a). Considered APs are installed in B4 building as follows: one on first floor, one on second floor, two on third floor, five on fourth floor, and two on fifth floor. Projection of APs in building is presented in Fig. 1b. According to technical information PWR-WiFi network using frequency 2.4 GHz in IEEE 802.11b/g/n standards and APs with using 5 GHz frequency in IEEE 802.11a/n standards. APs are wirelessly connected to switch. They are configured to get IP address from network and connecting to WiFi controller. All APs use communication protocols LWAPP (Light Weight Access Point Protocol) which allows communication between APs and controllers.

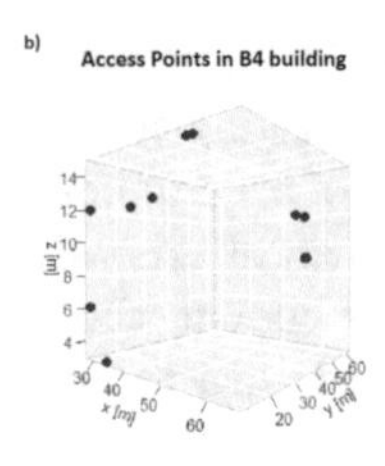

Fig. 1. B4 building with 11 APs inside, located at main campus of WUST and Projection on APs localization in this building

Data for analysis were taken from two periods of time 5th - 25th of May 2014 and 4th - 24th of May 2015. Data collected every hour between 7:00 AM and 9:00 PM are considered in research. This part of day was chosen because of specific behavior of network users. Because of PWR-WiFi is an open university network thus most of users that are students and employees of WUST like lecturers or administrative workers. The B4 building is closed by night and main traffic in network is during office hours and lectures. Mainly, the first classes are started at 7:30 AM and the last one finished at 8:30 PM. Generally, during the day, classes are started quarter after an odd hour. Moreover this parts of month May 2014 and 2015 are a period in semester where students coming regularly on classes, thus regularly need to access to the WiFi network. Experiment relies on analyses on expanse two years to better show the dynamic grow of number of PWR-WiFi network users and indicate periodical users behavior as well as pay attention on this wireless network performance. The most adequate parameter of PWR-WiFi network which was obtained from measurement is Number of Users (NoU) and Load Channel Utilization (LCU), which are investigated during research and presented in this paper. Author analyzed also another periods of time of PWR-WiFi located in B4 building [9] and prepare spatial model of prediction wireless network efficiency using estimation kriging method [10].

Author has performed all research presented in this paper in R language under R environment in version 3.4.1 which is available as Free Software under GNU licence [11]. Author was used also RGeostats package version 11.2.1 [12] to made geostatistical co-simulation prediction model with Turning Bands method. RGeostats is the Geostatistical Package (under R platform) developed by the Geostatistical Team of the Geosciences Research Center of MINES ParisTech.

5 Preliminary and Structural Analysis of Data

Fundamental statistics are presented in Table 1 shows increasing trend of usage PWR-WiFi network within two years 2014–2015. The maximum number of users is equal 72 in 2014 and 81 in 2015. Mean value of both NoU and also LCU is higher in 2015 than 2014. Variance of load channel utilization in 2014 is much lower and equal more than 250 and NoU almost 90 than variance in 2015 which

Table 1. Basics statistics of network parameters for all considered APs located in B4

Network parameter	Number of users	Number of users	Load channel utilization	Load channel utilization
Research period	05-25.05.2014	04-24.05.2015	05-25.05.2014	04-24.05.2015
Minimum value $\mathbf{Z}_{min}$	0	0	0	0
Maximum value $\mathbf{Z}_{max}$	72	81	96	115
Mean value $\mathbf{Z}$	7	8	20.66	26.20
Standard deviation $\mathbf{S}$	9.46	11.86	15.85	19.00
Variance $\mathbf{V}$	89.53	140.61	251.19	360.85

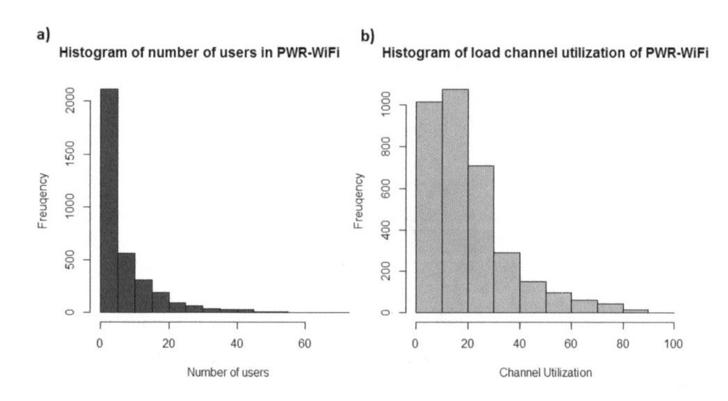

Fig. 2. Histogram: (a) NoU and (b) LCU parameters of PWR-WiFi network in 2014

equals more than 360 and 140 respectively what could means significant data differentiation. Furthermore dispersion of data characterized by standard deviation is also the largest in 2015 for both network parameters.

The highest frequencies of users in 2014 are in between 0 to 10 and 10 to 20 users which confirm histogram presented in Fig. 2a. The histogram is skewed left. Additionally, in Fig. 2b the level of load channel utilization is the highest up to 40%. It means that currently PWR-WiFi network is not overloaded. This histogram is also asymmetric.

The first step in structural analysis in geostatistical approach to create spatial prediction model is calculation of variogram. Two variograms for 2014 and 2015 were calculated in four different directions: 0, 45, 90, 135°. As example variogram of NoU in 2015 is presented in Fig. 4. Distance lags for each calculation direction equals 5. The number of lags for each calculation direction equals 15. For all directions it could be seen the nugget effect. The range of variogram function for two directions was equal about 30 m and for two others direction almost 60 m. In the next step, variogram function was approximated by theoretical functions: nugget effect, exponential, and spherical.

6 Spatial Prediction of PWR-WiFi Network Efficiency with the Co-Simulation Turning Bands Method

Two models of spatial prediction were prepared for two consecutive years 2014 and 2015. Models have 3 dimensions (x and y geographical coordinates and z the altitude coordinate) and prediction cover the space, where there is present the signal from APs belonging to PWR-WiFi wireless network in building B4. Space prediction was performed by geostatistical co-simulation method named Turning Bands described in third chapter. Model of spatial prediction contains theoretical model of variogram approximation and moving neighborhood. The moving neighborhood search is performed by angular sectors and the neighborhood ellipsoid is anisotropic. The search ellipsoid has three dimensions. The main variable in co-models is NoU and second variable LCU.

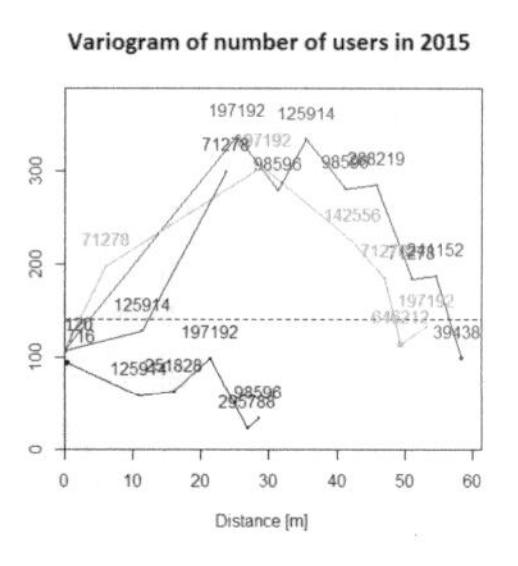

Fig. 3. Variogram calculated in four directions for number of PWR-WiFi wireless network users in 2015

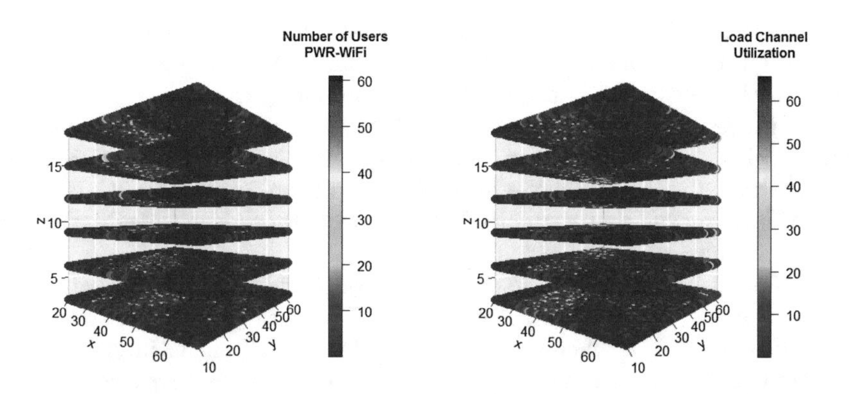

Fig. 4. Scatter 3D plot of: (a) Nou (b) LCU of PWR-WiFi wireless network users on each floor in B4 building at WUST campus in 2014.

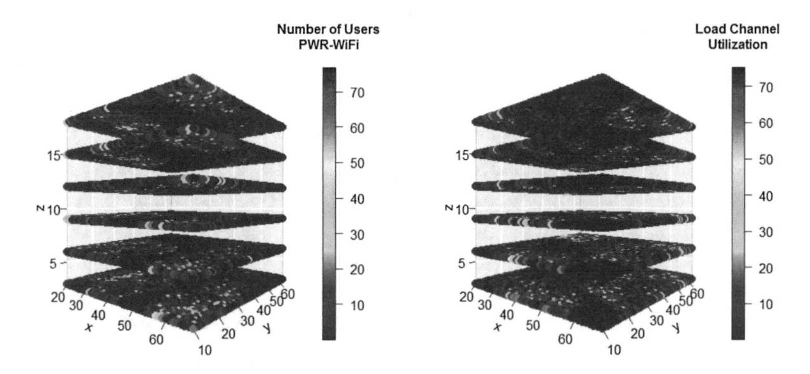

Fig. 5. Scatter 3D plot of: (a) Nou (b) LCU of PWR-WiFi wireless network users on each floor in B4 building at WUST campus in 2015.

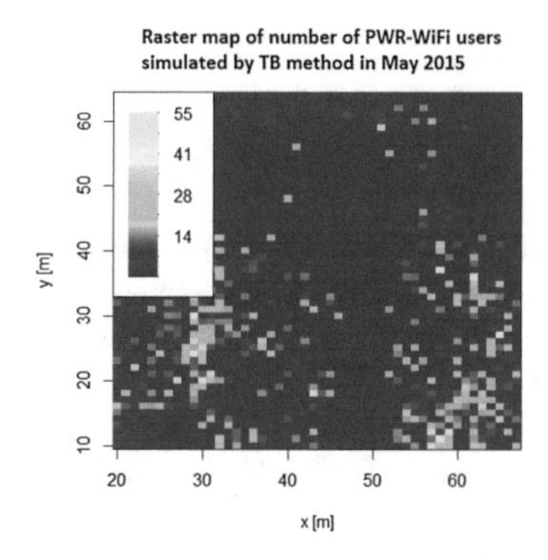

Fig. 6. Raster map of number of PWR-WiFi users simulated by TB method in B4 building in May 2015

In Figs. 4 and 5 there are presented raster maps for each floor of NoU and LCU of PWR-WiFi co-simulated by TB method in years 2014–2015. Localization of more concentrations of users such as students, lecturers or administrative workers are similar for all years. Probably it is related to the fact that it is nearby (depends of floor in the building): lecture hall, library or deanery.

Raster map of spatial wireless network efficiency prediction of PWR-WiFi on each level on the floor in B4 building in 2015 is presented in Fig. 6. Depending on localization of APs on each floor the spread of users are different. The main concentration of users is on the first two floors. On the fifth floor there are the least number of users because on this floor there is no lecture hall or classroom where students could generate the largest number of users during the day.

7 Conclusions

Spatial prediction models of PWR-WiFi wireless network efficiency with using co-simulation Turning Bands method were presented in this paper. There were two models of spatial prediction for two consecutive years: 2014, and 2015. Before prediction, preliminary and structural analysis of data had been discussed. Results of predictions confirm the rising trend of PWR-WiFi network loading. Generally, this kind of spatial prediction could be very helpful for administrators of networks. Raster maps show the good spread of signal from Access Points because there are more users connected to the wireless network. It could be valuable to see efficiency and capacity of network on this kind of map.

In future, Author plan to perform prediction of PWR-WiFi wireless network in 4D (space-time) model of wireless network prediction.

References

1. Cisco Visual Networking Index: Forecast and Methodology, 2016–2021, White paper, Cisco public, 6 June 2017
2. Pan, D.: Analysis of Wi-Fi performance data for a Wi-Fi throughput prediction approach, MSc Thesis, KTH Royal Institute of Technology, School of Information and Communications Technology (ICT), Stockholm, Sweden, June 2017
3. Rattaro, C., Belzarena, P.: Throughput prediction in wireless networks using statistical learning. In: LAWDN - Latin-American Workshop on Dynamic Networks, Buenos Aires, Argentina, 4 p., November 2010. (inria-00531743)
4. Sassi, I., Gouin, A., Thiriet, J.-M.: Wireless network performance evaluation for networked robots. In: 22nd IEEE International Conference on Emerging Technologies and Factory Automation (ETFA), pp. 1–5. IEEE (2017)
5. Herzen J., Lundgren H., Hegde N.: Learning Wi-Fi performance. In: 2015 12th Annual IEEE International Conference on Sensing, Communication, and Networking (SECON). IEEE (2015)
6. Matheron, G.: Quelques aspects de la montée. Internal Report N-271, Centre de Morphologie Mathematique, Fontainebleau (1972)
7. Matheron, G.: The intrinsic random functions and their applications. JSTOR Adv. Appl. Probab. **5**, 439–468 (1973)
8. Lantuejoul, Ch.: Geostatistical Simulation: Models and Algorithms. Springer, Heidelberg (2002)
9. Kamińska-Chuchmała, A.: Performance analysis of access points of university wireless network. Rynek Energii **1**(122), 122–124 (2016)
10. Kamińska-Chuchmała, A.: Spatial prediction models of wireless network efficiency estimated by Kriging method. Rynek Energii **2**(135), 89–94 (2018)
11. R Core Team: R: A language and environment for statistical computing. R Foundation for Statistical Computing, Vienna, Austria (2017). https://www.R-project.org
12. Renard, D., Bez, N., Desassis, N., Beucher, H., Ors, F., Freulon, X.: RGeostats: The Geostatistical R package 11.2.1 MINES ParisTech/ARMINES. http://cg.ensmp.fr/rgeostats

Industry 4.0

Neural Visualization for the Analysis of Energy and Water Consumptions in the Automotive Industry

Raquel Redondo[1(✉)], Álvaro Herrero[1], Emilio Corchado[2], and Javier Sedano[3]

[1] Grupo de Inteligencia Computacional Aplicada (GICAP), Departamento de Ingeniería Civil, Escuela Politécnica Superior, Universidad de Burgos, Av. Cantabria s/n, 09006 Burgos, Spain
{rredondo,ahcosio}@ubu.es

[2] Departamento de Informática y Automática, Universidad de Salamanca, Salamanca, Spain
escorchado@usal.es

[3] Instituto Tecnológico de Castilla y León, Pol. Ind. Villalonquejar, C/López Bravo 70, 09001 Burgos, Spain
javier.sedano@itcl.es

Abstract. This study presents the application of neural models to a real-life problem in order to study the energy and water consumptions of an automotive multinational company for resources saving and environment protection. The aim is to visually and naturally analyse different consumptions data for a whole year, month by month, from factories and locations worldwide where different kinds of products are produced. The data are studied in order to see whether the geographical location, the month of the year or the technology used in each factory are relevant in terms of consumptions and then take actions for a greener production. The consumptions dataset is analysed using different neural projection models: Principal Component Analysis and Cooperative Maximum-Likelihood Hebbian Learning. This unsupervised dimensionality reduction techniques have been applied, and subsequent interesting conclusions are obtained.

Keywords: Soft computing · Artificial neural networks · Exploratory projection pursuit · Industrial applications · Energy consumption

1 Introduction

In recent years, more and more companies and governments are concerned about the environment. Being respectful with the environment and consuming the minimum fossil energy and water is widely accepted as a strategy for greener industrial production as it greatly contributes to reduce atmospheric pollution [1].

Because of this commitment, companies carry out studies on their environmental impact. In order to do so, it is necessary to have relevant information and appropriate analysis tools to obtain indicators for taking decisions. In keeping with this idea,

M. Graña et al. (Eds.): SOCO'18-CISIS'18-ICEUTE'18, AISC 771, pp. 167–176, 2019.
https://doi.org/10.1007/978-3-319-94120-2_16

present study proposes soft computing techniques (more precisely, artificial neural networks) for the analysis of energy and water consumption of several factories from a multinational company in the automotive industry sector.

The consumption dataset is analysed using two projection methods: Principal Component Analysis (PCA) [2, 3] and Cooperative Maximum-Likelihood Hebbian Learning (CMLHL) [4], to identify the dataset structure and clearly identify the status of each one of the factories regarding environmental issues. If the initial collected dataset, once analysed shows a certain degree of clustering, it can be seen as a sign of a representative data set (there is not a single problem related when collecting the information and the process is well defined by such data set).

The data are provided by Grupo Antolin [5]. In order to characterize the different situations in which each one of its factories is, from an environmental point of view, it is necessary to analyse if it is a representative and informative enough dataset.

The remaining sections of this study are structured as follows: Sect. 2 presents the techniques and methods that are applied to analyse the data. Section 3 details the real-life case study, while Sect. 4 describes the experiments and results. Finally, Sect. 5 sets out the conclusions and future work.

2 Data Structure Analysis Using Connectionist Techniques

Soft Computing is a set of several technologies whose goal is to resolve inexact and complex problems [6]. It investigates, simulates, and analyses very complex issues and phenomena in order to solve real-world problems [7]. Soft Computing has been successfully applied for data analysis, and plenty of algorithms are reported in the literature [8, 9].

The unsupervised learning is useful to explore a dataset when there is no a specific goal or it is not clear what information the data contains. It is also a way to reduce the dimensionality of a given dataset.

Following subsections describe the techniques applied in present study.

2.1 Principal Component Analysis

Principal Component Analysis (PCA) [2] is a well-known method that provides the best linear data compression in terms of least mean square error by addressing the data variance.

PCA generates new variables by linear combinations of the original variables so that small numbers of new variables captures most of the information. When the datasets have many variables, often they are redundant. PCA takes advantage of the redundancy of information on them.

With PCA, it is possible to discover a smaller group of underlying variables that define the data. PCA has been the most frequently reported linear operation involving unsupervised learning for data compression and feature selection.

Though it was proposed as a statistical method, it has been proven that it can be implemented by several ANNs [10, 11].

2.2 Cooperative Maximum Likelihood Hebbian Learning

Exploratory Projection Pursuit (EPP) [12–14] is a more recent statistical method designed to resolve the difficult problem of identifying structure in high dimensional data. It projects the data onto a low dimensional subspace in which the data structure is searched by eye. However, not all projections will reveal this structure equally well. It therefore defines an index that measures how "interesting" a given projection is, and then represents the data in terms of projections that maximize the index.

Cooperative Maximum Likelihood Hebbian Learning (CMLHL) [4] is an extended version of Maximum Likelihood Hebbian Learning (MLHL) [15], an EPP connectionist model. CMLHL incorporates lateral connections [4, 16] derived from the Rectified Gaussian Distribution (RGD) [17]. The RGD is a modification of the standard Gaussian distribution in which the variables are constrained to be non-negative, enabling the use of non-convex energy functions. The CMLHL architecture is represented in Fig. 1, where lateral connections are shown.

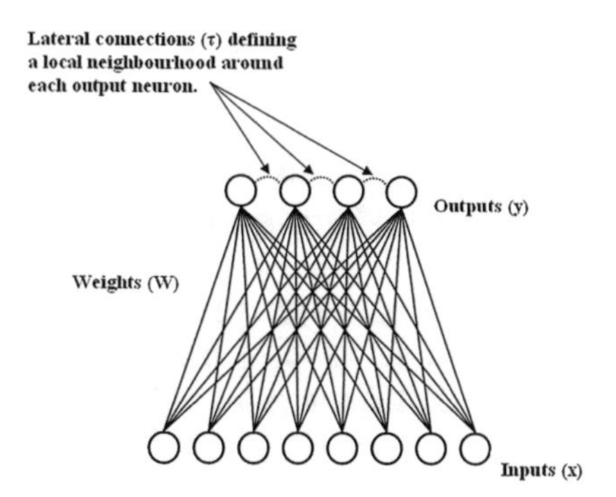

Fig. 1. CMLHL: lateral connections between neighbouring output neurons

The resultant net can find the independent factors of a data set but does so in a way that captures some type of global ordering in the data set.

Considering an N-dimensional input vector (x), an M-dimensional output vector (y) and with W_{ij} being the weight (linking input j to output i), CMLHL can be expressed as:

Feed-forward step:

$$y_i = \sum_{j=1}^{\mathbf{N}} W_{ij}x_j, \forall i \tag{1}$$

Lateral activation passing:

$$y_i(t+1) = [y_i(t) + \tau(b - Ay)]^+ \tag{2}$$

Feedback step:

$$e_j = x_j - \sum_{i=1}^{M} W_{ij} y_i, \forall j \tag{3}$$

Weight change:

$$\Delta W_{ij} = \eta . y_i . sign(e_j) |e_j|^p \tag{4}$$

Where: η is the learning rate, τ is the "strength" of the lateral connections, b the bias parameter and p is a parameter related to the energy function [4, 15].

A is a symmetric matrix used to modify the response to the data, the effect of which is based on the relation between the distances between the output neurons. It is based on Cooperative Distribution, but to speed up the learning process, it can be simplified to:

$$A(i,j) = \delta_{ij} - \cos(2\pi(i-j)/M) \tag{5}$$

where, δ_{ij} is the Kronecker delta.

3 A Real Case Study: Energy and Water Consumptions Analysis of an Automotive Multinational Company

As previously stated, in present research neural techniques are applied to analyse the energy and water consumptions of an automotive multinational company. To do this analysis, some data have been selected and gathered.

The data used in this study are taken from year 2016 and it comprises energy and water consumptions of all the worldwide factories and locations of Grupo Antolin [5]. This multinational company from the automotive industrial sector, is one of the largest players in the car interiors international market and the number 1 worldwide supplier of headliner substrates. It has presence in 26 countries with 167 production plants & Just in Time centers and 29 technical-commercial offices. Grupo Antolin manufactures products that are technologically sustainable, based on two premises: light and green, thereby contributing to lower CO_2 emissions. Grupo Antolin is strongly committed to the environment so this study is a first step in order to analyse with soft computing techniques the different consumptions in the company.

The dataset under analysis comprises a total of 1090 samples. The data are from year 2016, and there are a sample for each month of the year from 91 different factories. The following consumption parameters (features) were gathered for each one of the factories:

- Electricity (KWh)
- Natural gas (m^3)
- Propane (m^3)
- Fuel (l)
- Gasoil (l)
- Water (m^3)

As explained before, this dataset gathers data from different factories or centers, in different countries worldwide, where several kinds of technologies and products are manufactured. Each center could has several kinds of consumes, for example, not all of the locations consume propane or fuel but all the factories have electricity consumptions.

4 Results

The dimensionality reduction techniques have been applied in order to analyse and find the characteristics that best describe the collected data. The principal aim is to detect if a clear internal structure can be identified, this means that the selected data is informative enough. Otherwise, further data must be properly collected [18].

The above mentioned neural projection models have been applied to the data described in Sect. 3, whose results are compiled in present section. PCA is the first technique that has been applied in order to find a possible structure in the data. In next figures, data samples are depicted in the PCA projection (2 first principal components), according to different criteria: continent were the factory is located, production technology, business unit and the month of the year.

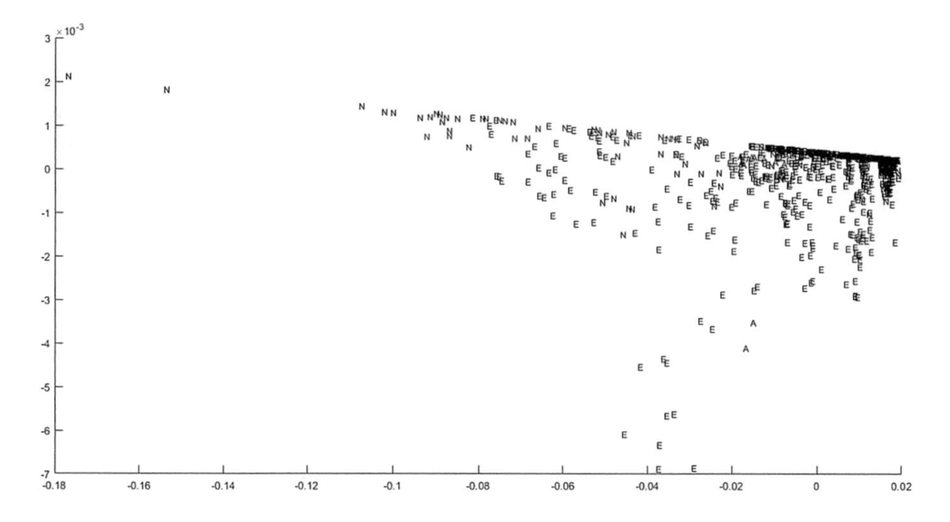

Fig. 2. PCA Projection–Continent: E: Europe, N: North America, A: Asia, F: Africa, S: South America.

Analysing the projection obtained with PCA (from Figs. 2, 3, 4 and 5) it can be concluded that PCA shows a clear internal structure in the dataset by identifying the consumption of electricity and natural gas as the most relevant features according to the data ordering, as stated in Fig. 5. On the other hand, different data groups can be identified in the projections, the samples located on the left of the projections are from big factories that consume more electricity, and the samples at the bottom are from big

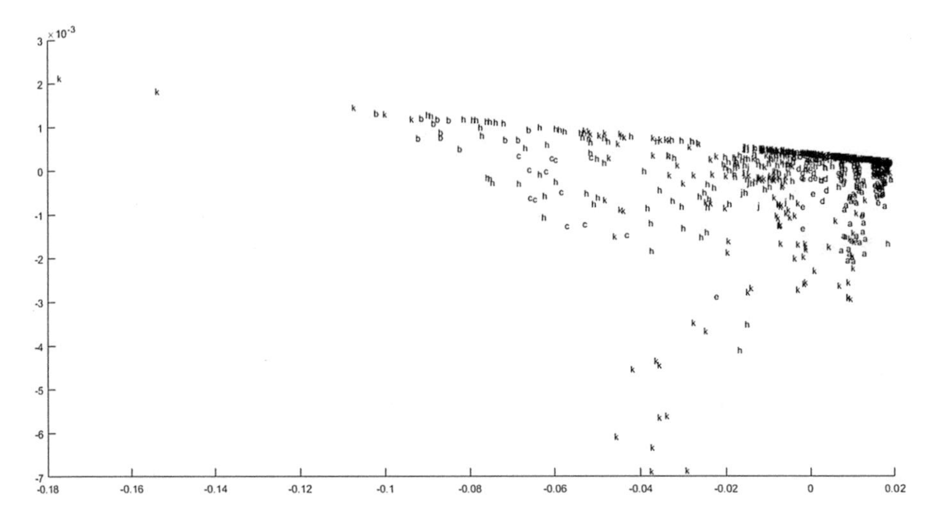

Fig. 3. PCA Projection–Technology.

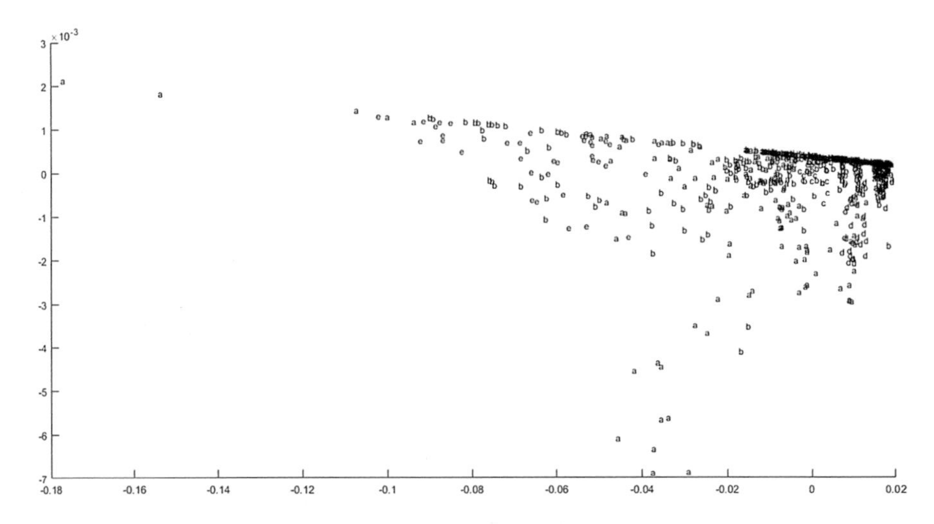

Fig. 4. PCA Projection–Business Unit.

factories that consume more natural gas. These factories, as can be seen in Fig. 2, are from different continents; as a result, they use different types of energy.

CMLHL is the second technique that has been applied. In the following figures CMLHL results are shown, where data are depicted according to four criteria as in the case of PCA (continent, technology, business unit and month of the year).

An analysis of the results obtained with the CMLHL model (from Figs. 6, 7, 8 and 9) leads to the conclusion that this method reveals the dataset structure in a clearer way than PCA.

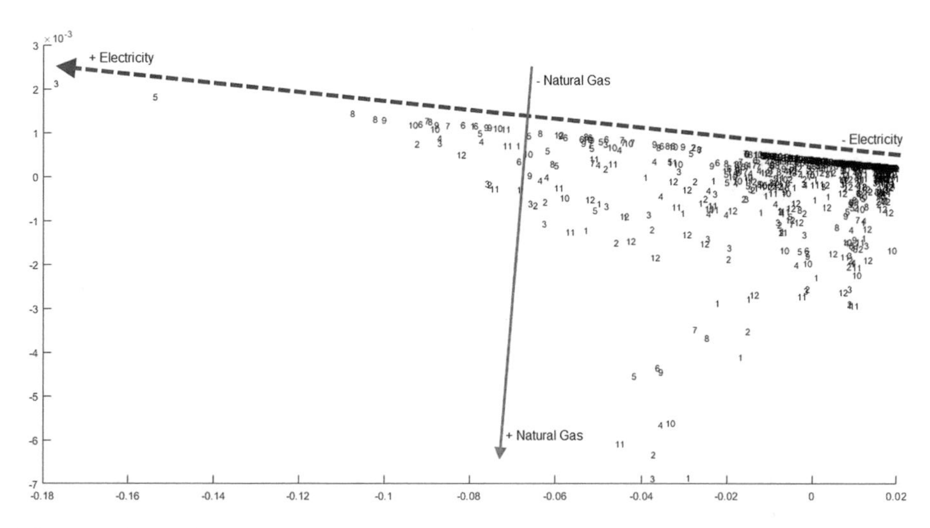

Fig. 5. PCA Projection–Month: one number per each month in the year (1-January, 2-February…)

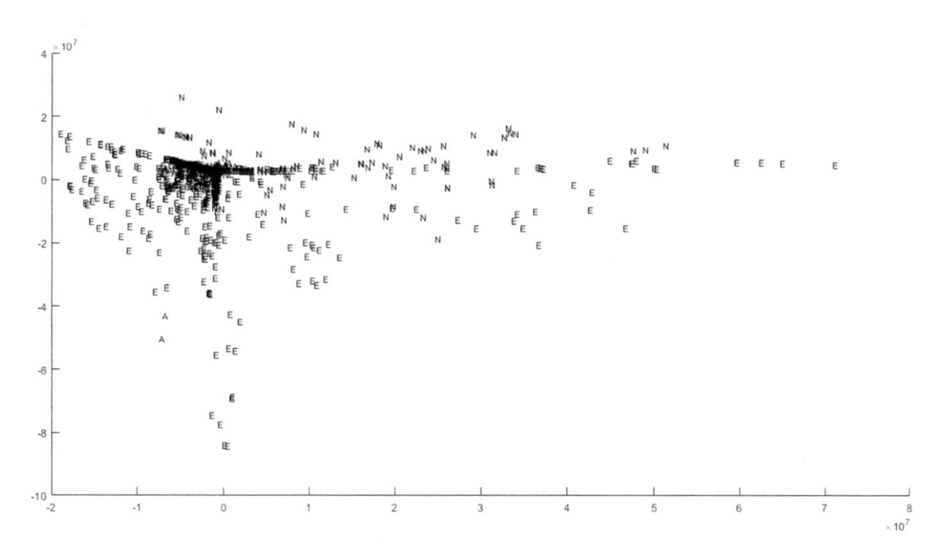

Fig. 6. CMLHL Projection–Continent: E: Europe, N: North America, A: Asia, F: Africa, S: South America

In Fig. 6, it is clearly identified the grouping according to the continent where the factory is located. There are two main groups in this projection; one gathering factories from Europe, Asia and South America and another one those from North America. In Fig. 7, different groups of data can be identified, there are several clusters were the technology '*k*' is clearly identified. In Fig. 8, groups by business unit are visibly represented, the different business units usually group similar technologies, and

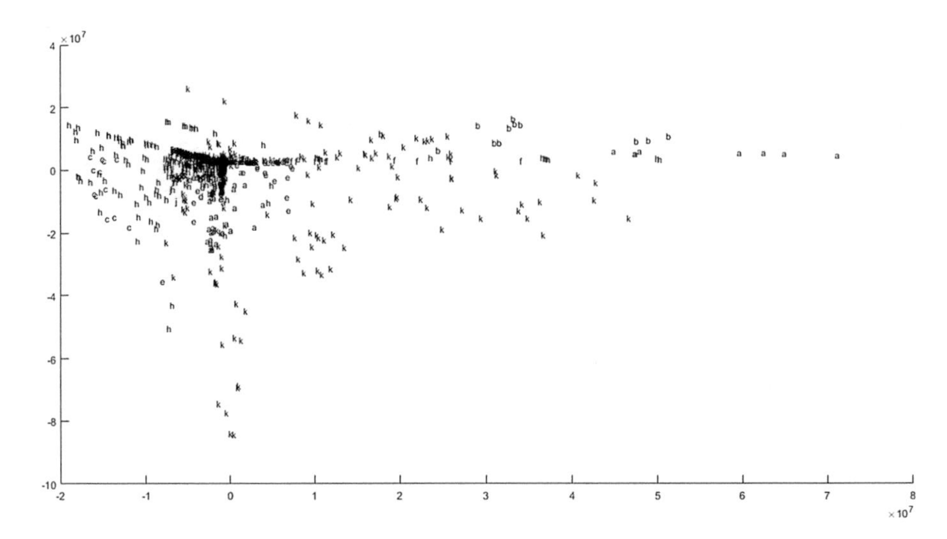

Fig. 7. CMLHL Projection–Technology.

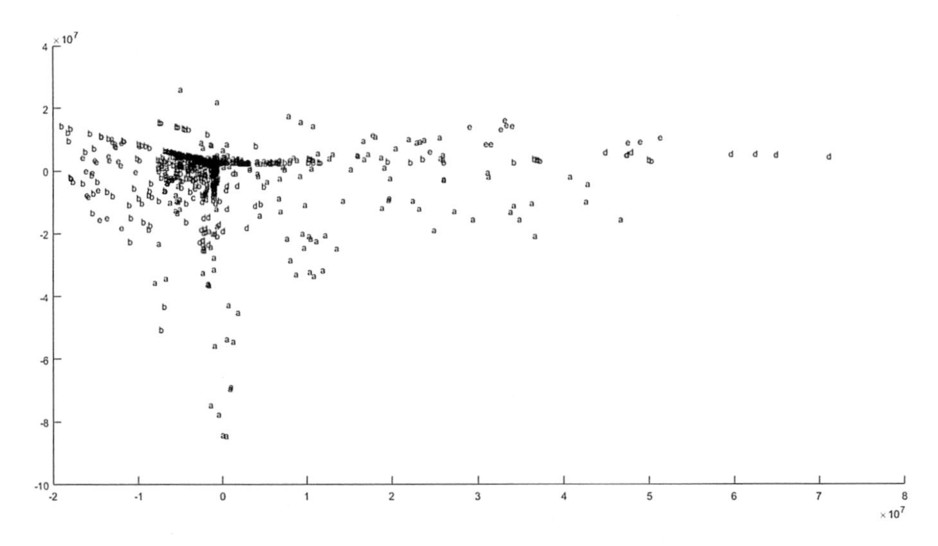

Fig. 8. CMLHL Projection–Business Unit.

because of that the projection for technologies are very similar to the business unit one; and in Fig. 9, it can be concluded that there are small clusters grouped by season, points from winter are closer between them and in the same way summer points. One reason for the difference between summer and winter is that in winter it is necessary to use heating. Also some factories in summer have vacations period.

In this study PCA and CMLHL are applied in order to bear out that the dataset is representative enough. By comparing two different neural projection techniques, more evidences supporting conclusions are provided.

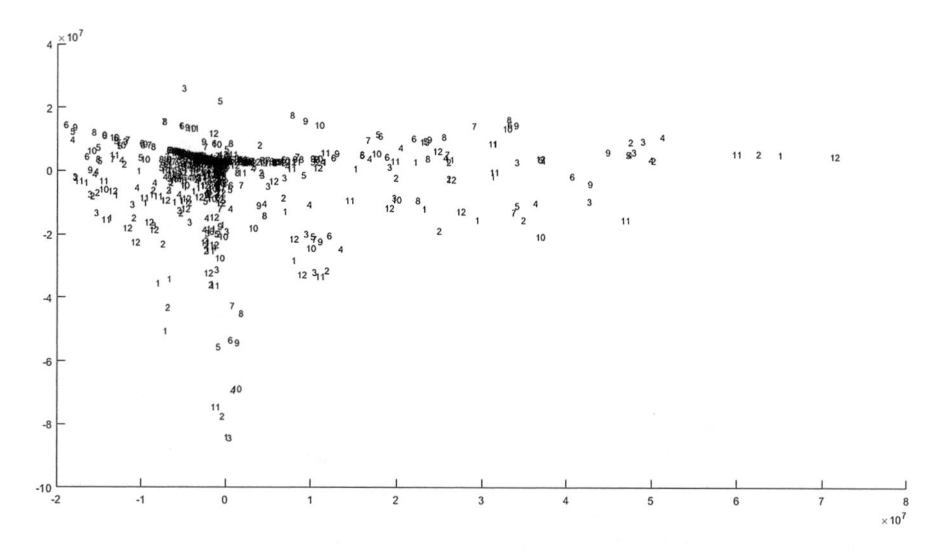

Fig. 9. CMLHL Projection–Month: one number per each month in the year

An analysis of the results obtained with the CMLHL model leads to the conclusion that this technique shows several different clusters in a clearer and sparser way than PCA; the points from the same factory or locations are grouped together.

5 Conclusions and Futures Lines of Work

Once the dataset has been analysed, it can be concluded that a certain structure has been revealed and thus, it can be seen as a representative and informative dataset. PCA projections show samples ordered by electricity and natural gas consumption. In addition, with CMLHL it is obtained more information because it is observed that the data are grouped by continent, technology, business unit and there are even clusters by factory.

These results are relevant for the Grupo Antolin company. The obtained projections show certain degree of clustering, either by type of manufacturing technology, business unit, etc., which will allow the company to make decisions, taking into account, for example, the number of pieces produced in each factory. The company could study which kind of energy used in certain factories is the most economic or the most environment friendly or even if one factory has anomalous consumptions when compared with another similar factories.

This is a first study to analyse if the data have a structure or information that could be relevant. In the future, this study will go on with data from additional years and also it would be possible to work with more variables such as CO2 emissions, hazardous waste, number of pieces produced, etc. With a future analysis the company would be able to detect what topics could be developed in order to improve its commitment to be environmentally-friendly.

Acknowledgments. The authors would like to thank the vehicle interior manufacturer, Grupo Antolin, for its collaboration in this research.

References

1. San José, R., Pérez, J.L., González, R.M.: An operational realtime air quality modelling system for industrial plants. Environ. Model Softw. **22**, 297–307 (2007)
2. Jolliffe, I.T.: Principal Component Analysis, 2nd edn. Springer, New York (2002)
3. Platon, R., Dehkordi, V.R., Martel, J.: Hourly prediction of a building's electricity consumption using case-based reasoning, artificial neural networks and principal component analysis. Energy Build. **92**, 10–18 (2015)
4. Corchado, E., Herrero, Á.: Neural visualization of network traffic data for intrusion detection. Appl. Soft Comput. **11**(2), 2042–2056 (2011)
5. Grupo Antolin. Accessed 03 Apr 2018. http://www.grupoantolin.com
6. Corchado, E., Baruque, B.: Wevos-visom: An ensemble summarization algorithm for enhanced data visualization. Neurocomputing **75**(1), 171–184 (2012)
7. Araya, D.B., Grolinger, K., ElYamany, H.F., Capretz, M.A.M., Bitsuamlak, G.: An ensemble learning framework for anomaly detection in building energy consumption. Energy Build. **144**, 191–206 (2017)
8. Garg, H., Rani, M., Sharma, S.P.: An approach for analyzing the reliability of industrial systems using soft-computing based technique. Expert Syst. Appl. **41**(2), 489–501 (2014)
9. Gitinavard, H., Mousavi, S.M., Vahdani, B.: Soft computing based on hierarchical evaluation approach and criteria interdependencies for energy decision-making problems: a case study. Energy **118**, 556–577 (2017)
10. Oja, E.: Principal components, minor components, and linear neural networks. Neural Netw. **5**, 927–935 (1992)
11. Oja, E.: Neural networks, principal components, and subspaces. Int. J. Neural Syst. **1**, 61–68 (1989)
12. Friedman, J.: Exploratory projection pursuit. J. Am. Stat. Assoc. **82**(397), 249–266 (1987)
13. Corchado, E., MacDonald, D., Fyfe, C.: Maximum and minimum likelihood Hebbian learning for exploratory projection pursuit. Data Min. Knowl. Disc. **8**(3), 203–225 (2004)
14. Herrero, Á., Corchado, E., Sáiz Bárcena, L., Abraham, A.: DIPKIP: a connectionist knowledge management system to identify knowledge deficits in practical cases. Computational Intelligence **26**(1), 26–56 (2010)
15. Krömer, P., Corchado, E., Snášel, V., Platos, J., García-Hernandez, L.: Neural PCA and Maximum Likelihood Hebbian Learning on the GPU. ICANN **2**, 132–139 (2012)
16. Corchado, E., Han, Y., Fyfe, C.: Structuring global responses of local filters using lateral connections. J. Exp. Theor. Artif. Intell. **15**(4), 473–487 (2003)
17. Seung, H., Socci, N., Lee, D.: The Rectified Gaussian Distribution. Adv. Neural. Inf. Process. Syst. **10**, 350–356 (1998)
18. Vera, V., Sedano, J., Corchado, E., Redondo, R., Hernando, B., Camara, M., Laham, A., Garcia, A.E.: A hybrid system for dental milling parameters optimisation. In: 6th International Conference on Hybrid Artificial Intelligence Systems. Wroclaw, Poland HAIS 2011, Part II, LNAI 6679, 2011, pp. 437–446 (2011)

A MAS Architecture for a Project Scheduling Problem with Operation Dependant Setup Times

Daniel Mota(✉), Constantino Martins, João Carneiro, Diogo Martinho, Luís Conceição, Ana Almeida, Isabel Praça, and Goreti Marreiros

GECAD – Research Group on Intelligent Engineering and Computing for Advanced Innovation and Development, Polytechnic of Porto, Porto, Portugal
{drddm,acm}@isep.ipp.pt

Abstract. The manufacturing industry is an ever-changing environment with companies facing increasing external and internal challenges, such as economic crises, technological development and global competition. These challenges create the need for companies to constantly adapt as the environment around them changes. As such, companies are adopting a more proactive approach to manufacturing rather than the usual reactive process, by taking advantage of the ongoing move towards automation and system interconnectivity in the context of Industry 4.0.

In this work, we propose an agent-based architecture that presents a solution to project scheduling problems, with operation dependent setup time that is resource, material and human-resource constrained.

Keywords: Industry 4.0 · Resource constrained project scheduling problem Multi-agent systems · Jade

1 Introduction

In the context of what has been identified as the 4th industrial revolution, with the advent of a new paradigm that revolves around the digitalization in the industrial sector, it's important to supply existing companies with the ability to survive the emergence of the denominated Industry 4.0. This new paradigm combines existing production systems with advanced information and communication technologies, promoting the integration between human and machine resources to achieve an overall improved manufacturing performance [1].

The implementation of the underlying principles of Industry 4.0 relies on incorporating appropriate sensorial mechanisms that intelligently perform acquisition, transmission and analysis of data at the industry's equipment level. Extracting knowledge from that data provides an appropriate support to reach the best possible industrial decisions [2].

M. Graña et al. (Eds.): SOCO'18-CISIS'18-ICEUTE'18, AISC 771, pp. 177–186, 2019.
https://doi.org/10.1007/978-3-319-94120-2_17

Efforts in the technological and scientific communities have been made to face the increasing challenges that emerge, as Industry 4.0 takes a center place in the manufacturing industry. It is important to improve the automation levels of the production system, but also to adopt a predictive perspective in the field of industrial maintenance [3].

The Resource Constrained Project Scheduling Problem (RCPSP) is a recurring subject within operational research and optimization, particularly in scheduling theory [4]. The RCPSP consists of finding a schedule of activities with limited resource availability, known duration, and precedence constraints that is optimized regarding certain objectives [5]. Although advances made in the 50's presented solutions to production scheduling, the study focused mainly on time as a constraint to the problem and resource limitations were not considered [6]. Resource constraints are identified as being the main problem when considering project planning in the real world [6].

In recent years, various RCPSP techniques have been developed and applied in the manufacturing industry, several of which have employed Multi Agent Systems (MAS). A MAS can separate a whole complex problem into smaller simpler problems and provide solutions to the problem based on the ability to solve the individual problems [7]. MAS supports decision making and the development of personalized agents representing the knowledge of experts in different areas to create a cohesive intelligent system [8]. The performance of MAS in the context of RCPSP have been greatly enhanced with the continuous improvement of multi-agent optimization models. These models combine the traditional project scheduling methods with the learning and collaborative ability of the agents to improve the quality of project scheduling [7].

This paper describes the research work that is being developed in the scope of the NIS (Núcleo de Investigação SISTRADE – SISTRADE Research Group). The project focuses on researching and developing solutions that have added value in the context of the new paradigm, Industry 4.0. In terms of specific objectives, it focuses on three primary research areas: (1) development of intelligent algorithms that are applicable in the context of industrial process scheduling; (2) research in the field of augmented reality in an industrial context; (3) development of usability solutions applied in the context of Industry 4.0. The paper focuses on the first primary objective, as it tackles the project scheduling problem in the manufacturing industry.

This document is organized as follows: (2) Reviewed Literature, presents an overview of existing work in the fields; (3) Architecture, presenting and exploring the details of the approach presented in the paper; (4) Conclusion, presents a summarized view of the document and other considerations.

2 Reviewed Literature

This section presents an overview on the subjects that have relevance to the problem that is being addressed and the resulting solution.

2.1 Scheduling

Scheduling deals with the allocation of resources over time to perform a collection of tasks [9]. Scheduling problems arise in domains as diverse as manufacturing, computer processing, transportation, education, etc. A resource can represent a machine that performs tasks, but it can also represent other entities such as consumables or even human skills [10]. Scheduling problems fit either in Constraint Satisfaction Problem (CSP) or Constraint Optimization Problem (COP) categories [10]. In a CSP reaching a solution that bypasses the problem constraints is considered as satisfactory [10]. A COP is a scheduling problem that requires providing an optimized output regarding a determined set of variables but still satisfying the problem constraints [10].

Dispatching rules (DR) are often used in project scheduling and provides solutions for CSP. A DR is a heuristic that consists of processing the jobs that are waiting on a machine based on a priority mechanism. Whenever a job is finished, the next one is selected based on the assigned priority [11]. Some dispatching example rules are:(1) Shortest Processing Time; (2) First-Come-First-Served; (3) Earliest Due Date; (4) Earliest Completion Time. According to Michael Pinedo, DR can be fast and simple to implement and may find a reasonably good solution in a relatively short time but can sometimes produce extremely bad solutions (often a combination of more than one DR can perform significantly better) [11].

For COP solutions, various exact optimization algorithms have been applied to the field but are faced with an exponential effort as the problem size increases and may not provide a near-optimal solution in an acceptable time frame [12]. That is why soft computing methods, such as meta-heuristics are often used. Meta-heuristics are high-level strategies that help other heuristics in the search for admissible solutions in domains where the problem is complex [13]. Very popular meta-heuristic algorithms include: (1) simulated annealing; (2) tabu search; (3) genetic algorithms; (4) scatter search; (5) filter-and-fan local search.

2.2 MAS Approaches to Scheduling

In the mid 90's, the Gartner Group used the term ERP (Enterprise Resource Planning) to establish a difference between the former generation of resource planning (MRP - Manufacturing Resource Planning) [14]. One of the main purposes of an ERP system consists in automating business processes [15]. In the field of automation and production it is common practice to map physical resources, such as tools, machines, robots, and logical objects such as schedulers and job orders as agents in a MAS. The study of MAS has its origin in the field of distributed artificial intelligence and may be defined as a population of agents that have specific social behaviors [16]. According to Lesser a MAS consists of an environment that is inhabited by several agents that interact cooperatively in order to overcome the restrictions imposed by their surroundings. Adopting cooperative mechanisms imbues the MAS of capabilities that are superior to the sum of the agent's individual skills. Using adequate distribution algorithms, the agents may be able to make their own decisions regarding resource consumption and coordination [17]. That superior performance is achieved from the dynamic interactions that occur naturally between the agents inhabiting that environment [18]. Gu et al. describes a MAS that uses a bidding-based approach to the integration between real-time scheduling, process planning and computer-aided design [19]. The Magenta Technology corporation developed a MAS for scheduling and optimizing production that resulted in improved service level, reduced delivery time and cost and improved their production system's scalability, performance and reliability [20]. Shpilevoy et al. presents an adaptive resource scheduling based on multi-agent technology. The solution presents a real-time, event-driven reaction to changes in the factory environment, resource control, allocation, scheduling and optimization. The authors highlight some of the observed results: (1) increase in productivity (between 10–15%); (2) reduced task scheduling, coordination and monitoring efforts; (3) improved resource efficiency (by more than 15%); (4) reduced response times to unexpected events; and (5) increased the number of orders executed within the defined due time (between 15–30%) [21]. Martin et al. proposes an agent-based distributed framework where each agent implements a different metaheuristic to solve the scheduling problem. In the search for the solution, the agents in the system continuously adapt using a direct cooperative protocol based on reinforced learning and pattern matching [22].

3 Architecture

In this section we present a proposal of an agent architecture for a resource/material/workforce constrained project scheduling problem with operation-dependent setup times. The problem consists of production orders (PO) that are to be scheduled. One PO results in the manufacture of n items of a certain product. The manufacture of a product requires a predetermined number of operations that need to be accomplished by a resource. Each resource can perform a determined number of operations. While there might be a relationship of precedence between some operations for the same product, other operations may be processed simultaneously. Given a PO with 5 operations, the weighted precedence graph is shown in Fig. 1. The vertices identify the operation and the edges between them are the time consumed between operations. In Fig. 1 it is possible to identify two separate operation paths. Path A = $\{1, 2, 4, 5\}$ and path B = $\{1, 3, 5\}$ are executed concurrently. The output from operation 2 is the input required for operation 4 to be executed. Operation 5 requires the outputs from Operation 3 and 4. Although the execution of A takes 11 units of time to reach from operation 1 to 5, operation 5 must wait until the output from 3 is available. These are restraints to the system that needs to be addressed because the machine performing operation 5 may be scheduled to another PO to minimize machine idle time.

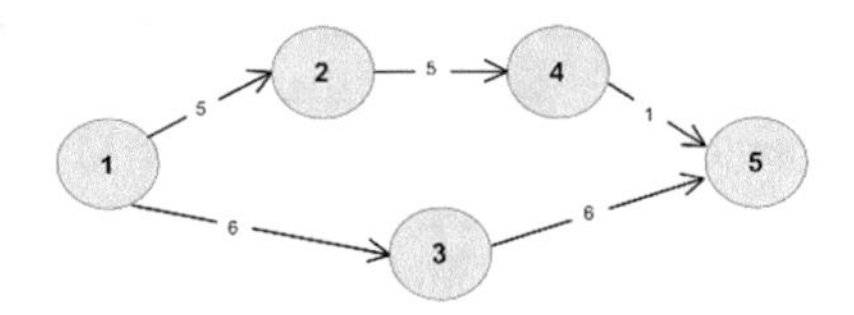

Fig. 1. Weighted precedence graph

The proposed solution presents a MAS with a hierarchical structure where each layer of the hierarchy has its own local responsibilities. The decision process flows from the bottom layer to the upper layer with a chain of responsibilities that produces

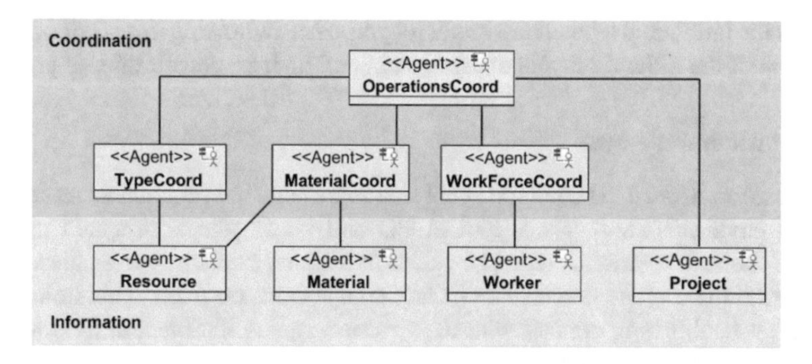

Fig. 2. Agent architecture

one or many solutions to be chosen when the top of the hierarchy is reached. The architecture diagram that depicts the relationship between the agents within the same layer, as well as the relationship between agents of the adjoining layer, is represented in Fig. 2.

The architecture contains two layers with different responsibilities: (1) the **Coordination** layer encapsulates the agents that coordinate specific groups of agents, which belong to the information layer. The Coordination layer is where the production objectives are set and the final decision is ultimately reached; (2) in the **Information** layer reside the agents that represent the information that constrains the production system and whose schedules need to be defined/optimized.

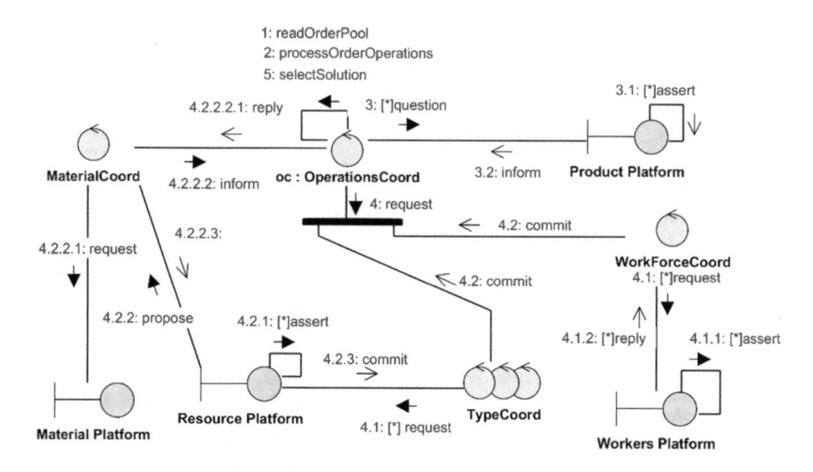

Fig. 3. Agent communication diagram

Figure 3 represents the flow of communication between the agents that are involved in the scheduling process. The boundary symbols that are depicted in the diagram represent different platforms where the information layer agents may be deployed and substitute the individual agents for simplicity purposes. Following this section, a detailed description of the different layers of the system and their responsibilities is presented.

3.1 Coordination Layer

The **OperationsCoord (OC)** is the agent that represents the production manager in an industrial environment. It has a pool of PO and knows the cost/benefit metric that should be minimized/maximized. The agent divides the POs into the smallest possible set of operations and the precedence (if any apply) between them. This information is obtained by establishing contact with the **Product** agents stationed in the lower layer. The final decision on which production schedule to adopt belongs to the OC and depends on pre-determined objectives. There may be more than one solution if more than one objective has been defined. Examples of objectives may be: (1) most profitable schedule; (2) fastest schedule; (3) biggest revenue. The OC collaborates with the **MaterialCoord (MC)** agent in the management of materials' stock for the production system's workflow to function without any delay. The same flow of events and responsibilities happen between the OC and the **WorkForceCoord (WFC)**. The WFC, entity that maps to the Human Resources manager of the company, coordinates the **Worker** agents.

The **TypeCoord (TC)** agent coordinates the efforts of all resources that can perform a determined operation and that belong to certain resource types. The system requires as many TC agents as there are operations. At this stage of the decision process, the TC handles the communications from the OC. The OC solicits the work of a TC, which coordinates the behavior of the resources that perform an operation and provides the metric that should be observed when creating the resources' schedules. The TC then creates a collection of all possible combinations of operations of the type it represents, sending that collection and metric down to the Information layer.

3.2 Information Layer

The **Resource** agents that inhabit the lower layer are entities that are mapped directly from resources existing in the physical industrial facility, such as machines, tools and other equipment that may perform specific operations. The knowledge these agents hold, correspond to the constraints that the production system needs to overcome. In the system that is being developed, a resource may be able to perform more than one operation and, as such, may be contacted by more than one TC. If a resource agent that can perform more than one operation receives instructions from one TC, it will ask the other TCs that work with it if it should expect something from them before providing the costs of performing the received operation. This happens because resources only execute one operation at a time, independently of which type of operation it may be. The resource will provide costs regarding all combinations of possible operations. Each resource agent has knowledge regarding the duration required to perform an operation, and the setup time required to change operations. The setup time of changing from operation i to operation j may be different from the time required to change from operation j to operation i. Setup time between operation i that corresponds to one PO to operation i corresponding to another PO is also being considered. The costs provided by the resource will take all that information into consideration and will therefore be included in the decision process up stream.

The resource agents' performance relies on knowledge that is not its responsibility to provide. The resources need to know if the materials, required in the production workflow, will be available when the operation is scheduled to begin production. As such, the resource agents collaborate with the MC agent to achieve an effective and reliable cost assumption. With this collaboration it is also possible to provide suggestions on how to improve production, e.g. if a resource is idle because of inexistent stock of materials at a certain date, it is suggested to produce/order more. This is possible by establishing a relation between the OC and the MC. The MC coordinates the **Material** agents. The material agents provide information on stock, stock location, expected delivery of stock, as well as various features of the material. It's also capable of managing its own stock, notifying the OC whenever the stock is running out.

As previously mentioned, this layer also holds the **Worker** and the **Product** agents.

Each product is responsible for knowing which operations are included in the manufacturing process and the precedence between these operations. It is the information that is required by the OC to identify the operations that are executed during the product manufacturing workflow. The agent also informs, preemptively, whenever a change in workflow is made to the product.

The worker agents represent the human resources that comprise the company's production workforce. The worker agents respond directly and exclusively to the WFC. Their collaboration provides the knowledge regarding the company's work force with information regarding the workers work days, shifts, vacation time, operations that may be performed by the worker, etc. Each worker has its own schedule and contributes to a level of uncertainty in the production planning.

3.3 Implementation

The proposed solution is being implemented using the JADE framework [23]. JADE is an open-source middleware, implemented in Java, that streamlines the process of creating distributed, FIPA-compliant [24] agent systems. Figure 4 depicts the representation of a snapshot of the scheduling for a manufacturing facility. In this example, all coordinator agents are running in the same platform, but JADE allows for the deployment of several platforms and permits agent mobility, as per FIPA specifications, even if only within the same platform [25], and other agents are distributed in different platforms.

A possible scenario that could result in the depicted representation of the system is when a printing company, that has the necessary resources to print different types of magazines, newspapers and books, receives a production order for the printing of 100 books that is due in a week. Let's consider the company has 3 resources, 3 workers capable of handling each resource and the required material to produce the intended order. The operations required in the printing process of a book are: (1) printing the interior pages; (2) printing the cover; (3) cutting the interior pages; (4) cutting the cover; (5) Gluing the book. Operation (1) precedes (3), operation (2) precedes (4) and operation (5) may only be started when operation (3) and (4) are finished. Operations (1) and (3) may be done in parallel with operations (2) and (4). Since each Resource

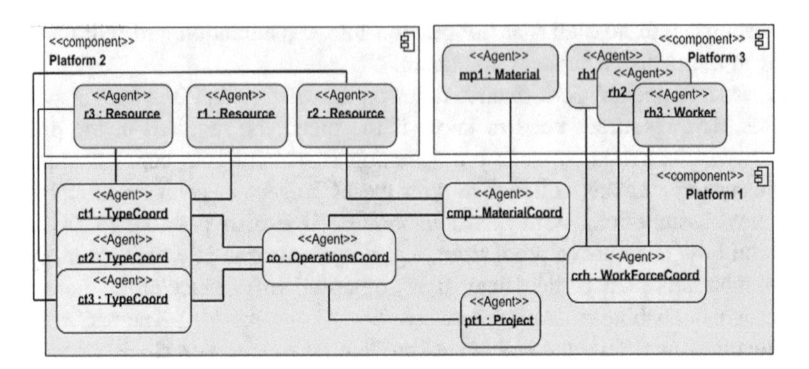

Fig. 4. System example

agent can present the costs for performing various combinations of operations, the OperationsCoord agent will rely on that expertise to receive the necessary individual solutions for each resource and select the individual solutions that also result in a solution to the overall problem that satisfies precedence relations, restrictions and results in the most maximized/minimized benefit/cost.

4 Conclusion

In this paper we presented an agent-based architecture for a project scheduling problem that has multiple constraints, such as resource, materials and workforce. The MAS approach was considered as the appropriate solution as we can represent the individual entities, both physical and logical, and the distributed nature of the agents easily supports the deconstruction of the problem into smaller, more contained issues that are then directed towards more specialized agents. The MAS system also allows the deployment of the agents in the exact physical location of the entity it represents, resulting in a real-time local update of the resource/material status.

The next stage of this project is to finish implementation of the MAS and includes benchmarking the solution against other existing project scheduling solutions. Deploying the system in an industrial environment to test the distributed nature of the system is also in the future of our research.

The NIS project also includes an approach to augmented reality and usability applied in the context of Industry 4.0. These topics will drive the focus of our research as soon as the project scheduling problem reaches satisfactory results.

Acknowledgments. This work was supported by NIS Project (ANI|P2020 21958) and has received funding from FEDER Funds through P2020 program and from National Funds through FCT – Fundação para a Ciência e a Tecnologia (Portuguese Foundation for Science and Technology) under the project UID/EEA/00760/2013.

References

1. Lasi, H., Fettke, P., Kemper, H.G., Feld, T., Hoffmann, M.: Industry 4.0. Bus. Inf. Syst. Eng. **6**(4), 239–242 (2014)
2. Rudtsch, V., Gausemeier, J., Gesing, J., Mittag, T., Peter, S.: Pattern-based business model development for cyber-physical production systems. Procedia CIRP **4**, 313–319 (2014)
3. Kolberg, D., Zühlke, D.: Lean automation enabled by I4.0. IFAC-PapersOnLine **48**(3), 1870–1875 (2015)
4. Krichen, S., Chaouachi, J.: The resource constrained project scheduling problem. Graph-Relat. Optim. Decis. Support Syst. 69–82 (2014). John Wiley & Sons. https://onlinelibrary.wiley.com/doi/book/10.1002/9781118984260. ISBN 9781848217430
5. Ling, W., Huan-yu, Z., Xiao-long, Z.: Survey on Resource-Constrained Project Scheduling Under Uncertainty. Tsinghua Tongfang Knowledge Network Technology Co., Ltd, Beijing (2014)
6. Abdolshah, M.: A review of RCPSP approaches and solutions. Int. Trans. J. Eng. Manag. Appl. Sci. Technol. **5**(4), 253–286 (2014)
7. Ren, H., Wang, Y.: A survey of multi-agent methods for solving resource constrained project scheduling problems. In: International Conference on Management and Service Science, Wuhan (2011)
8. Wooldridge, M., Jennings, N.R.: Agent theories, architectures, and languages: a survey. In: International Workshop on Agent Theories, Architectures, and Languages, pp. 1–39. Springer (1994)
9. French, S.: Sequencing and Scheduling: An Introduction to the Mathematics of the Job-shop. Wiley, Ottawa (1982)
10. Sadeh, N.: Look-Ahead Techniques for Micro-Opportunistic Job Shop Scheduling, Piitsburg (1991)
11. Pinedo, M.: Scheduling: Theory, Algorithms, and Systems. Springer (2012)
12. Myszkowski, P., Skowronski, M., Podlodowski, L.: Novel heuristic solutions for multi–skill RCPSP. In: Federated Conference on Computer Science and Information Systems (2013)
13. Reis, L.P.: Coordination in Multi Agent Systems: Applications in College Management and Robotic Football. Faculdade de Engenharia da Universidade do Porto, Porto (2003)
14. Dahlén, C., Elfsson, J.: An Analysis of the Current and Future ERP Market. Kungl Tekniska Hogskolan, Stockholm (1999)
15. Ahituv, N., Neumann, S., Zviran, M.: A system development methodology for ERP systems. J. Comput. Inf. Syst. **42**, 56–67 (2002)
16. Durfee, E.H., Rosenschein, J.: Distributed problem solving and multi-agent systems: Comparisons and examples. AAAI Technical report, pp. 52–62 (1994)
17. Leitão, P.: Agent-based distributed manufacturing control: a state-of-the-art survey. Eng. Appl. Artif. Intell. **22**, 979–991 (2009)
18. Lesser, V.R.: Cooperative multiagent systems: a personal view of the state of the art. IEEE Trans. Knowl. Data Eng. **11**, 133–142 (1999)
19. Gu, P., Balasubramanian, S., Norrie, D.: Bidding based process planning and scheduling in MAS. Comput. Ind. Eng. **32**, 477–496 (1997)
20. Andreev, M., Ivaschenko, A., Skobelev, P., Tsarev, A.: A multi-agent platform design for adaptive networks of intelligent production schedulers. IFAC Proc. Vol. **43**, 78–83 (2010)
21. Shpilevoy, V., Shishov, A.: Multi-agent system “Smart Factory”. In: 11th IFAC Workshop on Intelligent Manufacturing Systems, S. Paulo (2013)
22. Martin, S., Ouelhadj, D., Deullens, P., Ozcan, E., Juan, A., Burke, E.: A multi agent based cooperative approach to scheduling and routing. Eur. J. Oper. Res. **254**, 169–178 (2016)

23. Bellifemine, F., Caire, G.: JADE administrator's guide. 08 April 2010. http://jade.tilab.com/doc/administratorsguide.pdf. Accessed 22 Feb 2018
24. IEEE FIPA: The Foundation for Intelligent Physical Agents. http://www.fipa.org. Accessed 20 Jan 2018
25. Vaucher, J., Ncho, A.: JADE tutorial and primer. April 2004. https://www.iro.umontreal.ca/~vaucher/Agents/Jade/Mobility.html. Accessed Feb 2018

Collision Detector for Industrial Robot Manipulators

C. H. Rodriguez-Garavito[1], Alvaro A. Patiño-Forero[1(✉)], and G. A. Camacho-Munoz[2]

[1] Automation Engineering, La Salle University, Carrera 2 # 10-70, Bogotá, Colombia
{cerodriguez,alapatino}@unisalle.edu.co
[2] Industrial Engineering, La Salle University, Philadelphia, USA
gacamacho@unisalle.edu.co

Abstract. The increasingly complex tasks require an enormous effort in path planning within dynamic environments. This paper presents a efficient method for detecting collisions between a robot and its environment in order to prevent dangerous maneuvers. Our methods is based upon the transformation of each robot link and the environment in a set of bounding boxes. The aim of this kind of prismatic approximation is to detect a collision between objects in the workspace by testing collision between boxes from different objects. The computational cost of this approach has been tested in simulations, thus we have set up our environment with a HP20D robot and an obstacle, both represented by their corresponding chain of bounding boxes. The experiment implies to move the robot from an initial position, on the right of the obstacle, to a final position, on the left side of the obstacle, along a straight-line trajectory. The probe enabled us to check the correct behavior of a collision detector in a real situation.

Keywords: Collision avoidance · Robot manipulation
Bounding boxes

1 Introduction

The industrialization as a basis to sustain a consumption society requires automation of manufacturing techniques and methods, which in turn use manipulator robots to solve complex tasks in ever-dynamic environments.

Manipulator robots are highly popular tools in industry, for their ability to perform complex, efficient and highly accurate movements. One of the problems associated to this technology is the so-called autonomous planning. The problem consists of calculating a continuous trajectory for a set of possibly interconnected objects, with the purpose of enabling them to move from one position to the other, thus avoiding colliding with other static or mobile objects. In the case of a fixed standar industrial robot, the problem can be formulated as the calculation of a trajectory for the robot arm, not interfering with existing items

M. Graña et al. (Eds.): SOCO'18-CISIS'18-ICEUTE'18, AISC 771, pp. 187–196, 2019.
https://doi.org/10.1007/978-3-319-94120-2_18

in the working space. The literature has reported several efficient algorithms to solve this problem; those can be grouped in four categories: (i) Collision based on convex polytopes [4,5] are strategies founded on the distance between two convex polytopes. (ii) Collision based on bounding volumes hierarchies (Bounding Volumes Hierarchy-BVH) [6–8] follows the definition of geometric structures to represent objects in the robot working area, the most popular structures are boxes and spheres. For irregular shapes the use of deformable models is preferred as kd-trees and octrees [3]. (iii) Broad Phase Collision Detection (BPCD) is the name given to the collision detection techniques concerned with the question: which objects have a high collision possibility?. In order to prevent complexity $O(n^2)$, the BPCD technique is responsible for ruling out pairs of objects that are far apart from each other [3]. (iv) Point Clouds Collision Detection are algorithms that arises for the wide use of time-of-flight cameras as the main sensor in motion planning applications. Those applications use high update frequencies and deal with high volume of data, some level of noise and uncertainty that prevent a fully environment representation. In [9] they propose an algorithm that deals with those applications in two phases: converting from point cloud to octree, processing of collision and distance queries based on two techniques (a not-initialization and a dynamic AABB initialization).

This paper provides a method different from the so-called Ray-Tracing [10], for the collision calculation between two prismatic objects, the spatial definition of which results from the adjustment of their volume to the original solid. Therefore, this method can be classified within the group known as Bounding Volume Hierarchy-BVH. This proposal is aimed at obtaining a collision detector that can be used by algorithms of automatic trajectory planning at a reasonable computational cost.

Section 2 presents the strategy used to transform the structure of the working environment objects to bounding boxes, wherein an HP20D type Motoman manipulator robot interacts with other objects, which is in this case a repository of box type pieces. Section 3 below, describes systematically geometric concepts that allow to infer the objects collision through the intersection of planes in the space. Section 4 further makes the validation of collisions detection, wherein points are detected showing the occurrence of collision for a known trajectory. Finally, in the Conclusions section, a summary of the development is presented.

2 Environment Modelling

Objects making up the working environment: HP20D Motoman manipulator robot, box and floor, are built as sets of rectangles related to each other by homogeneous transformation matrices, as seen in Fig. 1. There each link is represented through a four faces hollowed prism (yellow, red, blue and green) referred to the reference system of its corresponding articulation. In home position, the robot features an alignment of colors by faces, which enables to visually detect articular rotations in a simple way: by colors misalignment between faces of consecutive links. The rotation of each link occurs about Z-axis for the associated reference system.

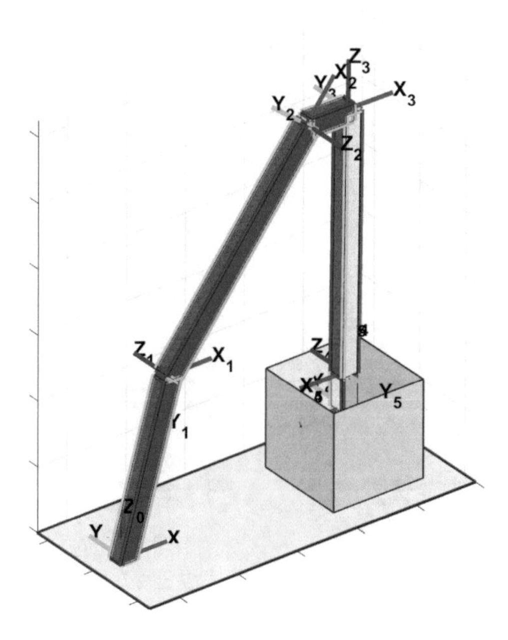

Fig. 1. Representation by rectangles of the HP20D Motoman environment.

2.1 HP20 Motoman Robot Model

The robot kinematic modelling is made using the homogeneous matrixes description by Denavit-Hartenberg [1] for each link and then obtaining the transformation matrix T_i^{i-1}, which represents the coordinate frame fixed at each link i with respect to the coordinate frame of the previous link $i-1$. Their sequential multiplication results in the representation of the robot end-effector relative to the base frame. $T_e^b = T_0^b T_1^0(q_1) T_2^1(q_2) \ldots T_6^5(q_6) T_e^6$; where the sub index e represents end-effector and the super-index b represent the base frame. Parameters defining the direct kinematic modelling were taken from [2] and they are presented on Table 1.

Table 1. Denavit-Hartenberg parameters for HP20D Motoman robot.

Link	d_i in mm	a_i in mm	α_i in degrees	θ_i in degrees
1	0	150	90	θ_1
2	0	760	0	$\theta_2 + 90$
3	0	140	90	θ_3
4	795	0	−90	θ_4
5	0	0	90	$\theta_5 - 90$
6	105	0	0	θ_6

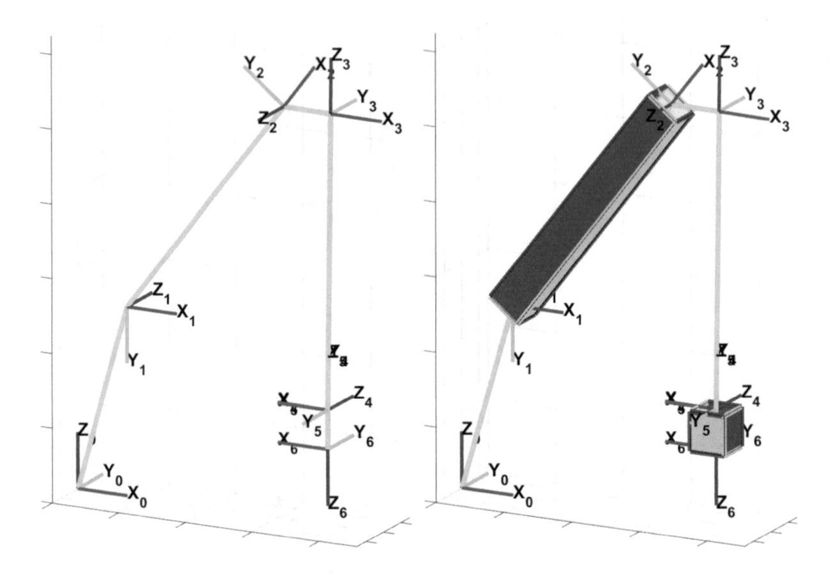

Fig. 2. Representation by wires and prismatic representation of the HP20D Motoman robot

With each parametric definition corresponding to link i-th the homogeneous transformation matrix is built which carries points in its $i-1$ reference system to the i reference system, as seen in Eq. (1).

$$T_i^{i-1} = \begin{bmatrix} c(\theta_i) & -s(\theta_i)c(\alpha_i) & s(\theta_i)s(\alpha_i) & a_i c(\theta_i) \\ s(\theta_i) & c(\theta_i)c(\alpha_i) & -c(\theta_i)s(\alpha_i) & a_i s(\theta_i) \\ 0 & s(\alpha_i) & c(\alpha_i) & d_i \\ 0 & 0 & 0 & 1 \end{bmatrix} \tag{1}$$

The following is a construction of the robot skeleton by joining wires between the T_i^0 y T_{i+1}^0 articulation systems, as seen in Fig. 2.

Finally, each wire is transformed into a four-face prismatic link. Note that only rectangles along the wire are kept and ending lids are not considered, since the kinematic restrictions of each articulation, prevent shocks between consecutive links. Prisms construction from the initial model wires are seen in Fig. 2. For clarification purposes, only links 2 and 5 are drawn over the wires model, as the intention here is to show how the bounding box representation is built from the simple wire representation.

2.2 Floor and Box Modelling

The floor on which the objects are supported is built as a rectangular face affine to thee X-Y plane of the reference system, M. The box seen in Fig. 1 (whose function is to serve as an obstacle in the robot working space) is build as a set of rectangular faces assembled; like a five-faced hollowed prismatic link, where just

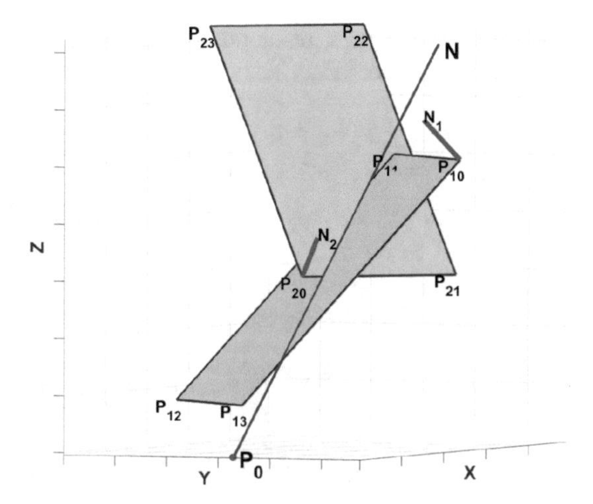

Fig. 3. A 3D representation of two planes intersection.

the volume upper face is non-existent. The box definition is built in solidarity to a C reference system.

3 Collisions Detection

The problem of detecting collisions in the space between objects built from rectangles, is then reduced to detecting the intersection between sets of rectangles. This implies to resolve in the first place the problem of determining when two rectangles are spatially crossed, a matter that will be focused on in the following section.

3.1 Intersection of Rectangles in the Space

The method used to assess the intersection of rectangles in the space starts with the intersection of planes where the rectangles are located. The analysis below requires that each rectangle be identified as the set of i-th of 3D $P_{ij} \in \mathbb{R}^3, i = 1, 2, \ldots, n \; j = 0, 1, 2, 3$ points. Likewise, the plane where each rectangle exists will be parameterized by vectors as a $P_{ij} \equiv (N_i, P_{i0})$ pair, where $N_i = (P_{i1} - P_{i0}) \times (P_{i2} - P_{i0})$, is the normal direction to the plane and P_{i0} is a point of origin, as seen in Fig. 3.

The first step of algorithm consists of finding the L_i line. This line is defined by the intersection between a director vector N and a reference point P_0 over L_i, where planes P_{1j} and P_{2j} are crossed in the space. The definition of the intersection line in the space is presented in Eq. (2) as a function of parameter τ.

$$L_i = (x, y, z) | \begin{cases} x = P_{0x} + N_x \tau \\ y = P_{0y} + N_y \tau \\ z = P_{0z} + N_z \tau \end{cases} \quad (2)$$

All points existing on Line L_i simultaneously comply with the equations of each of the two intersecting planes, Eq. (3).

$$\begin{aligned} xN_{1x} + yN_{1y} + zN_{1z} &= -d_1 \\ xN_{2x} + yN_{2y} + zN_{2z} &= -d_2 \end{aligned} \tag{3}$$

$N_1 \cdot P_{10} = -d_1$ and $N_2 \cdot P_{20} = -d_2$. Now, it is well known, by geometric inspection, that N is normal to N_1 and N_2. Since N is located on the planes and N_1, N_2 are perpendicular to the planes respectively, therefore, $N = N_1 \times N_2$, or by components of N are described in Eq. (4).

$$N = \begin{vmatrix} N_{1y} & N_{1z} \\ N_{2y} & N_{2z} \end{vmatrix} \cdot \boldsymbol{i} - \begin{vmatrix} N_{1x} & N_{1z} \\ N_{2x} & N_{2z} \end{vmatrix} \cdot \boldsymbol{j} + \begin{vmatrix} N_{1x} & N_{1y} \\ N_{2x} & N_{2y} \end{vmatrix} \cdot \boldsymbol{k} \tag{4}$$

Substitution (2) in (3) and simplifying in the resulting expression with the definition (4) and knowing that $N \cdot N_1 = 0$ and $N \cdot N_2 = 0$, an intersection line point is obtained which shall be taken as a reference thereof, P_0, Eq. (5).

$$P_0 = (P_{0x}, P_{0y}, P_{0z}) | \begin{cases} \text{if } P_{0x} = 0 \\ P_{0y} = \frac{d_2 N_{1z} - d_1 N_{2z}}{N_x} \\ P_{0z} = \frac{d_1 N_{2y} - d_2 N_{1y}}{N_x} \end{cases} \tag{5}$$

After the intersection line has been parametrically calculated, the next step consists of defining the conditions of intersection between rectangles. This procedure requires proving two conditions. The first one is revising if any of the two rectangles is not intersected to L_i, as seen in Fig. 4, if so, it is concluded that there is no intersection between rectangles. Otherwise, the second condition is confirmed as both rectangles are in this case, in contact with L_i.

The second condition firstly consists of obtaining the cutting points pc_{ij} between each line making up each rectangle, $Ln_{ij} \equiv (Nln_{ij} = (P_{i(j)} - P_{i(j-1)}), P_{i(j-1)})$ and the intersection line of L_i inter-planes. After the cutting points are defined, the segment S_i is extracted; this segment is made by the two points of cutting, which are located within the rectangle lines. The endings of the S_i segment will be hereinafter identified as S_{iA} and S_{iB}, as seen in Fig. 5. There, $S_{1A} = pc_1$ and $S_{1B} = pc_4$. The intersection of lines in the space occurs in conformity with Eq. (6).

$$\begin{vmatrix} t \\ s \end{vmatrix}^T = \begin{vmatrix} P_{(j-1)x} - P_0x \\ P_{(j-1)y} - P_0y \\ P_{(j-1)z} - P_0z \end{vmatrix}^{-1} \begin{vmatrix} N_x & NLn_{jx} \\ N_y & NLn_{jy} \\ N_z & NLn_{jz} \end{vmatrix} \tag{6}$$

where t is the value assumed by the parameter; this value meets the vector equation requirements of the inter-planes intersection straight line, on the point of cutting with each straight line of the rectangle Ln_j. The variable s is the value corresponding to the straight line on the rectangle. The values of S_{iA} and S_{iB} are found by verifying the values of s of each cutting point, and those comprised between 0 and 1 are filtered, then $S_{iA} = \min_{0 \leq s \leq 1}(pc_{ij})$ and $S_{iB} = \max_{0 \leq s \leq 1}(pc_{ij})$.

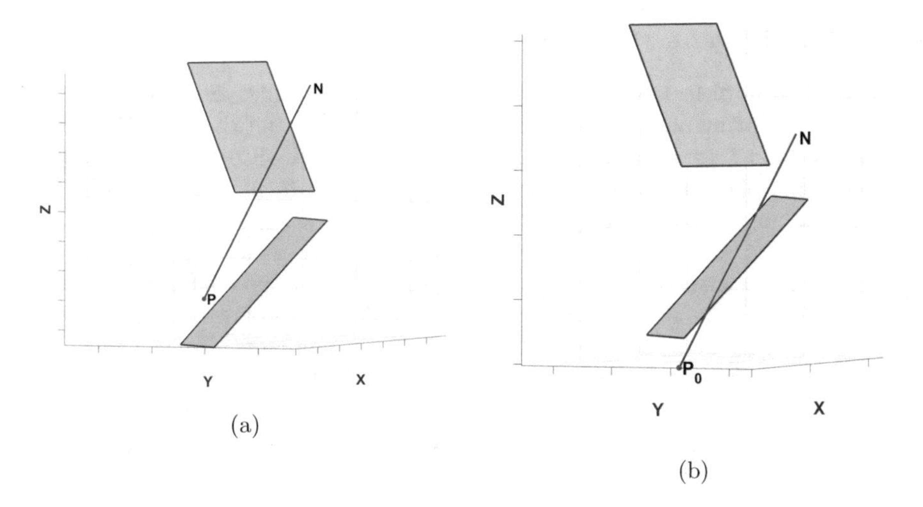

Fig. 4. (a) No intersection, plane 1. (b) No intersection plane 2. First case to evaluate in the rectangles intersection in the space.

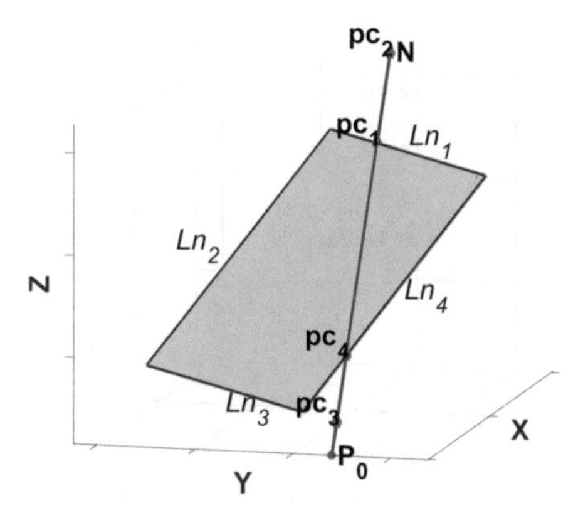

Fig. 5. Cutting points between one rectangle and a line define on the plane of the space where the rectangle exists.

Since there exists a matching between s and t, each segment $[S_{iA}, S_{iB}]$ will have a segment $[s_{iA}, s_{iB}]$ which is a range measured on line Ln_j of the rectangle and $[t_{iA}, t_{iB}]$ is a range measured on the intersection line Ln_i, therefore, to estimate if there is an overlapping between rectangles, it is sufficient to determine if two segments corresponding to two different planes $[t_{1A}, t_{1B}]$ and $[t_{2A}, t_{2B}]$ are separated to each other or not. If there is no segments intersection in t, then there is no intersection between rectangles as per condition 2.

3.2 Algorithm for Prismatic Bodies Intersection

The algorithm initial conditions are three Boolean variables enabling to identify collisions. A principal algorithm is available, which comprises decision structures, which are responsible for calling the functions. Function one is responsible for assessing the kinematic restrictions by articulation, in order to assure the non-contacting between consecutive links. In case of identifying a collision, the FCollisionLink variable becomes true. Therefore, within the decision structure, CollisionFinish assumes the true value as identified in the collision. It assesses if there is any collision between links. These links are distant in more than one position in the kinematic chain. The four faces are compared from i-th, i-th +2, until reaching the end of the chain. Function 2 on its part, is intended to evaluate the intersection among all the rectangles making up the kinematic chain, to all the rectangles making up all other objects of the HP20D robot environment.

Algorithm 1. Detection algorithm

Precondition:
1: $CollisionFinish \leftarrow False$
2: $FCollisionLink \leftarrow False$
3: $FCollisionPlane \leftarrow False$ **Result:** $CollisionFinish$
4: **if** $(CollisionArm)$ **then**
5: $\quad CollisionLink \leftarrow True$
6: $\quad Break;$
7: **else if** $(CollisionEnviroment)$ **then**
8: $\quad CollisionFinish \leftarrow True$
9: **else**
10: $\quad CollisionFinish \leftarrow False$
11: **end ifFunction 1:** $CollisionArm$
12: **for** $(link_i \in kinematicChain)$ **do**
13: $\quad$ **for** $(link_{j=i+2} \in kinematicChain)$ **do**
14: $\quad\quad FCollisionLink \leftarrow FCollisionLink\ or\ ColissionLink(Link_i, Link_j)$
15: $\quad$ **end for**
16: **end for**
17: **Return** $FCollisionLink$ **Function 2:** $CollisionEnviroment$
18: **for** $(link_i \in kinematicChain_n)$ **do**
19: $\quad$ **for** $(plane_j \in Link_i)$ **do**
20: $\quad\quad$ **for** $(Plane_k \in Environment)$ **do**
21: $\quad\quad\quad FCollisionPlane \leftarrow FCollisionPlane\ or\ CollisionPlane(Link_i.Plane_j, Plane_k)$
22: $\quad\quad$ **end for**
23: $\quad$ **end for**
24: **end for**
25: **Return** $FCollisionPlane$

---

4 Tests and Results

To test the performance of the collisions detector, a linear interpolation was made between two positions in the robot workspace, in the initial position to the right of the box and the final position to the left of the box. In each intermediate position, a calling was made to the collisions detector and the result was stored along the entire trajectory. It was made with the purpose of drawing in green the collision free positions and in red, the positions in which the robot was in contact with itself or with the box. The result of the interaction of HP20D robot to the working environment along the defined trajectory can be seen in Fig. 6. There, it is possible to see a green cylinder inside the box, in collision free positions and red on the points of interference with the box.

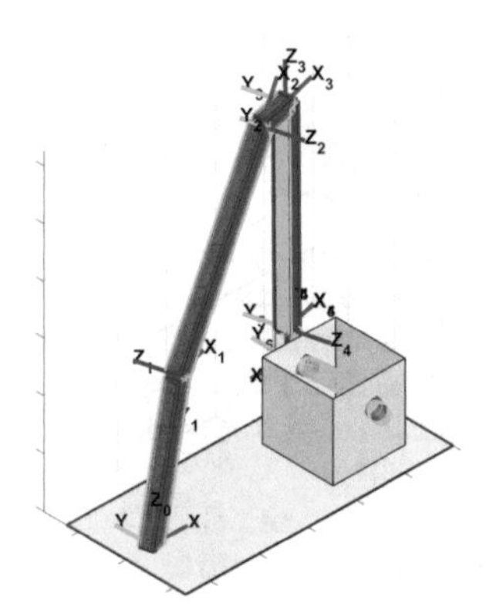

Fig. 6. A path described by HP20D Motorman robot. Free positions if collision are depicted in green, while the spatial interference points are red colored.

5 Conclusions

The collisions detector presented herein is a fast proposal, which solves the trajectories planning problem.

The encapsulation of links of a kinematic chain in the smallest prisms that can contain them is the foundation of the presented method. This foundation combined with planes intersection in the space has been used to resolve the collision detection. During this process, while empty spaces are assumed as part of the robot, such empty spaces are not significant, and instead, a saving of computational costs of the algorithm is obtained.

The method used herein was compared to a method based on the meshing of the objects surface, where each meshing is a sampling of the object, which produces a 3D points cloud. The objects sampling enables to estimate the distance by means of measuring distances between points using metrics, in order to conclude about the presence of collision condition. This alternative requires several orders of magnitude of computational time with respect to the collisions detector based on the intersection of rectangles.

As a further work, it is expected to implement the algorithm proposed using faster platforms using the parallel software. Additionally, this module will be integrated into the development of trajectories planners to solve the task of automated stowing of merchandise; this research project is in process at University of La Salle School of Engineering.

Acknowledgment. The authors acknowledge the collaboration of the Automation Engineering Programs and Industrial Engineering Programs by providing the materials to achieve the development of this work. Likewise, we also appreciate the active participation of the students in the development of the base algorithm.

References

1. Paul, R.P.: Robot manipulators: mathematics, programming, and control: the computer control of robot manipulators (1981)
2. Camacho-Munoz, G.A., Rodriguez, C., Alvarez-Martinez, D.: Modelling the kinematic properties of an industrial manipulator in packing applications. In: Proceedings of the 14th IEEE International Conference on Control and Automation. IEEE Control Systems Society, Washington, DC, USA (2018)
3. Pan, J., Chitta, S., Manocha, D.: FCL: a general purpose library for collision and proximity queries. In: 2012 IEEE International Conference on Robotics and Automation (ICRA). IEEE (2012)
4. Gilbert, E.G., Johnson, D.W., Keerthi, S.S.: A fast procedure for computing the distance between complex objects in three-dimensional space. IEEE J. Robot. Autom. **4**(2), 193–203 (1988)
5. Lin, M.C., Canny, J.F.: A fast algorithm for incremental distance calculation. In: Proceedings of the 1991 IEEE International Conference on Robotics and Automation. IEEE (1991)
6. Gottschalk, S., Lin, M.C., Manocha, D.: OBBTree: a hierarchical structure for rapid interference detection. In: Proceedings of the 23rd Annual Conference on Computer Graphics and Interactive Techniques. ACM (1996)
7. Klosowski, J.T., et al.: Efficient collision detection using bounding volume hierarchies of k-DOPs. IEEE Trans. Vis. Comput. Graph. **4**(1), 21–36 (1998)
8. Larsen, E., et al.: Fast proximity queries with swept sphere volumes. Technical report TR99-018, Department of Computer Science, University of North Carolina (1999)
9. Pan, J., et al.: Real-time collision detection and distance computation on point cloud sensor data. In: 2013 IEEE International Conference on Robotics and Automation (ICRA). IEEE (2013)
10. Rubin, S.M., Whitted, T.: A 3-dimensional representation for fast rendering of complex scenes. In: ACM SIGGRAPH Computer Graphics, vol. 14. No. 3. ACM (1980)

MASPI: A Multi Agent System for Prediction in Industry 4.0 Environment

Inés Sittón Candanedo(✉), Sara Rodríguez González, Fernando De la Prieta, and Angélica González Arrieta

BISITE Digital Innovation Hub, University of Salamanca, Edificio Multiusos I+D+I, 37007 Salamanca, Spain
{isittonc, srg, fer, angelica}@usal.es

Abstract. Prediction is the way to optimize the maintenance task by determining the correct moment to interview, repair or replace equipment which the most difficult decision for companies in Industry 4.0 environment. This research present MASPI. I is a multiagent system based on advantages of virtual organization. The goal of MASPI is to be a reference model for making predictions about data captured by sensors installed in equipment or industrial machines. The capability of MASPI is evaluated by applying it to SCANIA trucks dataset, using machine learnings algorithms to obtain the prediction and compare their accuracy.

Keywords: Virtual organization · Multiagent system · Predictive maintenance Failure diagnosis · Industry 4.0

1 Introduction

Prediction is an important task in Industry 4.0 environment, especially for Condition based Maintenance, also named Predictive Maintenance (PdM), Reliability Centered Maintenance (RCM), Plant Asset Management System (PAM) and Total Productive Maintenance (TPM), because this strategy reduce maintenance cost by applying actions only when machine failure occurs. Predictive Maintenance (PdM) will be used in this research as a concept to measure the condition of the equipment and establish if it will fail in a future period [9].

In Condition based maintenance or Predictive Maintenance (PdM) there exists major elements: sensors, signal converters and processors. Sensors are installed to the equipment to evaluate selected physical parameters such as temperature, pressure or vibration. Signal converters change the sensor signals into an electrical or digital form. Finally, processors provide the user with the data and strategies to make decisions [9, 19].

Predictive maintenance in Industry 4.0 environment is a difficult problem due to uncertainties involved in failure to detection the following: large volume of data, prediction accuracy, necessity to process the data and make decision in real time.

In this problem, the multiagent approach can to be an appropriate method to help reduce the complexity of the predictive maintenance and increase the prediction accuracy by creating autonomous agents, which can model independent subtasks that together comprise the overall goal.

M. Graña et al. (Eds.): SOCO'18-CISIS'18-ICEUTE'18, AISC 771, pp. 197–206, 2019.
https://doi.org/10.1007/978-3-319-94120-2_19

In recent decades, some research has been done on how to use multiagent systems in prediction problems, such as: (i) Wu *et al.,* in [22] used a multiagent system based on neural network model for failure prediction and (ii) Cerrada *et al.,* proposed a system for failure management in industrial processes based on agents [3]. Their experimental results show that prediction with multiagent was more accurate than prediction based only on data. In this sense, there are few papers about the use of intelligent multiagent approach for the prediction problems in Industry 4.0 environment, due to the fault that it is a new paradigm in this research field.

Therefore, this paper presents a MASPI: a multi agent system to develop a hybrid artificial intelligence model, based on intelligent agents for performing: processing data, train data, evaluate models, test data, make prediction and create a report to facilitate making decisions for Predictive Maintenance (PdM) in an Industry 4.0 environment through fault prediction. The capability of this proposal was evaluated by applying it to SCANIA trucks dataset. It consists of data collected from heavy SCANIA trucks in everyday usage, then in the dataset there exists a subset of data selected by experts to make a prediction model from possible failures of this specific component.

This paper is organized as follows: multiagent concept is described on Sect. 2. In Sect. 3 the proposed multiagent system is presented. Discussion and results are included in Sect. 4 and the last section contains conclusions and future work.

1.1 Predictive Maintenance (PdM) in Industry 4.0

For Dunn [6] Predictive Maintenance (PdM) estimates the condition of the equipment by measuring whether it will fail in an established future period. In Industry 4.0 paradigm, the fault prediction is an important aspect for Predictive Maintenance field because it is providing several methods to companies to perform time, cost and other resources.

In [20] the author provides a description of PdM. He defines Predictive Maintenance as a continuous monitoring machinery conditions, equipment or infrastructure to detect failures and make timely decisions. According to [9] some advantages of PdM are:

(i) the maintenance is only for the equipment with an imminent failure;
(ii) PdM reduces the total equipment shutdown time;
(iii) It can help to prevent catastrophic damage;

PdM increases machines life and their reliability [10]. However, for the same author, in Industry 4.0 environment, Predictive Maintenance faces several challenges [10]:

(1) Data: high demands on access, quality, fusion of data that comes from heterogeneous environment, therefore, integration between this data and systems is difficult.
(2) Industrial big data: industries need the ability to collect, process, support and store massive data sets to make historical analysis of critical trends which enable real-time predictive analysis.

(3) Accuracy: the prediction accuracy can prevent unnecessary maintenance. On the other hand the accuracy gives time to prepare machines or equipment for a maintenance operation.

In this research, fields of predictive maintenance will be studied from the perspective of accuracy, using machine learning algorithms, evaluating and comparing them in MASPI.

2 Multi Agent System

This section describes the principal concept about Agent, Virtual Organizations (VOs) and Multiagent System (MAS). The agent technology emerged like a field of artificial intelligence. The agent theory is present on a several research and it has been used to describe an autonomous entity capable of interacting with other agents. The main agent's properties as autonomy, reactivity, mobility, rationality, sociability, allowing them to offer intelligent solutions and to build autonomous computational systems [15, 18].

A Multiagent System (MAS) was defined by Jennings and Wooldridge in [21] and cited by [13], where they said that "MAS are based on individual agents who have common objectives within the system", and the main characteristics of these are organizations and the collaborative work through roles. Therefore, the role describes the responsibilities and benefits associated to an agent. For [13] a role "describes the constraints (obligations, requirements and skills) that an agent will have to satisfy to obtain a role".

The MAS stands out for their capacities: autonomy, reactivity, pro-activity, learning, ubiquitous and distributed communication and also for the intelligence among their elements [5].

Therefore, Virtual organizations is a group of agents that can be considered as a set of entities whose association or relationship is established by inheritance and aggregation with a structure of social relationships and workflows [4, 14]. The agents group work in turn of this organization structure and they can be associated by roles, resources or applications to facilitate their coordination and communication. The similarity between human organizations and agent organization are several, which is logical since agent organizations are born from human organizations. The aims of the organizations are reflected in their final objective, then it is translated into results in the form of products or services. Therefore, the objectives are the general aims that organizations pursue, these are expressed in a quantitative way: missions, purposes, goals, quotas and deadlines [16].

Knowing the services offered by an organization allows the agents that integrates it to be discovered, invoked, monitored and even make associations that give rise to new more complex services. Applying organizations to multiagent systems is very useful, because it controls agent's interaction, provides them a series of norms focusing on good coordination and achieving the general objectives of the organization itself [1, 2].

Multiagent system gives a new way to build distributed, robust and intelligent applications and to see the distributed systems. MAS may have interdependent actions by the use of shared resources between organizations or agents that share them, also a MAS developed with different languages, applications or characteristics are named heterogeneous [7, 8].

In these systems, the agents can be developed outside of an organization and at the same time it may have all its functionalities. This heterogeneous software platform contains tasks and roles that have a frequent interaction between the users (human) and the rest of systems process (algorithm, techniques, predictions, reports and others) which is appropriate to this research [12, 14]. Some examples of this heterogeneous software platform are: online auctions, predictions, collaborative filtering and others. With this platform it's possible that an agent exists to process a dataset, others can be an algorithm to obtain a model to make predictions, such as in our proposal named MASPI.

3 Proposed Multiagent System

In the proposed multiagent system there are three Virtual Organizations (VOs) with different roles and other agents. Every VOs synchronize the tasks, coordinate and supervise task execution and utilization of resources.

MASPI was designed as a heterogeneous system with a diversity of languages (for example some agents were developed on Python and Java), applications and characteristics. This makes it possible to combine systems based in virtual organizations with the use of machine learning algorithms in the field of Artificial Intelligence and some other distributed designs with the purpose of obtaining a multiagent system capable of allowing future modifications and extensions. Their Virtual Organizations are presented in Fig. 1, with the interaction between them. In point 3.1 to 3.4 the details of each VOs are described.

3.1 Data VOs

it manages Preprocessing and DataTrain agents. Their aim is to prepare the initial data for modeling. (i) Preprocessing agent select from the dataset a portion of the data i.e.: 70%, which will be sent to DataTrainAgent. The preprocessing includes a wide range of techniques that can group in two areas: data preparation and data reduction, both with the objective to provide data that will serve as input to the algorithms (k-nearest neighbors and Naïve Bayes in this case) For this research, the data preparation techniques are used with the aim to normalize, reduce noise and clean the data. The preprocessing is complete by DataTrain agent. It selects are potentially useful variables which may be available in prediction processes.

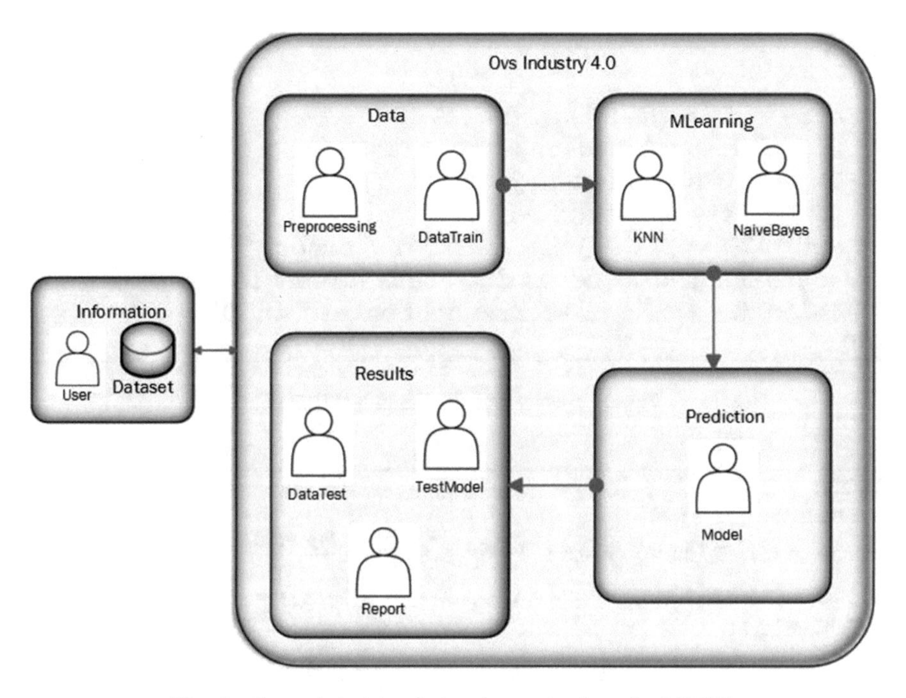

Fig. 1. General design of virtual organizations for MASPI

3.2 MLearning VOs

This VOs are integrated by two agents that represent the use of Machine Learning algorithms. For this research, only two algorithms were used: (i) k-nearest neighbors (KNN) and (ii) Naive Bayes (NB).

- **KNNAgent:** This agent used a k-nearest neighbors (K-NN) algorithm to obtain a model that can be proved first with DataTrainAgent and after with DataTestAgent. In Fig. 2, a Pseudocode for the basic K-NN Classifier is presented, it is based on Marsland, S., work [11]. Distances for all the cases to x are calculated, when the selected cases K are classified, D_x^K more near to the new case $(x1\ldots,)$, this will be assigned the class (variable c1, c2...) with more frequency between K and D_x^K objects.
- **NaïveBayesAgent:** Naïve Bayes Classifier algorithm is used by this agent. It requires the DataTrainAgent to estimate parameters. One of the advantages of Naïve Bayes is that its fast and it can be independently estimated as a one dimensional distribution. [11]. Figure 3 presented a Pseudocode for Naïve Bayes Classifier, in this, y is a class variable and x is a dependent feature vector, where: x_1 through x_n.

BEGIN

Entry: $D = \{(x1, c1), \ldots, (xN, cN)\}$
$x = (x1, \ldots, xn)$ new case to classify
FOR any classified object (xi, ci)
Calculate $di = d(xi, x)$
Sort $di(i = 1, \ldots, N)$ in ascending order
Stay with the K cases D_x^K classified closer to x
Assign to x the most frequent class in D_x^K
END

Fig. 2. Pseudocode for the K-NN Classifier

BEGIN
Entry: y and x,
The probability y is: $P(y|x_1, \ldots, x_n) = \frac{P(y)P((x_1, \ldots, x_n|y)}{P(x_1, \ldots, x_n)}$
Assumption that: $P(x_i|y, x_1, \ldots, x_{i-1}, x_{i+1}, \ldots, x_n) = P(x_i|y)$
FOR all i:
Simplified to $P(y|x_1, \ldots, x_n) = \frac{P(y)\prod_{i=1}^{n} P(x_i|y)}{P(x_1, \ldots, x_n)}$
Since $P(y|x_1, \ldots, x_n)$ is constant given the input
Where classifications rules
are: $P(y|x_1, \ldots, x_n) \propto P(y)\prod_{i=1}^{n} P(x_i|y)$
And $y = arg\max_y P(y)\prod_{i=1}^{n} P(x_i|y)$,
$P(x_i|y)$ is the relative frequency of class y in training and test data.
END

Fig. 3. Pseudocode for Naïve Bayes Classifier

3.3 Prediction VOs

In this virtual organization, the ModelAgent will create the prediction with the result of KNNAgent and NaïveBayesAgent. The Prediction VOs communicates with the Results VOs.

3.4 Results VOs

Three agents compose this VOs: DataTest, TestModel and Report agents. First, DataTestAgent uses 30% of dataset and sends this to TestModelAgent. The evaluation of a prediction model can be done according to different aspects. Therefore, the TestModelAgent uses the accuracy to evaluate the best model. The accuracy is the percentage of cases correctly classified by the model. Finally, the model with the best accuracy is sent to ReportAgent, then it has the task to organize the interaction with the user and create textual, graphical and other types of documents.

4 Discussion and Results

MASPI was implemented using a dataset consisting of data collected from heavy Scania trucks used everyday. For this research, the authors select this dataset because it is divided in two files that corresponding with TrainTestAgent (there is a file named Trainingset) and the other file is Test_Set for the DataTest agent.

MASPI uses SCANIA trucks dataset to obtain predictions about possible failures in Air Pressure system (APS) which generates pressurized air that is utilized in various functions in a truck, such as braking and gear changes.

SCANIA Trucks dataset only provide information about Air Pressure System (APS), one attribute describes the System Age, in other variable exist one positive class corresponding to failures in the APS and a negative class to equipment without failures or trucks with failures in components that not corresponded with APS.

The Data virtual organization via Preprocessing and TrainTestAgent, process the first file (Trainingset) sending this to Mlearning virtual organization, that contains the machine learnings algorithms used in this research (KNN and Naive Bayes). In MASPI each algorithm is an agent, which is an advantage because it facilitates the inclusion or the use of new agents in this virtual organization with other roles or task.

The Mlearning virtual organization send their results to Prediction VOs., then it has the task of creating models and sending them to Results virtual organization. For example the model obtained is received by TestModel then it uses the information and it works with DataTest agent to obtain prediction. Also it has the function of comparing models by their accuracy. The reports of best model by accuracy is elaborated by ReportAgent.

The proposed MAS – based MASPI allow the description of the main characteristics of the agents and virtual organizations with their respective objectives, services, capabilities, roles and tasks. Most of then has been programed on Python and Java to obtain results according to size of actual data set.

The k-nearest neighbors (KNN) prediction is presented in Fig. 4. A 0 is assigned for each value considered optimal and each value outside the normal range is assigned a 1 to perform the classification. In the graph of the figure, the test values are represented in green and the prediction values in red. For this research, the variable K was assigned the value of 10, with this number of neighbors the algorithm made the prediction with 92% accuracy. During the test process the user can change the number of neighbors and obtain different prediction percentages.

In Fig. 5, the Naïve Bayes results are presented. As in the same case of KNN, in this algorithm: for each value considered optimal is assigned a 0 and each value outside the normal range is assigned a digit 1 to perform the classification. This algorithm is used in sequential processes of maintaining a probabilistic model within a state that evolves over time, while being monitored by sensors [17]. In it a graph is observed that shows the values of the APS systems (green color) and the prediction made by the algorithm (red color). Evidence that the prediction of values performed by this algorithm is 90% accuracy.

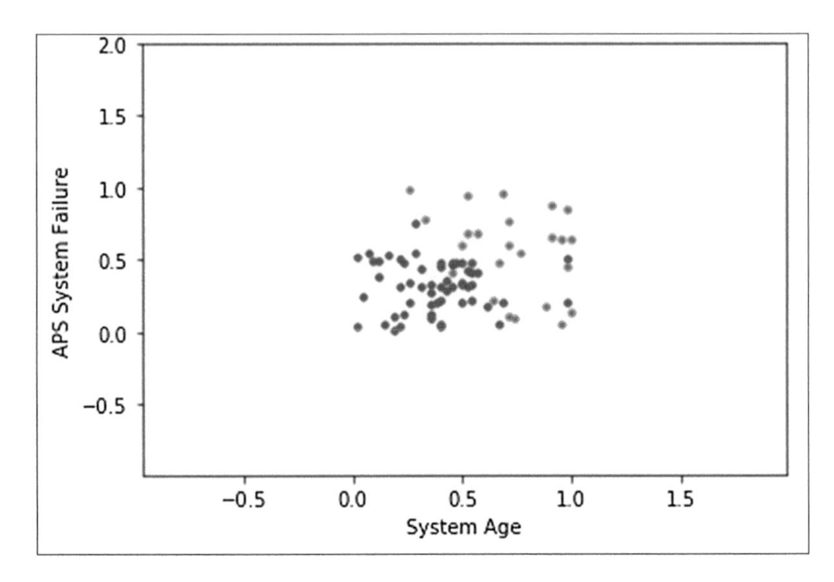

Fig. 4. Results with K-NN algorithm

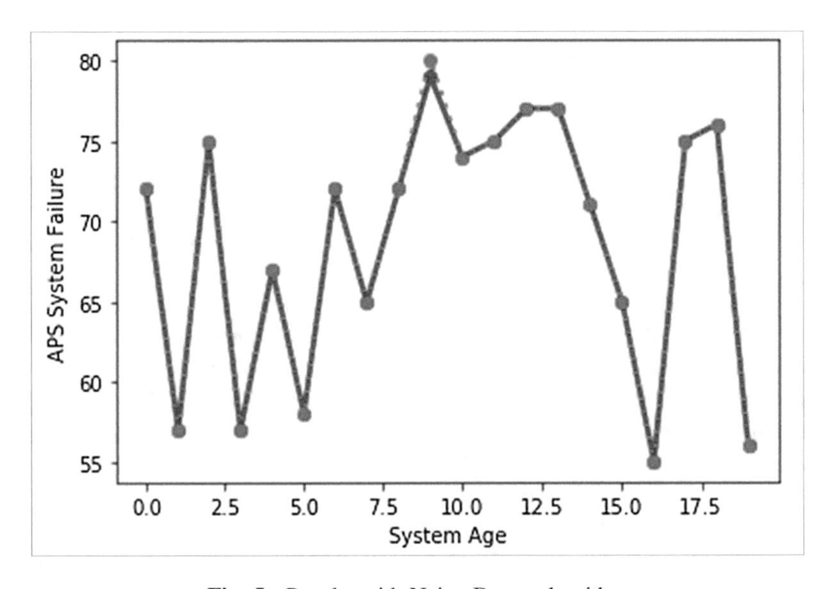

Fig. 5. Results with Naive Bayes algorithm

5 Conclusions and Future Work

MASPI uses the multiagent approach to reduce the complexity of the prediction problem by applying agents capable of processing data and using them to evaluate models by their prediction accuracy. To evaluate MASPI, it was applied to SCANIA Trucks dataset using machine learning algorithms, in order to provide a comprehensive form to see predictive maintenance. With the goal to use it in any industrial environment and integrate it in to a more complex architecture. The prediction accuracy improves over time in the whole prediction process with MASPI.

In future research the objective is to evaluate the effectiveness of MASPI using other case studies from real Industry 4.0 and integrate them into an architecture for Iot and Big Data platforms.

Acknowledgments. This work was supported by the Spanish Ministry, Ministerio de Economía y Competitividad and FEDER funds. Project: SURF, Intelligent System for integrated and sustainable management of urban fleets TIN2015-65515-C4-3-R.

I. Sittón has been supported by IFARHU – SENACYT scholarship program (Government of Panama).

References

1. Casado-Vara, R., Corchado J.M.: Blockchain for democratic voting: how blockchain could cast off voter fraud. Orient. J. Comp. Sci. and Technol **11**(1). http://www.computerscijournal.org/?p=8042
2. Cardoso, R.C., Bordini, R.H.: A multi-agent extension of a hierarchical task network planning formalism. ADCAIJ Adv. Distrib. Comput. Artif. Intell. J. **6**(2), 5–17 (2017)
3. Cerrada, M., Cardillo, J., Aguilar, J., Faneite, R.: Agents-based design for fault management systems in industrial processes. Comput. Ind. **58**(4), 313–328 (2007)
4. Chamoso, P., Rivas, A., Rodríguez, S., Bajo, J.: Relationship recommender system in a business and employment-oriented social network. Inf. Sci. (2017)
5. Corchado, J.M., Bajo, J., de Paz, Y., Tapia, D.: Intelligent environment for monitoring Alzheimer patients, agent technology for health care. Decis. Support Syst. **34**(2), 382–396 (2008). ISSN 0167-9236
6. Dunn, R.L.: Benchmarking maintenance. Plant Eng. **55**(3), 68–70 (2001)
7. García, E., Gallego, V., Rodríguez, S., Zato, C., de Paz, J.F., Corchado, J.M.: Simulation and analysis of virtual organizations of agents. In: Trends in Practical Applications of Agents and Multiagent Systems, pp. 65–74. Springer, Heidelberg (2012)
8. González, S.R., de la Prieta Pintado, F., Peña, E.G., Domínguez, C.Z., Rodríguez, J.M.C., Pérez, J.B.: Multiagent systems and self-organizative virtual organizations, a step ahead in adaptive MAS. In: 2011 11th International Conference on Intelligent Systems Design and Applications (ISDA), pp. 36–41. IEEE (2011)
9. Kumar, A., Shankar, R., Thakur, L.S.: A big data driven sustainable manufacturing framework for condition-based maintenance prediction. J. Comput. Sci. (2017)
10. Li, Z., Wang, K., He, Y.: Industry 4.0: potentials for predictive maintenance. In: International Workshop of Advanced Manufacturing and Automation, Manchester, October 2016
11. Marsland, S. (2015). Machine learning: an algorithmic perspective. CRC press
12. Peñaranda, C., Agüero, J., Carrascosa, C., Rebollo, M., Julián, V.: An Agent-based approach for a smart transport system. ADCAIJ Adv. Distrib. Comput. Artif. Intell. J. **5**(2), 2255–2863 (2016)
13. Rivas, A., Chamoso, P., Rodríguez, S.: An agent –based Internet of Things platform for distributed real time machine control. In: 2017 IEEE 17th International Conference on Ubiquitous Wireless Broadband (ICUWB), Salamanca, Spain, 12–15 September 2017 (2018)

14. Rodríguez, S., de la Prieta Pintado, F., Peña, E.G., Domínguez, C.Z., Rodríguez, J.M.C., Pérez, J.B.: Multiagent systems and self-organizative virtual organizations, a step ahead in adaptive MAS. In: 2011 11th International Conference on Intelligent Systems Design and Applications (ISDA), pp. 36–41. IEEE (2011)
15. Rodríguez, S., De La Prieta, F., Tapia, D.I., Corchado, J.M.: Agents and computer vision for processing stereoscopic images. LNCS (LNAI and LNBI), vol. 6077 (2010). https://doi.org/10.1007/978-3-642-13803-4_12
16. Román, J.A., Rodríguez, S., de da Prieta, F.: Improving the distribution of services in MAS. In: Communications in Computer and Information Science, vol. 616 (2016). https://doi.org/10.1007/978-3-319-39387-2_4
17. Siciliano, B.: Handbook of Robotic. Springer, Heidelberg (2008)
18. Tapia, D.I., Corchado, J.M.: An ambient intelligence based multi-agent system for Alzheimer health care. Int. J. Ambient Comput. Intell. **1**(1), 15–26 (2009). https://doi.org/10.4018/jaci.2009010102
19. Tian, Z., Wu, B., Chen, M.: Condition-based maintenance optimization considering improving prediction accuracy. J. Oper. Res. Soc. **65**(9), 1412–1422 (2014)
20. Wang, K.: Intelligent predictive maintenance (IPdM) system–Industry 4.0 scenario. WIT Trans. Eng. Sci. **113**, 259–268 (2016)
21. Wooldridge, M., Jennings, N.R.: Intelligent agents: theory and practice. Knowl. Eng. Rev. **10**(2), 115–152 (1995)
22. Wu, W., Liu, M., Liu, Q., Shen, W.: A quantum multi-agent based neural network model for failure prediction. J. Syst. Sci. Syst. Eng. **25**(2), 210–228 (2016)

Data Mining and Optimization

FEA Structural Optimization Based on Metagraphs

Diego Montoya-Zapata[1,2], Diego A. Acosta[3], Oscar Ruiz-Salguero[1(✉)], and David Sanchez-Londono[1]

[1] Laboratory of CAD CAM CAE, Universidad EAFIT, Medellín, Colombia
{dmonto39,oruiz,dsanch30}@eafit.edu.co
[2] Vicomtech, San Sebastián, Spain
[3] Process Development and Design Research Group (DDP), Universidad EAFIT, Medellín, Colombia
dacostam@eafit.edu.co

Abstract. Evolutionary Structural Optimization (ESO) seeks to mimic the form in which nature designs shapes. This paper focuses on shape carving triggered by environmental stimuli. In this realm, existing algorithms delete under - stressed parts of a basic shape, until a reasonably efficient (under some criterion) shape emerges. In the present article, we state a generalization of such approaches in two forms: (1) We use a formalism that enables stimuli from different sources, in addition to stress ones (e.g. kinematic constraints, friction, abrasion). (2) We use metagraphs built on the Finite Element constraint graphs to eliminate the dependency of the evolution on the particular neighborhood chosen to be deleted in a given iteration. The proposed methodology emulates 2D landmark cases of ESO. Future work addresses the implementation of such stimuli type, the integration of our algorithm with evolutionary based techniques and the extension of the method to 3D shapes.

Keywords: Topology optimization
Evolutionary structural optimization · Mathematical graph

Glossary

- AM: Additive manufacturing.
- BESO: Bidirectional evolutionary structural optimization.
- ESO: Evolutionary structural optimization.
- FEA: Finite element analysis.
- GA: Genetic algorithms.
- Ω_0: Compact and Bounded subset of $\mathbb{R}^2$, that represents an initial material stock from which to carve the part.
- Ω_i: The part after the i-th step of the evolution.
- f: Scalar function $f : \Omega_i \rightarrow \mathbb{R}$ that expresses how much is the neighborhood of a point $x \in \Omega_i$ being demanded by the stimuli (e.g. stress) being considered.

M. Graña et al. (Eds.): SOCO'18-CISIS'18-ICEUTE'18, AISC 771, pp. 209–220, 2019.
https://doi.org/10.1007/978-3-319-94120-2_20

g: Scalar function $g : \Omega_i \rightarrow \mathbb{R}$ that expresses the permissible level of demand f that the neighborhood of a point $x \in \Omega_i$ may stand (e.g. permissible stress allowable). In mechanical design, g is usually a constant for the whole domain Ω_i.

$G = (V, A)$: The Finite Element - based graph in which a vertex $v_i \in V$ is a finite element. An arc $(v_i, v_j) \in A$ means that finite elements v_i and v_j are neighbors.

$G_M = (V_M, A_M)$: A meta-graph built on G, in which a meta - vertex $V_i \in V_M$ is a connected set of finite elements of V. A meta - arc $(V_i, V_j) \in A_M$ means that meta - vertices V_i and V_j are neighbors.

1 Introduction

Shape evolution usually obeys the stimuli that such a shape receives. Examples of stimuli are heat, friction, stress/strain, etc. The underlying criterion may be reduction of weight, optimization of dimensional ratios, optimization of weight/inertia, minimization of drag, re-location of natural frequencies, etc.

Shape evolution, in nature, demands that the shape continue to execute the function while evolving. This circumstance obliges each evolution individual to transmit force, torque, motion, etc. as per the constraints imposed to it by the stimuli and demands that it receives. This stability in function designates neighborhoods of the shape as "indispensable", while others are "expendable". This manuscript focuses on the synthesis, classification and administration of indispensable vs. expendable shape neighborhoods of a given shape, via graph modeling.

The manuscript presents a methodology for structural optimization in which the exact nature of the stimulus may be generic. At the same time, the criterion of material removal may be also generic. Examples of such a criterion are under-demand, high exposure to friction, maximization of wave reflection (e.g. sound), etc. In this particular article, the material removal obeys to under-demand to stand stress/strain stimuli. Notice that, once the stimuli are calculated (by specialized outsourced software), the particular criteria for material removal can be applied in a generic manner.

In this manuscript, the Finite Element set is partitioned according to their status under the physical demand. Sets of the partition are then eliminated according to the rule imposed (e.g. under-demand) after guarantee that such eliminations do not affect the boundary conditions of the problem or render a disconnected piece. The meta-graph formalism supports in generic manner the application of such criteria. It is our purpose to inform the testing of this methodology in several manuscripts (each for different stimulus scenario), from which this is the first one.

2 Literature Review

2.1 Evolutionary Structural Optimization

Xie and Steven [16] introduce an structural optimization method called Evolutionary Structural Optimization (ESO). ESO technique removes progressively the low stressed portions of a structure by carrying out iterative FEA simulations. Therefore, the weight of the structure is reduced without affecting its functionality. Bidirectional ESO (BESO) [12] is an extension of ESO in which new material can be added in high-stressed zones. One of the main drawbacks of ESO and BESO is the formation of non-valid configurations as a result of material removal, which cause the end of the optimization process.

Recent publications on improvements of ESO/BESO techniques [6, 11] and on review articles focused on ESO/BESO [4, 10, 15] prove that the development of these algorithms is a matter of interest for the academic community.

One of the reasons of the popularity of ESO and BESO is the wide range of engineering problems that can be addressed with these methods. Some examples are: aeronautics [3], biomedicine [2], materials design [8].

Likewise, much of the research efforts in topology optimization are focusing on additive manufacturing (AM) [1, 14]. The recent advances in AM allow the exploitation of the full capacities of ESO/BESO techniques. Hence, it is necessary to extend the current topology optimization algorithms.

2.2 Graphs Representations for Connectivity Check in ESO

Reference [13] uses graphs as a representation of FEA meshes, in which each graph vertex represents a FEA element. In this paper, graphs are generated to verify that all FEA nodes subjected to a given stimulus remain connected. Structures that do not meet this requirement are discarded. In a similar fashion as [11, 13] uses a graph-based connectivity checker to extend BESO.

The reader may notice that graph representations have been used to find valid (or non-valid) configurations of FEA meshes while using ESO algorithms. Additionally, when a non-valid configuration is found, this branch of the optimization process is not taken into account anymore.

On the other hand, neither graphs abstractions have been integrated to the element-deletion process nor graph-based deletion criteria (e.g. vertex degree) have been used.

2.3 Graph Representations in Other Structural Optimization Algorithms

Graph representations have been used in conjunction with other structural optimization techniques, apart from ESO. For instance, [7] focuses on the topology optimization of trusses, using genetic algorithms (GA) as the basis to remove the useless material of the truss structure. Graph representations are used to establish a criterion to test if an individual is structurally valid.

Another example of the use of GA and graph representations in structural optimization is presented in [9]. In this work, each graph vertex represents a finite element, and a graph arc between two vertices indicates that the elements the vertices represent are neighbors. These graphs are created and modified using GA to find an optimal solution.

2.4 Conclusions of the Literature Review

Structural optimization and specifically ESO/BESO are topics of interest because of the multiple application fields in which they can be used. In particular, the use of topology optimization in AM is establishing as a necessity. Therefore, it is necessary to improve current topology optimization algorithms.

The use of graph abstractions is common in structural optimization, specially with truss structures. Likewise, graphs are used in ESO/BESO to check the validity of material configurations. However, invalid structures are usually discarded, not fixed.

The present manuscript intends to illustrate a methodology to use graphs to administrate the information of neighborhood and static connectedness to support an independent shape evolutionary strategy. We do not try in the present status of our manuscript to evaluate or apply alternative evolutionary strategies (e.g. mutation and crossover operators [7,9]). Our future work seeks to widen the variety of stimuli (kinematics, abrasion, temperature) that drive evolution in the nature domain.

3 Methodology

3.1 Problem Statement

Given

1. Let $\Omega_0 \subset \mathbb{R}^2$ be a compact and bounded domain that represents an initial oversized domain.
2. A stimulus function S that acts over $\Omega_S \subset \Omega_0$.
3. FEA mesh $M_0 = (N_0, E_0)$ for Ω_0, where N_0 is the set of nodes and E_0 is the set of elements.

Goal

To obtain the design domain $\Omega_F \subset \Omega_0$ that solves the optimization problem:

$$\begin{aligned} \min_{\Omega} \quad & A(\Omega) \\ s.t. \quad & f(x) \leq g(x), \qquad \text{for all } x \in \Omega, \Omega \subset \Omega_0 \\ & \Omega_S \subset \Omega \end{aligned}$$

where $A(\Omega)$ is the area of Ω, $f(x)$ is the response for $x \in \Omega$ to the stimuli S and $g(x)$ expresses the permissible level of demand f that the neighborhood of a point $x \in \Omega_i$ may stand (e.g. permissible stress allowable).

3.2 Implemented Optimization Algorithm

The implemented optimization algorithm follows the procedure described in Fig. 1(a). First, a FEA simulation is carried out, given the initial FEA mesh $M_0 = (N_0, E_0)$ and the stimulus function S. The FEA simulation allows to find the domain response f to the stimuli S. If f exceeds the permissible limit g, then the algorithm stops. Otherwise, the algorithm proceeds to delete the under-demanded FEA elements. The sub-algorithm that performs the deletion of the FEA elements is presented in Fig. 1(b). It is described in detail in Sect. 3.3.

Finally, another FEA simulation is performed with the resultant domain after the FEA elements deletion, and the cycle is repeated until no more elements can be deleted.

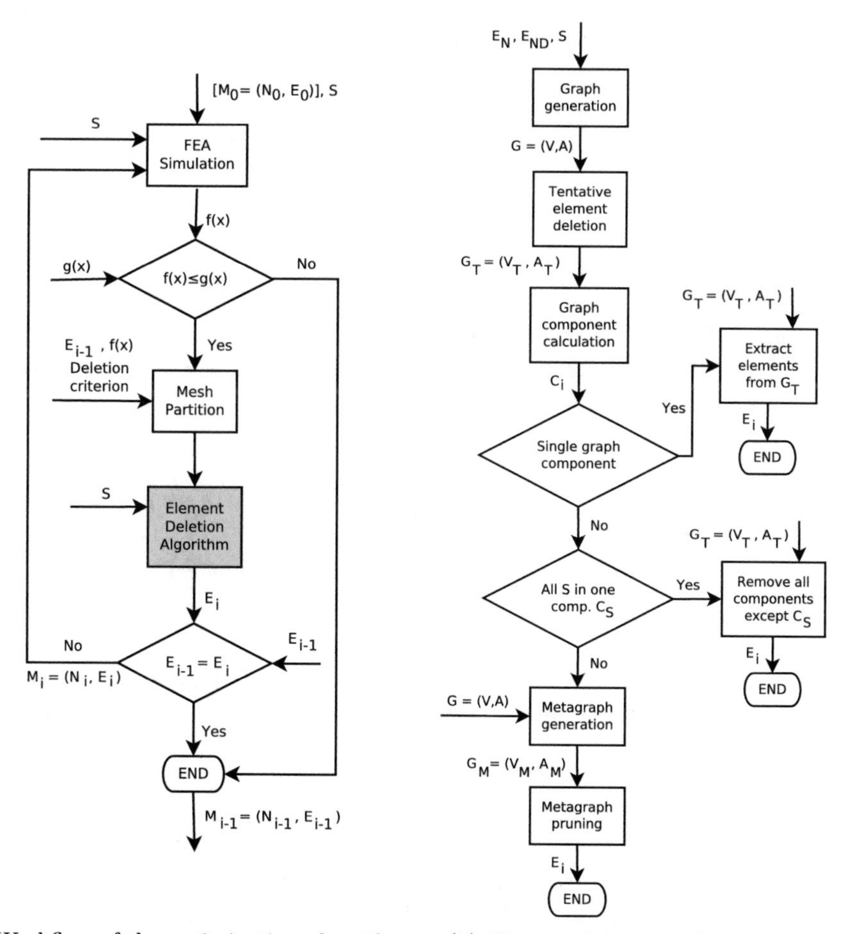

(a) Workflow of the optimization algorithm (b) Element deletion sub-algorithm

Fig. 1. Data flow of the implemented optimization procedure.

Stimuli Independence: The algorithm presented in this article reduces Ω_0 by removing under-demanded material (i.e. elements from E_0) given a deletion criterion. The stimulus function S and the deletion criterion are user-defined properties and f is calculated by using FEA software. Therefore, the presented algorithm is independent to the kind of stimuli (forces, friction, abrasion, humidity, etc.) to which the domain is subjected.

3.3 Element Deletion Algorithm

The element deletion algorithm is based on a graph abstraction of the design domain. The main objective of this algorithm is to assure that the resultant configuration (after removing the unnecessary FEA elements) is valid from a structural point of view. The algorithm is presented in Fig. 1(b). The main stages of the algorithm are discussed below.

Domain Graph Abstraction: For every FEA mesh $M = (N, E)$, a graph $G = (V, A)$ can be generated with the following procedure:

1. Assume that $E = \{e_1, e_2, \ldots, e_k\}$. Then, for every $e_i \in E$ create a graph vertex $v_i \in V$.
2. A graph arc $(v_i, v_j) \in A$ exists if and only if the corresponding FEA elements e_i, e_j are adjacent.

Different adjacency relations between elements can be defined for a 2D FEA mesh. In Fig. 2 are shown the results when the graph arcs are produced by FEA nodes adjacency or FEA edges adjacency.

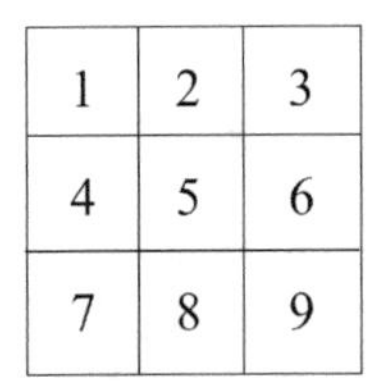

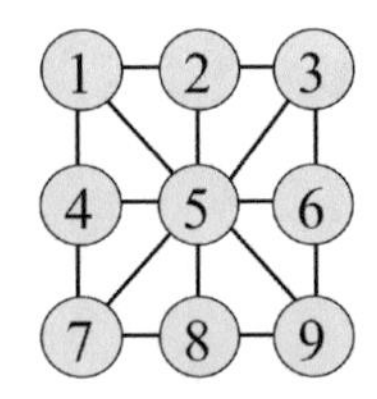

1 2 3
4 5 6
7 8 9

(a) FEA mesh

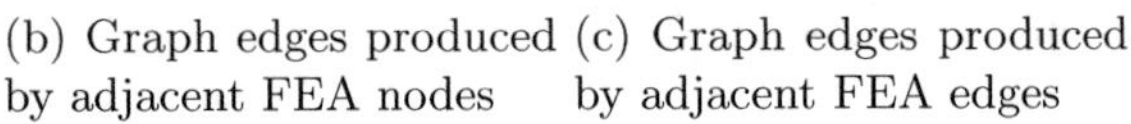
(b) Graph edges produced by adjacent FEA nodes (c) Graph edges produced by adjacent FEA edges

Fig. 2. FEA mesh to graph conversion using FEA node adjacency vs. FEA edge adjacency criteria.

Domain Metagraph Abstraction: The procedure to generate the metagraph G_M associated to (1) the graph $G = (V, A)$ and (2) the set of candidate elements to be deleted $E_D \subset E$ is as follows:

1. Divide G into two subgraphs: the first contains the graph vertices associated to E_D (G_{SD}). The second contains the graph vertices associated to $E_{ND} = E - E_D$ (G_{SND}).

2. Find the connected components of G_{SD} and denote them as $\{c_1, c_2, \ldots, c_p\}$.
3. Find the connected components of G_{SND} and denote them as $\{c_{p+1}, c_{p+2}, \ldots, c_{p+r}\}$.
4. Each connected component of G_{SD} and G_{SND} becomes a vertex (metavertex) of the metagraph.
5. Two metavertices c_i, c_j are adjacent if and only if exist vertices $v_i, v_j \in V$ such that: (a) the arc $(v_i, v_j) \in A$ exists, (b) vertex v_i belongs to the connected component c_i, and (c) vertex v_j belongs to the connected component c_j.

Metagraph Pruning Strategy: In order to preserve the connectivity of the stimulated subdomain Ω_S, a metavertex c_i is deleted if meets these conditions: (1) c_i is a connected component of G_{SD} and (2) c_i has degree 1.

The second condition assures that deleting c_i will not disrupt the connectivity between FEA elements that lie on Ω_S. When a metavertex c_i is deleted, all the elements of the connected component c_i are deleted.

Figure 3 depicts an example of the generation of the metagraph and the corresponding metavertex deletion. Marked squares represent the set of candidate elements to be deleted and the triangles represent elements under stimuli.

In Fig. 3(b) can be seen that the only metavertex that can be deleted is c_i, since c_2 fulfills the first condition but not the second (the degree of c_2 is 3).

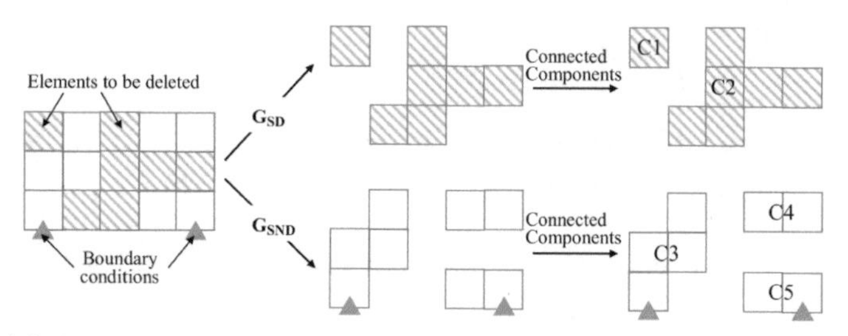

(a) Connected components in the two subgraphs G_{SD} and G_{SND}

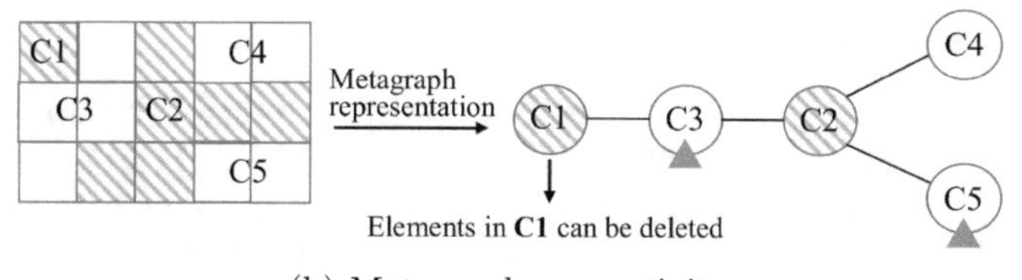

(b) Metagraph connectivity

Fig. 3. Metagraph associated to a FEA mesh and the candidate elements to be deleted

3.4 Boundary Synthesis

Since the presented algorithm works with FEA meshes, the final shapes obtained with the algorithm tend to be rough and difficult to manufacture. For this reason, the boundary of the final designs must be smoothed. In Fig. 4 is shown the process to obtain the smoothed boundary of a given FEA mesh.

4 Results

Section 4.1 reports the accomplished results with the implemented algorithm for two well-known problems: (1) a Michell structure and (2) a two bar frame. Section 4.2 reports the results for a Cantilever problem.

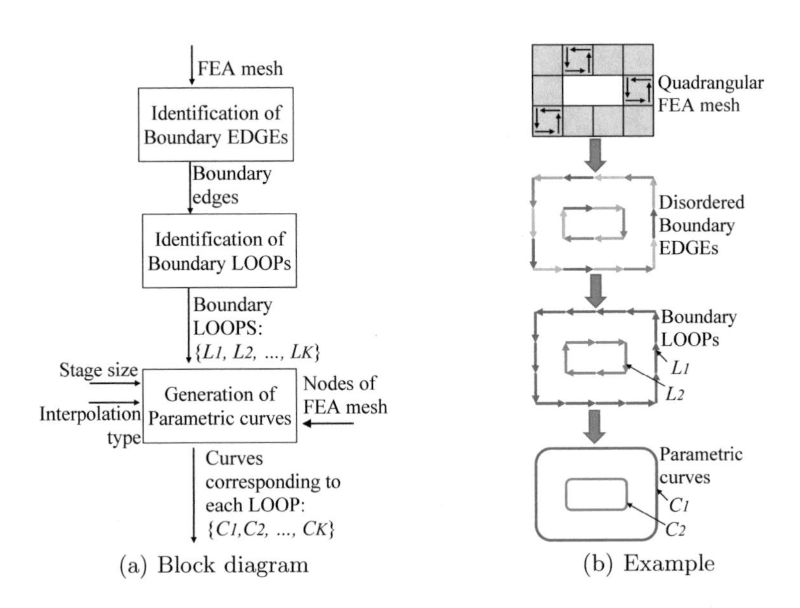

(a) Block diagram (b) Example

Fig. 4. Post-processing for part boundary smoothing.

4.1 Benchmarking Cases

Michell Structure: The design problem and theoretical solution of a Michell structure are depicted in Figs. 5(a) and (b). In Figs. 5(c)–(h) can be seen the evolution of the structure when the presented algorithm is applied. Notice that the final design (Fig. 5(h)) resembles the theoretical solution. It shows the capacity of the algorithm to replicate classical examples in 2D topology optimization.

Two Bar Frame: The initial domain and boundary conditions for the design of a two bar frame are shown in Fig. 6(a). Likewise, the theoretical solution is illustrated in Fig. 6(b). The evolution of the design is shown in Figs. 6(c)–(f). As in the Michell structure example, it can be seen that the design obtained with the metagraph abstraction is very similar to the theoretical solution. This result supports the validity of the proposed algorithm.

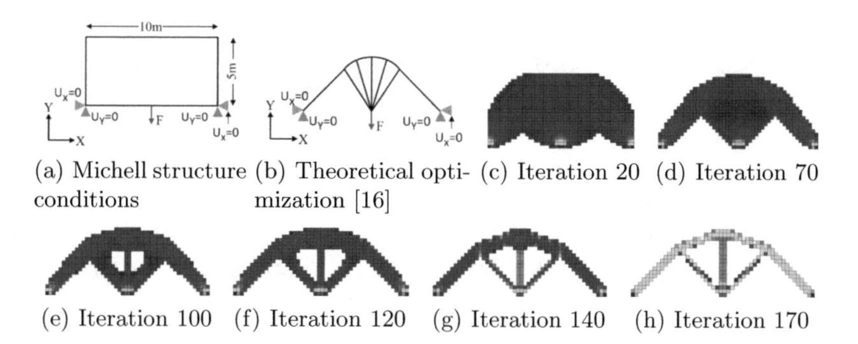

(a) Michell structure conditions (b) Theoretical optimization [16] (c) Iteration 20 (d) Iteration 70

(e) Iteration 100 (f) Iteration 120 (g) Iteration 140 (h) Iteration 170

Fig. 5. Michell structure. Evolution using our metagraph approach (figures (c) to (h)).

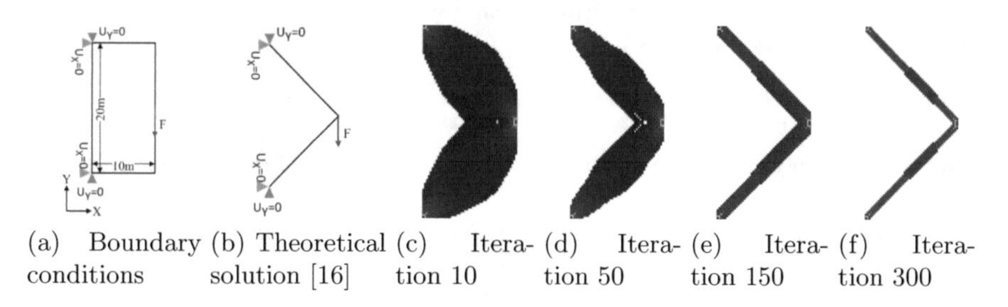

(a) Boundary conditions (b) Theoretical solution [16] (c) Iteration 10 (d) Iteration 50 (e) Iteration 150 (f) Iteration 300

Fig. 6. Two bar frame design problem. Evolution using our metagraph approach (figures (c) to (f)).

4.2 Cantilever

Another scenario that was studied involves a cantilever design. In this case, the domain is supported on one end and is vertically loaded on the other end, as can be seen in Fig. 7(a). In Figs. 7(b)–(d) can be seen the evolution of the domain using the metagraph approach. This example shows that the algorithm presented in this article can be used in complex problems, different from those found in the literature.

4.3 Computational Demands of the Proposed Algorithm

Due to the material removal procedure associated to the presented algorithm, given the mesh $M_i = (N_i, E_i)$ at iteration i we can say: $|N_0| \geq |N_i|$ and $|E_0| \geq |E_i|$, where $M_0 = (N_0, E_0)$ is the initial mesh. In addition, $|N_0| > |E_0|$. Therefore, the computational demands of one iteration of the algorithm can be expressed as a function of N_0 and the bandwidth W of the stiffness matrix calculated during the FEA simulation [5].

Table 1 presents the computational expenses of our algorithm. Notice that the time complexity and memory complexity of an iteration of our algorithm is dictated by the term $O(N_0^2)$. This term corresponds to the dominant generation of the graph and the metagraph associated to the FEA mesh.

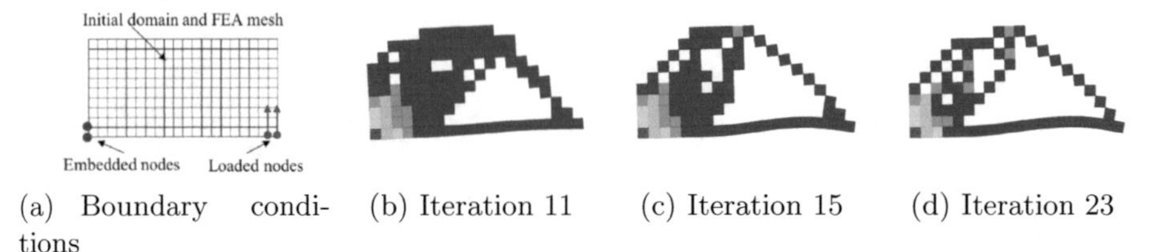

(a) Boundary conditions (b) Iteration 11 (c) Iteration 15 (d) Iteration 23

Fig. 7. Cantilever design problem. Evolution using our metagraph approach (figures (b) to (d)).

A comparison of the computational resources used vs. the efficiency of evolution is beyond our capabilities. One reason for this limitation is that the measure of the quality or efficiency of an evolution is itself an open research question at this time.

Table 1. Analysis of the computational costs of an iteration of the proposed algorithm

Process	Time expenses	Memory expenses
FEA simulation	$O(N_0 W^2)$ [5]	$O(N_0 W)$ [5]
Mesh partition	$O(N_0)$	$O(N_0)$
Element deletion	$O(N_0^2)$	$O(N_0^2)$
Graph generation	$O(N_0^2)$	$O(N_0^2)$
Graph components calculation	$O(N_0)$	$O(N_0)$
Metagraph generation	$O(N_0^2)$	$O(N_0^2)$
Metagraph pruning	$O(N_0)$	$O(N_0)$

4.4 Boundary Synthesis

The simulations carried out with the metagraph based algorithm showed that the resultant shape is very rough. Following the procedure described in Sect. 4.4, one of the designs obtained with the algorithm was smoothed. The results of the application of the smoothing algorithm using a mean filter of order 4 are shown in Fig. 8. Notice that the boundary of the shape was corrected without losing sensitive information of the design.

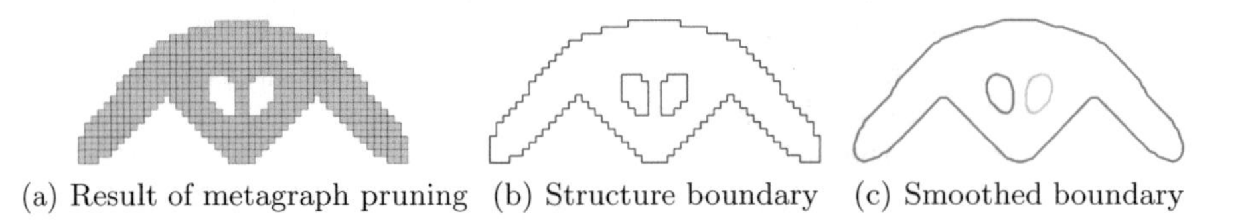

(a) Result of metagraph pruning (b) Structure boundary (c) Smoothed boundary

Fig. 8. Michell structure. Post-processing for border smoothing.

5 Conclusions

This article presents a methodology to perform 2D structural optimization using a graph abstraction in conjunction with ESO. The algorithm was tested with well-known examples in 2D topology optimization. The presented methodology yielded similar results to those found in [16]. The utilization of this method assured that all calculated solutions in the optimization cycle were valid from a structural point of view. Furthermore, the optimization process did not stop when non-valid structures could arise as the result of the FEA elements deletion operation.

As long as the effects of a set of stimuli can be described as a scalar field f, structural optimization of such a problem should be compatible with the presented algorithm.

5.1 Limitations and Shortcomings

In cases where some metanodes of degree 2 (or more) can be deleted the optimization process must be divided in multiple paths. However, since only metanodes of degree are deleted, the algorithm does not remove any of those metanodes. Thus, other paths are not considered.

5.2 Future Work

Future work should address the utilization of stimuli other than stress/strain: friction, abrasion, heat, humidity, etc. Future research should also address the extension of the algorithm to $\mathbb{R}^3$ and the integration of our algorithm with other evolutionary based techniques (e.g. mutation) to allow the generation of multiple feasible solutions that explore a wider region of the solution space.

References

1. Chen, J., Ahmad, R., Suenaga, H., Li, W., Sasaki, K., Swain, M., Li, Q.: Shape optimization for additive manufacturing of removable partial dentures - a new paradigm for prosthetic CAD/CAM. PLOS ONE **10**(7), 1–17 (2015)
2. Chen, Y., Schellekens, M., Zhou, S., Cadman, J., Li, W., Appleyard, R., Li, Q.: Design optimization of scaffold microstructures using wall shear stress criterion towards regulated flow-induced erosion. J. Biomech. Eng. **133**(8), 081008–081448 (2011)
3. Das, R., Jones, R.: Topology optimisation of a bulkhead component used in aircrafts using an evolutionary algorithm. Procedia Eng. **10**, 2867–2872 (2011). 11th International Conference on the Mechanical Behavior of Materials (ICM 2011)
4. Deaton, J.D., Grandhi, R.V.: A survey of structural and multidisciplinary continuum topology optimization: post 2000. Struct. Multidiscip. Optim. **49**(1), 1–38 (2014)
5. Farmaga, I., Shmigelskyi, P., Spiewak, P., Ciupinski, L.: Evaluation of computational complexity of finite element analysis. In: 2011 11th International Conference the Experience of Designing and Application of CAD Systems in Microelectronics (CADSM), pp. 213–214, February 2011

6. Ghabraie, K.: An improved soft-kill beso algorithm for optimal distribution of single or multiple material phases. Struct. Multidiscip. Optim. **52**(4), 773–790 (2015)
7. Giger, M., Ermanni, P.: Evolutionary truss topology optimization using a graph-based parameterization concept. Struct. Multidiscip. Optim. **32**(4), 313–326 (2006)
8. Huang, X., Xie, Y.M., Jia, B., Li, Q., Zhou, S.W.: Evolutionary topology optimization of periodic composites for extremal magnetic permeability and electrical permittivity. Struct. Multidiscip. Optim. **46**(3), 385–398 (2012)
9. Madeira, J.F.A., Pina, H.L., Rodrigues, H.C.: GA topology optimization using random keys for tree encoding of structures. Struct. Multidiscip. Optim. **40**(1), 227 (2009)
10. Munk, D.J., Vio, G.A., Steven, G.P.: Topology and shape optimization methods using evolutionary algorithms: a review. Struct. Multidiscip. Optim. **52**(3), 613–631 (2015)
11. Munk, D.J., Vio, G.A., Steven, G.P.: A bi-directional evolutionary structural optimisation algorithm with an added connectivity constraint. Finite Elem. Anal. Des. **131**, 25–42 (2017)
12. Querin, O., Steven, G., Xie, Y.: Evolutionary structural optimisation (ESO) using a bidirectional algorithm. Eng. Comput. **15**(8), 1031–1048 (1998)
13. Stojanov, D., Falzon, B.G., Wu, X., Yan, W.: Implementing a structural continuity constraint and a halting method for the topology optimization of energy absorbers. Struct. Multidiscip. Optim. **54**(3), 429–448 (2016)
14. Tang, Y., Dong, G., Zhou, Q., Zhao, Y.F.: Lattice structure design and optimization with additive manufacturing constraints. IEEE Trans. Autom. Sci. Eng. **PP**(99), 1–17 (2017)
15. Xia, L., Xia, Q., Huang, X., Xie, Y.M.: Bi-directional evolutionary structural optimization on advanced structures and materials: a comprehensive review. Arch. Comput. Methods Eng. **25**(2), 437–478 (2018). https://doi.org/10.1007/s11831-016-9203-2
16. Xie, Y., Steven, G.: A simple evolutionary procedure for structural optimization. Comput. Struct. **49**(5), 885–896 (1993)

Case-Based Support Vector Optimization for Medical-Imaging Imbalanced Datasets

I. A. Illan[4(✉)], J. Ramirez[4], J. M. Gorriz[4], K. Pinker[2,3], and A. Meyer-Baese[1]

[1] Scientific Computing Department, Florida State University, Tallahassee, FL 32306, USA
[2] Department of Radiology, Memorial Sloan-Kettering Cancer Center, New York, USA
[3] Department of Biomedical Imaging and Image-guided Therapy, Division of Molecular and Gender Imaging, Medical University of Vienna/AKH Wien, Vienna, Austria
[4] Department of Signal Theory, Networking and Communications, University of Granada, Granada, Spain
illan@ugr.es

Abstract. Imbalanced datasets constitute a challenge in medical-image processing and machine learning in general. When the available training data is highly imbalanced, the risk for a classifier to find the trivial solution increases dramatically. To control the risk, an estimate on the prior class probabilities is usually required. In some medical datasets, such as breast cancer imaging techniques, estimates on the priors are intractable. Here we propose a solution to the imbalanced support vector classification problem when prior estimations are absent based on a case-dependent transformation on the decision function.

Keywords: Medical image processing · Support vector machines DCE-MRI · Imbalanced datasets

1 Introduction

In the majority of the cases in statistical learning, the available training data set has few samples in a high dimensional space, allowing very poor estimations on probability distribution functions. Non-parametric approaches based on statistical learning theory, such as neural networks or support vector machines (SVMs), have been proven to be very successful solving classification problems. However, in the case of unbalanced training datasets, some difficulties arise if the learning algorithms are straightforwardly applied. For example, the soft margin solution in SVM [2] for non-separable classes, includes a term in the Lagrangian that accounts for the classification error rate together with the structural risk minimization. If the learning algorithm is optimized to minimize the risk of misclassifying samples, some additional constraints must be imposed to avoid the trivial solution in unbalanced datasets. The trivial solution is achieved when all

M. Graña et al. (Eds.): SOCO'18-CISIS'18-ICEUTE'18, AISC 771, pp. 221–229, 2019.
https://doi.org/10.1007/978-3-319-94120-2_21

samples are classified as the dominant class. In that undesirable case, the misclassification error can be very small if the dominant class outnumbers the scarce class in several orders of magnitude. A common solution is to apply penalties to the classification errors on the scarce class [1], so that the risk of classification errors is weighted. Usually, no other method or theoretical ground for class weight estimation but trial-and-error is proposed, prone to over-fitting or costly grid search algorithms.

Medical image dataset constitute an example of imbalanced data in which priors are hard to estimate. The disease incidence representation is seldom realistic, as well as limitations on the database size and dimension on the feature space make those estimations also unrealistic. In dynamic contrast enhancing magnetic resonance imaging (DCE-MRI) for breast cancer diagnosis, CAD systems have reported a high false positive rate [6]. The tumor regions in DCE-MRI are usually small with respect to the whole image and therefore the data is highly imbalanced. Thus, controlling the false positive rate is challenging since the size of the tumors can not be predicted in advance, and therefore the priors on training data do not provide accurate estimates for unseen data. Posterior probability estimates have been used in the literature to propose confidence based SVM [3,8], but not extended and studied in the unbalanced case. In order to maintain the non-parametric nature of SVM, we propose here a solution to the imbalanced data problem with intractable priors as a case-based hyperplane translation. Based on our previous works [4,5], we define a function from the estimation on the posterior probabilities that translates the hyperplane decision function, that controls the false positive rate.

2 Methods

2.1 Bayes Formulation of the Classification Problem

The binary classification problem in statistical learning theory can be stated as: given a set of N vectors $\{\mathbf{x}_i\} \in \mathbb{R}^d$, the d-dimensional feature space, together with its N corresponding class labels $y_i \in \{w_0, w_1\}$, find the function $f : \mathbb{R}^d \rightarrow \{w_0, w_1\}$ that correctly assigns their label to each sample. This function is usually called the classifier, and the process is named learning. The feature space is separated into two regions R_0 and R_1 by the classifier, each of them fulled by the set of vectors that are assigned with the class w_0 and w_1, respectively. Linear classifiers will divide the feature space with a hyperplane, and non-linear classifiers will take arbitrary shapes. Once the classifier has learned or is trained, it can be used to predict the unknown labels of new unseen data $\hat{\mathbf{x}}$.

In Bayes formulation, the classes in unseen data can be predicted from their posterior probabilities. Beginning with the probability of the event $\mathbf{x}$:

$$p(\mathbf{x}) = \sum_{i=1}^{s} p(\mathbf{x} \mid w_i) \cdot p(w_i) \tag{1}$$

where $p(w_i)$ is the prior probability of class w_i, and s is the number of different classes. By Bayes theorem, the posterior probability can be expressed as:

$$p(w_i \,|\, \mathbf{x}) = \frac{p(\mathbf{x} \,|\, w_i) \cdot p(w_i)}{p(\mathbf{x})} \tag{2}$$

and a sample $\hat{\mathbf{x}}$ will be assigned with the maximum a posteriori estimation $\hat{w}_{MAP}$:

$$\hat{w}_{MAP} = \arg\max_i p(w_i \,|\, \hat{\mathbf{x}}) \tag{3}$$

where $i = 0, 1$ in the binary case. In this formulation, the priors and conditional probability distribution functions must be known in order to solve the classification problem in unseen data. Given a binary learning function f as defined above, the classifying error E_i of f on the class w_i can be obtained by integrating the probability the sample is assigned with the class w_j over the region R_i.

Explicitly:

$$E_i = \int_{R_i} p(w_j|\mathbf{x})p(\mathbf{x})d\mathbf{x} \quad i,j = 0,1 \quad i \neq j \tag{4}$$

2.2 False Positive Rate Control by SVM Hyperplane Translation

SVM is a machine learning algorithm that separates a given set of binary labeled training data with a hyper-plane that is maximally distant from the two classes (known as the maximal margin hyper-plane). In the C-SVM formulation, the problem of finding the maximal margin hyperplane is solved by quadratic programming algorithms that try to minimize the dual of the cost function J:

$$J(\boldsymbol{w}, w_0, \xi) = \frac{1}{2}||\boldsymbol{w}||^2 + C\sum_{i=1}^{l} \xi_i, \tag{5}$$

subject to the inequality constraints:

$$y_i[\boldsymbol{w}^T x_i + w_0] \geq 1 - \xi_i, \quad \xi_i \geq 0 \quad i = 1, 2, ..., l. \tag{6}$$

where the slack variables ξ_i incorporate to the optimization those feature vectors that are not separable (details can be found in [2,12]). The solution to that problem can be expressed by a linear combination of a subset of vectors, called support vectors:

$$d(\mathbf{x}) = \sum_{i=1}^{N_S} \alpha_i y_i K(\mathbf{s}_i, \mathbf{x}) + w_0 \tag{7}$$

where $K(.,.)$ is the kernel function, α_i is a weight constant derived form the SVM process and $\mathbf{s}_i$ are the N_S support vectors [12]. Common kernels that are used by SVM practitioners for the nonlinear feature mapping are:

- Polynomial

$$K(\boldsymbol{x}, \boldsymbol{y}) = [\gamma(\boldsymbol{x} \cdot \boldsymbol{y}) + c]^d. \tag{8}$$

– Radial basis function (RBF)

$$K(\boldsymbol{x},\boldsymbol{y}) = \exp(-\gamma||\boldsymbol{x}-\boldsymbol{y}||^2). \quad (9)$$

as well as the linear kernel. Taking the sign of the function $d(\mathbf{x})$ leads to the binary classification solution.

In the imbalanced case, the misclassification part on the cost function J in C-SVM is modified to a class-weighted sum of errors [1], leading to the following optimization problem:

$$\begin{cases} \min \frac{1}{2}||\boldsymbol{w}||^2 + C\sum_{i\in+1}\xi_i + C^*\sum_{i\in-1}\xi_i \\ y_i[\boldsymbol{w}^T x_i + w_0] \geq 1-\xi_i, \quad \xi_i \geq 0 \quad i=1,2,...,l. \end{cases} \quad (10)$$

where the parameters C and C^* are fixed and obtained experimentally. For later convenience we define $\rho = C^*/C$.

The output of the SVM is not a probability but measures the distance from the test sample to the hyper-plane. There has been several proposals to transform the SVM output value to a probability. Platt [10] proposes a transformation to posterior probabilities based on a sigmoid function, while Zadrozny and Elkan [13] use isotonic regression to perform the task. However, they also report that histogram matching outperforms isotonic regression in some cases, provided that a proper bin selection is made. Lin and Sethi [8] propose a dynamic bin width allocation for histogram matching to estimate the posteriors, and empirical cumulative density (ECD) function to estimate the empirical error:

$$F_i(t) = \frac{\#(\mathbf{x} : d(\mathbf{x}) \leq t, \mathbf{x} \in w_i)}{\#\mathbf{x} \in w_i} \quad (11)$$

where $\#A$ denotes the cardinal of A, and t is a threshold on the output value distance of the SVM.

Tao et al. [11] define the empirical posterior probability as:

$$p(w_i \,|\, x_j) = \frac{l_{x_i}}{l_j} \quad j = 1,2 \quad (12)$$

where $r > 0$ is a real number, l_j is the number of elements in the class w_j and l_{x_i} is the number of elements in $\{x : x \in w_j, ||x - x_i|| \leq r\}$. In Tao's definition, the problem of bin allocation is translated to the selection of r. Both Tao's and Lin's empirical approaches, with an adequate selection of r, lead to the same estimation for the empirical error in a region $t_1 < d(\mathbf{x}) \leq t_2$:

$$\begin{aligned} E_i(t_1,t_2) &= \int_{t_1}^{t_2} p(w_j \,|\, x)p(x)dx \\ &= p(w_j)(F_j(t_2) - F_j(t_1)) \quad i \neq j,\, i,j = 1,2 \end{aligned} \quad (13)$$

where we used the Bayes formula. It is interesting to note that the empirical error E_1 on the class w_1 is proportional to the prior on the other class. Substituting the ECD function:

$$E_i(t_1,t_2) = \frac{\#\{\mathbf{x} : t_1 \leq d(\mathbf{x}) \leq t_2, \mathbf{x} \in w_j\}}{N} \quad (14)$$

Therefore, the magnitude $E_i(0,1)$ corresponds to the error inside the margin. The ratio ϵ between errors in the two classes is:

$$\begin{aligned}\epsilon &= \frac{E_1}{E_2} = \frac{p(w_2)(F_2(t_2) - F_2(t_1))}{p(w_1)(F_1(t_2) - F_1(t_1))} \\ &\approx \frac{\#\{\mathbf{x} : t_1 \leq d(\mathbf{x}) \leq t_2, \mathbf{x} \in w_2\}}{\#\{\mathbf{x} : -t_2 \leq d(\mathbf{x}) \leq -t_1, \mathbf{x} \in w_1\}}\end{aligned} \tag{15}$$

In balanced datasets the ratio ϵ takes values close to 1. However, in unbalanced datasets, the value of ϵ departs severely from 1. The undesirable scenario that leads to the trivial solution is the one with $\epsilon \to 0$. This is usually avoided by controlling the empirical risk separately on both classes. For instance, a weighted sum of errors (see Eq. 10) might control the risk, but a reasonable estimate on ρ must be given. The estimation can be either derived from an estimate on the priors, obtained through cross-validation, or even implemented through a fuzzy membership function on the errors. In some cases, the estimate of ρ is intractable even through cross validation, and some alternative approach is necessary to control the value of ϵ.

The empirical approximation to the error ratio ϵ of Eq. 15 serves as motivation to implement a translation on the decision function to control the error ratio on imbalanced datasets when the ρ estimation is intractable. We add a new uniform term $g(\mathbf{s}_i, \xi_i)$ to the hyperplane defining function $d(\mathbf{x})$, in terms of a subset of slack variables ξ_i and support vectors $\mathbf{s}_i$, so that the classification rule is now defined by:

$$f(\mathbf{x}) = \text{sign}\{d(\mathbf{x}) + g(\mathbf{s}_i, \xi_i)\} \tag{16}$$

With this term, the regions R_0 and R_1 change towards the translation of the hyperplane, and consequently the ECD function values are modified. If the function g is conveniently chosen, the ratio ϵ will acquire a non-zero value.

2.3 Estimation on the g Function

Consider the case in which the fraction of positive vs. negative samples in the test set is unknown, so the factor ρ can never be well estimated from the training data. In this work, following previous results in [4,5], the proposed solution to that case is to under-sample the dominant class to produce a balanced training set, and afterwards optimize the g function to a constant value that minimizes the false positive rate while keeps the ratio ϵ non-zero. If $g \geq d(\mathbf{s}_L)$, where $\mathbf{s}_L$ stands for the support vector with highest distance to the hyperplane then, from Eq. 15, $\epsilon \to 0$. But if we choose the set $A = \{\mathbf{s}_i : \xi_i \geq 1\}$ then:

$$g(\mathbf{s}_i, \xi_i) = \frac{1}{N_A} \sum_{i \in A} d(\mathbf{s}_i) \tag{17}$$

will guarantee a non-zero value for ϵ.

3 Results

We performed simulations of real case scenarios in which the data is imbalanced, and the unseen cases have a different imbalanced proportion than the training data. Such an example can be realized in dynamic-contrast-enhancing magnetic-resonance-imaging (DCE-MRI) for breast cancer diagnosis. Consider a DCE-MRI patient image with N voxels. In such case, the classification problem reduces to separate healthy tissues from malignant ones. A set of voxels m belonging to a malignant region constitutes a tumor. It is expected that the number of voxels b belonging to benign regions outnumber the voxels in malignant regions, so that $n/b \ll 1$ in either the training set or the unseen case. However, the size of the tumor is unknown in unseen cases, and do not have to match necessarily the size of tumors in the training set. Therefore, the proportion n/b is variable, but small. We will model that case in the following example:

Example 1: Let $p(w_1) = b/N$ and $p(w_2) = m/N$. Let

$$p(w_1 \,|\, \mathbf{x}) = \frac{1}{2\pi|\Sigma_1|^{\frac{1}{2}}} \exp\left(-\frac{1}{2}(\mathbf{x}-\mu_1)^T \Sigma_1^{-1}(\mathbf{x}-\mu_1)\right) \tag{18}$$

$$p(w_2 \,|\, \mathbf{x}) = \frac{1}{2\pi|\Sigma_2|^{\frac{1}{2}}} \exp\left(-\frac{1}{2}(\mathbf{x}-\mu_2)^T \Sigma_2^{-1}(\mathbf{x}-\mu_2)\right) \tag{19}$$

with $\mu_1 = 0$, $\mu_1 = 2$ and $\Sigma_1 = \Sigma_2 = \begin{pmatrix} 2 & 0 \\ 0 & 1 \end{pmatrix}$. The training set is built by joining $b = 10000$ samples of w_1 and $m = 100$ samples of w_2 (see Fig. 1). 20 different testing sets are built by modifying the proportion $\rho' = b'/m'$ ranging

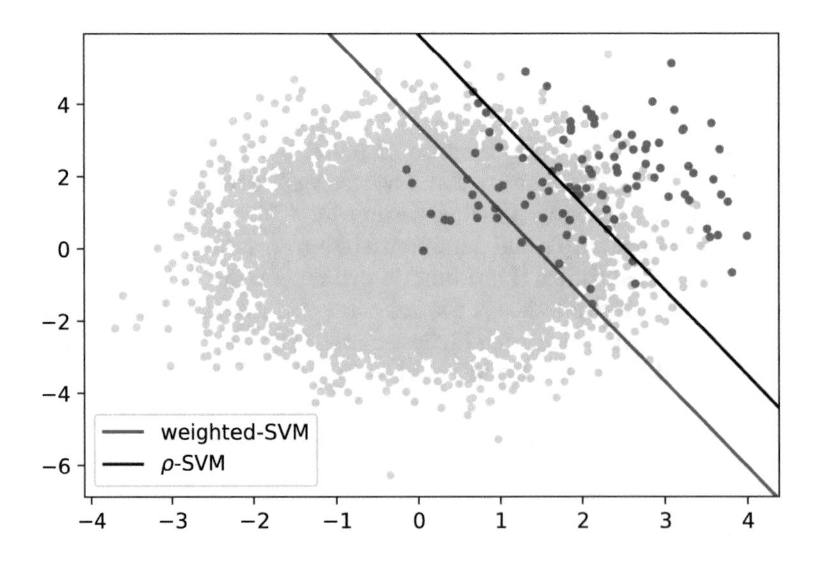

Fig. 1. Scatter plot of training set of Example 1. $\rho = 100$

Table 1. Average performance parameters on validating data with varying ρ'

	Test				
	False positives	ϵ	Accuracy	Specificity	Sensitivity
SVM		≈ 0	≈ 1	≈ 0	≈ 1
Weighted-SVM	8480 ± 30	0.013 ± 0.008	0.9149 ± 0.0003	0.889 ± 0.006	0.9152 ± 0.0003
ρ-SVM	1670 ± 15	0.2 ± 0.1	0.980 ± 0.001	0.713 ± 0.008	0.9833 ± 0.0002

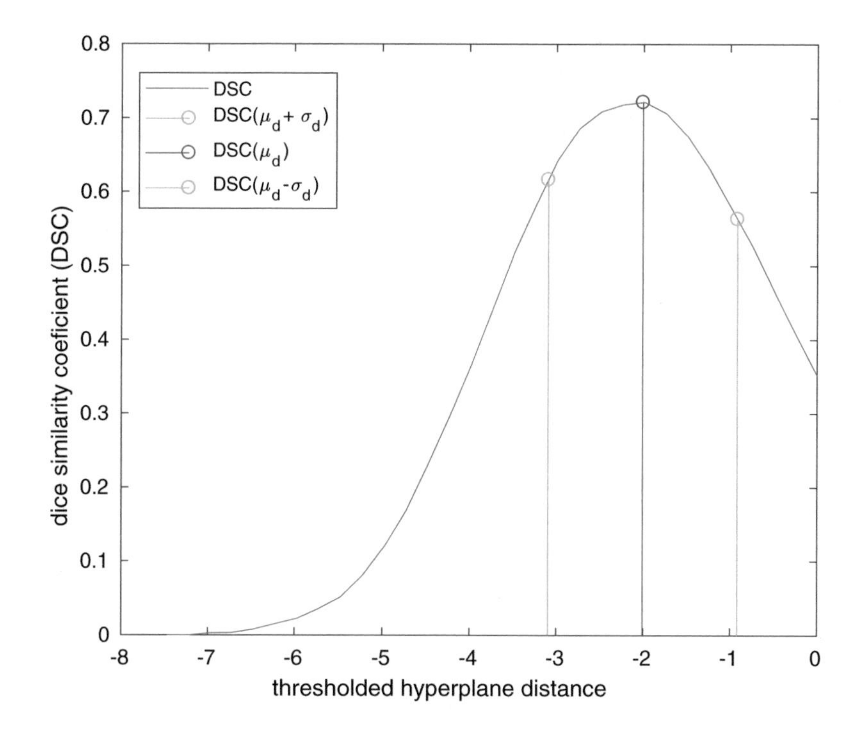

Fig. 2. Dice similarity coefficient for the validation data obtained by varying the decision defining value of the SVM hyperplane.

form $\rho' = 0.1\rho$ to $\rho' = 2\rho$ in 0.1 increments. The obtained results on the test sets are summarized on Table 1, where the parameters are averaged through all the possible values of ρ'. The results of weighted SVM are obtained following the equations on problem 10. The results of ρ-SVM are obtained training a SVM in a balanced set by downsampling the dominant class, and the applying the translation on the hyperplane decision function defined in Eq. 16.

Example 2: We collect a set of patients of a DCE-MRI database for breast cancer diagnosis, described in detail in [9]. We apply a feature selection process that combines PCA or ICA, described in [7], to classify benign and malignant regions on the images. We train a SVM in a balanced dataset obtained by downsampling the dominant class of benign voxels, and apply the translation on the hyperplane decision function defined in Eq. 16 to validate on a test set. We calculate the Dice similarity coefficient (DSC) between the manually delineated tumors by the radiologist and the voxels classified as malign by the hyperplane translated SVM. Figure 2 shows the results of the DSC for varying the hyperplane translation, including the proposed g value.

4 Discussion

We have provided evidence in Examples 1 and 2 that the false positive rate in medical imaging can be controlled by the proposed approach. In Table 1, it is shown that the false positives are reduced in almost one order of magnitude with the proposed ρ-SVM approach, while the values of accuracy, specificity and sensitivity maintain acceptable ranges if compared with the trivial solution provided by simple SVM. Moreover, the value of ϵ defined in Eq. 15 is shown to be higher for ρ-SVM, while in weighted SVM approximates to the trivial solution $\epsilon \to 0$. Figure 2 shows the DCS coefficients obtained by applying the false positive control defined previously. The validation results (in blue) are shown together with the DSC at mean distance g of the $\xi_i > 1$ support vectors (red), and its corresponding standard deviation interval (green). The DSC coefficient reaches a maximum by shifting the SVM hyperplane location approximately to $d = -2$. Although not displayed, the outlier support vectors defining the margin are at a maximum distance of $d = -4$, where the DSC is less than in $d = 0$. Therefore, locating the hyperplane at the mean distance μ_d duplicates the DSC, and prevents the effect of outliers. Moreover, the proposed value DSC(μ_d) is achieved within the margins of the actual empirical maximum, proving that the controlling strategy improves the DSC coefficient, reaching a peak value of 0.7215.

5 Conclusion

The problem of imbalanced datasets can be critical in medical imaging. Controlling the false positive rate is challenging when the imbalance between classes takes several orders of magnitude and the priors are not estimable. Here we propose a successful solution to that problem, and test it in synthetic and real data. The solution is based in the modification of the SVM decision function, by translating it to conservative values that avoid false positive rate increase. The solution is tested satisfactorily and compared to common weighted-SVM solutions.

Acknowledgment. This work has received funding from the European Unions Horizon 2020 research and innovation programme under the Marie Skodowska-Curie grant agreement No. 656886.

References

1. Akbani, R., Kwek, S., Japkowicz, N.: Applying support vector machines to imbalanced datasets. In: Machine Learning: ECML 2004. Lecture Notes in Computer Science, pp. 39–50. Springer, Berlin (2004)

2. Cortes, C., Vapnik, V.: Support-vector networks. Mach. Learn. **20**(3), 273–297 (1995)
3. Gammerman, A., Vovk, V., Vapnik, V.: Learning by transduction. In: Proceedings of the Fourteenth Conference on Uncertainty in Artificial Intelligence, UAI 1998, pp. 148–155. Morgan Kaufmann Publishers Inc., San Francisco (1998)
4. Gorriz, J.M., Ramirez, J., Suckling, J., Illan, I.A., Ortiz, A., Martinez-Murcia, F.J., Segovia, F., Salas-Gonzalez, D., Wang, S.: Case-based statistical learning: a non-parametric implementation with a conditional-error rate SVM. IEEE Access **5**, 11468–11478 (2017)
5. Gorriz, J.M., Ramirez, J., Suckling, J., Martinez-Murcia, F.J., Illan, I.A., Segovia, F., Ortiz, A., Salas-Gonzalez, D., Castillo-Barnes, D., Puntonet, C.G.: A semi-supervised learning approach for model selection based on class-hypothesis testing. Expert Syst. Appl. **90**(Supplement C), 40–49 (2017)
6. Gubern-Merida, A., Marti, R., Melendez, J., Hauth, J.L., Mann, R.M., Karssemeijer, N., Platel, B.: Automated localization of breast cancer in DCE-MRI. Med. Image Anal. **20**(1), 265–274 (2015)
7. Illan, I.A., Ramirez, J., Gorriz, J.M., Pinker, K., Meyer-Baese, A.: Automated non-mass enhancing lesion detection and segmentation in breast DCE-MRI (2018, to appear)
8. Li, M., Sethi, I.K.: Confidence-based classifier design. Pattern Recogn. **39**(7), 1230–1240 (2006)
9. Pinker, K., Grabner, G., Bogner, W., Gruber, S., Szomolanyi, P., Trattnig, S., Heinz-Peer, G., Weber, M., Fitzal, F., Pluschnig, U., Rudas, M., Helbich, T.: A combined high temporal and high spatial resolution 3 tesla MR imaging protocol for the assessment of breast lesions: initial results. Invest. Radiol. **44**(9), 553–558 (2009)
10. Platt, J.C.: Probabilistic outputs for support vector machines and comparisons to regularized likelihood methods. In: Advances in Large Margin Classifiers, pp. 61–74. MIT Press (1999)
11. Tao, Q., Wu, G.W., Wang, F.Y., Wang, J.: Posterior probability support vector machines for unbalanced data. IEEE Trans. Neural Netw. **16**(6), 1561–1573 (2005)
12. Vapnik, V.N.: Statistical Learning Theory. Wiley, New York (1998)
13. Zadrozny, B., Elkan, C.: Transforming classifier scores into accurate multiclass probability estimates. In: Proceedings of the Eighth ACM SIGKDD International Conference on Knowledge Discovery and Data Mining, KDD 2002, pp. 694–699. ACM, New York (2002)

Modeling a Synaesthete System to Generate a Tonal Melody from a Color Distribution

María Navarro-Cáceres(✉), Lucía Martín-Gómez, Inés Sittón-Candanedo, Sara Rodríguez-González, and Belén Pérez-Lancho

BISITE Digital Innovation Hub, University of Salamanca, Edificio Multiusos I+D+I, Calle Espejo s/n, 37007 Salamanca, Spain
maria90@usal.es

Abstract. Synaesthesia is a neurological phenomenon in which stimulation of one sense triggers an involuntary experience in a secondary sense. This paper draws on this phenomenon to generate a system that creates music using colors as a synaesthete input. The system optimizes a function based on the Tonal Interval Space by applying the Particle Swarm Optimization algorithm to automatically generate a tonal melody. The final result is a musical line that is evaluated in terms of musical quality.

Keywords: Music generation · Synesthesia
Particle swarm optimization

1 Introduction

One of the most peculiar phenomena in sensory pathways is synaesthesia, a perceptual condition in which the stimulation of one sense triggers an automatic experience in another sense. A synaesthete person can taste colors, listen to colors or touch sounds.

Among the different sensory associations, the combination of color-sound is especially interesting in the art world. The link between color and sound can be combined with intelligent techniques to obtain a more creative product [8]. [18] aims to compose music whose aim is relaxation and its tune depends on the way in which a physical material is stimulated by a user. The composition could also be based on a set of musical features that can express different moods and feelings [17], or based on some kind of metacreation, such as genetic algorithms, machine-learning or artificial intelligence techniques [6]. [15] created paintings that were the artists' interpretations of a small set of instrumental pieces. Results showed that people were able to correctly identify which painting corresponded to a particular piece of music. [20] presents a study on the possibilities of applying low-level visual-auditory associations to music generation. We should also highlight the research of [19], due to its similarity with our proposal. In it, the

M. Graña et al. (Eds.): SOCO'18-CISIS'18-ICEUTE'18, AISC 771, pp. 230–240, 2019.
https://doi.org/10.1007/978-3-319-94120-2_22

authors present a software that is able to generate acoustic sounds and images following a particle swarm optimization algorithm with motion captured by the camera.

Our proposal aims to model a system capable of transforming the colors of a picture into a musical melody by applying a swarm algorithm. The synaesthesia is used here as a source of inspiration for the generation of music. The input of the system will be a color map of one picture or a digital image. Each color will be associated to a particle, and together these colors will be transformed into different musical sounds. Initially, all the particles are immersed in a 2D space, where a negotiation process is started in order to optimize the sounds that can be part of a melody according to an optimization function. The design of this function is based on the Tonal Interval Space (TIS) proposed by [2] and used successfully for chord progressions [13] and music composition [3]. TIS is a 12-D space that uses the Fourier Transform (DFT) of the musical notes to measure distances. The space reflects musical properties through geometrical measures, which is very useful for the purposes of this work. Once the swarm algorithm converges, the notes are ordered according to the musical and position criteria of the particles. The final result is a color map translated into a music line.

Section 2 briefly describes the Tonal Interval Space. Section 3 details the principles which are the basis of the model, whereas Sect. 4 presents the proposed prototype and some preliminary results. Finally, the last section discusses the implications of the proposal and future work.

2 Tonal Interval Space Description

The Tonal Interval Space is a 12-D space where any pitch configuration, namely individual notes, chords or keys, can be represented. Initially, each element should be encoded by the corresponding binary chroma vector c, and transformed using the Discrete Fourier Transform (DFT). The result is a Tonal Interval Vector (TIV) T with six different components $T(k)$ that can be represented in the TIS. For an extensive description of the TIS, we reference Bernardes' work [2].

In the TIS, different geometrical measures can be applied to evaluate musical properties. For the purposes of this approach, the main geometrical property considered is the Euclidean distance.

Euclidean distance is used to measure two musical components: similarity between chords and consonance. In particular, the Euclidean measure between two pitch configurations is interpreted by how similar they are, and the Euclidean measure between the origin and the pitch configuration is related to the consonance of the pitch.

By encoding this phenomenon mathematically, Sim is obtained. It represents the Euclidean distance between the Fourier transform of the present candidate (note) and the previous note given as the input of the system.

$$Sim = \sqrt{\sum_{k=1}^{N} \mid T_i(k) - T_{prev}(k) \mid^2} \tag{1}$$

where $T_i(k)$ is the k-component of the i-candidate and $T_{prev}(k)$ represents the k-component of the previous note.

On the other hand, the consonance Con is encoded as the Euclidean distance between the chord point in the Fourier transform and the origin (point $(0+0i, 0+0i, 0+0i, 0+0i, 0+0i, 0+0i)$). Mathematically, consonance can be measured in our space as follows:

$$Con = 1 - \frac{\sqrt{\sum_{k=1}^{N} T(k)^2}}{C_{max}} \quad (2)$$

where: C_{max} is a value used to normalize v_1. Empirically, $C_{max} = 0.431$; $T(k)$ represents the k-component (complex number) of the Fourier transform of the chroma vector; N is the number of components. In our case, $N = 6$.

In tonal music, the notes most commonly used to create melodies are those close to tonality or main key. This is the main factor that maintains the hierarchy composed of the harmonic functions in the tonality. The distance between the note and the tonality is also modeled as the Euclidean distance shown in Eq. 1. Additionally, for the optimization function used in this manuscript, different parameters are empirically set, always to balance the influence of the musical properties in the total generation process.

3 Modeling the Problem

The present proposal aims to generate melodies from images provided by the users. Figure 1 shows the overall working flow where humans and a machine collaborate to generate music. The uploaders provide the digital images, considered the input of the system. The Hue, Saturation, Luminosity (HSL) codification is considered for the colors. Each color is then transformed into musical notes and located randomly in a search space. This is the starting point for a swarm algorithm which searches for consonant sounds, where each sound is an individual particle moving through the space. The best sounds are selected and ordered following the current position of the particle and tonal music criteria. Finally, the sounds are played so that the experts listen to the sounds and rate them.

The global system can be divided into different intelligent modules. The uploaders and the experts are considered individuals who are external to the system. The color module maps the color distribution of the digital picture. The synaesthete module transforms the colors into musical notes. For each correspondence, the module creates particles which are plunged into a two-dimensional space and it starts a swarm algorithm that searches for the best particles in the space. The group of notes is then transformed into MIDI notation to be played and listened to by musical experts.

The next sections will describe the different modules in detail.

3.1 Mapping Colors

A digital image usually contains multiple colors. Some of them are almost imperceptible to the human eye. Therefore, we extract dominant colors that can be

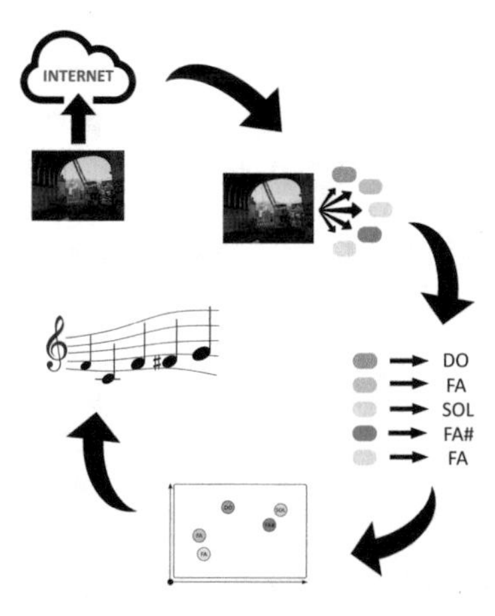

Fig. 1. Shows system workflow.

perceived from the image by applying color quantization, which reduces the number of colors used in an image. Among the most popular modern color quantization algorithms are the closest color algorithm (for fixed palettes), the Median cut algorithm, and an algorithm based on octrees. Due to its robustness, the Median cut algorithm is applied in our work [4, 11].

Although the Median-cut algorithm works with RGB codification, the final colors extracted are encoded in HSL, because the HSL colors can be automatically transformed into sound. In our study, a standard relationship following the Lagresille system is made [16]. According to this system, the Hue is associated with a specific note name. The saturation is related to loudness; the greater the color intensity, the louder the sound. However, the loudness will not be considered for the purposes of this work. Finally, luminosity represents the octave, when luminosity value is high, the sound should have a higher pitch, and viceversa. In order to preserve some coherence between different notes, luminosity is mapped between octave 2 and octave 6, according to MIDI codification (Fig. 2).

Fig. 2. Color extraction process in which the 70 most dominant HSL colors are considered.

3.2 Modeling Melody

To model the construction process of a melody, a particle swarm optimization is applied [5,7,10]. This algorithm has been successfully applied to generate music in different situations [14].

The fitness function F used by the particles to modify their paths considers the consonance properties between the i-particle a_i and the j-neighbor a_j according to Eq. 3.

$$F(a_i, a_j) = \sum_{i=0}^{M} (C(a_i, a_j)) \qquad a_i, a_j \subset A \tag{3}$$

C represents the tonal standards that the fitness function follows to evaluate the quality between two musical notes. This C function collects several properties in the TIS. As it is explained above, TIS is defined as a 12-D space, where geometrical distances of the Fourier Transform of notes captures musical properties. [2] demonstrate that Euclidean distances and other geometrical measures taken in such space capture some musical properties. In particular, the following measures were examined here:

- Consonance of the individual note a_i, measured as $Con(T(a_i))$ (Eq. 2).
- Similarity between two notes or particles a_1 and a_2, measured with $Sim(T(a_1), T(a_2))$ (Eq. 1).

C will be a linear combination between both previous measures, according to Eq. 4:

$$C(a_i, a_j) = \alpha \cdot Con(T(a_i)) + \beta \cdot Sim(T(a_1), T(a_2)) \tag{4}$$

where α and β are constant values empirically set to 1 and 0.5, respectively. In the main system, the particles must comply with some rules in order to prevent undesirable situations:

- There is a maximum distance between two particles. This prevents particles from going too far from the center of the system. To do so, the Euclidean distance between a_i and its nearest particle is calculated for each iteration. If the distance is above an established threshold, a temporal attraction force is created so that the particle can approach the other particles.
- Avoid collisions between particles. In this case, repulsive forces are needed if the distance between two particles is too small. If the Euclidean distance of a particle a_i and its nearest particle is below a threshold, a temporal repulsion force is generated to avoid possible collision.

Considering these two main rules, the position p_t of a particle a can be encoded as a function:

$$\boldsymbol{p}_t^i = f(\boldsymbol{p}_{t\text{-}1}^i, \boldsymbol{v}_t^i) \tag{5}$$

where p_{t-1}^i corresponds to its previous position and v_t^i the current velocity of the particle a_i.

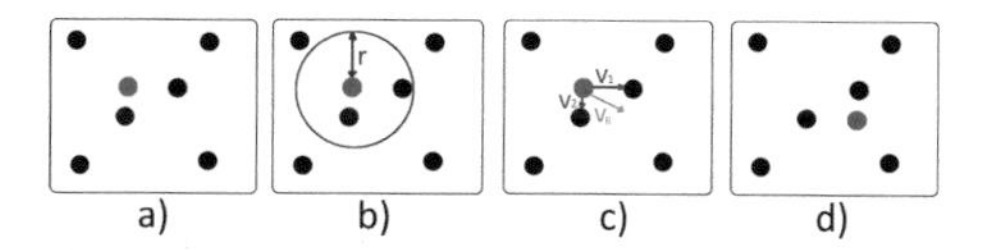

Fig. 3. PSO application. In each iteration, every particle is selected as shown in Fig. (a). Figure (b) examines the radius and its neighboring particles. Figure (c) shows the velocity between the main particle in the iteration and its neighbors, and the resultant velocity is calculated. Finally, Fig. (d) illustrates the new position of the main particle.

Figure 3 illustrates the process. In the first stage of the algorithm, the particles look for their neighbors Fig. 3a,b. One particle a_i can have several neighbors, and an attraction force will result in the comparison of a_i with each neighboring particle a_j, which will depend on the fitness function values obtained. Each attraction force between two particles creates a velocity vector, as shown in Fig. 3c.

Each vector will be influenced by the fitness function, the previous velocity or inertia phenomenon, and the direction. Likewise, the total velocity will be the vector sum of the velocities, and will depend on the different direction and modulus of each component. Figure 3c shows the total velocity according to the different components obtained from the comparison between particles a_1 and a_2, and a_1 and a_3. Equation 6 formalizes this operation.

$$\boldsymbol{v}_t^{total} = \sum_{k=1}^{N} F * \boldsymbol{v}_{t\text{-}1}^{i} + (\boldsymbol{p}_{t\text{-}1}^{k} - \boldsymbol{p}_{t\text{-}1}^{k}) \tag{6}$$

where k represents the k neighbor agent present in particle a_i. The fitness function F regulates the effect of the momentum (velocity) component. The vector $(p_{kt} - p_t))$ allows for the movement of particles towards the best position found by all the neighbor agents.

Finally, the particle takes a new position p_t^i which is a linear combination of the velocity and the previous position (Eq. 7).

$$\boldsymbol{p}_t^i = \boldsymbol{p}_t^i + \boldsymbol{v}_t^i \tag{7}$$

Within each iteration the particles are searching for better positions until they all converge. This convergence criterion occurs when the positions of the particles are not noticeably modified. At this stage, diverse groups of particles have been generated, which makes it easier for the ordering process to generate a melody.

3.3 Ordering Sounds

The notes grouped can now be ordered in a proper way. The first note will be the tonic note according to tonality. To select the tonality, Krumhansl's approach [12] is used, which presents an extraction method based on tonal hierarchies presented in the music composition. In our case, the appearances of each note are counted to extract a key profile and hereby extract the main tonality. Additionally, to avoid the repetition of notes, each time a note is added to the melody, the particle is discarded as candidate for the future.

Mathematically, to get the best note to follow the last note in the melody, the system searches for the best value of Eq. 8.

$$\begin{aligned} Comp = W_f \cdot F(a_i, a_j) + W_{key} \cdot Sim(T(a_1), T_{key} \\ + W_{at} \cdot Sim(T(a_i), T(a_j)) * e^{\gamma \cdot nSem} + W_c \cdot d(a_i, a_j) \end{aligned} \quad (8)$$

where $nSem$ represents the number of semitones between a_i and a_j, and γ is a real value empirically set to 0.3. W_f,W_{key}, W_{at} and W_c are empirically set to 1.1, 1.2, 7 and 1.3, respectively.

3.4 Modeling Rhythm

Rhythm assignation was not the main goal of our work. However, it was necessary to generate some simple rhythms for each note. That depends on the Euclidean distance between the notes. The distances between the maximum and the minimum obtained and considered durations of 2 pulses, 1 pulse and half pulse were framed. With these frames, each note a_i is given the corresponding duration according to the distance between a_i and the previous note in the melody. Once this task is accomplished, the MIDI information is extracted and transformed into audio information so that the computer can reproduce it.

4 Results and Discussion

This aim of this proposal is to design a creative system that transforms colors into sounds and generates a consonant melody from a tonal point of view. This system will be evaluated in two parts; first, the musical quality of the generated melodies will be assessed and then the correlation between colors and sounds will be analyzed.

Evaluation will consist of checking whether the fitness function values captures the musical quality of the sounds. Consequently, a preliminary experiment is designed with different musical excerpts composed by the system, leaving apart the images. This first experiment was presented as a listening test, with musical fragments in different tonalities, in particular, three fragments in C Major, two in F# Major, one in Db Major, one in D Major and one excerpt in G Major. The melody always begins with the tonic note to establish the key because the tonic determines the tonal basis of the music [1]. The final notes of the melody are also related to the tonic to give the feeling of musical ending. The listening test can be found online on this link: https://form.jotformeu.com/70191628235354. The listeners were asked to evaluate the overall quality of the melody according to the tonal standards. Each musical excerpt could be rated from 1 (very bad) to 5 (very good) according to the listeners' opinion (the listeners are considered musical experts with more than 6 years of musical training). Finally, 35 musical experts took this listening test and evaluated the musical excerpts.

We expect the objective function to reflect the perceptual quality of the melodies with lower objective values corresponding to worse candidates than higher values independently of the key.

For such categorical data, the chi-square test is used to prove that the results reflect the quality of the melodies according to the subjective ratings given by the listeners. It is important to note that we can get very subjective ratings since all listeners can have different interpretations of the musical pieces and different musical tastes. Thus, we considered the analysis of the median and the mode a useful step.

Table 1 collects statistical values calculated from the data, namely the p-value for the chi-value χ^2 the median M_e and the mode M_d. The statistical analysis suggests that the objective function captures the perceptual quality of the melodies. In particular, p-value leads us to reject the null hypothesis, which implies the existence of a relationship between objective values and subjective scores. The majority of listeners think the values are "Good" (according to the M_d) and the median indicates at least half of the ratings obtained are scored as "Good" or even better. The statistical results lead us to conclude that the fitness function captures perceptual musical quality quite well.

Table 1. Statistics resulting from organizing the subjective ratings as a function of distance measures from the TIS representation.

χ^2	M_e	M_o
$6.8381e-04$	Good	Good

The objective of the second part of the experiment is to measure the correlation between the color composition of a digital image and the sounds. Here, the picture plays an important role because it is a source of inspiration that generates the music. A new listening test was designed to rate the musical perception of music and sound. In this case, the same musical excerpts were presented along with the digital image representing the origin of the composition. The experts should rate the musical fragments between 1 (no similarity at all) and 5 (very similar) according to their personal perception of similarity between the music and the image.

In this case, 32 musical experts were considered to rate the perception of similarity between music and sound. Figure 4 shows the final results obtained.

The plot shows the level of similarity between the melody and the color considered by the listeners as well as the standard deviation obtained for each musical excerpt. The values were also normalized for clarification purposes. All the mean rates are above 0.5 ("similar enough") according to the subjective rates. However, the standard deviation calculated shows that some opinions are below this threshold, considering the melody as not similar enough. It is important to note that the perception of similarity between colors and sound is very personal and depends on factors such as culture or mood. Therefore, very different results

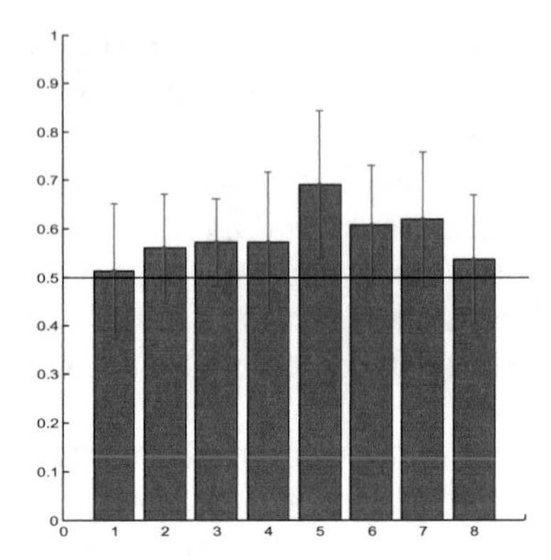

Fig. 4. Shows the similarity rates between colors and sound according to the listeners.

could be obtained for the same musical excerpt depending on the context of the listener. Additionally, telling experts that there is a connection between image and music can biased the analysis. As a future work it is proposed to develop a new analysis to compare the punctuations received by melodies associated with images and melodies that are not.

In general, the melody aims to follow the main tonal standards but is always inspired by a painting. Thus, the melody might not be a full tonal piece, it is not designated to be a classical melody. Although Western tonal standards are quite mainstream when asked to evaluate a melody, music is always a subjective art that depends on personal preferences and culture.

Another problem detected is that the number of colors determines the notes that will be used in the composition. When a picture has very uniform colors, the melody contains a low diversity of notes. For future work, we propose extracting musical motifs using the notes given in the system, and use them as a basis for a musical composition following other parameters that can be also extracted from the image, such as shapes or distribution of elements.

Eventually, the similarity between color and sounds was very difficult to measure, as personal perceptions play a crucial role in the final results. This topic needs a more indepth analysis in which psychological, social and cultural issues should be considered. In a future study, a new test will be developed to improve how the perception of correlation between music and image is collected.

5 Conclusions

The system proposed can transform colors into a melody, based on a particle swarm optimization algorithm (PSO).

From the particles to the generation of music, voice-leading criteria and Euclidean distances between particles were followed to select the next particle in the melody. The final results were presented to the experts, who evaluated the musical quality and the music's similarity to the digital pictures, which were the source of inspiration.

Specifically, two listening tests were designed with the same musical fragments, one to measure the musical quality, and the other to collect the overall perception of the experts on the similarity of colors and sound. Both tests gave encouraging results: the music quality was generally well rated and the association between music and sound was fairly well perceived among our users.

This work presents an innovation with respects to previous proposals, such as Monalisa App [9]. Monalisa is a standalone application that also converts images into sound and vice-versa, but using a fixed translation and rules. In this particular case, the particles are dynamically changing, producing different melodies following quality measures and optimizing a mathematical function which encodes musical properties.

Acknowledgments. This work was supported by the Spanish Ministry, Ministerio de Economía y Competitividad and FEDER funds. Project. SURF: Intelligent System for integrated and sustainable management of urban fleets TIN2015-65515-C4-3-R.

References

1. Agmon, E.: Functional harmony revisited: a prototype-theoretic approach. Music Theor. Spectr. **17**(2), 196–214 (1995)
2. Bernardes, G., Cocharro, D., Caetano, M., Guedes, C., Davies, M.E.: A multi-level tonal interval space for modelling pitch relatedness and musical consonance. J. New Music Res. **45**(4), 281–294 (2016)
3. Bernardes, G., Cocharro, D., Guedes, C., Davies, M.E.: Conchord: an application for generating musical harmony by navigating in a perceptually motivated tonal interval space. In: Proceedings of the 11th International Symposium on Computer Music Modeling and Retrieval (CMMR), pp. 71–86 (2015)
4. Casado-Vara, R., Corchado, J.M.: Blockchain for democratic voting: How blockchain could cast off voter fraud. Orient. J. Comp. Sci. Technol. **11**(1) (2018)
5. Chamoso, P., Prieta Pintado, F.d.l.: Swarm-based smart city platform: A traffic application (2015)
6. Eigenfeldt, A., Thorogood, M., Bizzocchi, J., Pasquier, P.: Mediascape: towards a video, music, and sound metacreation. J. Sci. Technol. Arts **6**(1), 61–73 (2014)
7. Faia, R., Pinto, T., Abrishambaf, O., Fernandes, F., Vale, Z., Corchado, J.M.: Case based reasoning with expert system and swarm intelligence to determine energy reduction in buildings energy management. Energ. Build. **155**, 269–281 (2017)
8. Horn, B., Smith, G., Masri, R., Stone, J.: Visual information vases: towards a framework for transmedia creative inspiration. In: Proceedings of the Sixth International Conference on Computational Creativity, p. 182 (2015)
9. Jo, K., Nagano, N.: Monalisa: see the sound, hear the image. In: Proceedings of the 8th International Conference New Interface for Musical Expression, NIME, vol. 8, pp. 315–318 (2008)
10. Kennedy, J.: Particle swarm optimization. In: Encyclopedia of Machine Learning, pp. 760–766. Springer (2010)
11. Kruger, A.: Median-cut color quantization. Dr Dobb's J. Softw. Tools Prof. Programmer **19**(10), 46–55 (1994)

12. Krumhansl, C.L., Cuddy, L.L.: A theory of tonal hierarchies in music. In: Music perception, pp. 51–87. Springer (2010)
13. Navarro, M., Caetano, M., Bernardes, G., de Castro, L.N., Corchado, J.M.: Automatic generation of chord progressions with an artificial immune system. In: Evolutionary and Biologically Inspired Music, Sound, Art and Design, pp. 175–186. Springer (2015)
14. Poli, R.: An analysis of publications on particle swarm optimization applications. Department of Computer Science, University of Essex, Essex, UK (2007)
15. Ranjan, A., Gabora, L., OConnor, B.: The cross-domain re-interpretation of artistic ideas. In: Proceedings 35th Annual Meeting of the Cognitive Science Society, pp. 3251–3256 (2013)
16. Sanz, J.C.: Lenguaje Del Color: Sinestesia cromática en poesía y arte visual. H. Blume (2009)
17. Scirea, M., Nelson, M., Cheong, Y.G., Bae, B.C.: Mood expression in real-time computer generated music using pure data (2014)
18. Sysoev, I., Chitloor, R.D., Rajaram, A., Summerlin, R.S., Davis, N., Walker, B.N.: Middie mercury: an ambient music generator for relaxation. In: Proceedings of the 8th Audio Mostly Conference, p. 20. ACM (2013)
19. Unemi, T., Matsui, Y., Bisig, D.: Identity sa 1.6: an artistic software that produces a deformed audiovisual reflection based on a visually interactive swarm. In: Proceedings of the 2008 International Conference on Advances in Computer Entertainment Technology, pp. 297–300. ACM (2008)
20. Wu, X., Li, Z.N.: A study of image-based music composition. In: 2008 IEEE International Conference on Multimedia and Expo, pp. 1345–1348. IEEE (2008)

Minimization of the Number of Employees in Manufacturing Cells

Wojciech Bożejko(✉), Jarosław Pempera, and Mieczysław Wodecki

Faculty of Electronics, Wrocław University of Technology, Janiszewskiego 11-17, 50-372 Wrocław, Poland
{wojciech.bozejko,jaroslaw.pempera,mieczyslaw.wodecki}@pwr.edu.pl

Abstract. In the paper we consider the problem of scheduling of technological operations implementation in productions cells in a company which produce steel structures of car seats. We propose an optimization algorithm based on the Branch and Bound method which determines minimal number of team members which operating production cells maintaining the maximum efficiency of the cells. The usefulness of the algorithm for practical purposes has been verified on real data.

Keywords: Optimization · Production cell · No-store constraint
Cyclic scheduling

1 Introduction

Robotization and automation of manufacturing systems enables relatively quick adaptation of production plans to the demand generated by contractors. This is particularly important in the case of production in large volumes, large sizes or with large masses. Then, storage costs can be a significant part of the cost of manufacturing of the final product. In this type of companies, a just-in-time or just-in-sequence strategies are used.

In the enterprise at work, the process of manufacturing seats and other car equipment includes product design, computer and real-life testing related to the safety of their use, and the production of finished products. Extensive experience in both designing and production of products allows the company to produce many products for many contractors.

Despite of the significant and continuous development of robots and automation devices, the most important element of the manufacturing system is man. In advanced production systems, its role is often to precisely place processed products in the holders of automatic welding machine modules, to make welding corrections or to weld in hard to reach places, to carry out product quality control and to level and polish the surface of products at various stages of production.

Nowadays a chronic shortage of highly-qualified workforce is observed in many countries. This situation does not change significantly despite an increase in financial incentives for employees. Therefore, effective management of the staff is very important.

M. Graña et al. (Eds.): SOCO'18-CISIS'18-ICEUTE'18, AISC 771, pp. 241–248, 2019.
https://doi.org/10.1007/978-3-319-94120-2_23

2 Manufacturing Cell

The task of a typical production cell is to weld previously embossed metal parts into a raw intermediate or end product. In later stages, it is varnished and packaged. The cell consists of several working places (here called machines). In each of the machines, specific technological activities are performed. Most often these are welding operations performed in welding modules or in a few cases by hand. Welding operations are preceded by the placement of the product in the holder that the operator performs and supervises. The machines in the cell are placed on the perimeter of the straight-angle. Metal parts produced on the press and other parts are delivered in containers placed in a separate place near the working place.

Cells are designed for the production of one product or several types of similar products. The change of production from one product to another requires in some cases the replacement of some tools. Products in the cell are produced in series. In one series, identical products are produced. There are no buffers (temporary storage places) between the positions where the partially processed product could be stored. It is caused by a limited surface allocated for the cell and additional activities and costs related to the operation of buffer devices. The operator may serve several machines, in particular, machines that are physically close to each other. The maximum productivity of the production cell is directly related to the time of performing the longest technological activity.

Cyclic production is required to increase the efficiency and simplify the operation of the operators. In cyclic production, the operator performs most of the working time in the same order. This rule does not apply for a short period at the beginning and the end of the production of series of products.

We want to set a minimum number of cell operators (employees, or – in the future – robots) ensuring production with its maximum efficiency. Determination of such a number is of special importance at the time of high absenteeism of employees or urgent execution of many orders for contractors.

3 Problem Description

The production cell consists of m production machines from the set $M = \{1, 2, \ldots, m\}$. In the cell, cyclically, n products should be made. Each product is processed in each machine in the order of $1, 2, \ldots, m$. Each product therefore generates a production task consisting of m operations. There are no buffers between stations [11,12] (so called *no-store* constraint). The processing time of the job at the position $k \in M$ is $p_k > 0$ [minutes]. The maximum cell productivity is defined as

$$P = 60/\max_{k \in M} p_k \tag{1}$$

pieces per hour.

Each employee is qualified to work in any machine, performs the tasks in a given machine or supervises them throughout the entire processing time in the position. Only one product can be made at a given time, the product can be processed on only one machine and the employee can work on only one machine. Performing tasks on machines cannot be interrupted.

Each technological operation performed by the operator in the system is represented by the pair (j, i), where $j = 1, 2, ..., n$ denotes the sequence number of the product, and $i \in M$ machine for the cell. Let us denote by α_k the order of performing technological activities by the operator $k = 1, 2, ..., o$. The sequence of performing all operations by operators in the production cell is described by $\alpha = (\alpha_1, \ldots, \alpha_o)$, $\alpha_k = (\alpha_k(1), \ldots, \alpha_k(n_k))$, where n_k is the number of operations performed by the operator k. Further, let us denote by $S_{j,i}$ and $C_{j,i}$ the start and completion time, respectively, of the task j on a machine i. The schedule of performing tasks for a given sequence α has to fulfill the following constraints:

$$S_{j,i} \geq 0, \quad j = 1, \ldots, n, \quad i = 1, \ldots, m, \tag{2}$$

$$S_{j,i} \geq C_{j,i-1}, \quad j = 1, \ldots, n, \quad i = 2, \ldots, m, \tag{3}$$

$$C_{j,i} = S_{j,i} + p_{j,i}, \quad j = 1, \ldots, n, \quad i = 1, \ldots, m, \tag{4}$$

$$S_{\alpha_k(s)} \geq C_{\alpha_k(s-1)}, \quad s = 2, \ldots, n_k, \quad k = 1, \ldots, o, \tag{5}$$

$$S_{\alpha_k(s)} \geq S_{A(\alpha_k(s-1))}, \quad s = 2, \ldots, n_k, \quad k = 1, \ldots, o, \tag{6}$$

where $A((j, i)) = (j, i + 1)$ denotes the technological successor of the operation (j, i). Inequality (2) anchors the task execution schedule at time 0. Inequality (3) means that the start time of the operation of a given task cannot be earlier than the end of the previous operation of this task. Equality (4) means that the operation cannot be interrupted. It results from the inequality (5) that the start time of the s-th operation performed by the operator k can only start after the previous operation has been completed. The limitation "no-store" models the constraint (6) and means that we can start operations only after the job has been released by the previous task, i.e. when the next operation of this task begins its execution.

The order of operations performed by the α operators is feasible if there is a solution of the inequalities (2–6). Checking the feasibility of α and the schedule of performing operations in the order α can be determined in time $O(nm)$ by constructing a directed graph based on the inequalities (2–6) (similar to work [8]).

The sequence α will be called the cyclic sequence if it is constructed on the basis of a cyclic core. The cyclic core consists of cyclically performed m operations for one or more tasks. The order of executing operations from the cyclic core will be denoted by symbolem $\pi = (\pi_1, \ldots, \pi_o)$, $\pi_k = (\pi_k(1), \ldots, \pi_k(\vartheta_k))$, $m = \sum_{s=1}^{o} \vartheta_s$ where ϑ_k is the number of operations performed by the operator k. Then the order α is a core assembly, i.e.

$$\propto = \pi^0 \pi\pi \ldots \pi\pi^*, \tag{7}$$

where π^0 - stands for the preliminary order, $\pi\pi ... \pi$ - repeatedly repeating the cyclic core, π^* - the ending order.

The most commonly considered literature in the literature are fully automated production cells in which the transport of products is carried out by an industrial robot [3,6,7,9].

3.1 Example

The production cell consists of $m = 6$ positions and is operated by $o = 4$ operators. The names of technological activities and machining times [in seconds] are given in Table 1. Operators 1, 2, 3 perform only one operation respectively 3, 4 and 6, while operator 4 operations 1, 2 and 5. Consider a cyclic core $\pi = (\pi_1, \ldots, \pi_4)$, in the following form: $\pi_1 = ((1,3))$, $\pi_2 = ((1,4))$, $\pi_3 = ((1,4))$, $\pi_4 = ((5,1),(1,3),(2,3))$.

The system should perform $n = 7$ products. For such a cyclic core, the order π^0 and π^* take the form $\pi^0 = (\pi_4^0)$, where $\pi_4^0 = ((1,1),(2,1),(1,2),(2,2),(1,3),(2,3))$ and $\pi^* = (\pi_1^*, \pi_2^*, \pi_3^*, \pi_4^*)$, where $\pi_1^* = ((5,3),(6,3),(7,3))$, $\pi_2^* = ((5,4),(6,4),(7,4))$, $\pi_3^* = ((5,6),(6,6),(7,6))$, $\pi_4^* = ((5,5),(6,5),(7,5))$.

Table 1. Technological activities performed in the production cell

Position	Operation name	Execution time [sec.]
1	Clinching	45
2	Welding	26
3	Welding	134
4	Manual welding	104
5	Gurtholma welding	34
6	Assembly	104

Figure 1 illustrates the schedule for performing operations in the form of Gantt chart. In the graph: dotted line, the schedule for performing operations resulting from the order π^0 and π^* is marked, the continuous operation schedule resulting from the first and last occurrence of the cyclic core is marked, the

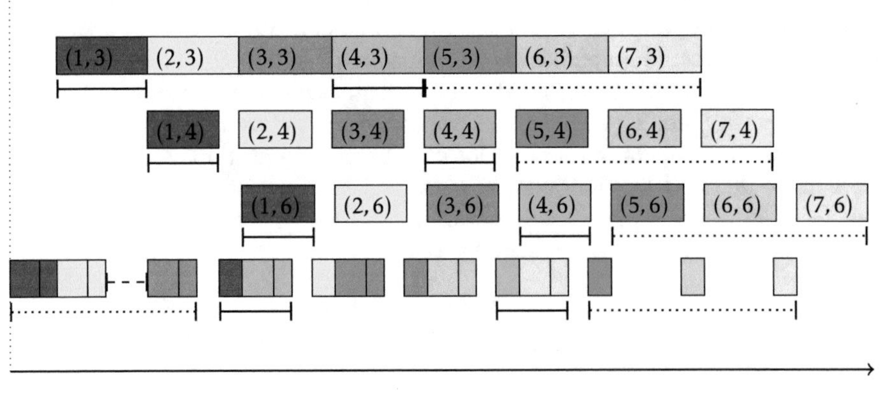

Fig. 1. The schedule of operations performed by operators

Table 2. Schedule of technological operations

Part		Operations					
		1	2	3	4	5	6
1	S_{1i}	0	45	71	205	309	343
	C_{1i}	45	71	205	309	343	447
2	S_{2i}	71	116	205	339	443	477
	C_{2i}	116	142	339	443	477	581
3	S_{3i}	205	250	339	473	577	611
	C_{3i}	250	276	473	577	611	715
4	S_{4i}	343	388	473	607	711	745
	C_{4i}	388	414	607	711	745	849
5	S_{5i}	477	522	607	741	845	879
	C_{5i}	522	548	741	845	879	983
6	S_{6i}	611	656	741	875	979	1013
	C_{6i}	656	682	875	979	1013	1117
7	S_{7i}	745	790	875	1009	1113	1147
	C_{7i}	790	816	1009	1113	1147	1251

dashed line the delay resulting from station blocking due to the lack of interstate buffer. In addition, the start and end times of all operations are summarized in Table 2.

Let us note that the operator 4 supports 3 positions: 1, 2 and 5. Therefore, they should be located nearby, eg stations 1 and 2 should be placed next to each other while 5 in front of positions 1 and 2. The work schedule of this operator is the most complicated, especially at the beginning of work. First, it performs the first two operations of the first task, then the first two operations of the second task, then it must wait for the second position to be released (i.e. until the task 2 in step 1 starts) and perform the first two operations of the third task. Subsequent production activities of the operator 4 are carried out cyclically and consist of three technological operations, successively in stand 5, 1 and 2, and a break related to the wait for the operator 2 to complete the operation.

4 Optimization Algorithm

At the very beginning, note that after a sufficient number of repeats of the cyclic core, the starting moments for the corresponding operations belonging to two successive cores $x-1$ and x are separated by the same period of time T, i.e.

$$S_{x,i} = S_{x-1,i} + T, \quad i = 2, \ldots, m. \tag{8}$$

The period T will be called the *cycle time*. It is easy to see that the production socket works at full productivity (its performance is limited by the time of the longest operation performed) if the cycle time is $\max_{k \in M} p_k$.

The length of the cycle time for the cyclic core π can be determined by generating the order α and determining T in the steady state. Unfortunately, this type of method requires the generation of α based on the large number of core repetitions. Cycle time can be determined much faster using a method similar to that used in [1,2,4,5,10].

In a natural way, the lower limit of the number of operators is FT/T, where $FT = \sum_{s=1}^{m} p_s$ is the time the task flows through the production system, while the upper limit is equal to the number of machines. The minimum number of operators can be determined by searching half of the minimum number of operators for which a schedule with cycle time $T = \max_{k \in M} p_k$ can be constructed.

Due to the small number of operations in the cyclic core, it was decided to construct a complete review algorithm, in which all possible operations assignments to operators and all possible operator execution orders are generated. During the operation of the algorithm, cyclic cores fulfilling the condition

$$\max_{k \in M} \sum_{i=1}^{\vartheta_k} p_{\pi_k(i)} > T, \tag{9}$$

were not generated.

Of course, among the cores that meet this condition may be those for which it is impossible to construct a cyclic schedule with a cycle time T. The algorithm ends when the core is found with the cycle time T or after generating all cyclic cores for a given number of operators.

5 Computational Experiments

Computer research was carried out on 20 products produced in the company. The following data was collected for each product:

1. m - the number of technological operations,
2. $FT = \sum_{s=1}^{m} p_s$- time of task flow through the production system,
3. $T = \max_{k \in M} p_k$ – cycle time.

Next, the minimum number of employees NO was determined by the optimization algorithm and the following were determined:

1. FT/T - lower bound estimation of the number of employees,
2. m/NO - the ratio of the number of positions to the number of employees.

The optimization algorithm has been implemented in C++ in the Visual Studio 2010 environment and launched on a computer with an Intel i7 2.4 GHz processor working under the Windows 8.1 operating system. The operation time of the optimization algorithm for each product did not exceed several dozen seconds. The analysis of the experimental results collected in Table 3 shows that in all 20 cases, the algorithm has determined the optimal number of operators equal to the lower bounds estimation. It should be noted that determining the lower

Table 3. Results of experimental research

Product	m	FT	T	NO	FT/T	m/NO
1	9	10.197	2.39	5	5	1.8
2	10	11.049	2.39	5	5	2
3	12	12.579	2.989	5	5	2.4
4	12	12.579	2.989	5	5	2.4
5	12	13.431	2.989	5	5	2.4
6	5	19.129	10.6	2	2	2.5
7	5	19.563	10.6	2	2	2.5
8	5	13.258	9.204	2	2	2.5
9	6	7.452	2.232	4	4	1.5
10	7	13.319	4.061	4	4	1.7
11	8	14.399	4.061	4	4	2
12	8	17.498	7.416	3	3	2.6
13	9	18.212	7.416	3	3	3
14	6	13.002	4.41	3	3	2
15	5	18.09	12.766	2	2	2.5
16	3	5.202	2.487	3	3	1
17	3	4.044	1.825	3	3	1
18	3	4.044	1.825	3	3	1
19	3	4.405	2.186	3	3	1
20	3	12.321	5.187	3	3	1

bound of the number of employees is trivial, but determining the allocation of operators to machines and technological activities and determining a schedule for their implementation for most products is very difficult and requires the use of the proposed algorithm.
The values of the m/NO coefficient range from 1 to 2.6. The value means that for certain products the number of employees is equal to the number of machines, so each employee only serves one machine. In the case of the highest value of this coefficient, the average employee supports as many as 2.6 machines, i.e. 3 employees operate 8 machines in the production cell with full efficiency.

6 Summary

The work considers the problem of determining the smallest number of operators in a multi-station manufacturing cell. The original concept of the so-called a cyclic core simplifying the sequence of servicing many positions by employees. The method of constructing the order of performing all technological activities and the schedule of their implementation has been proposed.

An optimization algorithm based on the method of an all-round review of all cyclic cores has been proposed. The algorithm eliminates the generation of cyclic cores for which the schedule would not guarantee operation of the cell with a full productivity. Research has been carried out on 20 products manufactured in a company that makes car seats. For all products, a minimum number of employees was set equal to the lower estimate and an acceptable work schedule.

The method of constraints modeling proposed in the work and the optimization algorithm can be used to design manufacturing cells and determine the number of employees, assign them to machines, set a cyclical schedule of their implementation for new production cells and improve the existing ones. The proposed exact algorithm is proposed, therefore the number of employees is determined optimally.

Acknowledgement. The paper was partially supported by the National Science Centre of Poland, grant OPUS no. DEC 2017/25/B/ST7/02181.

References

1. Bożejko, W., Gnatowski, A., Pempera, J., Wodecki, M.: Parallel tabu search for the cyclic job shop scheduling problem. Comput. Ind. Eng. **113**, 512–524 (2017)
2. Bożejko, W., Pempera, J., Wodecki, M.: A fine-grained parallel algorithm for the cyclic flexible job shop problem. Arch. Control Sci. **27**, 169–181 (2017)
3. Bożejko, W., Gnatowski, A., Klempous, R., Affenzeller, M., Beham, A.: Cyclic scheduling of a robotic cell. In: Proceedings of 7th IEEE International Conference on Cognitive Infocommunications CogInfoCom 2016, 16–18 October, Wrocław, Poland, pp. 379–384 (2016)
4. Bożejko, W., Uchroński, M., Wodecki, M.: Parallel metaheuristics for the cyclic flow shop scheduling problem. Comput. Ind. Eng. **95**, 156–163 (2016)
5. Bożejko, W., Uchroński, M., Wodecki, M.: Block approach to the cyclic flow shop scheduling. Comput. Ind. Eng. **81**, 158–166 (2015)
6. Che, A., Chu, C.: Multi-degree cyclic scheduling of a no-wait robotic cell with multiple robots. Eur. J. Oper. Res. **199**(1), 77–88 (2009)
7. Drobouchevitch, I.G., Sethi, S.P., Sriskandarajah, C.: Scheduling dual gripper robotic cell: one-unit cycles. Eur. J. Oper. Res. **171**(2), 598–631 (2006)
8. Grabowski, J., Pempera, J.: The permutation flow shop problem with blocking. A tabu search approach. Omega **35**, 302–311 (2007)
9. Levner, E., Kats, V., Levit, V.E.: An improved algorithm for cyclic flowshop scheduling in a robotic cell. Eur. J. Oper. Res. **97**(3), 500–508 (1997)
10. Pempera, J., Smutnicki, C.: Open shop cyclic scheduling. Eur. J. Oper. Res. **269**, 773–781 (2018)
11. Ronconi, D.P.: A branch-and-bound algorithm to minimize the makespan in a flowshop with blocking. Ann. Oper. Res. **138**, 53–65 (2005)
12. Ronconi, D.P.: A note on constructive heuristics for the flowshop problem with blocking. Int. J. Prod. Econ. **87**, 39–48 (2004)

Soft Computing Methods in Manufacturing and Management Systems

Rationalization of Production Order Execution with Use of the Greedy and Tabu Search Algorithms

Kamil Musiał, Joanna Kochańska, and Anna Burduk(✉)

Faculty of Mechanical Engineering, Wroclaw University of Science and Technology, Wybrzeze Wyspianskiego 27, 50-370 Wroclaw, Poland
{kamil.musial, joanna.kochanska, anna.burduk}@pwr.edu.pl

Abstract. In the paper, rationalization of production order execution in a large manufacturing company is suggested. Till now, the decision-making process was based on the human factor, which resulted in irregular utilization of manufacturing resources. The presented work was aimed at developing new order selection system. That would make it possible to utilize the admitted resources possibly best and thus to meet the deadlines and to adapt to specific production requirements. In the work, the greedy and Tabu Search algorithms were used. As a basis for the research employing, historical data were accepted. Each order were given a priority, production time, profit and penalty for failure. Additionally, the machine failure risk was calculated based on empirical measurements. Simulations of several subsequent working weeks were performed in order to analyse the results obtained thanks to the suggested methods and to compare them with the results presently reached by the company.

Keywords: Execution of production orders · Decision-making processes
Greedy algorithm · Tabu search · Meta-heuristics
Intelligent optimisation methods of production systems

1 Introduction

From the point of view of a company oriented to profit maximisation, execution of production orders is an important process [1, 12, 18]. Correct managing this process means striving for possibly high profits with use of possibly small (financial, temporal, material, human etc.) resources. At present, any algorithms leading to possibly best solutions of the set problems are most often used as supporting methods for decision-making processes [8, 9, 13, 14, 16, 19]. Their application reduces time, labour input and human errors in the process of making production decisions.

In this paper, application of the greedy and the Tabu Search algorithms for decision-making in managing execution of production orders is suggested. The greedy algorithm takes a locally optimal decision with no assessment of its effect on further steps [7, 15]. In turn, the Tabu Search algorithm searches the solution space by means of specific sequence of movements [3, 6, 11]. These two algorithms can be used at solving variable optimisation problems, both the simpler ones and the NP

M. Graña et al. (Eds.): SOCO'18-CISIS'18-ICEUTE'18, AISC 771, pp. 251–259, 2019.
https://doi.org/10.1007/978-3-319-94120-2_24

(nondeterministic polynomial) problems [2, 4, 5, 10, 17]. None of the above algorithms guarantees finding the optimum solution, but they make it possible to indicate the solution sufficiently good and acceptable from the company point of view. The research aimed at selecting the method to rationalise management of production orders was carried-out on the example of a large manufacturing company from the automotive industry.

2 Optimisation Algorithms

2.1 Greedy Algorithm

The greedy algorithm is a method of finding a solution by making a selection in each step [15]. This means that the chosen solution is the one that seems to be the best at the moment. The rating is locally optimal but does not analyse, whether the data in the following steps will make sense [7, 15]. This method is most often used for maximisation or minimisation of the considered value. So, it chooses the target function that locally maximises or minimises the target value. Main advantages of this type algorithms include fast action, moderate memory requirements and easy implementation [15, 20].

2.2 Tabu Search

Tabu Search is an algorithm that searches the space created by all possible solutions with a particular sequence of movements [3, 6, 11]. Among them, there are taboo (forbidden) movements [6, 11]. The algorithm avoids oscillation around the local optimum by storing information about proven solutions in form of a Taboos List (TL). The Tabu Search algorithm is deterministic, in contrast to a genetic algorithm. Treating the same data several times with the same Tabu Search algorithm gives identical results.

3 Case Study

The research aimed at improvement of order execution was carried-out on the example of a large manufacturing company from automotive industry, where a possibility was noticed to improve work performance by a change of planning the time schedule. So far, the decision-making process was based on the human factor only, which not always made it possible to fully utilize the available production resources, but also permitted generation of errors. In the face of complexity of the production process, "manual" planning based on experience appeared insufficient.

The plant realizes various size orders for many customers. For each order, production time and profit are estimated. In addition, the orders are given priorities related to their significance resulting from the concluded contracts. A failure to keep the order deadline entails variable financial penalties dependent of the contract terms. The plant works in three-shift system, five days a week. When necessary, the additional weekend shift is launched, which requires additional financial expenses. The additional problem related to production planning is retooling time of individual machines. Various groups of products need availability of various machines and retooling time of the machinery depends on the longest retooling time among all the machines. Moreover, for each machine there is a certain risk of failure, estimated on the grounds of previous failure history.

Therefore, production management requires handling the sequence of the processed orders, taking into consideration their priorities and outlays in form of both costs of production resources and time (dependent also on number of retoolings).

3.1 Basic Example for the Algorithms

As a basis for the research employing the above-mentioned algorithms, historical data of 2017 were accepted, when, within execution of 6 projects, 51 various products were manufactured in the plant. Each of them passed through 3 main production stages, namely die stamping, assembly and painting. Individual processes required suitable work stations. Therefore, material flow was various for each product. In addition, even if the main part of the machines is dedicated to specific groups of products, there are also some machines participating in manufacture of several products, or shared among various products.

3.2 Accepted Parameters

Within elaboration of the algorithm, the following parameters were accepted for each order:

- priority [prio]- [$prio_1$, $prio_2$, …, $prio_n$] - depends on terms of the contract with the customer and on the required order execution time;
- production time [pt]- [pt_1, pt_2, …, pt_n] - required for the order execution;
- profit [p]- [p_1, p_2, …, p_n] - that the company can gain thanks to its execution;
- penalty [pen]- [pen_1, pen_2, …, pen_n] - for not completing the order;
- risk [r]- [r_1, r_2, …, r_n] - of the machine failure is accepted as the factor predetermining the order execution, calculated on the grounds of empirical data based on history of failures and repairs.

The purpose is selection of orders in the way guaranteeing possibly highest profit, at the same time meeting the order execution deadlines and properly making provision against potential risk of the machine failures.

4 Rationalization with Use of Tabu Search and Greedy Algorithms

Effectiveness of the selected algorithms was examined by simulations of subsequent working weeks. For the examined case, the Tabu Search and greedy algorithms were considered. The aim was to find the calculation method giving a solution in maximum

5 s using a mid-range tablet. This would allow the company to modify orders and parameters easily, as well as to obtain calculated solutions in nearly real-time.

4.1 Solutions of the Greedy Algorithm

Greedy:

Sort in descending order:
In this step, the algorithm multiplies profit of each order by its priority value and production time, and sorts the products in descending sequence:

$$prio_{1'}p_{1'}pt_{1'} \geq prio_{2'}p_{2'}pt_{2'} \geq prio_{3'}p_{3'}pt_{3'} \geq \ldots \geq prio_{n'}p_{n'}pt_{n'} \quad (1)$$

Add highest orders:
Thereafter, the algorithm adds production times from the highest sorted values until it reaches the maximum value, not exceeding the maximum available time:

$$pt_{1'} + pt_{2'} + \ldots + pt_{m'} \leq T_{max}. \quad (2)$$

Add next:
When adding another order results in exceeding the maximum volume (maximum available time), this solution is skipped, the next order is added and the criterion is checked again.

If:

$$pt_{1'} + pt_{2'} + \ldots + pt_{m'} + pt_{m+1'} > T_{max}, \quad (3)$$

then:

$$pt_{1'} + pt_{2'} + \ldots + pt_{m'} + pt_{m+2'} \leq T_{max}. \quad (4)$$

The algorithm will terminate, when the available time is used:

$$pt + pt_{2'} + \ldots + pt_{m'} = T_{max} \quad (5)$$

or when all the orders are checked:

$$pt_{1'} + pt_{2'} + \ldots + pt_{m'} + pt_{m+2'} + pt_{m+5'} + pt_{n-1'} + pt_{n'} \leq T_{max}. \quad (6)$$

A result of the algorithm action is total profit from execution of the orders, i.e. sum of profits from all the accepted orders:

$$p_{sum} = p_{1'} + p_{2'} + \ldots + p_{m'} + p_{m+2'} + p_{m+5'} + p_{n-1'} + p_{n'} \quad (7)$$

4.2 Solutions of the Tabu Search Algorithm

Tabu Search:

Algorithm 1. Tabu Search pseudo-code for the considered problem

```
Algorithm Tabu Search (S0, var S, max_m, max_it)
 Set S = S0 and n_iter = 0
 Repeat
          m = 0
          best = 0
          it = it + 1
               Repeat
                      m = m+1
                      Execute Check_the_neighboring_solution (S,Sm)

                      Execute Check_Tabu_list (S,Sn)
                             if (f(Sm) > best then (best = f(Sm) and (m2= m))
               until (m = max_m)

          Execute Add_to_Tabu_list (S,Sm2)

          S= Sm2

          If best > solution then solution = best
 Until n_iter = max_it
```

The analysed problem may be presented as a binary sequence:

1 1 1 0 0 1 0 0 1 . . .,

in which each bit corresponds to the order assigned to the working week. The value “1” means that the order is accepted and proceeded, the value “0” means that the order is rejected. Implementation of five orders can be described as a bit sequence, e.g.:

1 1 0 0 1.

Check the neighbouring solution:
In this step, the algorithm checks all of the neighbouring solutions (differing in one bit):

1 1 0 0 **0**
1 1 0 **1** 1
1 1 **1** 0 1
1 **0** 0 0 1
0 1 0 0 1.

Check Tabu List:
The algorithm checks, whether the transition between previous and actual solutions has been already made. If so, this solution is omitted and then the best solution is chosen, according to the specified criterion:

11101.

Add to Tabu List:
The algorithm adds it to the Tabu List, i.e. the list of prohibited movements (TL). The algorithm remembers, which solution is the best one and repeats the steps the specified number of times.

In the case when each bit represents an order [o]- [o_1, o_2…..o_n], its parameters are:

production time [pt]- [pt_1, pt_2…..pt_n]
profit [p]- [p_1, p_2…..p_n]
priority [prio]- [$prio_1$, $prio_2$…..$prio_n$]
risk [r]- [r_1, r_2…..r_n]
penalty [pen]- [pen_1, pen_2…..pen_n].

According to the assumptions, the total operational time is strictly defined. The algorithm has a limit: total time of the orders multiplied by the bit value representing the order can not exceed the maximum operational time T_{max}.

$$\mathbf{0}pt_1 + \mathbf{1}pt_2 + \mathbf{1}pt_3 + \ldots + \mathbf{0}pt_n \leq T_{max} \tag{8}$$

In addition, risk is taken under consideration. Probability of a failure varies according to the machine operational time. The highest failure probability occurs at the beginning of production process and after some hours, depending on the machine. Following the empiric company data, quadratic function of failure risk is calculated for each machine and specific time buffers are assumed. The total of time buffers related to the machines in a specific process is added to the equation.

Final version of the algorithm can be presented as:

$$\mathbf{0}(pt_1 + r_1) + \mathbf{1}(pt_2 + r_2) + \mathbf{1}(pt_3 + r_3) \ldots + \mathbf{0}(pt_n + r_n) \leq T_{max}. \tag{9}$$

This means that total time of the selected orders and risk buffers can not exceed the maximum available operational time of the week.

According to the adapted objective function, the criterion for choosing the best solution is profit of the orders multiplied by priority values and by values of bits representing the order data.

$$F(c) = \mathbf{0}p_1prio_1 + \mathbf{1}p_2prio_2 + \mathbf{1}p_3prio_3 + \ldots + \mathbf{0}p_nprio_n \rightarrow MAX \tag{10}$$

Penalties for non-performance of the order, reducing final profit of the company, are also taken into consideration. A penalty is charged for each non-performed order. Therefore, the final target function is as follows:

$$\begin{aligned} F(c) = \mathbf{0}p_1prio_1 - \mathbf{1}pen_1 + 1p_2prio_2 - \mathbf{0}pen_2 + 1p_3prio_3 - \mathbf{0}pen_3 \\ + \ldots + \mathbf{0}p_nprio_n - 1pen_n \rightarrow MAX \end{aligned} \tag{11}$$

In the case of an executed order, where profit and the order priority are multiplied by "1", penalty for non-performance of the order is multiplied by "0" and vice versa. This means that either profit (for performance) or penalty (for non-performance) is considered for each order.

The algorithm implements also calculation of the weekend shift. If the calculated time is longer then T_{max} – which means that the weekend shift works – total profit is decreased by 15 000 euro.

The algorithm was set to 100 iterations.

4.3 Comparison of Algorithms Results

The results obtained with use of the suggested algorithms are given in the table and illustrated in the diagram. Potential profits obtained with each of the methods for the analysed working weeks are shown in Fig. 1. Amounts of the borne penalties are also marked. Information concerning work time for each week, together with numbers of executed orders and numbers of activated weekend shifts is collected, see Table 2. The obtained results are compared with actual historical data of the company.

Analysis of the obtained results indicates that the greedy algorithm in the presented form is not suitable for solving the considered type of production problems. The reason is that, in each of five examined cases, real profit of the company was higher than that calculated with the greedy algorithm.

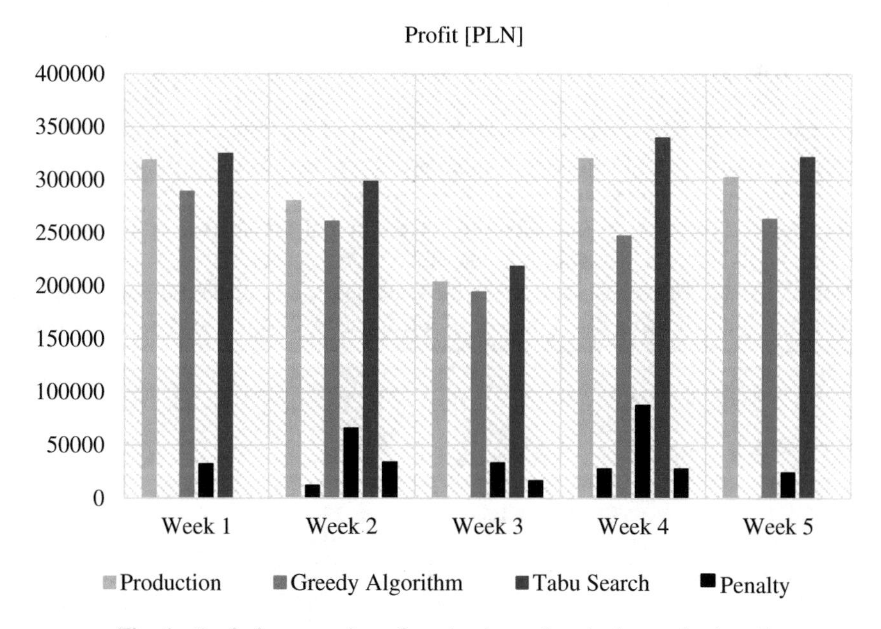

Fig. 1. Profit for execution of production orders in the weeks 1 to 5

Table 2. Production data in the weeks 1 to 5

		Week 1	Week 2	Week 3	Week 4	Week 5
Work time [h]	Production	102	99	110	118	109
	Greedy Alg.	101	103	101	112	99
	Tabu Search	103	112	112	118	104
Number of executed orders	Production	18	12	23	15	26
	Greedy Alg.	19	9	20	8	13
	Tabu Search	19	16	24	16	27
Number of activated weekend shifts	Production	0	0	1	2	1
	Greedy Alg.	0	0	0	2	0
	Tabu Search	0	2	2	2	0

However, in each of the analysed cases, the Tabu Search algorithm suggests a solution giving the company higher profit than that really reached. It can be also noticed that, for the Tabu Search algorithm, the penalties that should be potentially paid are identical or even higher than those to be paid in reality. Therefore, the results demonstrate participation of human factor in the decision-making process and namely natural tendency to avoid penalties and additional costs. The examination results show that, in some cases, it can appear profitable to pay higher penalty or to activate two additional weekend shifts. The algorithm was set to 100 iterations. The best solution appeared the one No. 17-32, depending on the test. This can mean that it is possible to reduce the number of iterations with maintained accuracy. This can also make a ground for the examinations focused on use of the Tabu Search algorithm in analysis of more complex questions.

5 Summary

The obtained results demonstrate also high potential of the Tabu Search algorithm that can be successfully used for solving similar or even more difficult problems. In relation to the above, further works on this algorithm are planned. These works would support production management by consideration of additional variables like advanced risk analysis, availability of personnel or retooling times depending on sequence of the performed orders. The measurements in the production bay of the analysed plant are continued, which should permit the algorithm to be developed in the direction both practical and useful for the business operation.

References

1. Betz, F.: Strategic business models. Eng. Manage. J. **14**(1), 21–28 (2002)
2. Bożejko, W., Uchroński, M., Wodecki, M.: Parallel tabu search algorithm with uncertain data for the flexible job shop problem. In: Rutkowski, L., Korytkowski, M., Scherer, R., Tadeusiewicz, R., Zadeh, L.A., Zurada, J.M. (eds.) ICAISC 2016. LNCS, vol. 9693, pp. 419–428. Springer, Cham (2016)

3. Brandão, J.: A tabu search algorithm for the open vehicle routing problem. Eur. J. Oper. Res. **157**(3), 552–564 (2004)
4. Burduk, A., Musiał, K.: Genetic algorithm adoption to transport task optimization. In: Graña, M., López-Guede, J.M., Etxaniz, O., Herrero, Á., Quintián, H., Corchado, E. (eds.) ICEUTE/SOCO/CISIS -2016. AISC, vol. 527, pp. 366–375. Springer, Cham (2017)
5. Burduk, A., Musiał, K.: Optimization of chosen transport task by using generic algorithms. In: Saeed, K., Homenda, W. (eds.) CISIM 2016. LNCS, vol. 9842, pp. 197–205. Springer, Cham (2016)
6. Cordeau, J.F., Gendreau, M., Laporte, G.: A tabu search heuristic for periodic and multi-depot vehicle routing problems. Networks **30**(2), 105–119 (1997)
7. DeVore, R.A., Temlyakov, V.N.: Some remarks on greedy algorithms. Adv. Comput. Math. **5**, 173–187 (1996)
8. Gola A., Kłosowski G., Application of fuzzy logic and genetic algorithms in automated works transport organization. In: Advances in Intelligent Systems and Computing, vol. 620, pp. 29–36 (2018)
9. Górnicka, D., Markowski, M., Burduk, A.: Optimization of production organization in a packaging company by ant colony algorithm. In: Intelligent Systems in Production Engineering and Maintenance, ISPEM 2017. Advances in Intelligent Systems and Computing, pp. 336–346. Springer (2018)
10. Grabowski, J., Pempera, J.: New block properties for the permutation flow shop problem with application in tabu search. J. Oper. Res. Soc. **52**(2), 210–220 (2001)
11. Grabowski, J., Wodecki, M.: A very fast tabu search algorithm for the permutation flow shop problem with makespan criterion. Comput. Oper. Res. **31**(11), 1891–1909 (2004)
12. Guillén, G., Badell, M., Espuña, A., Puigjaner, L.: Simultaneous optimization of process operations and financial decisions to enhance the integrated planning/scheduling of chemical supply chains. Comput. Chem. Eng. **30**(3), 421–436 (2006)
13. Ho, W., Xu, X., Dey, P.K.: Multi-criteria decision making approaches for supplier evaluation and selection: a literature review. Eur. J. Oper. Res. **202**(1), 16–24 (2010)
14. Jones, D.F., Mirrazavi, S.K., Tamiz, M.: Multi-objective meta-heuristics: an overview of the current state-of-the-art. Eur. J. Oper. Res. **137**(1), 1–9 (2002)
15. Kahraman, C., Engin, O., Kaya, I., Öztürk, R.E.: Multiprocessor task scheduling in multistage hybrid flow-shops: a parallel greedy algorithm approach. Appl. Soft Comput. **10**, 1293–1300 (2010)
16. Kotowska, J., Markowski, M., Burduk, A.: Optimization of the supply of components for mass production with the use of the ant colony algorithm. In: Intelligent Systems in Production Engineering and Maintenance, ISPEM 2017. Advances in Intelligent Systems and Computing, pp. 347–357. Springer (2018)
17. Musiał, K., Kotowska, J., Górnicka, D., Burduk, A.: Tabu search and greedy algorithm adaptation to logistic task. In: Saeed, K., Homenda, W., Chaki, R. (eds.) Computer Information Systems and Industrial Management, CISIM 2017. Lecture Notes in Computer Science, vol. 10244, pp. 39–49. Springer, Cham (2017)
18. Papageorgiou, L.G.: Supply chain optimisation for the process industries: advances and opportunities. Comput. Chem. Eng. **33**(12), 1931–1938 (2009)
19. Pohekar, S.D., Ramachandran, M.: Application of multi-criteria decision making to sustainable energy planning - a review. Renew. Sustain. Energy Rev. **8**, 365–381 (2004)
20. Zhang, Z., Schwartz, S., Wagner, L., Miller, W.: A greedy algorithm for aligning DNA sequences. J. Comput. Biol. **7**, 203–214 (2000)

Dealing with Capacitated Aisles in Facility Layout: A Simulation Optimization Approach

Hani Pourvaziri[1(✉)] and Henri Pierreval[2]

[1] Université Clermont Auvergne, SIGMA-Clermont, LIMOS, UMR CNRS 6158, Campus des Cézeaux, CS 20265, 63178 Aubiere Cedex, France
hani.pourvaziri@sigma-clermont.fr

[2] SIGMA-Clermont, LIMOS, UMR CNRS 6158, Campus des Cézeaux, CS 20265, 63178 Aubiere Cedex, France

Abstract. In manufacturing systems layout of machines has a significant impact on production time and cost. The most important reason for this is that machine layout impact on transportation time and cost. When designing layout, the aisles structure has an effect on transportation. The aisles are paths that transporters go through them to move the materials between machines. The capacity of the aisles is not infinitive and there is a limitation for the number of transporters that can pass an aisle at the same time. This causes transporters wait for the aisle to be empty or even transporters crossing. Therefore, when optimizing the layout of machines, the aisles structure and their capacity must be investigated. This paper proposes an approach for layout design in manufacturing systems taking into account capacitated aisles structure. First, the aisle structure is determined. Then, the machines are assigned in the possible areas. A simulation optimization approach is proposed to solve the problem. This helps us to avoid unrealistic assumptions and consider realistic conditions such as stochastic characteristics of the manufacturing system, random process time and random breakdown. A genetic algorithm is used to search for the best position of machines. Finally, a numerical example is included to illustrate the proposed approach.

Keywords: Facility layout problem · Manufacturing systems
Capacitated aisles · Simulation

1 Introduction

Facility layout problem (FLP) is known as one of the difficult combinatorial problems in operational research. It is concerned with finding the most efficient position of facilities (*e.g.*, manufacturing cell, an office or a machine) in a system. According to Tompkins et al. [1], layout decisions in manufacturing systems can contribute to 15–70% of the total operational costs and a good layout can at least reduce those costs by at least 10–30%. The research questions are concerned with the placement of machines on the surface of the workshop to minimize production objective(s), under several types of constraints [2]. Most research articles dealing with layout problem do not consider aisle. In manufacturing systems aisle are paths used by transporters to transport the materials between machines. The aisles are very important in determining

M. Graña et al. (Eds.): SOCO'18-CISIS'18-ICEUTE'18, AISC 771, pp. 260–269, 2019.
https://doi.org/10.1007/978-3-319-94120-2_25

the machine layout because the distance between two machines is dependent on the aisles structure. The width of these aisles cannot be infinitive and there is a limitation for number of transporters that can pass an aisle at the same time. The wider aisle let more transporters to pass through. We call this character the aisle capacity. The aisle capacity may cause increase the transportation time. Since, minimizing transportation time is the most important objective of machine layout, paying attention to capacity of aisles is essential to find the best layout. This, considering capacitated aisles in layout design, is the main motivation of this research. We also consider dynamic and stochastic behaviour of the manufacturing system such as possible breakdowns etc. Due to the complexity of the problem, we propose an approach based on simulation and optimization. In the proposed approach first the structure if the aisles is determined and then the layout of machines is found. The article is organized as follows. We first analyze the related literature and highlight our contribution regarding the works published in this area. In Sect. 3 the general principals are presented. Then, we explain the suggested approach. Section 4 gives a simple example to illustrate the approach. Our conclusion and future research directions terminate the article.

2 Related Researches

Benson and Foote [3] was one of the first who studied the problem of determining locations of pickup/delivery (P/D) points when material transportation occurs in aisles. Bock and Hoberg [4] introduced an approach to simultaneously determine machine layout and transportation paths. A path is determined between each pair of machines and the transporters are only allowed to move through the paths. Hamzeei et al. [5] studied the problem of designing flow path and the P/D points in a block layout in which AGVs were responsible for transportation. In order to optimize the problem two cutting-plane and annealing algorithms were presented. Mohammadi and Forghani [6] presented an integrated model to design both inter and intra cell layout in cellular manufacturing system. They considered two types of intra-cell aisle and inter-cell aisle. The intra-cell aisle was the spacing between machines inside a cell and inter-cell aisle was the spacing between cells. Based on bay structure layout, Wang and Chang [7] presented a two-stage model for FLP. Their proposed layout problem includes two subsystems of inter-bay and intra-bay. AGV was considered as transportation system. They used two types of aisles. The first one was to show the AGV path between intra-bays and the second one was to show the AGV path for inter-bay movements. Klausnitzer and Lasch [8] proposed a mixed integer programing model to optimize layout arrangement and aisle structure simultaneously. The aisles are placed around the departments. Friedrich et al. [9] presented a slicing tree based parallel tempering heuristic to solve plant layout problem. They considered each facility as a rectangular area which has one P/D point. The P/D points are next to the aisle. The material handling system was only allowed to move through the aisles. Zafar Allahyari and Azab [10] developed a mathematical model for unequal-area FLP. Their model accommodates for aisles structure.

Based on this review, it can be noted that the literature is relatively limited in terms of methods addressing the layout design considering aisles. Unfortunately, the few articles that have addressed this, have not specifically addressed aisles structure and aisles capacity. This has been pointed out as one literature gap in [9]. As a consequence, this paper presents an approach to address layout design considering capacitated aisles.

3 General Principles

3.1 Layout and Aisle Network

In manufacturing systems the products require to be processed on different machines and this causes flow of materials between machines. Different types of transporters, such as industrial trucks, conveyors, hoists, and cranes are used to handle the materials between machines. In certain manufacturing systems, producing small batches of a variety of products, industrial trucks such as forklift, hand lift and side loaders are often used for transportation. In these manufacturing systems, the transporters are only allowed to move on specified paths, named aisles. By crossing the aisles, a network is generated. The transporters are only allowed to move through this network. Designing the aisles network is important to determine the machine layout [11]. The machines and aisles must be place in the shop floor so that (1) all the machines have access to the aisles network, (2) no overlap between machines and aisles is occurred and (3) a path between each pair of machines via the aisle network is available. When the transportation system receives to the movement request, a transporter is dispatched to the origin machine and handles the material to the destination machine. Sometimes the received requests are more than the capacity of transporters and we need to prioritize the requests. This leads to bring up phenomena such as dispatching policy and queuing (e.g., parts waiting for the availability of a transporter).

3.2 Aisle Capacity

Each aisle is characterized by its width, length and location. The width of each aisle a, wa_a, induces a capacity related to the transporters. When a transporter tr is going to pass aisle a, obviously its width, wth_{tr}, must be less than the width of the aisle, $w_a \geq wth_{tr}$. When two transporters tr and tr' are passing aisle a in opposite direction, they definitely will go across each other. Therefore, the width of the aisle must be larger than their total width, $w_a \geq wth_{tr} + wth_{tr'}$. Otherwise, transporters would collide with each other. In general it can be said that the width of an aisle must be larger than the total width of transporters that go across each other in that aisle.

4 Proposed Approach

4.1 Determining the Aisle Network

The shop floor of the manufacturing system has a rectangular shape with length L and width W. The machines are equal and have rectangular shapes with predetermined orientation. There are vertical and horizontal aisles. The width of the aisles is equal to wa.

The aisles divide the shop floor into sections, called position. The length of vertical aisles is equal to W and the length of horizontal aisles is equal to L and so the aisles are extended to the boundaries of shop floor. With such a consideration we avoid creating triangular or odd-shaped positions. Therefore, each possible position of machine will have a rectangular shape. Machines are assigned to the possible positions. The aisle network must be so that there is enough space for machines to be assigned inside the positions. To satisfy this first we determine the maximum number of vertical and horizontal aisles. By knowing the length and width of machines, (lm, wm), the number of vertical (R) and horizontal (S) aisles must be so that Eqs. (1) and (2) is satisfied.

$$\lfloor (L - lm)/(lm + wa) \rfloor \leq R \leq \lfloor (L + lm)/(lm + wa) \rfloor \tag{1}$$

$$\lfloor (W - wm)/(wm + wa) \rfloor \leq S \leq \lfloor (W + wm)/(wm + wa) \rfloor \tag{2}$$

Therefore, there is a choice in determining the number of aisles. Since at last one machine can be placed in each position, the number of vertical and horizontal aisles must be determined so that the possible number of position is greater than number of machines. This is guaranteed through the following equations.

$$M \geq N \tag{3}$$

$$N = R' \times S' \tag{4}$$

$$R' = \begin{cases} R+1, & if\ R = \lfloor (L - lm)/(lm + wa) \rfloor \\ R-1, & if\ R = \lfloor (L + lm)/(lm + wa) \rfloor \\ R, & else \end{cases} \tag{5}$$

$$S' = \begin{cases} S+1, & if\ S = \lfloor (W - wm)/(wm + wa) \rfloor \\ S-1, & if\ S = \lfloor (W + wm)/(wm + wa) \rfloor \\ S, & else \end{cases} \tag{6}$$

M is number of machines and N is number of possible positions generated by aisles. This guarantee that the number of generated positions is at least equal to the number of machines.

We start by placing vertical aisles. If we had used the maximum number of allowable vertical aisles, we start by placing the aisle. Then, the next vertical aisle is placed so that the distance between it and the left-side vertical aisle is equal to the length of a machine. According to the same procedure other vertical aisles are placed one by one until it is not possible to add new aisle or machine. If we had used the minimum allowable aisle, then we start by placing the machine. After that aisles and machines are placed one by one according to the described procedure. After placing the vertical aisles, there may be empty space between the boundaries of shop floor and last machine or aisles. These empty spaces can be used to increase the width of the most used aisles. The horizontal aisles are placed just like the way we placed vertical aisles.

4.2 Assignment of Machines to Possible Positions

After determining the aisles structure and positions, the machines must be placed in the shop floor in a manner that no overlap between machines and aisle occurs. Considering the length and width of the positions and machines dimension, by placing one machine in each position no overlap will occurs. Therefore, we have a set of *M* processing machines must be assigned to *N* positions. The aim is to assign the machines to the positions so the MFT is minimized. Before characterizing the problem, let us first introduce the new notations:

Indices

j Index for machines, $j = 1,\ldots,M$
l Index for positions, $l = 1,\ldots,N$
tr Index for transporters, $tr = 1,2,\ldots,TR$
a Index for aisles, $a = 1,2,\ldots,R + S$

Parameters

wth_{tr} The width of transporter tr
wa_a The width of aisle a

Dependent decision variable

$M_{tr,tr',a}$ 1 if transporter tr and tr' go across each other in aisle a
0 otherwise

Independent decision variable

$X_{j,l}$ 1 if the machine j is placed in the position l
0 otherwise

$$\min E(Mean\ Flow\ Time) = \mathbb{E}\left[f\left(X_{j,l}, \xi\right)\right] \tag{7}$$

$$\sum_{l=1}^{N} X_{j,l} = 1 \qquad \forall j \tag{8}$$

$$\sum_{j=1}^{M} X_{j,l} \leq 1 \qquad \forall l \tag{9}$$

$$M_{tr,tr',a}(wth_{tr} + wth_{tr'}) \leq wa_a \qquad \forall tr, tr', a \tag{10}$$

$$X_{j,l}, M_{tr,tr',a} \in \{0, 1\} \qquad \forall i, k \tag{11}$$

Mean flow time (MFT) is selected as performance measure. It can be seen as a measure of the operational performance of the system and almost all simulation studies include MFT as a major performance measure [12]. MFT can reflect the impact of position of machines and capacitated aisles. Therefore, the objective of the mathematical model is to minimize the MFT. Formally, the binary decision variable $X_{j,l}$ takes value 1 if the machine j is placed in the position l, and 0 otherwise. Constraints set (8) guarantee that each machine is assigned to one position. Constraints set (9) guarantee that in each position at most one machine is placed. Constraints set (10) control that at each time the number of transporters in an aisle is less than the aisle capacity. The constraints set (11) control the decision variables.

A major difficulty to solve the model comes from calculating the objective function. The stochastic aspects of manufacturing system and capacitated aisles make it difficult to calculate the value of MFT. To deal with these difficulties, we propose a simulation optimization approach to search the best assignment of machines to possible positions. Even if several researchers have used simulation optimization approach in FLPs [7, 13–15] they have not specifically addressed aisles structure and their capacity.

Due to the problem complexity of the problem and in order to have a realistic estimation of the performance criterion, the performance of each layout solution in terms of MFT is evaluated by simulation in a simulation optimization procedure (Fig. 1). To search for the best layout, we apply a genetic algorithm (GA). The GA iteratively searches to find the best layout while simulation is used to evaluate each layout. For a better understanding of simulation optimization readers are referred to [16].

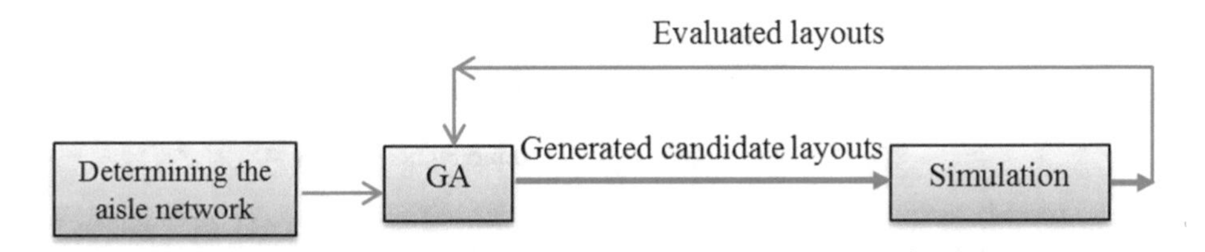

Fig. 1. Simulation optimization approach

The simulation optimization approach is able to satisfy the constraints. The way we deal with satisfying the constraints is described below.

Since, the aisle network is determined based on the method describe in Sect. 4.1, no overlap between machines and aisles will occur. Constraints (8) and (9) are satisfied by the way we define the solution in the GA. To satisfy constraint (10), when a transporter is going to enter an aisle, the simulation model verifies two things. (1) The capacity of that aisles and (2) if any other transporter is in that aisle or not. According to these two matters the simulation model let the transporter to go across the aisle or stop the transporter until the aisle gets empty. By this way the constraint (10) is satisfied.

4.3 Genetic Algorithm

Genetic algorithms (GA) have been successfully used for layout design [17]. The GA starts with a set of chromosomes. In order to produce new chromosomes, two operators cross over and mutation are used in GA. Then, based on the fitness function values, some of the chromosomes with better values of the fitness function are saved. By this way, better chromosomes can participate to generate new chromosomes in the next generations. The algorithm continues this procedure until some termination criteria is satisfied.

Fitness Function
Our fitness function is the MFT that should be minimized (Eq. 7). The MFT of jobs is defined by Eq. (12).

$$MFT = \frac{\sum_{i=1}^{I} f_{ji}}{I} \tag{12}$$

Where f_{ji} is the flow time of job i and I is number of jobs.

Chromosome Structure
One of the most important issues in applying any meta-heuristic algorithm is to develop an effective solution representation. The solution encoding of the proposed problem is permutation of N numbers. By scanning from left to right, the first number shows the position wherein the machine 1 must be assigned; the second number shows the position wherein the machine 2 must be assigned and so on. Since the number of each position is appeared only one time and as many as machines, the number is generated, the constraint set (8) and (9) will be satisfied. Figure 2 shows an example of chromosome with 6 machines and 6 positions.

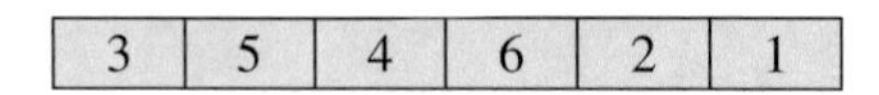

Fig. 2. An example of chromosome representation

Crossover Operator
The crossover operator combines two chromosomes to produce two new chromosomes. To apply the proposed crossover operator first a cross point is randomly selected between 1 to N. Then, the machine numbers before the cross point of Parent 1 are directly copied in the offspring. The remaining machine numbers are put into empty positions according to their relative positions in Parent 2. By this way we avoid of duplicating numbers and so constraint set (8) and (9) are satisfied. According to experimental result the cross over percent is considered equal to 0.6.

Mutation Operator
The purpose of the mutation is to prevent of population from being too similar and getting trapped into local optima. According to Smullen et al. [18], it is interesting to apply the mutation only when the similarity solutions exceed a predefined value. In order to find the similarity coefficient (SC) between each pair of chromosomes Eqs. (13) and (14) are used.

$$SC_{ab} = \frac{\sum_{j=1}^{M} \partial\left(X_{ja}, X_{jb}\right)}{M} \tag{13}$$

$$\partial\left(X_{ja}, X_{jb}\right) = \begin{cases} 1 & \textit{if } X_{ja} = X_{jb} \\ 0 & \textit{otherwise} \end{cases} \tag{14}$$

In which SC_{ab} are the positions of machine in the chromosomes a and b is the number of machines. $\partial(X_{ja}, X_{jb})$ is the similarity between two especial genes. Then, the average similarity coefficient of the population is calculated by Eq. (15).

$$\overline{SC} = \frac{\sum_{a=1}^{Np-1} \sum_{b=a+1}^{Np} SC_{ab}}{\binom{Np}{2}} \tag{15}$$

In which Np is the number of chromosomes in the population. Finally, when the similarity coefficient is greater than pre-defined threshold (0.9) the mutation is incorporated into the GA loop. To perform the mutation operator first a chromosome is randomly selected. Then to create a great change the genes of the selected chromosome are arranged reversely. We note that because we just changed the sequence of genes, the chromosomes remain feasible.

Stopping Criterion
The algorithm generates new populations until the total number of iterations reached to a predefined number (in our experimental results this value is set to 30).

5 Application Example

In this section, the application of the proposed approach is illustrated by an example. The example is a 83 × 39 manufacturing system with 10 machines with dimension 17 × 9. The width of the aisles and transporters are respectively 3 and 2. Therefore, the capacity of the aisle will be equal to 1. According to the Eqs. (1) and (2) the number of vertical aisle must be between 3 and 5 and the number of horizontal aisle must be between 2 and 4. We select 5 vertical and 4 horizontal aisles. By this way, 12 possible positions are generated. The machine can be placed inside each position, except positions 1 and 12 which are used as input and output. The path of movement the transporters between the aisles is determined by Dijkstra is algorithm. Dijkstra algorithm selects the nodes which lead to shortest path from a starting point to a target point. We consider the capacity of all aisles equal to one. The greatest-queue-length (GQL) is selected as dispatching policy. According to Ho et al. [19], GQL is the best dispatching policy for minimizing the time-based indexes. In GQL, a transporter gives priority to the machine that has the greatest number of waiting parts. The number of simulation replications for each layout is set at 30 and the warm-up period is 1000 s.

The GA search for the best layout and uses the simulation to calculate the MFT corresponds to each layout. Figure 3 presents the convergence diagram of the GA. The mean curve shows the average values objective function of chromosomes in each generation and best is the objective function of best chromosome. The MFT for the best obtained layout is 5807 s.

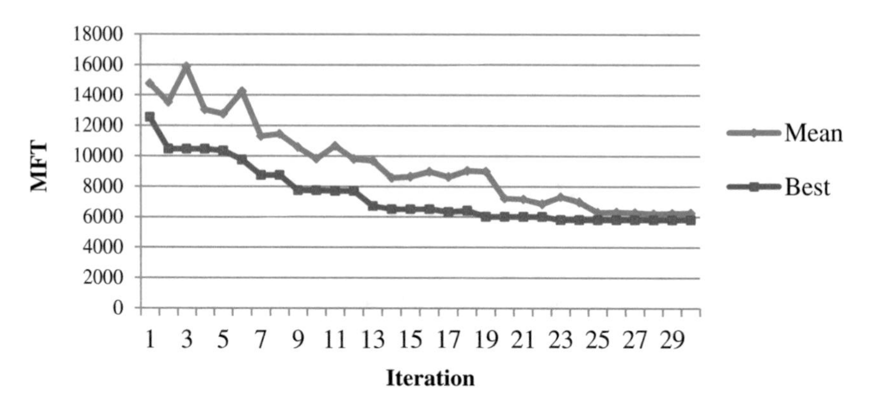

Fig. 3. The convergence diagram of GA

6 Conclusion and Future Research

We presented an approach to solve FLP in presence of capacitated aisles and stochastic characteristics. The goal was to search for finding the best layout of machines so that MFT is minimized. This approach can deal with capacitated aisles, stochastic transportation time, stochastic demand of products and machine breakdowns. Our approach could tackle these realistic considerations and search for the best layout. This is possible thanks to the use of simulation optimization approach. The applicably of the proposed approach was tested by an example.

The next step of this research is to be able to assign more than one machine to a possible position e.g. using a continual layout approach. Other future research directions are concerned with reducing computing costs by formulating the problem based on stochastics mathematical programming or queueing theory. The problem of simultaneously determining the aisles structure and machines layout can also be investigated.

References

1. Tompkins, J.A., White, Y.A., Bozer, E.H., Frazelle, J.M.A., Tanchoco, J.: Facilities Planning. Wiley, New York (1996)
2. Drira, A., Pierreval, H., Hajri-Gabouj, S.: Facility layout problems: a survey. Ann. Rev. Control **31**, 255–267 (2007)
3. Benson, B., Foote, B.L.: DoorFAST: a constructive procedure to optimally layout a facility including aisles and door locations based on an aisle flow distance metric. Int. J. Prod. Res. **35**, 1825–1842 (1997)
4. Bock, S., Hoberg, K.: Detailed layout planning for irregularly-shaped machines with transportation path design. Eur. J. Oper. Res. **177**(2), 693–718 (2007)
5. Hamzeei, M., Farahani, R.Z., Rashidi-Bajgan, H.: An ant colony-based algorithm for finding the shortest bidirectional path for automated guided vehicles in a block layout. Int. J. Adv. Manuf. Technol. **64**(1–4), 399–409 (2013)

6. Mohammadi, M., Forghani, K.: A novel approach for considering layout problem in cellular manufacturing systems with alternative processing routings and subcontracting approach. Appl. Math. Model. **38**, 3624–3640 (2014)
7. Wang, G., Yan, Y., Zhang, X., Shangguan, J., Xiao,Y.: A simulation optimization approach for facility layout problem. In: IEEE International Conference on Industrial Engineering and Engineering Management, pp. 734–738 (2008)
8. Klausnitzer, A., Lasch, R.: Extended model formulation of the facility layout problem with aisle structure. In: Mattfeld, D., Spengler, T., Brinkmann, J., Grunewald, M. (eds.) Logistics Management. Lecture Notes in Logistics. Springer, Cham (2016)
9. Friedrich, C., Klausnitzer, A., Lasch, R.: Integrated slicing tree approach for solving the facility layout problem with input and output locations based on contour distance. Eur. J. Oper. Res. (2018). https://doi.org/10.1016/j.ejor.2018.01.001
10. Zafar Allahyari, M., Azab, A.: Mathematical modeling and multi-start search simulated annealing for unequal-area facility layout problem. Expert Syst. Appl. **91**, 46–62 (2018)
11. Wu, Y., Appleton, E.: The optimization of block layout and aisle structure by a genetic algorithm. Comput. Ind. Eng. **41**, 371–387 (2002)
12. Bokhorst, J.A.C.: Shop floor design: layout, investments, cross-training, and labor allocation, Doctor of Philosophy, University of Groningen (2005)
13. Chan, F.T.S., Chan, H.K.: Design of a PCB plant with expert system and simulation approach. Expert Syst. Appl. **28**(3), 409–423 (2005)
14. Gyulai, D., Szaller, A., Viharos, Z.J.: Simulation-based flexible layout planning considering stochastic effects. Procedia CIRP **57**, 177–182 (2016)
15. Prajapat, N., Waller, T., Young, J., Tiwari, A.: Layout optimization of a repair facility using discrete event simulation. Procedia CIRP **56**, 574–579 (2016)
16. Wang, L.F., Shi, L.Y.: Simulation optimization: a review on theory and applications. Acta Autom. Sin. **39**(11), 1957–1968 (2013)
17. Pourvaziri, H., Naderi, B.: A hybrid multi-population genetic algorithm for the dynamic facility layout problem. Appl. Soft Comput. **24**, 457–469 (2014)
18. Smullen, D., Gillett, J., Heron, J., Rahnamayan, SH.: Genetic algorithm with self-adaptive mutation controlled by chromosome similarity. In: IEEE Congress on Evolutionary Computation (CEC), Beijing, China (2014)
19. Ho, Y.C., Liu, H.C.: A simulation study on the performance of pickup-dispatching rules for multiple-load AGVs. Comput. Ind. Eng. **51**(3), 445–463 (2006)

Unlocking Augmented Interactions in Short-Lived Assembly Tasks

Bruno Simões(✉), Hugo Álvarez, Alvaro Segura, and Iñigo Barandiaran

Vicomtech, Paseo Mikeletegi 57, 20009 Donostia/San Sebastián, Spain
{bsimoes,halvarez,asegura,ibarandiaran}@vicomtech.org

Abstract. Augmented Reality (AR) has evolved over the past years, but before it is widely adopted and used in manufacturing industry, it has to overcome a number of technological challenges. Although new advancements in tracking and display technology have been a priority in recent research works, the use of accurate registration methods is not fundamental for users to understand the intent of the augmentation. Moreover, interactive visualization of contextual data in augmented spaces did not receive enough attention from the research community. In this paper, we investigate the creation of AR workspaces focused on interaction and visualization modes rather than on the registration accuracy, and how to provide more effective means to support assembly tasks in hybrid human-machine manufacturing lines. In particular, we focus on short-lived assembly tasks, i.e. manufacturing of limited batches of customized products, which do not yield significant returns considering the effort necessary to adapt AR systems and the production time frame.

Keywords: Augmented Reality · Interaction · Assembly
Manufacturing

1 Introduction

Manufacturing assembly lines are subject to a continuous fluctuation in production demands such as customization level and quantity [1]. While many continuous and repetitive actions in assembly tasks can be automated to improve production efficiency, the biggest challenge to automation is when a new product variant is introduced into the production line.

The manufacturing of products variants that change in relatively short intervals is efficient when processes integrate human workers, as they adapt to new manufacturing operations without interrupting the production process. However, human workers are more prone to errors than machines. Replacing the human factor with an automated process could, perhaps, decrease the complexity of the solution. Nonetheless, maintaining the presence of human workers in the assembly chain can create a competitive advantage. This advantage comes from the fact that humans can rely on their natural senses to form very complex and

M. Graña et al. (Eds.): SOCO'18-CISIS'18-ICEUTE'18, AISC 771, pp. 270–279, 2019.
https://doi.org/10.1007/978-3-319-94120-2_26

intuitive, yet instant, solutions. In contrast, automated pipelines require reprogramming to a new product family or manufacturing problem. A technological solution that provides training and assistance to workers in assembly tasks could mitigate the issue of human errors and be more cost-effective that reprogramming the entire system for a new product family.

In recent years, manufacturing industries have been pursing solutions based on augmented reality (AR) to empower their workers, actively and passively, with new possibilities of perceiving information in a timely manner, e.g. about assembly processes, without altering their established work routines. In this paper, we propose an AR framework based on low cost technology that provides effective means to support assembly tasks, in particular in short-lived assembly tasks, i.e. manufacturing of very limited batches of customized products. Manufacturing of products that are limited to a few units do not yield significant returns to justify the effort required to adapt the AR system. We also describe how to create AR user assembly manuals with minimal effort and adapt them to multiple context and user profiles, e.g. different levels of worker disabilities. The AR framework automatically interprets assembly diagrams and creates superimposed task-related content for assembly tasks. In our test case, we evaluate the use of projection-based AR and we compare it to AR display-based approaches, that often divert workers' hands and attention between digital instructions and the assembly process.

The rest of the paper continues as follows. Section 2 provides an overview of related works of AR in manufacturing processes. Section 3 describes the proposed AR framework, including the required setup (Sect. 3.1), the procedure to create AR multimedia content (Sect. 3.2), the tracking methodology (Sect. 3.3) and the calibration that is required to project the assembly instructions on the workspace (Sect. 3.4). Finally, Sect. 4 provides a discussion and enumerates future works.

2 State of the Art

The integration of AR in industrial manufacturing processes, namely in assembly tasks, has proved to reduce manufacturing errors and to easy cognitive load of participants by 82% when compared to paper-based instructions, on a monitor, or steadily displayed on an HMD (Head Mounted Display) [2]. Previous studies also demonstrated an improvements in the performance of untrained workers [3–5]. In these studies it was also observed that proposed AR solutions failed to address the existent know-how of expert workers and attempted to impose new routines, which led to a deterioration in their performances [3].

Assembly operations typically require workers to locate assembly parts and then to use their both hands either to hold or to perform the assembly task. Consequently, the interaction with visual interfaces through gestures is inconvenient and indirectly shifts their attention from assembly parts. Projection-based solutions have been increasingly regarded as an alternative for ubiquitous visual output that can replace displays and AR smart glasses. One of the advantages is that it can overcome the pitfalls of limited field-of-view found in AR glasses (which makes harder to guide the worker to theses parts) and can augment physical objects with digital content, without the need to hold any accessory or shift attention between a display and the workspace.

In the field of industrial manufacturing, Sand et al. [6] developed a prototype to project AR instructions into the physical workspace, which enabled the worker not only to find the pieces but also to assemble products without prior knowledge. Rodriguez et al. [7] proposed a similar solution in which instructions are directly overlaid with the real world using projection mapping. Petersen et al. [8] projected video overlays into the environment at the correct position and time using a piecewise homographic transform. By displaying a color overlay of the user's hands, feedback could be given without occluding task-relevant objects.

Challenges in the visualization and interaction with projected-based systems have been widely investigated. Tang et al. [2] found that occlusions of objects by digital content or presenting information over a cluttered background can decrease task performance. Funk et al. [9] experiments demonstrated that users tended to prefer instructions to not be directly projected onto the actual assembly situation but peripherally beside the target region, thus not occluding relevant objects. Feiner et al. [10] proposed a rule-based prototype to create different types of graphics for the user. However, very little efforts has been taken to investigate AR systems that embed task know-how and facilitate personalization towards different worker skills and disabilities.

3 Solution Fundamentals

Many projector-based prototypes have been created to project content on arbitrary surfaces. However, most of these works failed to consider the interaction with assembly parts that pose challenges to computer vision algorithms. Other works have solved the problem by employing expensive technology such as 3D cameras to track and interact with objects in projected spaces. In this section, we introduce the concepts that enable us to transfer the interaction space from the conventional desktop space to the direct interaction in the physical space, i.e. where the projected displays are located. Our solution aims to deliver a robust and low cost tracking system for the assembly of electrical cabinets, which involves assembly parts that pose great challenges to modern computer vision algorithms. Another advantage in our solution is the minimal amount of time required to deploy new digital assembly manuals that take into account worker skills and disabilities.

3.1 Enabling Technology and Setup

Assembly operations often require workers to hold assembly parts or to use both hands to perform assembly sub-tasks. Consequently, hand-based interaction with visual interfaces is inconvenient and affects their attention. In this section, we describe an AR framework that provides graphical instructions projected or overlaid on top of physical components. The proposed use case requires workers to assembly an electrical cabinet. In particular, we focus on the task of cable wiring the cabinet.

Our framework supports assembly of electrical cabinets with two different applications. The first application helps users to design AR assembly manuals for new product families, see workflow in Fig. 1.

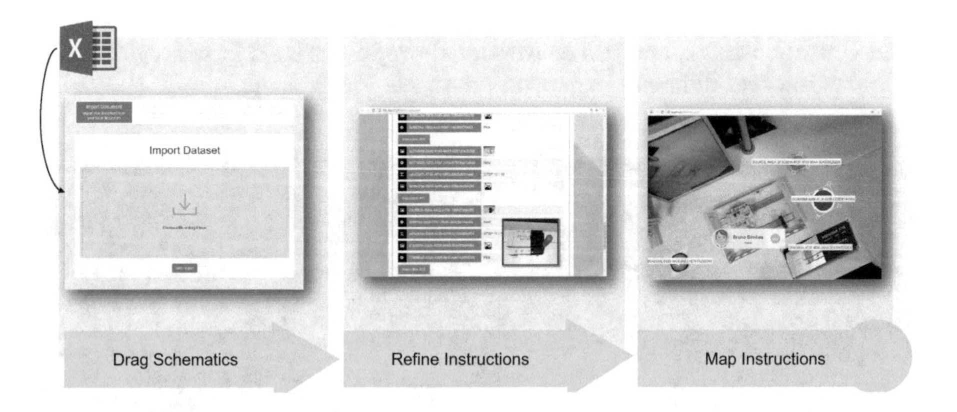

Fig. 1. Ingestion and creation of AR applications

The application consists of a web-based interface that can ingest electrical schematics or drawings in DWG or DXF format, as well Excel files describing the assembly process of a given product. In real assembly processes, this material is printed to hard copies and delivered to workers. Hence, the effort required to create new AR applications is limited to assigning a location to the AR content. The task can be performed in-situ using the projectors as display or remotely through a web browser that depicts a video camera streaming of the physical location and the digital content, see last step in Fig. 1.

The second application provides real-time projection-based instructions to the worker. The proposed framework also relies on a rule-based decision system to adapt the AR experience to workers with disabilities. The use of a rule-based decision system simplifies significantly the design of digital assembly manuals because it reduces human-machine interaction to an event-consequence metaphor, which in turn reduces the complexity of programming workflow. Moreover, it allows digital assembly manuals to be automatically customized by external sensors and worker processes. As part of this application, we have developed a robust approach for the detection of assembly parts that lack of textural information. Tracking assembly parts is required to superimpose content and ensure that certain elements are rendered at appropriated locations. The user interaction, requires workers to use bracelets with printed markers to enable the framework to estimate their 3D hands pose without the requirement of a 3D camera, see Fig. 2. Our approach has the advantage that it works from a single camera perspective and benefits from usability at the workspace and tracking speed, hence maintaining good levels of interaction feedback.

To summarize, our framework is built on top of a distributed network architecture, which defines mechanisms to automatically adapt the AR experience to different hardware configurations, e.g. the use of multiple projectors to reduce projection occlusions by physical materials or irregular surfaces, see Fig. 2. The setup consists of a table, which corresponds to the workspace, fixing brackets that hold a camera in zenith position (it is a low cost web camera, configured to capture at 640×480 resolution), bracelets with printed markers to track user hand with robustness, and two conventional projectors fixed in strategical aerial positions to cover different projection areas, see Fig. 2.

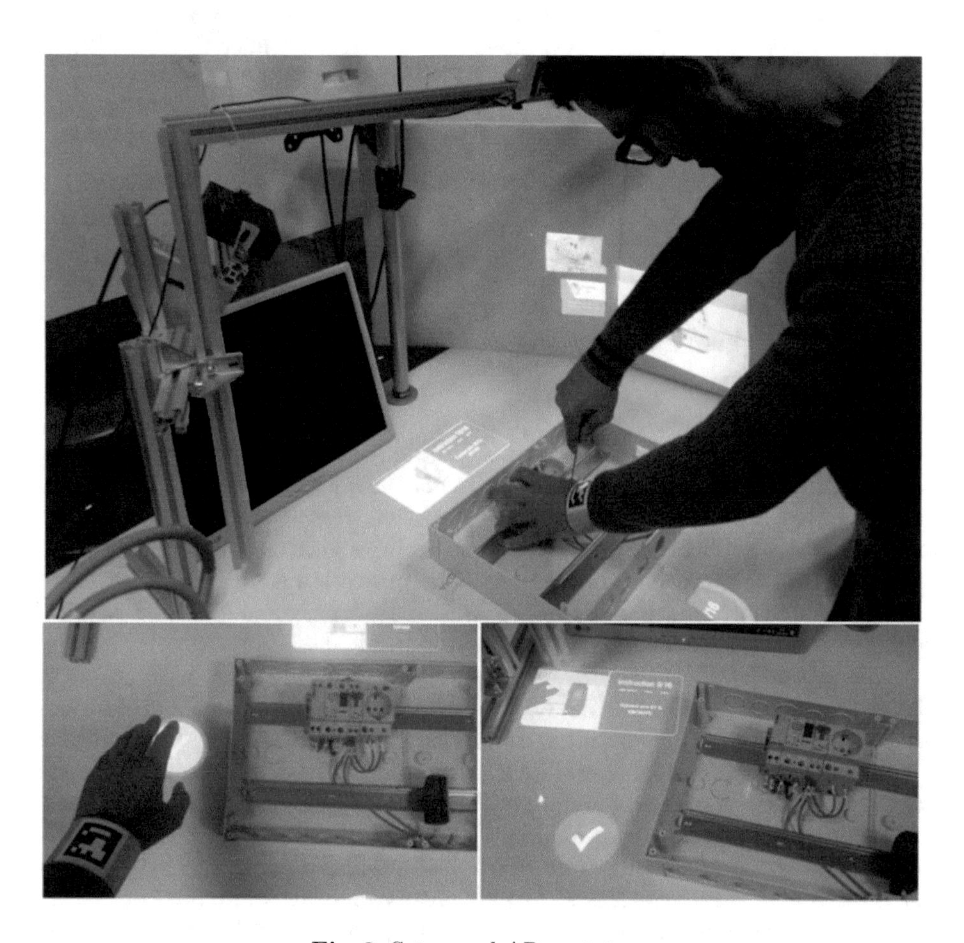

Fig. 2. Setup and AR prototype

3.2 Interaction Context and Focus

To obtain an easy-to-use interaction concept, we kept the set of visual interaction elements and gestures as limited as possible. Interaction gestures were constrained to touch of projected UI and the tracking technology limited to low cost cameras, as described in the previous section.

When the operator places the base of an electrical cabinet on the table, the system automatically identifies the object by processing the images captured by the camera and starts projecting AR instructions on the table to help the worker with the assembly task (installation of wiring in the electrical cabinet). Tracking assembly parts is very important because as observed in previous studies, workers might prefer certain types of content not to be projected over the components. Hence, we need to ensure that projected content is re-adapted whenever assembly parts are moved. In the following sections, we describe the techniques used for the identification of the object, detection of gestures, and for the projection of multimedia content.

The detection of users gestures, e.g. touches in projected elements, requires the worker to wear bracelets to estimate the 3D hand pose. Our approach mitigates existing limitations in tracking approaches, linked to issues in industrial environments like magnetism and luminosity, and can estimate the probability of touching projected elements on the physical surface with a low cost solution. In Fig. 2 we can find the design of the reusable bracelets, which are designed to not affect workers' ergonomics. In future work, bracelets could be used also to encode the user identity, which would simplify aspects in collaborations that involve multiple workers with different disabilities.

As mentioned in previous sections, interaction context and focus is inferred by a rule-based engine that analyses the state of each device and sensor participating in the AR experience (e.g. two computers that control two projectors, tracking module, worker skills and disabilities, and service providing AR instructions). The main advantage of our configuration is that it simplifies the initial design of the digital assembly manual, which is automatically enriched with content that depends on sensor feedback and worker profiles. For example, our framework can interact with third-party workload schedulers to generate rules that reprogram workflow of projected-based content to a specific worker, i.e. task context. The same approach can be applied to personalize projected elements to different worker's profiles. The rule-based model is build separately from the system that provides AR instructions and can be personalized through a web interface or controlled with an API. Eventually, rules can be managed in real-time by computational modules that observe the assembly process at different granularities. Rules use the syntax form:

$$\text{Rules} \begin{cases} r_1\text{: Device.UserID = `bsimoes'} \rightarrow \text{Execute.AppID = `App1'} \\ r_2\text{: Text.Color = red \& User.Protanopia} \rightarrow \text{Text.Color = cyan} \end{cases}$$

The entire framework is event-driven and uses JSON notation for all communications. The rule syntax takes the type of messages exchanged through the network and asserts on their JSON attributes.

In Fig. 2 we can visualize in the lower left corner, the worker interacting with projected elements and in the lower right corner the content adopting a new location due to a translation of the electrical cabinet.

3.3 Tracking of Assembly Parts

The setup of the system, in which the camera is in a zenith position and the object (the electrical cabinet) moves in a table whose plane is parallel to the camera's image, enable us to simplify the problem of detection and tracking to an image warping (the idea is similar to the one applied when mapping aerial photographies). Instead of using the 3D model of the object as input (which is not always available and its creation is time consuming), we only take an aerial photo of it to extract all required data. This aerial image is used as a reference, hence the problem of the camera pose can be modeled by a homography, which is a transformation that moves points from one plane (camera image) to another plane (reference image).

The electrical cabinet that is shown in Fig. 2 is an object that is difficult to detect in an image, especially when the camera is low cost and does not offer a high quality image. The electrical cabinet has poor texture, and therefore, the well known texture-based techniques (particularly those that use feature points and descriptors [11]) are not well suited. However, it does have predominant edges that can be used to perform the detection based on its shape. Given that, the detection is solved by applying a similarity measure that matches the edges directions of the electrical cabinet extracted in a reference image with those extracted in the current input camera image. In an offline step, given a reference image of the electrical box, the edges and their directions are extracted. Thus, in an online stage, the reference edge model is tested throughout all the positions of the input camera image (in a coarse to fine mode to reduce the computational cost and make it close to real-time) and the position with the highest similarity is returned (Fig. 3). Moreover, to achieve rotation and scale invariance, the reference image is rotated and scaled (columns and rows of the offline step of Fig. 3, respectively) with different rotation and scale steps, and several reference models are created (one for each rotation-scale combination). See [12] for more details. Although this technique is not robust against viewpoint changes, we do not suffer this weakness, as the camera is in zenith position and the electrical cabinet only moves on a table, i.e., the viewpoint change is minimal.

Once the electrical cabinet has been detected in the current frame, its position has to be updated in subsequent frames, i.e., it has to be tracked. *Tracking*, unlike *detection*, exploits information from previous frames to update the current location of the object. Thus, for example, the positioning problem, instead of covering the entire image, is restricted to only an area close to the position of the previous frame. This *temporal coherence* assumption simplifies the problem, and due to this, techniques that we discarded for detection have been demonstrated suitable for tracking. This is the case of the optical flow of some feature points [13]. We have observed that it is possible to extract a few interest points on the surface of the electrical cabinet, enough to apply the optical flow and get an initial estimation of the position.

The optical flow technique is robust against fast movements, but it accumulates error and suffers from drift. To overcome this problem, the position estimation given by the optical flow is refined using an edge tracker [14] (Fig. 3).

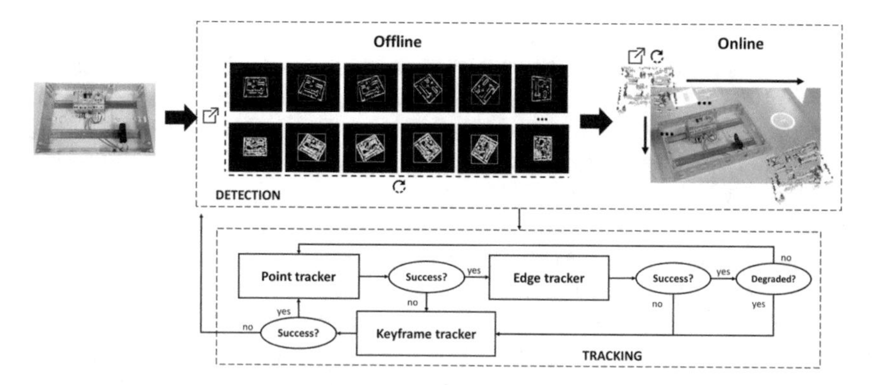

Fig. 3. Detection and tracking pipelines. The colors of the edge pattern represent different gradient directions. See text for more details.

The input reference image of the electrical cabinet is also processed here in an offline step to extract some reference edges (those that have strong response). Then, during the online execution, these reference edges are positioned using the initial estimation of the optical flow and their presence in the current frame is measured to minimize the error between prior and current positioning and achieve more accuracy.

Additionally, we have included a keyframe tracker to recover from errors of the other two trackers described above. When the electrical cabinet is detected, that frame is used as keyframe, extracting some feature points and descriptors. In this way, when a new camera frame is received, its feature points and descriptors are computed and matched with those of the keyframe (a similar approach is described in [15]). When good enough matches are found, the position of the electrical cabinet can be recovered in the current frame. The recovery strategy is applied only when we detect a long displacement of the box (10 pixels in our case) or when the optical flow has been degraded (20% of feature points have been lost) (Fig. 3). This keyframe tracker allows us to avoid having to switch to the *detection* state in some occasions, whose processing is more expensive computationally. Furthermore, we have noticed that this keyframe tracker grants more robustness in case of a partial occlusion of the electrical cabinet, especially when the operator uses his/her hand to place a component into the cabinet.

3.4 Projection Mapping and Spatial Calibration

The projection of multimedia content at a desirable location, when observed through a camera from a random angle, requires us to find a transformation that moves points from the camera's image to the projector's space. Although this task could be done manually, we have automatized the calibration process in our solution. Assuming the camera and projector are calibrated independently, it consists in a sequential projection of a chess pattern along the entire projection area (a table in our case), such that each projection of the pattern is detected in the image of the camera. Therefore, it is possible to find the camera-projector relation (or vice versa, since the problem is analogous). Our implementation is similar to the solution described in [16]. It is also noteworthy that we have modeled the transformation between the camera and the projector using a homography, which is a valid simplification because the multimedia content is projected on the table, which is a plane.

Once both devices are calibrated and the camera pose is known (Sect. 3.3), the projection of the multimedia content is straightforward. Let H_{ref}^{cam} be the homography that transform points from the reference image of the object to the current camera image, and let H_{cam}^{proj} be the homography that relates the camera image with the projector image that is projected. Then, the multimedia content that has been positioned in the reference image (M_{ref}) is projected in the desired location using the following transformation composition: $H_{cam}^{proj} * H_{ref}^{cam} * M_{ref}$.

4 Discussion and Conclusions

In this paper, we propose a projection-based AR framework to augment assembly workspaces that require the interpretation of electrical schematics. Workers' disabilities were considered in the definition of how the AR system interacts. The advantages of our work include fast creation and personalization of projection-based assembly instructions to different humans skills, which is critical to industries that have to deal with high product customization and short product life cycles. Our solution overcomes drawbacks in 2D tracking technology, and does not require multiple cameras to track and find 3D projection coordinates in arbitrary surfaces.

In our preliminary tests, the creation of digital manuals from electrical schematics took between 0 up to 10 s per instruction. The time required considered minor changes to projected elements and definition of relevant rules. The framework did not require adaptation of existent assembly materials to the framework and facilitated the adaptation of the manual to different worker's skills. Hence, improving the time needed to create manuals and to assemble electrical cabinets.

Acknowledgements. This project has received funding from the European Union's Horizon 2020 research and innovation programme under the grant agreement no 723711 and from the Ministry of Economy and Competitiveness of the government of Spain through a Torres Quevedo grant.

References

1. ElMaraghy, H.A.: Flexible and reconfigurable manufacturing systems paradigms. Int. J. Flex. Manuf. Syst. **17**(4), 261–276 (2005)
2. Tang, A., Owen, C., Biocca, F., Mou, W.: Comparative effectiveness of augmented reality in object assembly. In: Proceedings of the SIGCHI Conference on Human factors in computing systems, pp. 73–80. ACM (2003)

3. Funk, M., Bächler, A., Bächler, L., Kosch, T., Heidenreich, T., Schmidt, A.: Working with augmented reality?: a long-term analysis of in-situ instructions at the assembly workplace. In: Proceedings of the 10th International Conference on PErvasive Technologies Related to Assistive Environments, ser. PETRA 2017. New York, pp. 222–229. ACM (2017)
4. Hořejší, P.: Augmented reality system for virtual training of parts assembly. Procedia Eng. **100**, 699–706 (2015)
5. Peniche, A., Diaz, C., Trefftz, H., Paramo, G.: Combining virtual and augmented reality to improve the mechanical assembly training process in manufacturing. In: American Conference on Applied Mathematics, pp. 292–297 (2012)
6. Sand, O., Büttner, S., Paelke, V., Röcker, C.: smart. assembly-projection-based augmented reality for supporting assembly workers. In: International Conference on Virtual, Augmented and Mixed Reality, pp. 643–652. Springer, Cham (2016)
7. Rodriguez, L., Quint, F., Gorecky, D., Romero, D., Siller, H.R.: Developing a mixed reality assistance system based on projection mapping technology for manual operations at assembly workstations. Procedia Comput. Sci. **75**, 327–333 (2015)
8. Petersen, N., Pagani, A., Stricker, D.: Real-time modeling and tracking manual workflows from first-person vision. In: 2013 IEEE International Symposium on Mixed and Augmented Reality (ISMAR), pp. 117–124, October 2013
9. Funk, M., Kosch, T., Schmidt, A.: Interactive worker assistance: comparing the effects of in-situ projection, head-mounted displays, tablet, and paper instructions. In: Proceedings of the 2016 ACM International Joint Conference on Pervasive and Ubiquitous Computing, pp. 934–939. ACM (2016)
10. Feiner, S., Macintyre, B., Seligmann, D.: Knowledge-based augmented reality. Commun. ACM **36**(7), 53–62 (1993)
11. Hassaballah, M., Abdelmgeid, A.A., Alshazly, H.A.: Image Features Detection, Description and Matching, pp. 11–45. Springer, Cham (2016)
12. Steger, C.: Similarity measures for occlusion, clutter, and illumination invariant object recognition. In: Radig, B., Florczyk, S. (Eds.) Pattern Recognition, pp. 148–154. Springer, Heidelberg (2001)
13. Bouguet, J.Y.: Pyramidal implementation of the lucas kanade feature tracker. Intel Corporation, Microprocessor Research Labs (2000)
14. Drummond, T., Cipolla, R.: Real-time visual tracking of complex structures. IEEE Trans. Pattern Anal. Mach. Intell. **24**(7), 932–946 (2002)
15. Choi, C., Christensen, H.I.: Real-time 3D model-based tracking using edge and keypoint features for robotic manipulation. In: 2010 IEEE International Conference on Robotics and Automation, pp. 4048–4055, May 2010
16. Moreno, D., Taubin, G.: Simple, accurate, and robust projector-camera calibration. In: Proceedings of the 2012 Second International Conference on 3D Imaging, Modeling, Processing, Visualization & Transmission, ser. 3DIMPVT 2012. IEEE Computer Society, Washington, DC, pp. 464–471 (2012). https://doi.org/10.1109/3DIMPVT.2012.77

Deep Learning for Deflectometric Inspection of Specular Surfaces

Daniel Maestro-Watson(✉), Julen Balzategui, Luka Eciolaza, and Nestor Arana-Arexolaleiba

Department of Robotics and Automation, Mondragon University, 20500 Arrasate, Spain
{dmaestro,jbalzategui,leciolaza,narana}@mondragon.edu

Abstract. Deflectometric techniques provide abundant information useful for aesthetic defect inspection in specular and glossy/shinny surfaces. A series of light patterns is observed indirectly through their reflection on the surface under inspection, and different geometrical or texture information about the surface can be extracted. In this paper, we present a deep learning based approach for the automated defect identification in deflectometric recordings. The proposed learning framework automatically learns features used for classification. Although the method is in an early stage of development, the experiments with industrial parts show promising results, and a very direct application if compared to handcrafted feature definition approaches.

Keywords: Automated surface inspection · Defect detection Specular surfaces · Deflectometry · Convolutional Neural Networks

1 Introduction

In many industrial environments, such as automotive, home appliance or consumer electronic manufacturing, surface appearance affects consumers' perception of the quality. The human perception is especially sensitive to imperfections on surfaces that exhibit a specular or glossy appearance, as both local shape deviations and imperfections in the surface finish produce effects that can be perceived by the human eye, even in micrometer scale. Thus, a surface inspection process is often required in order to ensure that only unblemished surfaces are shipped out to customers.

Deflectometric techniques [23,24] are used to perform full-field optical measurements on surfaces that exhibit specular reflection behavior. A series of patterns are displayed on the LCD screen, and the camera observes them indirectly through their reflection on the surface. From the captured deflectometric measurements different geometrical information can be extracted. Depending on the employed setup, the normal vector field and the absolute shape [2,8,22], or the local curvature of a surface [3] can be obtained. Additionally, the intensity of the reflected patterns can also provide information on the texture of the surface

M. Graña et al. (Eds.): SOCO'18-CISIS'18-ICEUTE'18, AISC 771, pp. 280–289, 2019.
https://doi.org/10.1007/978-3-319-94120-2_27

finish [16]. Thus, deflectometric recordings provide abundant information useful for aesthetic defect inspection in specular and glossy/shinny surfaces.

Different approaches have been proposed for the exploitation of deflectometric data for automated defect identification, but they are often dependent on the particular application. Generally, the feature engineering required to design an inspection system is a tedious process that consists of selecting the parameters and pre-processing operations required for the classification [4,9,11,21,25].

The usage of Deep Neural Networks (DNN) for feature learning has become very popular in the last five years, due to their remarkable results in classification task [10]. More recently, the application of these techniques to surface inspection has been studied by many researchers, showing its potential in different scenarios such as rail surface defect detection, pavement crack detection, X-Ray inspection of metal pipes, or micro-structure defect detection in microscope images [5,6,17].

In this paper we present a deep learning based approach for the automated defect identification in deflectometric data.

2 Related Works

2.1 Deflectometry for Detection of Surface Imperfections

Although in some applications a spatial analysis of the pattern variations can suffice to fulfill the inspection task. E.g. Tarry et al. [21] describe an approach which detects paint defects in molded plastic parts (foreign contamination, cold-slug...), locally inferring anomalous regions with fixed thresholds. Usually, the task is more complex and some kind of pattern recognition method is required.

Few works have studied the use of machine learning techniques to automate the defect detection on industrial objects by analyzing deflectometric recordings, and most of them rely on shallow learning techniques:

Caulier et al. have worked in identification of textural (grease, scratch, roughness...) and geometrical (hit, wear, abrasion...) imperfections in long metallic surfaces, e.g. [4]. Their work is based on the intensity images of a reflected sinusoidal pattern. Different combination of features (both spatial, and frequential) and classification algorithms (NB, 1-NN, 3-NN, ...) are tested, concluding that the optimal setup is application dependent.

Ziebarth et al. have worked on the identification of geometric imperfections (dents and pimples) in lacquered metal sheets. They tackle a multi-scale analysis of height-maps [25] obtained from a deflectometric setup. They analyze several methods to extract geometric features, including high-pass, gradient, Laplacian-of-Gaussian (LoG), or wavelets filters, and consider several classification algorithms (NB, MB, SVM). The feature extraction based on bi-orthogonal wavelet filter banks designed to match characteristics of specific defects, and an SVM classifier seems to give them good results [11].

In [9] the authors deal with the detection of geometric imperfections (cavities and cracks) in silicon wafers for the semiconductor industry. They classify surface regions by analyzing the curvature obtained from a deflectometric registration, using different statistical properties of the curvature patch as a feature vector for classification algorithms (SVC, KNNC, Fisher LDC, and QDC).

All these methods employ hand-crafted features that need to be defined for a particular task.

2.2 Deep Learning

Deep learning is a subfield of automatic learning in which different architectures are built and trained in an iterative process using a large set of data, from which the model learns to extract meaningful characteristics. The extracted information is then used to make predictions for different types of tasks such as classification. During the training process, the accuracy of the predictions is improved through loss optimization algorithms, obtaining a better model for the assigned task in each iteration.

These techniques have been used for multiple purposes like natural language processing, image classification and object detection or automatic text translation, applied into different kinds of domains: human pose estimation, pedestrian detection [1], medical image analysis, parse scenes and natural language sentences, language translation. The examples above are few of the projects developed around these groups of methods.

Regarding the defect detection scenario, various researchers have applied a deep learning approach. Some implementations have been carried out directly using exclusively deep learning techniques, while others have been combinations of these with other techniques as support: Malekzadeh et al. [13] used a pre-trained Convolutional Network (CNN) as a feature extractor from keypoints proposals of aircraft fuselage images by SURF detector, and then classified them with a linear SVM classifier. Muniategui et al. [15] used a CNN approach to classify quality on welded parts and compress information through autoencoders within an industrial process. Fan et al. [6] used a CNN directly over pavement crack images without any kind of preprocessing. Other examples of CNN implementations in industry were by Soukup et al. [20] and Masci et al. [14], both for defect detection on steel surfaces.

3 Method

3.1 Data Acquisition

In this work we employ a sinusoidal pattern coding approach based on phase shifting [19] and temporal phase unwrapping algorithms [18], using two sets of orthogonal sinusoidal patterns of different frequencies. The pattern phase encodes information on the spatial location of the light source that generated them. Decoding the acquired images results in a mapping ($\boldsymbol{l}$) between each image pixel $p_{\mathrm{i}} = [u_{\mathrm{c}}, v_{\mathrm{c}}]$, and the position that it observes on the LCD screen $p_{\mathrm{l}} = [u_{\mathrm{l}}, v_{\mathrm{l}}]$, as shown in Eq. 1. This mapping is called a deflectometric registration or light map.

$$\boldsymbol{l} : p_{\mathrm{i}} \rightarrow p_{\mathrm{l}}, \quad \boldsymbol{l}\left(u_{\mathrm{c}}, v_{\mathrm{c}}\right) = \left[u_{\mathrm{l}}, v_{\mathrm{l}}\right] \tag{1}$$

Obtaining local curvature is straightforward, as the light-map itself already contains information directly related to surface slopes, and therefore, its partial derivatives (see Eq. 2) correspond to local curvature deviations [3,23] in orthogonal directions. Local curvature deviations have been proven very meaningful in aesthetic defect perception, and their computation is much simpler than absolute shape measurements, which require an accurately calibrated setup [8,22].

$$\tilde{\boldsymbol{C}}_{\mathrm{x}}\left(u_{\mathrm{c}}, v_{\mathrm{c}}\right)=\frac{\partial \boldsymbol{l}\left(u_{\mathrm{c}}, v_{\mathrm{c}}\right)}{\partial u_{c}}, \qquad \tilde{\boldsymbol{C}}_{\mathrm{y}}\left(u_{\mathrm{c}}, v_{\mathrm{c}}\right)=\frac{\partial \boldsymbol{l}\left(u_{\mathrm{c}}, v_{\mathrm{c}}\right)}{\partial v_{c}} \tag{2}$$

Textural or appearance defects cannot always be detected by analyzing the local curvature, because they do not alter the surface geometry, but rather the amount of reflected light, e.g. stains, scratches or texture variations. For this kind of defects the contrast of the captured patterns (γ) can provide valuable information, as it is directly related to the reflectance properties of the surface [16]. The contrast is usually high for non-defective areas, but textural imperfections reduce the amount of light that is reflected specularly, resulting in a contrast loss on the captured patterns. Contrast can be easily obtained from the employed phase shifting algorithm [19].

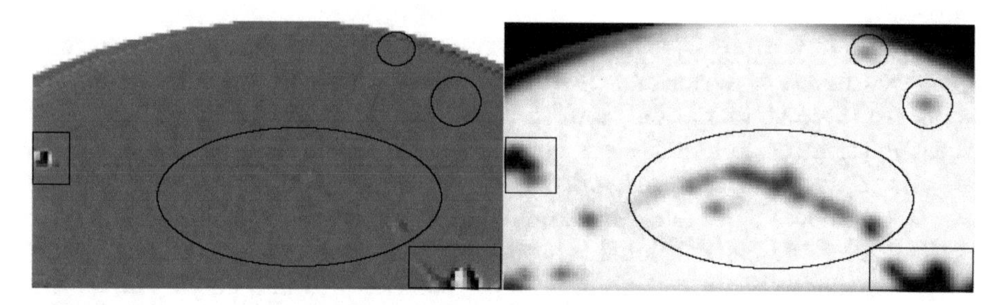

Fig. 1. Sample deflectometric registration with several visible textural (circles) and geometrical imperfections (squares): (a) Local curvature x direction - $\tilde{\boldsymbol{C}}_{\mathrm{x}}$ (b) Contrast x direction - γ_{x}

In Fig. 1 the curvature and contrast of a sample surface can be observed. Different imperfections are present on it, namely a slight scratch (big circle), two dents (squares), and two stains (small circles). It can be seen that geometrical defects (squares) are observed in both variables, but curvature provides more information. On the other hand, the textural defects (circles) can clearly be seen in the contrast information, but can only be barely perceived in the curvature.

In order to be able to detect textural and geometrical defects, we propose employing both variables, curvature and contrast, each calculated in orthogonal directions. Thus, for each measurement four data matrices will be produced and stacked together, namely $\tilde{\boldsymbol{C}}_{\mathrm{x}}$, $\tilde{\boldsymbol{C}}_{\mathrm{y}}$, γ_{x}, and γ_{y}.

3.2 Convolutional Networks

Deep convolutional neural networks (DCNN), proposed by LeCun et al. [12], are a specialized kind of neural network for processing data that has a known grid-like topology. Examples include time-series data, which can be thought of as a 1-D grid taking samples at regular time intervals, and image data, which can be thought of as a 2-D grid of pixels. Convolutional networks have been tremendously successful in practical applications. A convolution is a specialized kind of linear mathematical operation. Convolutional networks are simply neural networks that use convolution in place of general matrix multiplication in at least one of their layers. A convolution layer connects to the next layer in a similar manner as the traditional multi-layer neural network. However, weights are shared between sets of connections. Each set of weight sharing connections forms a filter that is convoluted with the input data. The filter is the same concept as that used in image processing: a filter, kernel or mask used for blurring, sharpening embossing, edge detection, etc. There are usually several such filters trained in parallel at each layer, and every filter acts as a feature detector.

Thus, the result of applying a convolution across an image forms a feature map. This feature map is automatically optimized directly from training data, and does not need of a feature selection process involved in a conventional image processing application.

Another important part of the deep neural networks is the activation function. This function performs a nonlinear transformation of a data space, and facilitates the discrimination between image classes. The Rectifier Linear Unit (ReLU) activation function is the default activation function recommended for use with most feedforward networks. This function is nearly linear, which makes easy to optimize neural networks through gradient descent based methods.

Typically, in a convolutional layer, there is also a pooling layer. The feature maps resulting from a convolution layer are sub-sampled in a pooling layer. In a max-pooling layer, the dimensions of the feature maps are reduced by merging local information (selecting maximum values) within a neighborhood window.

Convolutional layers and max-pooling layers are laid successively to create the DCNN architecture. DCNN has multiple layers that can represent complex functions, and compared to shallow architectures are more efficient and have greater generalization accuracy.

3.3 Training

A concept referred to as regularization is a central problem in machine learning. It deals with making an algorithm perform well not just on training data, but also on new inputs. Many strategies used in machine learning are explicitly designed to reduce the test error. A great many forms of regularization are available to the deep learning practitioner. In fact, developing more effective regularization strategies has been one of the major research efforts in the field [7].

A standard gradient descent method involves optimizing the error over the entire training set and can be extremely computationally expensive for a large network. Therefore, the training of a DCNN is made through a stochastic gradient descent method. This is an approximation method where the optimization is computed iteratively over small subsets of the training data (batches), which is computationally much cheaper than standard gradient descent. Although this method uses approximations of the gradient, it has been shown that the average or trend of the error converges correctly and provides good classification results. In this paper, we have used batch sizes of 64 images for the optimization.

Apart from the optimization method, and in order to avoid overfitting during training, we have also used dropout, batch normalization, and the application of Gaussian noise to the training samples.

Dropout provides a computationally inexpensive but powerful method of regularizing a broad family of models. This method turns off (drops out) a set of neurons to be optimized in each iteration of the optimization, helping to avoid overfitting by breaking-up non-relevant coincidental patterns.

Noise robustness, is applied to the inputs as a database augmentation strategy. For some models, the addition of noise with infinitesimal variance at the input is equivalent to imposing a penalty on the norm of the weights.

3.4 Architecture

In this paper we have implemented a DCNN network composed of four stages, as can be seen in Table 1:

(1) Input: The input layer takes a 4 channel image of size 256×256 per analyzed part, as provided by the deflectometry preprocessing stage.
(2) Convolutional stage: this stage is composed of 5 convolutional layers: the first two consist of 8 filters of size 5×5, and the rest consist of 8 filters of size 3×3. A Max-Pooling layer with a filter size of 2×2 is connected in between each convolutional layer, successively reducing the size of the feature maps. At the end of this stage, the feature maps are flattened to a vector of size 512.
(3) Dense stage: this stage is composed of 4 fully-connected layers, with 512, 256, 64, and 2 nodes respectively. The first and third layers use a 50% dropout during the training stage.
(4) Output: Finally, the output layer provides the binary classification, i.e. it labels each part as *ok* or *nok*.

4 Experiments

In order to evaluate the feasibility of the presented method, an experiment was carried out with parts from an automotive manufacturer. The goal is to test if the data generated from deflectometric registrations can be successfully exploited with deep learning techniques to perform an industrial surface inspection.

Table 1. Employed architectures

Convolutional layers Number of feature maps@Filter size					Fully Connected layers (Number of nodes)			
CL1	CL2	CL3	CL4	CL5	FCL1	FCL2	FCL3	FCL4
8@5 × 5	8@5 × 5	8@3 × 3	8@3 × 3	8@3 × 3	512*	256	64*	2

Notes: A Max-Pooling layer with a filter size of 2 × 2 is connected at the output of every convolutional layer. The fully connected layers marked with a star (*) use a 50% dropout during the training stage.

4.1 Employed Setup and Dataset

In total 512 sample parts are employed, with 40 of them being labeled as defect free (*ok*), and 472 as defective (*nok*), by an expert. In order to generate a larger dataset, and to account for the variability in the positioning of the parts, each part is recorded eight times in different orientations, resulting in a dataset composed of 4096 samples. The class distribution of the dataset is clearly unbalanced, but those where all the samples that we where able to gather at this stage. In order to alleviate this problem, we used data augmentation for the *ok* class. Out of each class, we reserve 20% of the samples for validation and use the remaining 80% for training.

The parts have an approximate diameter of 20 mm and a surface that presents a mirror-like reflective appearance. The sample contains a representative variety of surface imperfections that can be geometrical (anomaly in surface shape), textural (anomaly in surface finish), or a mixture of both. Some examples are shown in Fig. 2.

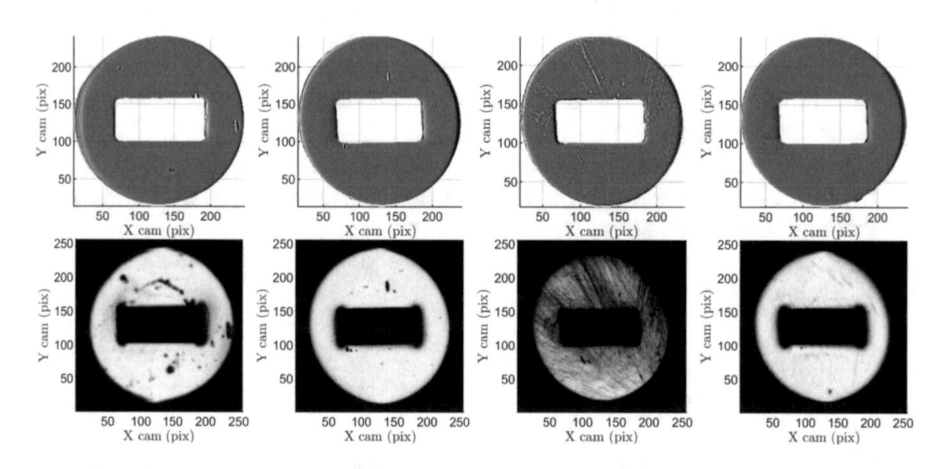

Fig. 2. Geometrical and textural defects can be observed in curvature and contrast

The employed recording setup consists on an off-the-shelf 24" LCD screen with a resolution of 2560 × 1440 pixel with a 0.2 mm pitch, and a 5 Mpixel gray-scale industrial camera with a f=25 mm lens. The setup has been designed to perform the simultaneous observation of 32 parts with an approx. lateral resolution of 0.1 mm/pixel, achieving an average inspection speed of 2.5 parts/s.

Before the classification stage, the 32 parts contained in a measurement must be segmented. Assuming no other reflective surface is in view, the background can be easily removed by thresholding γ_x and γ_y, and simple blob processing. Each off the remaining connected components corresponds to a sample, allowing us to mask and crop the measurement results to form $265 \times 256 \times 4$ pixel matrices, each containing a single part. Figure 2 shows several examples of input channels used in the classification stage.

For the implementation of the deep learning architecture, *Tensorflow* library with the *Keras* wrapper has been used. This stack is popular for DL practitioners.

4.2 Results

First, several model parameters such as the learning rate where adjusted during initial test runs. Afterwards, the model was trained for 30 epoch, obtaining an accuracy of 98.36%. In order to monitor the accuracy on unseen data, a test dataset was generated by setting apart a 20% of the original training data. The results for this dataset showed a maximum accuracy of 88.91%. The evolution of the accuracy over the training period can be seen in Fig. 3.

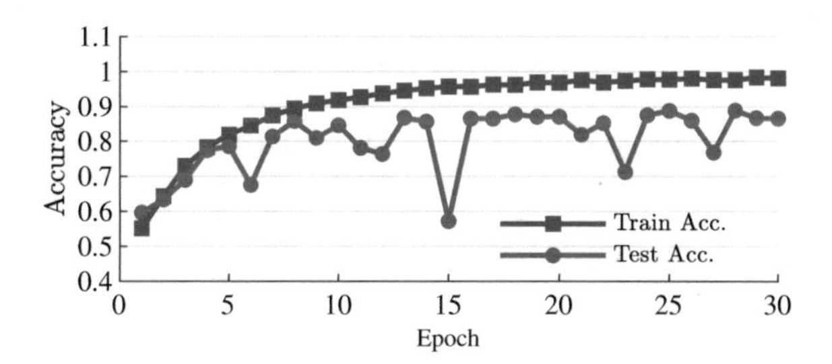

Fig. 3. Model performance: evolution of the accuracy over the training period.

Finally, the network performance was tested using the validation dataset, obtaining an accuracy of 89.57%. A deeper evaluation of the results shows a recall of 97.65% and a specificity of 81.04%, which means that the classifier favors the *nok* class. We suspect that this is caused by the use of an unbalanced class dataset, with much more samples belonging to the *nok* class. This hypothesis should be tested by repeating the experiment with a larger dataset.

In any case, in industrial inspection systems it is critical to assure no defective sample is shipped to a customer, so the recall measure is quite meaningful. Of course, the specificity is also important, but the samples classified as *nok* can be revised later by human operators, returning the incorrectly classified ones to the production line. For this reason, we consider these results good enough for our purposes at this initial stage, although they should be improved in order to apply the method in a real industrial application.

5 Conclusion

We have proposed a deep learning based approach for the automated defect detection in deflectometric data. By feeding local curvature maps and radiometric data to a DCNN classification architecture, surfaces containing geometric or textural imperfections are automatically identified from raw deflectometric data. Compared to shallow learning approaches previously employed in deflectometric defect detection literature, the employed method avoids hand-crafted feature selection work, considerably reducing the engineering efforts on deflectometric surface inspection system development.

We demonstrate the feasibility of the approach with an experiment performed with parts from an automotive industry manufacturer, obtaining classification rates around 89%, which shows that deep learning can be used to exploit deflectometric registrations to perform automated specular surface inspections.

Despite the promising results, the method is still in an early stage of development, and several improvements should be made. Apart from obtaining more data and repeating the experiments, a finner tunning of the hyperparameters should be performed, and different pre-processing and normalization techniques should be evaluated. Additionally, the application of transfer leaning techniques could be investigated, employing pre-trained networks to take advantage of previous knowledge. Also, instead of approaching the problem as a binary classification that just labels parts as *ok* or *nok*, more elaborated schemes that account for the identification of the defect types and their severity could be tested.

Acknowledgments. This work has been developed by the intelligent systems for industrial systems group supported by the Department of Education, Language policy and Culture of the Basque Government. The authors would also like to thank the collaboration partner Ekide S.L. for their contributions.

References

1. Angelova, A., Krizhevsky, A., Vanhoucke, V., Ogale, A.S., Ferguson, D.: Real-time pedestrian detection with deep network cascades. In: Proceedings of British Machine Vision Conference (BMVC), vol. 2, p. 4 (2015)
2. Burke, J., Li, W., Heimsath, A., von Kopylow, C., Bergmann, R.: Qualifying parabolic mirrors with deflectometry. J. Eur. Opt. Soc. Rapid Publ. **8** (2013)
3. Burke, J.: Inspection of reflective surfaces with deflectometry. In: Proceedings of the International Conference on Processes in Combined Digital Optical & Imaging Methods, pp. 108–111 (2016)
4. Caulier, Y.: Inspection of complex surfaces by means of structured light patterns. Opt. Express **18**(7), 6642–6660 (2010)
5. Faghih-Roohi, S., Hajizadeh, S., Núñez, A., Babuska, R., Schutter, B.D.: Deep convolutional neural networks for detection of rail surface defects. In: 2016 International Joint Conference on Neural Networks (IJCNN), pp. 2584–2589, July 2016
6. Fan, Z., Wu, Y., Lu, J., Li, W.: Automatic Pavement Crack Detection Based on Structured Prediction with the Convolutional Neural Network. ArXiv, February 2018

7. Goodfellow, I., Bengio, Y., Courville, A., Bengio, Y.: Deep Learning, vol. 1. MIT press, Cambridge (2016)
8. Häusler, G., Faber, C., Olesch, E., Ettl, S.: Deflectometry vs. interferometry. In: Proceedings of the SPIE 8788, Optical Measurement Systems for Industrial Inspection (2013)
9. Kofler, C., Spöck, G., Muhr, R.: Classifying defects in topography images of silicon wafers. In: 2017 Winter Simulation Conference (WSC), pp. 3646–3657, December 2017
10. Krizhevsky, A., Sutskever, I., Hinton, G.E.: Imagenet classification with deep convolutional neural networks. In: Proceedings of the 25th International Conference on Neural Information Processing Systems, NIPS 2012, USA, pp. 1097–1105 (2012)
11. Le, T.T., Ziebarth, M., Greiner, T., Heizmann, M.: Systematic design of object shape matched wavelet filter banks for defect detection. In: 39th International Conference on Telecommunications and Signal Processing (TSP), pp. 470–473, June 2016
12. LeCun, Y., et al.: Generalization and network design strategies. Connectionism in perspective, pp. 143–155 (1989)
13. Malekzadeh, T., Abdollahzadeh, M., Nejati, H., Cheung, N.M.: Aircraft Fuselage Defect Detection using Deep Neural Networks. ArXiv e-prints, December 2017
14. Masci, J., Meier, U., Ciresan, D., Schmidhuber, J., Fricout, G.: Steel defect classification with max-pooling convolutional neural networks. In: 2012 International Joint Conference on Neural Networks (IJCNN), pp. 1–6. IEEE (2012)
15. Muniategui, A., Hériz, B., Eciolaza, L., Ayuso, M., Iturrioz, A., Quintana, I., Álvarez, P.: Spot welding monitoring system based on fuzzy classification and deep learning. In: IEEE International Conference on Fuzzy Systems (FUZZ-IEEE), pp. 1–6 (2017)
16. Nagato, T., Fuse, T., Koezuka, T.: Defect inspection technology for a gloss-coated surface using patterned illumination. In: Proceedings of the SPIE, vol. 8661 (2013)
17. Ren, R., Hung, T., Tan, K.C.: A generic deep-learning-based approach for automated surface inspection. IEEE Trans. Cybern. **48**(3), 929–940 (2018)
18. Saldner, H.O., Huntley, J.M.: Temporal phase unwrapping: application to surface profiling of discontinuous objects. Appl. Opt. **36**(13), 2770–2775 (1997)
19. Schreiber, H., Bruning, J.H.: Phase Shifting Interferometry. In: Optical Shop Testing, pp. 547–666. John Wiley & Sons (2007)
20. Soukup, D., Huber-Mörk, R.: Convolutional neural networks for steel surface defect detection from photometric stereo images. In: International Symposium on Visual Computing, pp. 668–677. Springer (2014)
21. Tarry, C., Stachowsky, M., Moussa, M.: Robust detection of paint defects in moulded plastic parts. In: Canadian Conference on Computer and Robot Vision, pp. 306–312, May 2014
22. Tutsch, R., Petz, M., Fischer, M.: Optical three-dimensional metrology with structured illumination. Opt. Eng. **50**, 101507–101510 (2011)
23. Werling, S., Mai, M., Heizmann, M., Beyerer, J.: Inspection of specular and partially specular surfaces. Metrol. Measur. Syst. **16**(3), 415–431 (2009)
24. Zhang, Z., Wang, Y., Huang, S., Liu, Y., Chang, C., Gao, F., Jiang, X.: Three-dimensional shape measurements of specular objects using phase-measuring deflectometry. Sensors **17**(12), 2835 (2017)
25. Ziebarth, M.: Empirical comparison of defect classifiers on specular surfaces. In: Proceedings of the 2013 Joint Workshop of Fraunhofer IOSB and Institute for Anthropomatics, Vision and Fusion Laboratory (2014)

Special Session: Optimization, Modeling and Control by Soft Computing Techniques

Disturbances Based Adaptive Neuro-Control for UAVs: A First Approach

J. Enrique Sierra and Matilde Santos(✉)

Computer Science Faculty, Complutense University of Madrid, Madrid, Spain
{jesier01,msantos}@ucm.es

Abstract. In this work an adaptive neuro-control is proposed to cope with some external disturbances that can affect unmanned aerial vehicles (UAV) dynamics, specifically: the variation of the system mass during logistic tasks and the influence of the wind. An intelligent control strategy based on a feedforward neural networks is applied. In particular, a variant of the generalized learning algorithm has been used. Simulation results show how the on-line learning increases the robustness of the controller, reducing the effects of the changes in mass and the effects of wind on the UAV stabilization, thus improving the system response. It has been compared with a PID controller obtaining better results.

Keywords: Neuro-control · Adaptive control · Disturbances rejection
Online learning · Neural networks · Unmanned Aerial Vehicle (UAV)
Quadrotor

1 Introduction

In recent years, new and valuable applications of unmanned aerial vehicles (UAV) have appeared in different sectors such as defense, security, construction, agriculture, entertainment, shipping, etc. These and other applications demand the design of efficient and robust controllers for those autonomous vehicles. Thus, the modeling and control of these complex and unstable systems still motivate the research and the interest of the scientific community [1–3].

Nevertheless, the modeling and control of unmanned aerial vehicles are not easy tasks. Their complexity comes from the randomness of the airstreams and of the exogenous forces, the high non-linearity of the dynamics, the coupling between the internal variables, the uncertainty of the measurements… These factors make the techniques based on artificial intelligence a promising approach for the identification and control of these systems.

These intelligent strategies are especially interesting when the model parameters vary while the system is working. For example, the total mass will undergo variations when the vehicle is performing logistic tasks, since the mass depends on the packages that are shipped.

There are several studies where neural networks are applied to model these systems [4, 5] and to control them [6–9]. But these intelligent strategies are usually applied without considering external disturbances, an issue that deserves further research.

M. Graña et al. (Eds.): SOCO'18-CISIS'18-ICEUTE'18, AISC 771, pp. 293–302, 2019.
https://doi.org/10.1007/978-3-319-94120-2_28

In this work an adaptive neuro-control strategy is proposed to stabilize an unmanned aerial vehicle. A variant of the generalized learning algorithm has been applied. Changes in the system dynamics due to mass variation and external disturbances (wind) are taken into account. The results show how the online learning increases the robustness of the controller, reducing the effect of the mass variations and the wind on the UAV's stabilization, and notably improving the system response.

The paper is organized as follows. In Sect. 2 the equations which describe the dynamic behavior of the system are presented. The Sect. 3 describes the adaptive neuro-control strategy that has been implemented. Simulation results are presented and discussed in Sect. 4. The document ends with the conclusions and future works.

2 System Description

A quadrotor vehicle is composed by four perpendicular arms, each one with a motor and a propeller (Fig. 1, left). The four motors drive the lift and direction control.

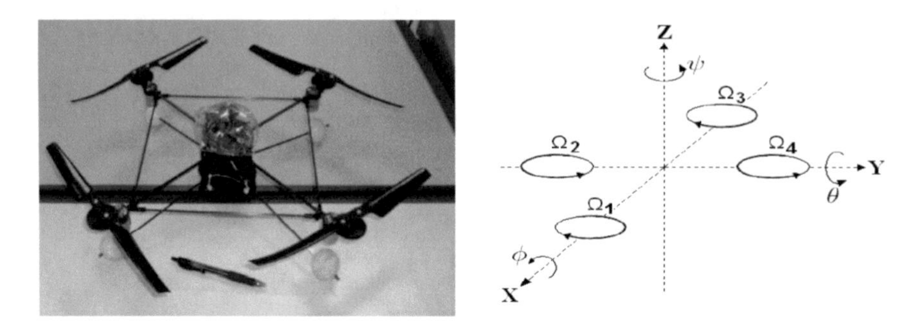

Fig. 1. Example of a quadrotor vehicle (left) and UAV's coordinate system (right)

The UAV absolute position is described by three coordinates, (x, y, z), and the attitude is given by the three Euler's angles (ϕ, θ, ψ), under conditions: $(-\pi \leq \psi < \pi)$ for the yaw angle, $(-\pi/2 \leq \phi < \pi/2)$ for the roll, and $(-\pi/2 \leq \theta < \pi/2)$ for the pitch. In order to avoid the potential instability of the UAV near the boundary conditions, the range of the pitch, roll and yaw angles is reduced to [−1.22, 1.22] radians (70º), of course velocity dependent.

The system is based on two couples of propellers which are opposed each other (1, 3) y (2, 4) (Fig. 1, right). In order to balance the system, one pair of motors turns clockwise while the other one spins counterclockwise. The increment of the speed of rotor 3 respect to rotor 1 produces a positive pitch $(\theta > 0)$, while increasing the speed of rotor 4 regarding rotor 2 produces a positive roll $(\phi > 0)$. The increment of the speeds of rotors 1 + 3 respect to rotors 2 + 4 produces a positive yaw $(\psi > 0)$.

Using the Newton-Euler's method, the angular dynamic of the system is represented as follows (1) [10]:

$$\tau = J\dot{\omega} + \omega \times J\omega \tag{1}$$

$$J = \begin{pmatrix} I_x & 0 & 0 \\ 0 & I_y & 0 \\ 0 & 0 & I_z \end{pmatrix} \tag{2}$$

Where τ is the vector of torques in the three axes, J is the inertia tensor (2), ω is the angular velocities vector, and × represents the vector product.

The translational dynamic is given by (3):

$$m\dot{v} = RT - mge_3 \tag{3}$$

Where m is the mass of the quadrotor, R is the rotation matrix, g is the gravitational acceleration, T is a vector of forces, and $e_3 = [0, 0, 1]^T$ is a unit vector which describes the rotor orientation.

The vectors τ (4) and T (5) are a function of the velocities of the propellers:

$$\tau = \begin{pmatrix} bl(\Omega_4^2 - \Omega_2^2) \\ bl(\Omega_3^2 - \Omega_1^2) \\ d(\Omega_2^2 + \Omega_4^2 - \Omega_1^2 - \Omega_3^2) \end{pmatrix} \tag{4}$$

$$T = \begin{pmatrix} 0 \\ 0 \\ b(\Omega_1^2 + \Omega_2^2 + \Omega_3^2 + \Omega_4^2) \end{pmatrix} \tag{5}$$

In Eqs. (4–5), b is the thrust coefficient, d is the drag coefficient, l is the longitude of each arm, and $\Omega_1, \ldots, \Omega_4$, are the velocities of the rotors 1 to 4, respectively.

In order to simplify the calculations, instead of using the speed of the rotors, it is possible to define a set of control signals $u_1, u_2, u_3\, y\, u_4$ as follows (6):

$$\begin{bmatrix} u_1 \\ u_2 \\ u_3 \\ u_4 \end{bmatrix} = \begin{bmatrix} 1 & 1 & 1 & 1 \\ 0 & -1 & 0 & 1 \\ -1 & 0 & 1 & 0 \\ 1 & -1 & 1 & -1 \end{bmatrix} \begin{bmatrix} \Omega_1^2 \\ \Omega_2^2 \\ \Omega_3^2 \\ \Omega_4^2 \end{bmatrix} \tag{6}$$

This matrix is invertible, so it is possible to generate speed references for the rotors from a set of control signals.

Finally, from Eqs. 1 to 6, the following system of equations is derived (7–12):

$$\ddot{\phi} = \dot{\theta}\dot{\psi}(I_y - I_z)/I_x + (lb/I_x)u_2 \tag{7}$$

$$\ddot{\theta} = \dot{\phi}\dot{\psi}(I_z - I_x)/I_y + (lb/I_y)u_3 \tag{8}$$

$$\ddot{\psi} = \dot{\phi}\dot{\theta}(I_x - I_y)/I_z + (d/I_z)u_4 \tag{9}$$

$$\ddot{X} = -(sin\theta cos\phi)(b/m)u_1 \tag{10}$$

$$\ddot{Y} = (sin\phi)(b/m)u_1 \tag{11}$$

$$\ddot{Z} = -g + (cos\theta cos\phi)(b/m)u_1 \tag{12}$$

The constants of Eqs. 7 to 12 that have been used during the simulations are listed in Table 1. The values have been extracted from [10].

Table 1. Constants of the model

Parameter	Description	Value/Units
l	Longitude of an arm	0.232 m
m	Mass of the quadrotor	0.52 kg
d	Drag coefficient	$7.5e^{-7}$ N.m.s^2
b	Thrust coefficient	$3.13e^{-5}$ N.s^2
I_x	Inertia in X	$6.228e^{-3}$ kg.m^2
I_y	Inertia in Y	$6.225e^{-3}$ kg.m^2
I_z	Inertia in Z	$1.121e^{-2}$ kg.m^2
ρ_{air}	Density of the air	1.2 kg/m^3
A	Area in the direction of the wind	0.0186 m^2
Cd	Wind drag coefficient	1

3 Description of the Neuro-Controller

3.1 Control Strategy

There are different control strategies with neural networks [5–9]. In our case, a variant of the generalized learning algorithm (GLA) has been used. The modification consists of the refinement of the network during the execution of the controller by adaptive learning.

The first step is the application of the GLA algorithm to off-line training the neural network in order to identify the inverse dynamic of the plant (Fig. 2).

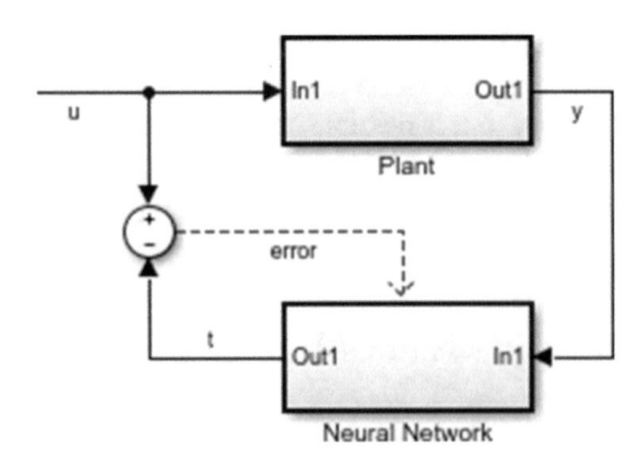

Fig. 2. Off-line training to identify the inverse plant dynamic

Once the network has been off-line trained, it is placed in cascade connection with the plant and a PID controller. In Fig. 3, this control strategy is shown for the altitude variable. Then, the configuration of the network is on-line refined. In order to do this, during each control interval two processes are sequentially applied to the network (first the simulation, later the on-line learning):

1. Simulation: The output of the PID, $PID_{OUT}(t)$, feeds one of the inputs of the artificial neural network; the rest of the inputs are past values of the plant output, $PLANT_{OUT}(t - i * Ts)$. The network generates the control input, u_1 (13), which is the input of the plant (Fig. 3, switch in the upper position).

$$\begin{aligned} u_1(t) &= f_{NET}(PID_{OUT}(t), PLANT_{OUT}(t - i * Ts), par_{NET}(t - Ts)) \\ i &= 1\ldots(N_{inputs} - 1) \end{aligned} \tag{13}$$

Where Ts is the sampling time, par_{NET} denotes the configuration parameters of the network, and N_{inputs} means the number of inputs of the neural network.

2. On-line learning: The neural network is trained again with the current and previous outputs of the plant, in order to generate the control output, u_1, obtaining the new configuration parameters par_{NET} (14). The network input dataset is made up of the past values of the plant output, $PLANT_{OUT}(t - i * Ts)$. The output dataset is simply the current value of the plant input $u_1(t)$ (Fig. 3, switch in the lower position).

$$par_{NET}(t) = f(PLANT_{OUT}(t - i * Ts), u_1(t), par_{NET}(t - Ts)) \quad i = 0\ldots(N_{inputs} - 1) \tag{14}$$

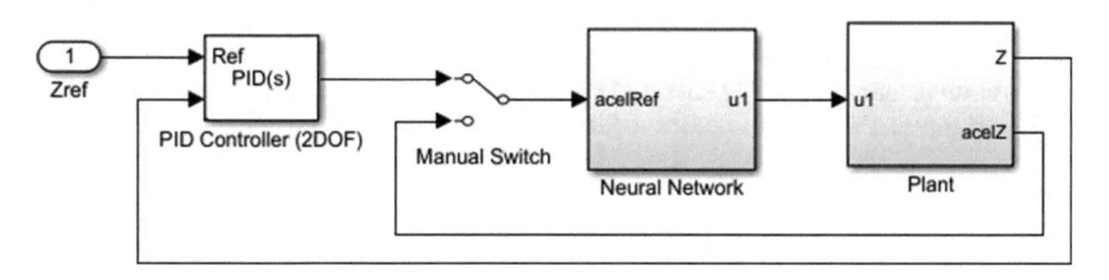

Fig. 3. Adaptive altitude neuro-control strategy

3.2 Altitude Control

In order to test the validity of this proposal we have firstly focused on the altitude control. UAVs are normally provided with accelerometers, so it is assumed that the acceleration in the z-axis $(\ddot{Z})$ is available. The network must be able to simulate the control signal u_1 by using acceleration measurements.

In this example, PID_{OUT} (13) is the reference of the acceleration in the z-axis, $\ddot{Z}_{PID}$; $PLANT_{OUT}$ is the acceleration in the z-axis, $\ddot{Z}$, and u_1 is the control signal. The sampling time Ts has been fixed to 10 ms.

The tuning parameters of the PID are set to Kp = 2, Kd = 2, Ki = 0, obtained by trial and error. Thanks to the artificial neural network, the PID does not need to include the plant gain. The network is able to learn the plant gain and work with it. In other words, with this approach it is not necessary to know the system parameters to control it. Another advantage of this approach is that a PD controller is enough because the network is able to reduce the stationary error.

The selected network is a multilayer perceptron (MLP), with a hidden layer with 5 neurons. The number of inputs of the network, N_{inputs}, is fixed to 10. The Levenberg-Marquardt algorithm with $\mu = 0.001$ has been used.

A train of pulses with variable amplitude is generated during the off-line phase to initially train the network. The duration of the train of pulses is 2 s. Before the training, the input and output signals are normalized to the interval [0–1].

4 Simulation Results

The simulation results have been obtained with Matlab/Simulink software. The duration of each simulation is 15 s. The controller is off-line trained during the first 2 s. Then the on-line learning algorithm is applied the remaining 13 s by introducing the reference Zref to the controller. In this preliminary work, only the influence of the system mass variation and the wind are considered

The performance of the proposed approach is compared with the application of a PID regulator. In order to do a fair comparison, the PID output has been multiplied by a factor given by *m/b* to include the plant gain. Otherwise the system becomes unstable. It is also necessary to add the integral term in the PID to eliminate the stationary error. During the simulations, Ki is set to 0.9.

In Fig. 4 (left), the behavior of the controller during the different phases is shown. Time (sec) is represented in the x-axis; the acceleration (m/s^2) in z (blue line), the acceleration reference (m/s^2) generated by the controller once it has been trained (green line), and the UAV altitude (m) (red line) are represented in the y-axis.

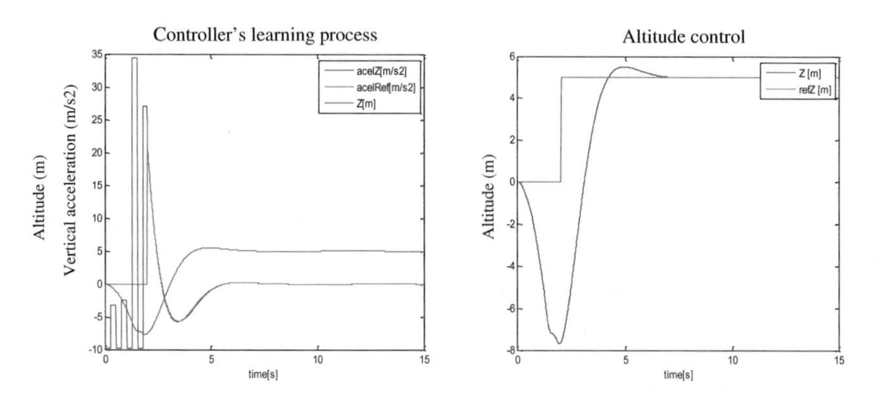

Fig. 4. Learning of the neural controller (left) and altitude with online learning (right)

Figure 4 (right) shows the altitude (m) in blue and the reference (m) in green. The controller is trained from t = 0 s up to t = 2 s. For this reason, the reference (green line) is 0 in that interval. The control signals used to off-line training the network produce changes in the altitude during this interval. The control phase begins at t = 2 s, when the Z reference is set to 5 m. From then, acceleration references are continuously generated to stabilize the Z over this value. It is also possible to see how the signals acelRef (green) and acelZ (blue) match better over time. The controller is able to stabilize the altitude to the desired value.

These results can be extended to the UAV Euler's angles control for path following. Figure 5 shows the result of the application of this technique to the Euler's angles, when the reference of every angle is set to $\pi/4$ radians.

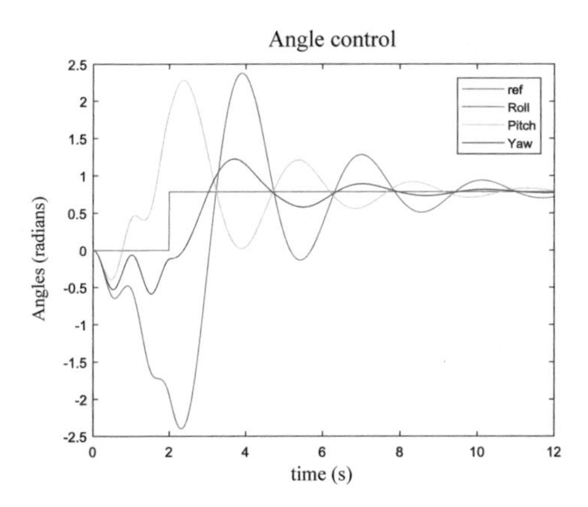

Fig. 5. Application of the control strategy to the UAV Euler's angles

Although the system response can be improved, the neuro-control strategy stabilizes the UAV.

4.1 Robustness When Varying the Mass

In this experiment the effect of the mass variation is simulated by adding a new term, $dist_m$ (16) to Eq. 12, resulting (15).

$$\ddot{Z} = -g + (cos\theta cos\phi)(b/(m * dist_m))u_1 \tag{15}$$

$$dist_m = 1 + step(t - 4) \tag{16}$$

Where *step* denotes the unit step function. That is, the mass of the quadrotor is duplicated from t = 4 s on. Figure 6 shows the reference (green line), the altitude with online-learning (blue line), without it (red line), and only with PID (purple line), in m. The error at the output of the neural network worsens the system response. Indeed, the stationary error when on-line learning is not applied is significant, due to the fact that after the perturbation the network needs to learn the new stationary error.

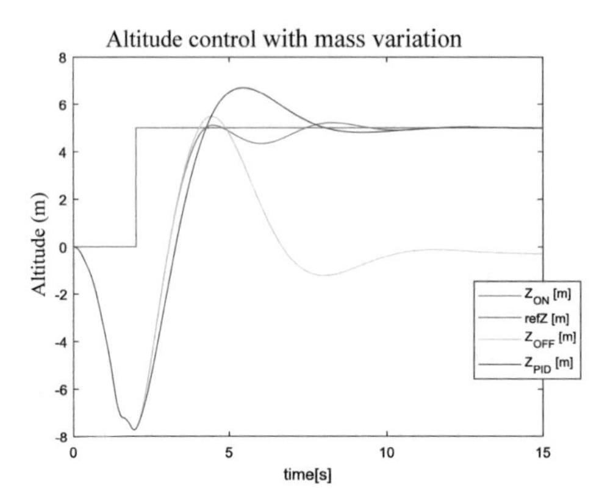

Fig. 6. Effect of the variation in mass

The performance of the controllers with and without online learning in this experiment is compared in Table 2. It is noticeable how the overshoot and the steady state error are clearly reduced by the online learning. The settling time is also slightly reduced. As expected, the learning makes the response a bit slower. It is significant how this control approach improves the response respect to the PID, specially the overshoot.

Table 2. Performance of the controllers when varying the mass

Parameter	Without online learning	With online learning	Only with PID
Rise time (s)	**2.05**	2.24	2.27
Settling time (s)	9.42	**7.44**	8.56
Overshoot (%)	9.66	**2.26**	34.04
Steady state error	5.31	**0.015**	**0.015**

4.2 Robustness with Wind Disturbance

A new term $(dist_w)$ is added to Eq. 12 to model the wind disturbances (17).

$$\ddot{Z} = -dist_w - g + (cos\theta cos\phi)(b/m)u_1 \tag{17}$$

The disturbance represents the variation of the acceleration, modeled by (18), caused by the external wind in the movement direction [11]:

$$dist_w = sgn(v_w) \cdot \rho_{air} \cdot A \cdot Cd \cdot \left(\dot{Z} - v_w\right)^2 / (2m) \tag{18}$$

Where v_w is the wind speed, ρ_{air} is the air density, A is the area of the quadrotor, Cd is the drag coefficient respect to the wind, $\dot{Z}$ is the velocity in the z-axis and sgn denotes the sign function.

In this experiment, the wind speed is simulated by a step with Gaussian noise at t = 4 s. The SNR between the average wind and the noise is 10 dB. The average wind speed is 12 m/s. This value matches number 6 in the Beaufort's scale (strong breeze).

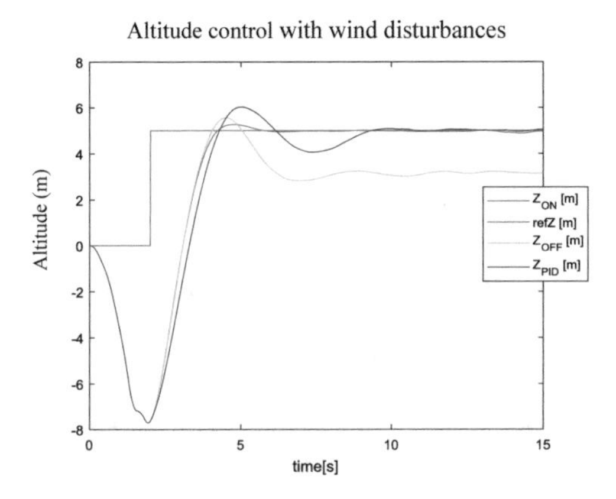

Fig. 7. Influence of wind disturbance

Figure 7 shows the altitude reference (red line) and the height of the system with and without on-line learning and only with PID (blue, yellow and purple lines, respectively), in m. As in Fig. 6, the controller without on-line learning is not able to cope with this disturbance, but in this case the stationary error is not as large.

The performance of the controllers with and without online learning is compared in Table 3. The overshoot and the steady state error have been reduced or eliminated with the online learning. Indeed, there is not stationary error, and now the settling time is notably reduced, as it should correspond to an agile system like this UAV. Again it is also significant how the overshoot is reduced respect to the application of the PID.

Table 3. Performance of the controllers with external disturbance

Parameter	Without online learning	With online learning	Only with PID
Rise time (s)	**2.05**	2.22	2.27
Settling time (s)	9.42	**3.57**	7.24
Overshoot (%)	11.04	**5.6**	20.42
Steady state error	1.86	**0**	0.1

5 Conclusions and Future Works

Unmanned aerial vehicles are difficult systems to be modeled and controlled. The complexity comes from their strong non-linear and coupled dynamics and also from external disturbances that change the behavior of the system.

In this work an adaptive neuro-controller is proposed. A MLP neural network with a modified generalized learning algorithm has been applied to the UAV. Simulation results validate the effectiveness of this stabilization controller, even with disturbances such as mass variation and wind. Even more, the online learning improves the robustness of the neuro-controller, reducing the effects of the disturbances on the system response.

Among other possible future works we may highlight the study of the influence of other disturbances such as the ones generated by the engines, and the comparison between the performance provided by this approach and other controllers.

References

1. Ortega, J.J., Sigut, M.: A low-cost mobile prototype for high-realism flight simulation. RIAI - Revista Iberoamericana de Automática e Informática Industrial **13**(3), 293–303 (2016)
2. Fahmy, A.A., Ghany, A.A.: Adaptive functional-based neuro-fuzzy PID incremental controller structure. Neural Comput. Appl. **26**(6), 1423–1438 (2015)
3. Garcia-Auñón, P., Santos, M., Cruz, J.M.: A new UAV ship-tracking algorithm. Preprints of the 20th IFAC World Congress, pp. 13632–13637 (2017)
4. Sierra, J.E., Santos, M.: Modeling engineering systems using analytical and neural techniques: hybridization. Neurocomputing **271**, 70–83 (2018)
5. Bansal, S., Akametalu, A.K., Jiang, F.J., Laine, F., Tomlin, C.J.: Learning quadrotor dynamics using neural network for flight control. In: IEEE 55th Conference on Decision and Control (CDC), pp. 4653–4660 (2016)
6. Boudjedir, H., Bouhali, O., Rizoug, N.: Adaptive neural network control based on neural observer for quadrotor unmanned aerial vehicle. Adv. Robot. **28**(17), 1151–1164 (2014)
7. Bakshi, N.A., Ramachandran, R.: Indirect model reference adaptive control of quadrotor UAVs using neural networks. In: 2011 10th International Conference on Intelligent Systems and Control (ISCO), pp. 1–6 (2016)
8. Sierra, J.E., Santos, M.: Control de un vehículo cuatrirrotor basado en redes neuronales. Actas de las XXXVIII Jornadas de Automática, CEA, pp. 431–436 (2017)
9. Yañez-Badillo, H., Tapia-Olvera, R., Aguilar-Mejía, O., Beltran-Carbajal, F.: On line adaptive neurocontroller for regulating angular position and trajectory of quadrotor system. RIAI - Revista Iberoamericana de Automática e Informática Industrial **14**(2), 141–151, (2017). https://doi.org/10.1016/j.riai.2017.01.001, ISSN 1697-7912
10. Wise, K.A., Lavretsky, E., Hovakimyan, N.: Adaptive control of flight: theory, applications, and open problems. In: American Control Conference, 2006, 6-pp. IEEE (2006)
11. NASA Glenn Research Center. https://www.grc.nasa.gov/WWW/K-12/airplane/ldrat.html

Optimizing a Fuzzy Equivalent Sliding Mode Control Applied to Servo Drive Systems

Zhengya Zhang and Matilde Santos(✉)

Computer Science School, University Complutense of Madrid,
28040 Madrid, Spain
zhangzhengyachina@gmail.com, msantos@ucm.es

Abstract. Positioning accuracy of servo drive systems is very important for tasks that require precision. From the control point of view, servo drive systems are complex due to their non-linear time-varying dynamics. Most of the control strategies applied to these systems either introduce undesirable chattering in the response or suppress it at the cost of producing large tracking errors. In this work, an equivalent sliding mode control based on fuzzy logic is applied to a servo system. The fuzzy membership functions of the switching function are optimized in order to improve the control robustness and to obtain accurate tracking at the same time. Simulation results show that this soft computing control proposal can effectively eliminate the chattering and reduce the tracking error for servo drive systems. It has been compared to conventional sliding mode control and sliding mode control with boundary layer with encouraging results.

Keywords: Fuzzy logic · Sliding mode control
Non-linear time-varying systems · Membership function optimization
Servo drive system

1 Introduction

Servo drive systems are widely used in the industrial production processes. But they are complex systems due to their non-linear time-varying dynamics. Besides, undesirable vibrations, mechanical wear and lack of installation accuracy influence their behaviour. Hence, precise control of servo drive system under internal and external disturbances is highly valued [1].

Sliding mode control is capable of addressing the uncertainty caused by some of those undesirable effects that changes the dynamics of the system. It does not require an accurate mathematical model of those disturbances. As long as the upper bounds of the uncertain parameters of the model are known, an asymptotically stable controller can be designed. Particularly, once the sliding surface has been pre-designed, the system is unaffected by small parameter variations, showing robustness. However, the implementation of sliding mode control is often disturbed by high frequency oscillations known as "chattering" in system outputs issued by dynamics from actuators and sensors ignored in system modelling.

M. Graña et al. (Eds.): SOCO'18-CISIS'18-ICEUTE'18, AISC 771, pp. 303–312, 2019.
https://doi.org/10.1007/978-3-319-94120-2_29

Different solutions have been tried to eliminate sliding mode controller chattering [2]. One commonly used is boundary layer method which introduces a boundary layer implemented as a saturation function instead of the switching function of the controller. This method can effectively suppress chattering but it produces large tracking error [3]. To address this issue, researchers have worked hard on adjusting the boundary-layer thickness. For example, in [4], dynamic boundary layer based on neural network quasi-sliding mode control has been studied for soft touching down on asteroid and numerical simulation results demonstrate its effectiveness. In [5], the design and analysis of a new autonomous underwater vehicle sliding control system was investigated using the dynamic boundary layer method to ensure the stability; the system response does not exceed the amplitude constraint of the rudder actuator. Furthermore, other sliding control methods such as disturbance observer [6], gain switching [7], reaching law [8] and neural networks [9] have been applied to different non-linear systems. Each of these methods has its own advantages and drawbacks. Indeed, in general, any of the sliding control strategies applied to these complex systems either introduces undesirable chattering in the response or suppress it at the cost of producing tracking errors.

On the other hand, fuzzy logic is also a popular solution when we have to deal with uncertainty. Even more, it is well known that fuzzy logic smooths the signals as it works with flexible range of values.

Therefore, in this work we will take advantage of the combination of fuzzy logic and the effective sliding control strategy to control servo drive systems. That is, fuzzy logic will soften the control signal to reduce or avoid the chattering phenomenon of ordinary sliding mode control.

Sliding mode fuzzy control has already been applied to different complex systems. For instance, in [10], the depth control of a bioinspired robotic dolphin based on sliding-mode fuzzy control method is successfully applied to steer the robot toward and along the desired depth. In [11], composite fuzzy sliding mode control is tested on perturbed systems, achieving a good control response. In [12], an adaptive fuzzy back-stepping sliding mode control for MIMO uncertain nonlinear systems was designed and its effectiveness proved with simulation experiment of a simple numerical system. In [13], the equivalent sliding mode control based on fuzzy control is proposed for an ideal electrical servo system, which showed that chattering phenomenon is suppressed, but not well eliminated.

In our proposal, an equivalent sliding mode controller based on fuzzy logic is applied and the membership functions of the switching function are optimized for a servo drive system. It is compared with the results of applying a sliding mode control with and without boundary layer to the same servo drive system.

The paper is organized as follows: Sect. 2 introduces the servo drive system model. Section 3 describes the sliding mode control methods that will be applied for comparison purposes. In Sect. 4, the optimized fuzzy controller is described. Section 5 discusses the results of the simulation of the fuzzy control approach and compares it to the sliding mode and boundary layer sliding control methods. Conclusions and future work end the paper.

2 Servo Drive System Model

Figure 1 shows the system under study. The load is driven by the servo motor through the two-stage gear and the ball screw. The simplified mathematical model is obtained relating the input angle of the servo motor *θi(t)* and the displacement of the pushing mechanism $x_h(t)$ [14].

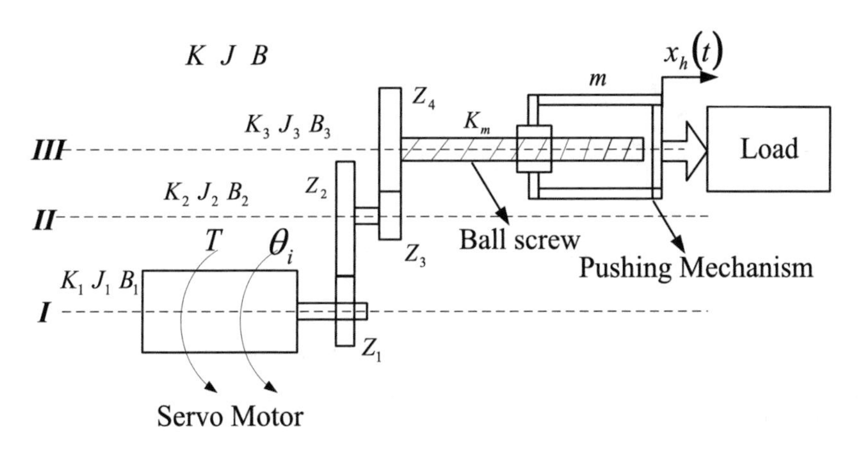

Fig. 1. Servo drive system

The equivalent torque balance equation for *I* axis (see Fig. 1) can be described as follows:

$$J\frac{d^2\theta_m(t)}{dt^2} + B\frac{d\theta_m(t)}{dt} + T_L + K[\theta_m(t) - \theta_i(t)] = 0 \tag{1}$$

where *J*, *B*, *K* are the equivalent total moment of inertia, viscous damping coefficient, and stiffness coefficient, respectively. Any of these parameters *J, B, K* can have a small perturbation, Δ, added to its initial value, J_0, B_0, K_0, which corresponds to the static values of the total moment of inertia, the viscous damping coefficient, and the total stiffness coefficient, respectively; T_L is the equivalent load torque; *θm(t)-θi(t)* is the shaft relative rotation angle.

The motor shaft equivalent angle for load displacement θ_m is related to the load displacement $x_h(t)$ as,

$$\theta_m(t) = \frac{1}{i_1 i_2}\frac{2\pi}{L}x_h(t) \tag{2}$$

With $i_1 = Z_1/Z_2$ and $i_2 = Z_3/Z_4$, being i_1 and i_2 the first and second stage of gear transmission ratio, respectively. *L* and K_{m} are the ball screw pitch and stiffness, respectively.

Substituting (2) in (1) and developing the mathematical expression, the model is now,

$$J\frac{d^2x_h(t)}{dt^2} + B\frac{dx_h(t)}{dt} + Kx_h(t) + \frac{L}{2\pi}i_1 i_2 T_L = \frac{L}{2\pi}i_1 i_2 K\theta_i(t) \tag{3}$$

This model can be expressed as a second order system, with the state variable $x_1 = x_h$, and,

$$\begin{cases} \dot{x}_1 = x_h \\ \dot{x}_1 = x_2 \\ \dot{x}_2 = [-a_1(t)x_1 - a_2(t)x_2] + g(t)u(t) - d(t) \end{cases} \tag{4}$$

where

$$\begin{cases} a_1(t) = \frac{K}{J}, & a_2(t) = \frac{B}{J} \\ g(t) = \frac{L}{2\pi J}i_1 i_2 K, & d(t) = \frac{L}{2\pi J}i_1 i_2 T_L \end{cases} \tag{5}$$

As the values of K, B and J are time variant, with an uncertain variation, the values of $a_1(t)$, $a_2(t)$ and $g(t)$ and the upper and lower bounds of $\Delta a_1(t)$, $\Delta a_2(t)$, $\Delta g(t)$ and $d(t)$ are given as follows:

$$\begin{cases} a_i(t) = a_{0i} + \Delta a_i(t) & |\Delta a_i(t)| \le \alpha_i \\ g(t) = g_0 + \Delta g(t) & |\Delta g(t)| \le \beta \\ |d(t)| \le D \end{cases} \tag{6}$$

where a_{0i} and g_0 are the non-variant part of a_i and g, respectively; Δa_i and Δg are the uncertain parts of a_i and g, respectively; $\alpha_i(t)$ and $\beta(t)$ are the upper limits of Δa_i and Δg, respectively. Finally, we obtain,

$$\begin{cases} \dot{x}_2 = f(t) + \Delta f(t) + g(t)u - d(t) \\ f(t) = -a_{01}\cdot x_1 - a_{02}\cdot x_2 = -\sum_{i=1}^{2} a_{0i}x_i \\ \Delta f(t) = -\Delta a_1(t)\cdot x_1 - \Delta a_2(t)\cdot x_2 = -\sum_{i=1}^{2}\Delta a_i(t)\cdot x_i \end{cases} \tag{7}$$

3 Equivalent Sliding Mode Control

The selected control strategy is defined as a control law and a switching function for systems with uncertainty parameters and external disturbances [15].

Firstly, the switching function is designed as:

$$s(t) = c.e_1 + e_2 \quad (8)$$

where c is a positive constant, e_1 is the tracking error, e_2 is the derivative of the tracking error.

The control law *u(t)* is designed as:

$$u(t) = u_{equ}(t) + u_n(t) \quad (9)$$

with

$$u_n(t) = \frac{1}{g(t)} \eta sgn(s(t)), \eta \geq \sum_{i=1}^{2} \alpha_i |x_i| + D \quad (10)$$

where $u_{equ}(t)$ is the equivalent control law and $u_n(t)$ is the switching control law.

Chattering is produced due to the existence of the discontinuous function *sgn(x)* (10) in the control law.

3.1 Equivalent Sliding Mode Controller with Boundary Layer

Equivalent sliding mode controller with boundary layer consists of modifying the switching control law $u_n(t)$ of the formal equivalent sliding mode control [16]. The new switching control law $u_{nw}(t)$ is now,

$$u_{nw}(t) = \frac{1}{g(t)} \eta sat(\frac{s(t)}{\Phi}), \; sat(\frac{s(t)}{\Phi}) = \begin{cases} \frac{s(t)}{\Phi} & \left|\frac{s(t)}{\Phi}\right| \leq 1 \\ sgn(\frac{s(t)}{\Phi}) & \left|\frac{s(t)}{\Phi}\right| > 1 \end{cases} \quad (11)$$

where Φ is the coefficient that determines the thickness of the boundary layer.

4 Equivalent Fuzzy Sliding Mode Control

Fuzzy sliding model control means implementing the switching function (8) with a fuzzy inference system (FIS).

Thus, the fuzzy inference system has the switching function *s(t)* as input, and the control law *u(t)* as output. For this Mamdani-type fuzzy system, three trapezoidal and triangular fuzzy sets are assigned to these variables, with labels Negative (N), Zero (Z) and Positive (P) (Fig. 2). In this figure, the value of *k* in the *s*-membership function will be furtherly determined.

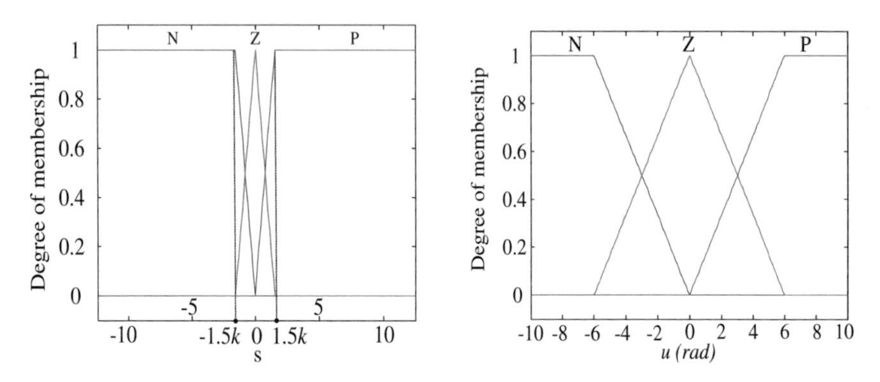

Fig. 2. Input (left) and output (right) variables fuzzy sets

Now, according to the fuzzy sliding mode control principle, the control law consists of the sum of the equivalent control and the switching action (9) [13, 15]; i.e., the control fuzzy rules are:

$$\text{If } s(t) \text{ is Z then } u \text{ is } u_{eq}$$
$$\text{If } s(t) \text{ is NZ then } u \text{ is } u_{eq} + u_n$$

where Z means around zero and NZ represents any other value (P or N).

That is, the fuzzy sliding mode controller is the equivalent control only when the switching function $s(t)$ is zero, and it is the sum of both actions when this switching control law is different to zero.

Then output is obtained with the centre of gravity defuzzification method, i.e.,

$$u = \mu_Z(s)u_{eq} + (1 - \mu_Z(s))(u_{eq} + u_n) = u_{eq} + (1 - \mu_Z(s))u_n \tag{12}$$

When $\mu_Z(s) = 0$, $u = u_{eq} + u_n$, and the control law is the equivalent sliding mode control. When $\mu_Z(s) \neq 0$, the chattering can be eliminated by adjusting the membership function $\mu_Z(s)$.

According to Lyapunov's Stability theory and sliding mode control principle, the basic reaching condition of sliding-mode control is met when $\dot{s}(t) \leq 0$ [17]. Then, the fuzzy rules are designed based on reaching this condition as follows:

$$\text{If } (s \text{ is N}), \text{ then } (u \text{ is N})$$
$$\text{If } (s \text{ is Z}), \text{ then } (u \text{ is Z})$$
$$\text{If } (s \text{ is P}), \text{ then } (u \text{ is P})$$

4.1 *s*-Membership Optimization Design

To explore the influence of the *s*-membership design on the control performance, an experiment is run. The comprehensive evaluation index f is designed so it includes the

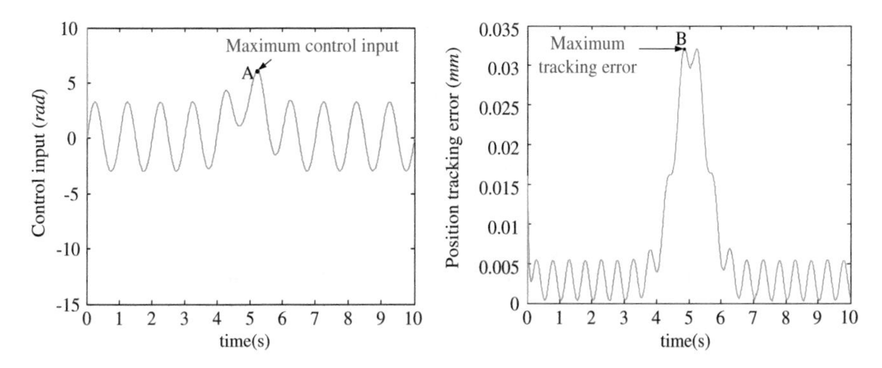

Fig. 3. Maximum control input (left) and maximum tracking error (right)

maximum control input, u_{max} (Fig. 3, left), and maximum tracking error, e_{max} (Fig. 3, right).

Then the goal is to minimize this index f (13).

$$\begin{cases} f = \eta_1 \widehat{e}_{max} + \eta_2 \widehat{u}_{max} \\ \eta_1 + \eta_2 = 1 \end{cases} \tag{13}$$

where η_1 and η_2 are the weight coefficients, $\hat{e}_{max}$ is the normalized $e_{max,}$ $\hat{u}_{max}$ is the normalized $u_{max.}$ In this case both, η_1 and $\eta_{2,}$ are set to 0.5, meaning that both indexes, control and error, are equally balanced but this can be changed.

5 Simulation Results Discussion

We are going to apply the three control strategies to the servo drive system described in Sect. 2. The system parameters are shown in Table 1.

Table 1. Servo drive system parameters

Parameters	Unit	Value
J_0	$kg.mm^2$	1.0×10^2
B_0	$N.mm.s/rad$	10.0
K_0	$N.mm/rad$	2.0×10^4
L	mm	5
i_1	1	0.4
i_2	1	0.8

In particular, the disturbances of relevant parameters are designed as below.

$$\begin{cases} \Delta a_i = 0.2a_{i0}\sin(2\pi t) \\ \Delta g = 0.2g_0\sin(2\pi t) \\ \alpha_i = 0.2a_{i0} \\ \beta = 0.2g_0 \end{cases} \quad (14)$$

The variable $d(t)$ varies with time following a Gaussian function, with $c_i = 5$, $b_i = 0.5$.

$$d(t) = 200\exp(-\frac{(t - c_i)^2}{2b_i^2}) \quad (15)$$

The reference signal is set to $x_d = \sin(2\pi t)$, and initial values of $c = 25$, $\Phi = 10$. Then simulations are carried out.

First, it is necessary to calculate f and to find the k value that minimizes it. The k value will vary from 0.5 to 5.5. The simulation time is 10 s. Results are shown in Table 2. As it is possible to see, f reaches the minimum value when k *is* equal to 4. Therefore, k is set to 4 for the control simulation experiments.

Table 2. Data processing for f calculation

k	e_{max}	$\hat{e}_{max}$	u_{max}	$\hat{u}_{max}$	f
0.5	0.00404	0	6.201	1	0.5
1.0	0.00807	0.10005	6.190	0.92810	0.51408
1.5	0.01210	0.20010	6.175	0.83007	0.51508
2.0	0.01611	0.29965	6.157	0.71242	0.50603
2.5	0.02012	0.39921	6.141	0.60784	0.50352
3.0	0.02412	0.49851	6.126	0.50980	0.50416
3.5	0.02812	0.59782	6.110	0.40523	0.50152
4.0	**0.03216**	**0.69811**	**6.094**	**0.30065**	**0.49938**
4.5	0.03621	0.79866	6.079	0.20261	0.50064
5.0	0.04026	0.89921	6.064	0.10458	0.50189
5.5	0.04432	1	6.048	0	0.5

Once k has been optimized, it is used to design the membership functions of the fuzzy switching function.

The control signal and the position tracking error of ordinary equivalent sliding mode controller, sliding control with boundary layer, and fuzzy sliding mode control are obtained and shown in Figs. 4 and 5, respectively.

As it can be observed in those figures, the ordinary equivalent sliding mode controller (green line) produces chattering, as expected. Nevertheless, the equivalent sliding mode controller with boundary layer (blue line) notably reduces the chattering,

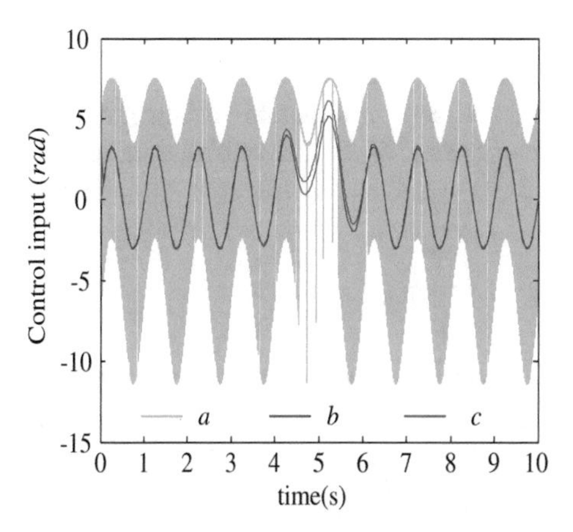

Fig. 4. Control signal of ordinary equivalent sliding mode controller (a, green line), with boundary layer (b, blue line), and fuzzy logic sliding mode control (c, red line)

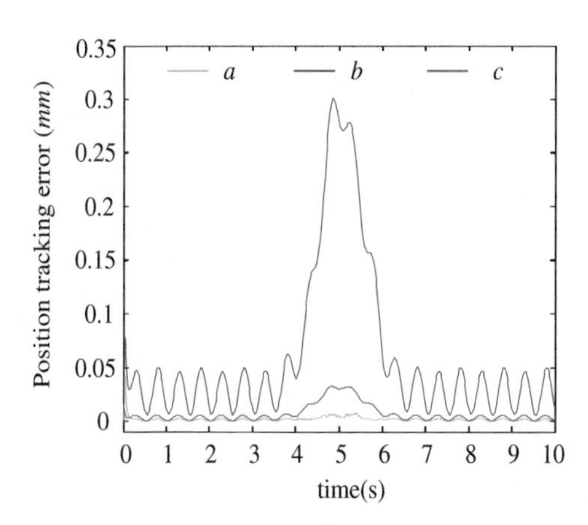

Fig. 5. Position tracking error of ordinary equivalent sliding mode controller (a, green line), with boundary layer (b, blue line), and fuzzy logic sliding mode control (c, red line)

yet it has large tracking error. Only the equivalent sliding mode controller based on fuzzy logic is able to eliminate the chattering (red line) and simultaneously effectively reduce the position tracking error.

6 Conclusions and Future Work

In this paper, a fuzzy approach of sliding mode control is applied to a servo drive system. The fuzzy memberships of the switching function of the controller are optimized in order to improve the robustness and accuracy of the control.

Simulations results are compared to those obtained by an ordinary equivalent sliding mode control and this controller with boundary layer. The optimized fuzzy sliding mode control can better satisfy the control requirements as it is able to soften the control chattering and reduce the tracking error.

Further study will be focused on dynamic sliding mode control using fuzzy logic and other intelligent techniques [18] for the implementation of the dynamic switching function of this system.

References

1. Zhang, Z., Yan, P., Hao, G.: A large range flexure-based servo system supporting precision additive manufacturing. Engineering **3**(5), 708–715 (2017)
2. Lee, H., Utkin, V.I.: Chattering suppression methods in sliding mode control systems. Annu. Rev. Control **31**(2), 179–188 (2007)
3. Sumantri, B., Uchiyama, N., Sano, S.: Least square based sliding mode control for a quad-rotor helicopter and energy saving by chattering reduction. Mech. Syst. Signal Process. **66**, 769–784 (2016)
4. Liu, X., Shan, Z., Li, Y.: Dynamic boundary layer based neural network quasi-sliding mode control for soft touching down on asteroid. Adv. Space Res. **59**(8), 2173–2185 (2017)
5. Gao, F., Pan, C., Han, Y.: Design and analysis of a new AUV's sliding control system based on dynamic boundary layer. Chin. J. Mech. Eng. **26**(1), 35–45 (2013)
6. Chen, M., Yu, J.: Disturbance observer-based adaptive sliding mode control for near-space vehicles. Nonlinear Dyn. **82**(4), 1671–1682 (2015)
7. Farhane, N., Boumhidi, I., Boumhidi, J.: Smart algorithms to control a variable speed wind turbine. Int. J. Interact. Multimedia Artif. Intell. **4**(6), 88–95 (2017)
8. Xu, B., Shi, G., Ji, W., Liu, F., Ding, S., Zhu, H.: Design of an adaptive nonsingular terminal sliding model control method for a bearingless permanent magnet synchronous motor. Trans. Inst. Meas. Control **39**(12), 1821–1828 (2017)
9. Joo, Y.H., Tien, L.Q., Duong, P.X.: Adaptive neural network second-order sliding mode control of dual arm robots. Int. J. Control Autom. Syst. **15**(6), 2883–2891 (2017)
10. Yu, J., Liu, J., Wu, Z., Fang, H.: Depth control of a bioinspired robotic dolphin based on sliding-mode fuzzy control method. IEEE Trans. Industr. Electron. **65**(3), 2429–2438 (2018)
11. Nagarale, R.M., Patre, B.M.: Composite fuzzy sliding mode control of nonlinear singularly perturbed systems. ISA Trans. **53**(3), 679–689 (2014)
12. Yoshimura, T.: Design of adaptive fuzzy backstepping sliding mode control for MIMO uncertain discrete-time nonlinear systems based on noisy measurements. J. Vib. Control **24** (2), 393–406 (2018)
13. Lin, Y., Liping, F.: The equivalent sliding mode control based on fuzzy control for an electrical servo system. In: Proceedings of the 27th IEEE Chinese Control Conference, CCC 2008, Kunming, Yunnan, pp. 118–121 (2008)
14. Wang, J.W., Wu, Z.S.: Mechanical Engineering Control Foundation, 2nd edn. Higher Education Press, Beijing (2010)
15. Liu, J.K.: Matlab simulation for sliding model control. Tsinghua University Press, Beijing (2005)
16. Slotine, J.J., Sastry, S.S.: Tracking control of non-linear systems using sliding surfaces, with application to robot manipulators. Int. J. Control **38**(2), 465–492 (1983)
17. Zhao, W.J., Liu, J.Z.: An improved method of sliding mode control with boundary layer. Acta Simulata Systematica Sinica **17**(1), 156–158 (2005)
18. Santos, M.: An applied approach of intelligent control. Revista Iberoamericana de Automática e Informática Ind. RIAI **8**(4), 283–296 (2011)

Genetic Simulation Tool for the Robustness Optimization of Controllers

Matilde Santos[(✉)] and Nicolás Antequera

Computer Architecture and Automatic Control Department,
University Complutense of Madrid, C/ Profesor García Santesmases 9,
28040 Madrid, Spain
msantos@dacya.ucm.es, nicolas.antequera@gmail.com

Abstract. When designing controllers for complex systems, it is not only necessary to stabilize the system but to improve the robustness in order to get a better response. Some indexes allow to measure this robustness of the system response, such as the gain and phase margins. In this paper a computational tool that implements a Multi-Objective Genetic Algorithm (MOGA) is designed and applied to optimize the robustness of different controllers. So, it is possible to analyse how the variation of the controller parameters influences the robustness of the system. The tool is applied to the optimization of a Linear Quadratic (LQ) and Eigenvalues assignment (EA) controllers for a MIMO autonomous vehicle, a helicopter, with satisfactory results.

Keywords: Intelligent control · Robustness · Gain margin · Phase margin
LQR · EA control · UAV

1 Introduction

Controlling a real system is never a simple task, especially for complex systems. Even more, when designing a controller, one of the most demanding issues is adjusting the configuration parameters. These tuning parameters are usually interrelated and have a strong influence on the system response. Besides, after the design of a controller and the initial tuning, it is usually necessary to change those values in order to improve the response and to get more demanding control specifications accurately [1].

Soft computing techniques have been applied to this control issue with encouraging results, not only to the determination of the gains, but to the qualitative tuning of these parameters, including expert knowledge [2].

One of the most important goals when designing the controller is the robustness. This characteristic broadens the range of application as it implies that a controller designed for a particular set of parameters will also work well under a different scenario. This is especially useful with systems that work under changing environments.

Some indexes allow measuring the robustness in the frequency domain, such as the gain and phase margins of the system response [3]. But it is not easy to analyse how those measurements of the response vary with the controller parameters and therefore how to get the best ones.

M. Graña et al. (Eds.): SOCO'18-CISIS'18-ICEUTE'18, AISC 771, pp. 313–323, 2019.
https://doi.org/10.1007/978-3-319-94120-2_30

In this paper, a computation tool based on Genetic Algorithms (GA) is implemented and applied not only to find good values of the controller parameters [4], but to analyse how to improve the robustness measured by the phase and gain margins. These evolutive strategies have been proved quite useful for complex applications in different engineering fields [5]. Moreover, in our work the evolutive tool has been applied to simultaneously optimize the two functions that determine the robustness of a controller in the frequency domain, that is, in a Multi-Objective Genetic Algorithm (MOGA) approach.

To test the developed tool, it has been applied to control a complex system, a MIMO unmanned aerial vehicle (UAV). Controlling a helicopter is difficult due to its strongly non-linear and highly unstable dynamics. Besides, it has many parameters to be fixed that are interrelated and the number of system states is greater than the number of inputs. For that, many of the controllers found in the literature are designed to only govern some specific outputs, such as position or velocity tracking [6, 7].

The proposed robustness optimization tool has been applied to two different non-linear controllers: a Linear Quadratic Regulator (LQR) and a Eigenstructure Assignment (EA) controller. Simulation results prove the efficiency of these control strategies and the responses of the UAV with both controllers are compared in terms of robustness. These experiments have allowed us to validate the proposal.

The article is structured as follows. In Sect. 2 the robustness measurements are defined. The computational tool that implements the Multi-Objective Genetic Algorithm (MOGA) to the robustness optimization of the controllers is presented in Sect. 3. Section 4 describes the model of the helicopter to be controlled. Sections 5 and 6 show the application of the optimization tool to a LQR and EA controllers, respectively. Conclusions end the paper.

2 Robustness: Gain Margin and Phase Margin

Robustness means to minimize the sensitivity of a system over its frequency spectrum when the system parameters change. Transfer functions that are commonly used in control design for robust performance include the complementary sensitivity function (T) for reference tracking and sensor noise rejection, and the sensitivity function (S), which is used for disturbance rejection analysis [12].

The difficulty in achieving good robust performance arises from the relationship between the two sensitivity functions, $S + T = 1$. Hence, the control system design problem is that of achieving better disturbance rejection or reference tracking despite the coupling between system transfer functions, S and T.

To obtain the robustness of MIMO systems, the following functions can be used: the sensitivity function $S = (I + L)^{-1}$, and the complementary sensitivity function $T = L(I + L)^{-1}$ or the balanced sensitivity function $(S + T)$, where L is the open loop gain matrix and I the identity matrix.

This sensitivity function can be calculated at the inputs and outputs of the actuator. At the inputs, $L = HG$ and at the outputs, $L = GH$, where H is the controller transfer function matrix and G is the plant transfer function matrix.

Phase margin (PM) and gain margin (GM) are measures of stability in closed loop. Phase margin indicates relative stability, the tendency to oscillate during its damped response to an input change such as a step function. Gain margin indicates absolute stability and the degree to which the system will oscillate without limit given any disturbance. Both of them must be positive, but a low phase or gain margins will lead to undesirable oscillations in the step response.

The maximum of the singular values (σ) of L will be used to obtain the gain and phase margins (1) and (2). The parameter a is calculated by the expression $a = 1/\sigma$, where $a \leq 1$. Then the Gain Margin (GM) and the Phase Margin (PM) are,

$$\mathrm{GM} = [1/(1+a), 1/(1-a)] \tag{1}$$

$$\mathrm{PM} = \pm 2 sin^{-1}(a/2) \tag{2}$$

3 Multi-objective GA Optimization Tool for Controller Robustness

The problem we want to solve is the optimization of the robustness of the controller, that is, of the gain and phase margins, by using a heuristic strategy, Genetic Algorithms. The GA multi-objective optimization tool is implemented using the epsilon-constraint method [8]. This firstly allows us to transform the multi-objective problem into a sequence of parameterized single-objective problems, where the optimum of each single-objective problem corresponds to a Pareto-optimal solution. Then, GAs are applied to each of these functions [9].

Given n functions to be optimized, this strategy first optimizes one of them until the result is satisfactory enough. Then it proceeds with the following function, taking into account that the previously optimized function should not be reduced in its optimum value in more than a percentage named epsilon (ε). Mathematically,

$$\text{maximize} \quad \Phi(f(x)) \equiv \Phi(f_1(x, \ldots, f_m(x))$$

$$\text{subject to} \quad f_i(x) \triangleright \varepsilon_i, \forall i \in \{1, \ldots, m\}, x \in X,$$

The fitness function is the robustness of the controller, i.e., the gain margin and the phase margin, or according to Sect. 2, the sensitivity functions.

The optimization tool developed has a graphic interface where the following values can be specified by the user (Fig. 1),

- Stability criteria to be optimized (GM, PM, both).
- State space matrixes of the system model (A, B, C, D).
- Limit values of GM or PM. This means that if only one of these variables is to be optimized, the desired value that we want to achieve will stop the algorithm.
- Differential (%), i.e., the epsilon constraint ε. Once the value of one margin has been obtained, it cannot be changed in more that this value when optimizing the other one.

- Configuration parameters of the genetic algorithm (population size, mutation probability, etc.)

Every iteration the algorithm shows the best individual. Once the procedure ends, the evolution of the individuals is displayed. The values of GM, PM, and the tuning parameters of the controllers are recorded in a file each iteration. So, at any time the user knows the values of the system that give certain stability values.

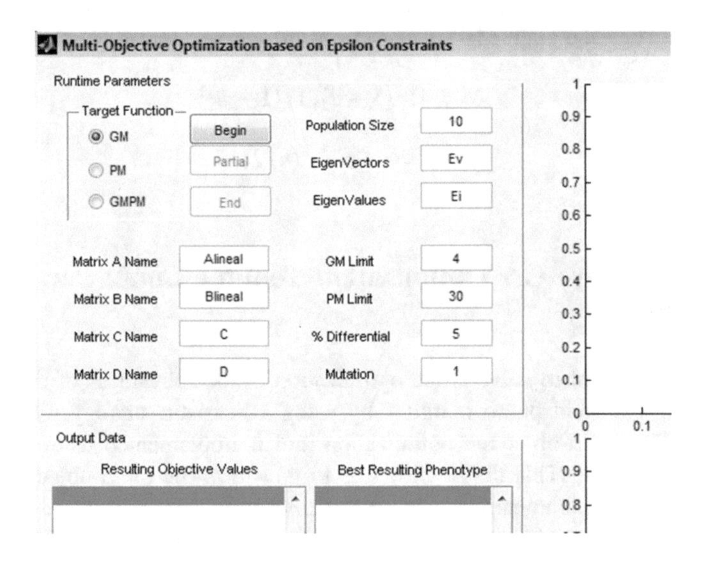

Fig. 1. Interface of the MOGA optimization tool

The size of the population is 10 individuals. Mutation probability is set by the user (%) and it is multiplied by the number of genes of the individual. In this case, Pm = 1 (1%). The tournament strategy is applied for the selections of the parents.

The limits for the GM and PM are fixed to 4 dB and 30°, respectively. The GA will not stop until it reaches those values. The tolerance is 5% (differential), meaning that once we have obtained a good value of GM, it cannot be reduced in more than 5% when optimizing the PM, and vice versa.

This is a quite general optimization tool for any system, as it takes as inputs the state space matrixes of the system, being only necessary to adapt the control logic under study, that is, control LQR, EA, etc.

4 A Case Studied: UAV Model

The autonomous vehicle to be controlled is a helicopter, as a benchmark that has been used in previous works [10]. It is a strongly non-linear and highly unstable system. It is an example of a MIMO (Multi-Input Multi-Output) system where the different inputs and the states are coupled.

The helicopter dynamics can be represented by the approximate non-linear model derived in [6]. The linearized model in hover and slow flight conditions can be expressed in the state space as,

$$\begin{aligned} \mathbf{x}' &= \mathrm{A}\mathbf{x} + \mathrm{B}\mathbf{u} \\ \mathbf{y} &= \mathrm{C}\mathbf{x} + \mathrm{D}\mathbf{u} \end{aligned} \tag{3}$$

where the matrix A represents the internal dynamic of the system. The B matrix expresses the relationship between the inputs and states of the helicopter. The C matrix describes the observable states. In this case, the D matrix is zero.

The state vector, **x**, of the model of the helicopter is,

$$\mathbf{x} = \begin{bmatrix} \mathrm{V}^{\mathrm{T}} & \Theta^{\mathrm{T}} & \omega^{\mathrm{T}} \end{bmatrix}^{\mathrm{T}} \tag{4}$$

with $\mathrm{V} = [\mathrm{Vx} \quad \mathrm{Vy} \quad \mathrm{Vz}]^{\mathrm{T}}$ the linear velocity vector; $\Theta = [\varphi \quad \theta \quad \psi]^{\mathrm{T}}$ (roll, pitch and yaw, respectively) the Euler angles vector (rotations), and the angular velocity $\omega = [\mathrm{p} \quad \mathrm{q} \quad \mathrm{r}]^{\mathrm{T}}$.

The inputs contain both the linear and the rotational forces; that means that the linear and the rotational dynamics of the helicopter are highly coupled [11]. The input vector **u** is defined as,

$$\mathbf{u} = [\mathrm{T_M}, \quad \mathrm{T_T}, \quad \mathrm{a_{1S}}, \quad \mathrm{b_{1S}}]^{\mathrm{T}} \tag{5}$$

where $\mathrm{T_M}$ and $\mathrm{T_T}$ (in N) are the normalized main and tail rotor thrust, respectively, and $\mathrm{a_{1s}}$ and $\mathrm{b_{1s}}$ (°) are the longitudinal and lateral tilt of the main rotor. This non-linear system is driven by the inputs.

The output vector **y** consists of the three velocities and the heading, ψ. That is,

$$\mathbf{y} = [\mathrm{Vx} \quad \mathrm{Vy} \quad \mathrm{Vz} \quad \psi]^{\mathrm{T}} \tag{6}$$

The control action is based on these outputs.

The position vector of the UAV, **P**, consists of the three components $\mathbf{P} = [\mathrm{Px} \quad \mathrm{Py} \quad \mathrm{Pz}]^{\mathrm{T}}$ where $[\mathrm{x} \quad \mathrm{y} \quad \mathrm{z}]$ are the scaled Cartesian coordinates of the helicopter's centre of mass. They are obtained by integrating the velocity vector (4).

5 Example 1: Optimization of a LQR

The LQR has been designed applying the expression,

$$J[u] = \int_0^{\infty} \left(x^T Q x + u^T R u\right) dt \tag{7}$$

where $Q \in \mathbb{R}^{n \times n}$, the quitar matrix, and $R \in \mathbb{R}^{m \times m}$ are symmetric weighting matrices. The control law is $\mathbf{u} = A + BK$, with K the controller gain vector that minimizes the quadratic cost function $J[\mathbf{u}]$. In the diagonal of matrix Q, the values of the velocities are weighted by 10 and the rest are equal to one. The weighting matrix R has been obtained taking into account the range of the control signal, i.e., T_M, T_T, a_{1S} and b_{1S}. The diagonal values of this R matrix are: $[0.011069, \quad 0.095057, \quad 1.146, \quad 1.432]$. The eigenvalues of the system obtained with the LQR are presented in Table 1.

Table 1. LQR eigenvalues (P_i, $i = 1 \ldots 9$)

−233.17	−137	−10.53	−2.21 + 2.25i	−2.21 − 2.25i	−2.25 + 2.22i	−2.25 − 2.22i	−1	−1.92

The system response when using the LQ controller is shown in Fig. 2. The states variables that are represented are Vx (solid), Vy (dashed), Vz (dots), and the yaw angle (dash-dot). It is difficult to see the yaw behaviour because of its order of magnitude. On the left side, Fig. 2 shows how the LQR stabilizes the system when it is disturbed from the initial conditions. On the right side it is possible to see how the system follows the references when they are abruptly changed. The response is satisfactory (the weighting matrix R has been rightly tuned) but it presents overshoot.

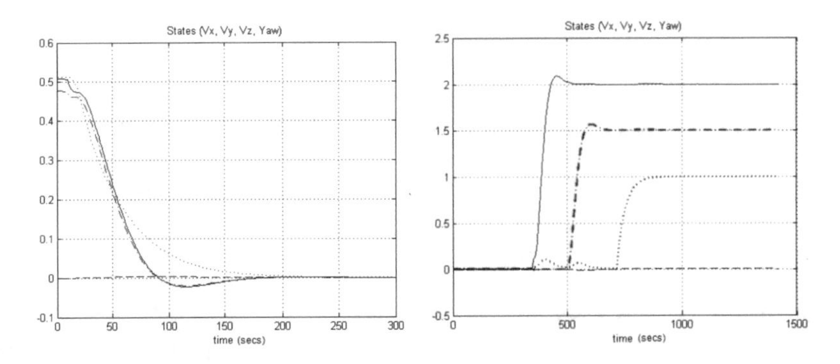

Fig. 2. System response (velocities and yaw) with the LQR, disturbance response (left) and reference step changes (right)

The robustness is computed for the system controlled with the LQR, and the gain and phase margins for inputs and outputs are calculated (Table 2). As it is possible to see, the GM is quite low, and the PM is lower than the required 30°. Although the response is not bad, it can be improved if the controller is better tuned.

Therefore, the genetic optimization tool has been applied to this problem. The initial individuals of the population of the GA are the values of the Q and R matrixes. In fact, the individuals are vectors with the diagonal values of Q (8 values) and R (4 values). Initially the diagonal values of the desired matrixes Q_d and R_d are set to 1. A hard constraint has been added as matrix R must be positive.

Table 2. System robustness with the LQ controller

Actuators	σ	$a = 1/\sigma$
Input	3.8922	0.2569
Output	3.9756	0.2515
	GM (dB)	PM (deg)
Input	$[-1.9862,\quad 2.5793]$	±14.7613
Output	$[-1.9489,\quad 2.5166]$	±14.4502

The evolution of the GM and PM is shown in Fig. 3. The initial GM for this LQR was 1.2414 dB but, as Fig. 3 (left) shows, after the first iteration, the GM is improved to 2.6 dB, and finally it reaches the value of 5.95 dB. The initial value of the PM was 8.1769°. After the first iteration it reaches a PM value of 20.5° and finally it reaches up to 38° (Fig. 3, right).

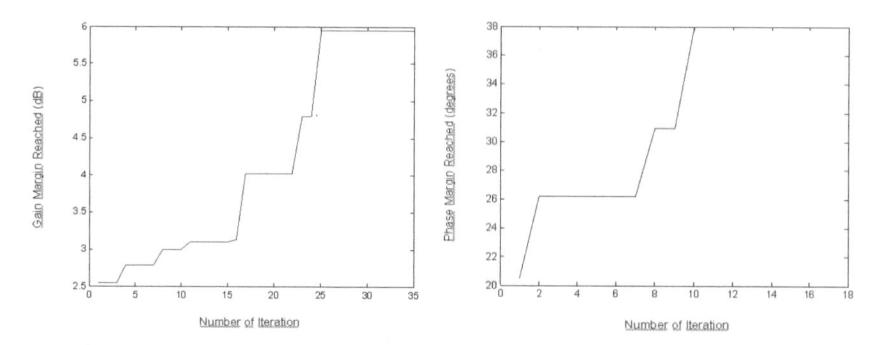

Fig. 3. Gain (left) and Phase (right) margins obtained with the LQR optimized by the MOGA.

The Q and R matrixes obtained by the GA have the following diagonal values: Qdiag $= [504 \quad 474 \quad 62 \quad 256 \quad 179 \quad 323 \quad 238 \quad 275]$, Rdiag $= [216.31 \quad 427.89 \quad 423.57 \quad 451.59]$.

The system response has improved as it is quicker and it does not have overshoot. In Fig. 4 it is possible to see the velocities of the system, Vx (blue), Vy (cyan), Vz (red) and the yaw angle (green) for some step changes in the references.

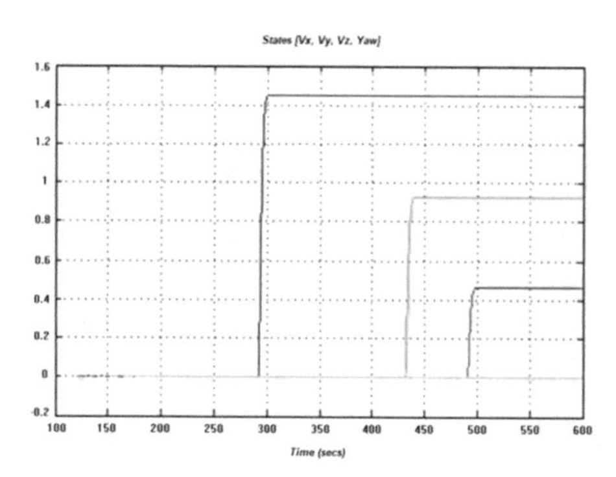

Fig. 4. System response with the LQ optimized controller.

6 Example 2: Optimization of a EA Controller

The eigenstructure assignment (EA) technique allows us the design of an initially robust controller and to solve the main problem of some complex systems such as the helicopter, the strong coupling between its variables. At the same time, it provides the engineer with the degree of freedom that he needs to place the poles in the closed-loop and to decouple some of the modes of the system.

Given a set of desired eigenvalues, Λ_d, and the corresponding desired eigenvectors, V_d, the control problem consists of finding the feedback matrix $\mathbf{K}$ ($m \times p$) of the control equation, $\mathbf{u} = -\mathbf{K}y$, being p the number of outputs and m the inputs, such that the eigenvalues of the closed loop system matrix, $(\mathbf{A} - \mathbf{BKC})$, include Λ_d as a subset.

Initially, the eigenvalues of the helicopter are zero, meaning that the system is completely unstable. Applying the EA controller designed in [10] for this UAV, the poles are now shifted to the left semi plane and the system is stable (Table 3).

Table 3. EA eigenvalues (P_i, $i = 1 \ldots 9$)

−14.9919	−10.0001	−8.0006	−1.7409 + 0.0090i	−1.7409 − 0.0090i	−1.5045	−1.0044	−1.0095	−1.2577

Even though the resulting gain matrix gives a good modal decoupling, the system response is unsatisfactory because of the input matrix B. So, a pre-compensation gain was added using the equation $N = -\mathrm{inv}[C(A - BK)^{-1}B$.

With this matrix the system response is now better and the steady-state error is zero.

The system may be now disturbed by any disturbance value Δ that satisfies $1/(1+a) < \Delta < 1/(1-a)$, or any initial conditions different from the operation point (Fig. 5, left). As it is possible to see, the EA controller stabilizes the system very quickly, especially the Vz state. The system also follows the references (Fig. 5, right). The response with the EA controller is smoother than with the LQR as it does not present overshoot yet it is still a quick response.

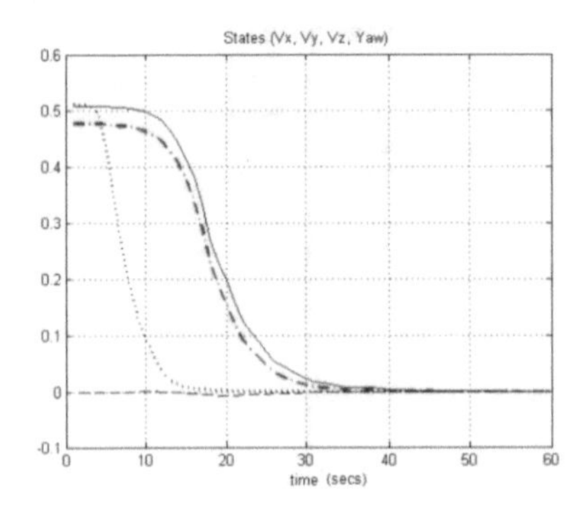

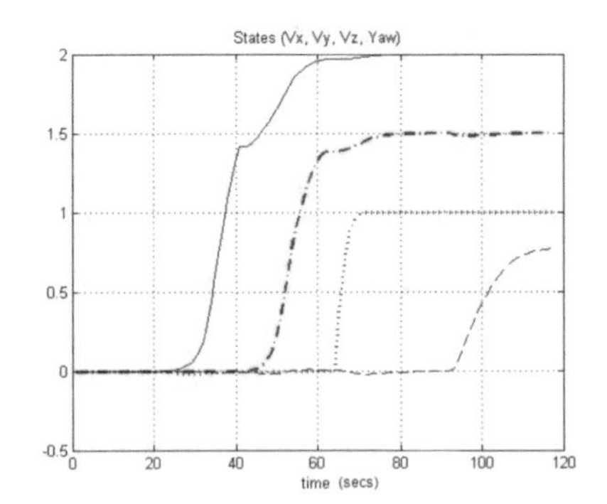

Fig. 5. System response (velocities and yaw) with the EA controller, disturbance response (left) and step (right)

The values of the gain and phase margins for the system controlled with the EA regulator are shown in Table 4.

Table 4. System robustness with the EA controller

Actuators	σ	$a = 1/\sigma$
Input	1.4473	0.6910
Output	4.7616	0.2100
	Gain margin (dB)	Phase margin (deg)
Input	$[-4.5626, \quad 10.1995]$	± 40.4216
Output	$[-1.6558, \quad 2.0476]$	± 12.0551

Now, the GA optimization tool is applied to the EA controller to improve the robustness. The desired eigenvalues and eigenvectors are introduced.

The population size is 10 (although a different size can be selected in the interface), each individual with 8 chromosomes (eigenvalues). Although it has been randomly initialized, one of the individual is replaced by the following one:

	Forward speed	Side speed	Heave speed	pitch/pitch	roll/roll	yaw
Pi, i = 1 … 8	–0.00199	–0.00526	–2.1	$-1.45 \pm 1.45i$	$-1.5 \pm 1.5i$	–2

After 103 iterations, the system reaches a GM = 4.0452 dB and PM = 26.26° values. After that, the algorithm selects the best individuals of the different runs. Based on those values, the GM is again optimized. The best value obtained has been 6.28 dB. The evolution of this value is shown in Fig. 6 (left). The same strategy has been applied to optimize the PM. The best value, conditional on the fixed GM value previously obtained, is 32° approximately (Fig. 6, right).

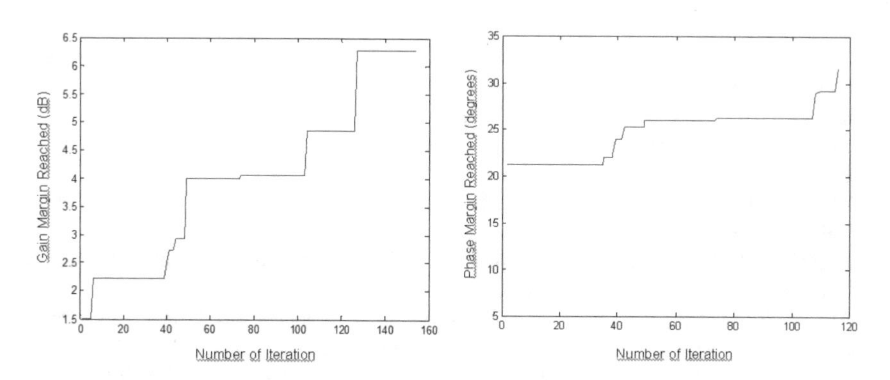

Fig. 6. Optimization of GM up to 6.28 dB (left) and of PM with GM fixed to 6.28 dB (right).

Figure 7 shows the system response using the EA controller optimized by the MOGA tool for some step changes in the reference.

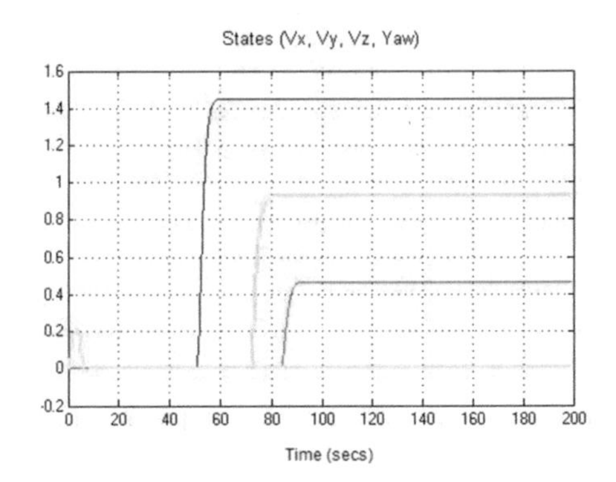

Fig. 7. System response using the EA controller optimized by GA (velocities and *yaw*)

7 Conclusions and Future Work

In this paper, an evolutive multi-objective optimization tool has been developed. Based on genetic algorithms, it optimizes the variables that determine the robustness of a system response, the gain and phase margins, or both simultaneously. It can be applied to any controller and to any system.

It has been proved on two different non-linear controllers, LQR and EA, for a MIMO system, an autonomous helicopter, with satisfactory results. It optimizes the parameters of the controller and gets a better system response in terms of stability.

References

1. Ribeiro, J.M.S., Santos, M.F., Carmo, M.J., Silva, M.F.: Comparison of PID controller tuning methods: analytical/classical techniques versus optimization algorithms. In: 18th International Carpathian Control Conference (ICCC), pp. 533–538. IEEE (2017)
2. Santos, M.: An applied approach of intelligent control. Revista Iberoamericana de Automática e Informática Industrial RIAI **8**(4), 283–296 (2011)
3. Zhou, K., Doyle, J.C., Glover, K.: Robust and Optimal Control, vol. 40, p. 146. Prentice Hall, New Jersey (1996)
4. Jamshidi, M., Krohling, R.A., Coelho, L.D.S., Fleming, P.J.: Robust control systems with Genetic Algorithms, vol. 3. CRC Press, Boca Raton (2002)
5. Alonso, F., Santos, M.: Heuristic optimization of interplanetary trajectories in aerospace missions. Revista Iberoamericana Automática e Informática Industrial **14**(1), 1–15 (2017)
6. Shim, H., Koo, T.J., Hoffmann, F., Sastry, S.: A comprehensive study of control design for an autonomous helicopter. In: Proceedings of the 37th IEEE Conference on Decision and Control, vol. 4, pp. 3653–3658. IEEE, December 1998

7. Nair, V.V., Jayasree, P.R., Parvathy, G.: Robust control of helicopter with suspended load. In: International Conference on Circuit, Power and Computing Technologies (ICCPCT), pp. 1–6. IEEE (2017)
8. Marler, R.T., Arora, J.S.: Survey of multi-objective optimization methods for engineering. Struct. Multidiscip. Optim. **26**(6), 369–395 (2004)
9. Santos, M., Cruz, J.M.: Algoritmos genéticos. Métodos de Procesamiento Avanzado e Inteligencia Artificial en Sistemas Sensores y Biosensores, Reverté **15**, 321–333 (2009)
10. Antequera, N., Santos, M., de la Cruz, J.M.: A helicopter control based on eigenstructure assignment. In: IEEE Conference on Emerging Technologies and Factory Automation, ETFA 2006, pp. 719–724. IEEE, September 2006
11. Yañez-Badillo, H., Tapia-Olvera, R., Aguilar-Mejía, O., Beltran-Carbajal, F.: On line adaptive neurocontroller for regulating angular position and trajectory of quadrotor system. RIAI **14**(2), 141–151 (2017)
12. Zimenko, K., Polyakov, A., Efimov, D., Kremlev, A.: Feedback sensitivity functions analysis of finite-time stabilizing control system. Int. J. Robust Nonlinear Control **27**(15), 2475–2491 (2017)

A PSO Boosted Ensemble of Extreme Learning Machines for Time Series Forecasting

Alain Porto[1(✉)], Eloy Irigoyen[2], and Mikel Larrea[2]

[1] IK4-Ideko, Arriaga Industrialdea 2, 20870 Elgoibar, Spain
aporto@ideko.es
[2] UPV/EHU, Alda. Urquijo S/N, Bizkaia, Spain
{eloy.irigoyen,m.larrea}@ehu.eus
http://www.ehu.eus/es/web/gici

Abstract. In this work, a first approach of using the Particle Swarm Optimization (PSO) as a method for optimizing an Ensemble Model built with Extreme Learning Machines is presented. The paper focuses on the obtaining of the parameters of a weighted averaging method for a Ensemble Model, using Extreme Learning Machines as models. The main contribution of this document is the use of the heuristic algorithm PSO for searching optimum parameters of the weighted averaging method. The experiments show that PSO is suitable for computing the parameters of the ensemble, obtaining an average improvement of 68% of the error comparing with an individual model. Also other comparisons have been made with basic combining methods of Ensemble Model fulfilling the expectations.

Keywords: PSO · ELM · ANN · Optimization · Ensemble model

1 Introduction

Systems modelling and prediction is one of the most challenging problems to address in engineering. In many different fields, an accurate prediction of the system behaviour is a very important factor which ensures success [1]. Prediction data is often based on a Time Series (TS) data points, which are a collection of values sequentially recorded over time. Different methods have been proposed in order to face TS, such as Auto-Regressive Integrated Moving Average [2] or Neural Networks [3,4]. In many studies carried out to date, in order to create an accurate system model, different candidate learning machines are trained for the same TS problem, and it is a standard practice to select and keep the best model, while discarding the rest [5]. Moreover, in Time Series modelling and prediction area, accuracy is one of the most important factor that should be guaranteed, and in most cases, it is difficult to reach certain accuracy ratios with a single model [6]. Ensemble Models (EMs) emerge as a method to overcome the

M. Graña et al. (Eds.): SOCO'18-CISIS'18-ICEUTE'18, AISC 771, pp. 324–333, 2019.
https://doi.org/10.1007/978-3-319-94120-2_31

waste of already trained models, combining different learning machines. The EMs improve the overall performance and the accuracy of the predictions compared to the best single learning machine [7]. EMs had been applied successfully for improving accuracy and performance in different scopes [8]. An EM consists of a set of models taken from one single class, such as neural networks [9] or support vector machines. Usually, each one of the models in the EM is referred as 'member' or 'expert'.

Due to the improvement that the EMs introduce, many authors had tried this technique [10]. Moreover, a lot of investigations have been done proposing combining methods that improves both accuracy and efficiency of the outputs of the set of experts. Some of these methods are Simple Averaging, Majority Voting, Ranking, or Weighted Averaging. It has been demonstrated that weighted methods yields to better results [11], due to the capability of emphasizing an expert among the others. The mayor issue involving weighted methods is the computing of the weights, because of the size of the solution space. In addition, the scientific publications that address this problem from an EM point of view is scarce [12].

The aim of this paper is to contribute in finding a suitable technique for computing the weights of a weighted averaging method for EM. As first approach, a modification of the PSO is proposed, as it has been used successfully in ensembles [13] and its suitability has been demonstrated. Also, a study comparing the performance of the proposed model with simple averaging methods has been done. In the second section, a background information about the different used tools are presented, such as the Extreme Learning Machines and Particle Swarm Optimization function. In the third section the methodology for obtaining the final EM is shown. Later on, in section four the results of the performance of each individual model and the performance EM itself are introduced. Finally, in sections five and six, the discussion of the performance reached as well as the conclusions of using this methodology are presented.

2 Background Information

2.1 Extreme Learning Machine

The Extreme Learning Machines (ELMs) are a specific type of Feed Forward Artificial Neural Networks (FFNNs) introduced in 2006 by Huang et al. [14]. Specifically, an ELM is a Single Layer FFNN (SLFN) with a random initialization of the weights and biases of the input layer. In this scenario, the output of the SLFN can be easily computed, hence, the output layer's weights can be found to minimize the output error [15]. In Fig. 1 the structure of a SLFN can be seen. Here, $\mathbf{X}_{M\times 1} = [x_1, \ldots, x_M]^T$ represents the inputs to the SLFN in an input layer of M nodes. $\mathbf{W}_{N\times M}$ is the matrix built with ω_{ij}, the weights between the input and the hidden layer (with N nodes). Analogously, $\mathbf{B}_{O\times N}$ represents the matrix constructed with β_{ij}, the weights between the hidden layer and the output layer (with O nodes). The biases of the hidden layer are shown as $\mathbf{b}_{N\times 1} = [b_1, \ldots, b_N]$. The activation function of the hidden layer (common for all neurons) is represented

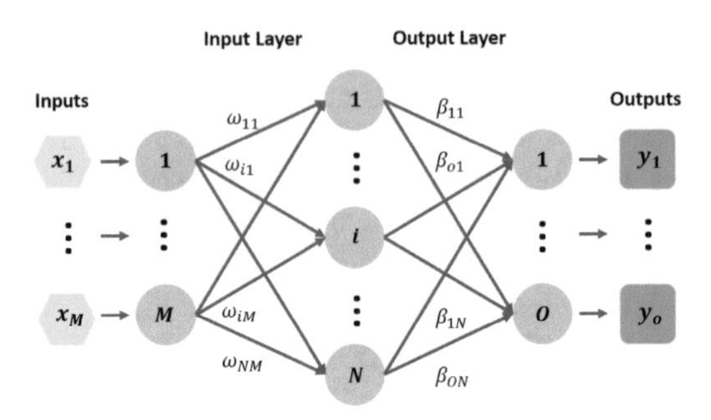

Fig. 1. ELM general scheme

with σ and the outputs of the SLFN are represented with $\hat{\mathbf{Y}}_{O\times 1}$. Using this compact notation, the output of the SLFN can be easily computed as

$$\hat{\mathbf{Y}}_{O\times 1} = \mathbf{B}_{O\times N} \cdot \left[\sigma\left(\mathbf{W}_{N\times M} \cdot \mathbf{X}_{M\times 1} + \mathbf{b}_{N\times 1}\right)\right], \tag{1}$$

If there is a set of training samples $(\mathbf{X}_i, \mathbf{Y}_i)|i \in (1, D)$, the output of the SLFN can also be computed, if we stack all inputs together as $X_{N\times D} = [X_1, \ldots, X_D]$ matrix and create the matrix $\mathbf{b}_{N\times D}$ repeating $\mathbf{b}_{N\times 1}$ D times,

$$\hat{\mathbf{Y}}_{O\times D} = \mathbf{B}_{O\times N} \cdot \left[\sigma\left(\mathbf{W}_{N\times M} \cdot \mathbf{X}_{M\times D} + \mathbf{b}_{N\times D}\right)\right]. \tag{2}$$

Usually, the resultant matrix of $\sigma\left(\mathbf{W}_{N\times M} \cdot \mathbf{X}_{M\times D} + \mathbf{b}_{N\times D}\right)$ is denoted as $\mathbf{H}_{N\times D}$. Now, the goal of training an artificial neural network is that its outputs fits at best the training set's outputs, i.e. $\hat{\mathbf{Y}}_{O\times D} = \mathbf{Y}_{O\times D}$, where $\mathbf{Y}_{O\times D} = [\mathbf{Y_1}, \ldots, \mathbf{Y_D}]$, so $\mathbf{B}_{O\times N}$ can be found using (1),

$$\hat{\mathbf{Y}}_{O\times D} = \mathbf{B}_{O\times N} \cdot \mathbf{H}_{N\times D} \rightarrow \mathbf{B}_{O\times N} = \hat{\mathbf{Y}}_{O\times D} \cdot \mathbf{H}^{\dagger}_{D\times N}. \tag{3}$$

where $\mathbf{H}^{\dagger}_{D\times N}$ denotes the Moore-Penrose generalized inverse matrix of $\mathbf{H}_{N\times D}$. This equation implies that by using (3) the weights of the output layer of the SFLN would be tuned at its best, given that is a least-squares approach.

2.2 Ensemble Model

As pointed before, an EM is a set of models that can be somehow combined in order to improve a final prediction. An EM can be built with different class of learning machines such as, Support Vector Machines, Decision Trees or Artificial Neural Networks [16]. EM has been applied to different problems with successful results as can be found in [17,18]. The aim of this work is to improve the combining methods of an EM, for this purpose a weighted averaging method has been used, and for computing the weights the PSO method has been chosen, as in other studies has shown better result than other heuristics algorithms [19]. The improvement of using EM is mathematically formalized in the work of Wichard et al. [20]. This work shows that the variance of the ensemble is lower than the average variance of all the individual models, and it demonstrates that the error of an ensemble is lower than an individual model.

$$E_{ens} = \frac{1}{n} E_{avg}. \quad (4)$$

Ensemble Model's Combining Methods. There are plenty of different combining methods to set an EM. Some of them are suitable for classification, like majority voting or weighted majority voting, and others for regression, such as weighted averaging or ranking [12]. In this section, some combining methods are explained in a brief description, as presented by Bishop in [21].

1. Weighted voting: It is based on majority voting, where the output is the most repeated value over the experts, but assigning weights (ω_j) to each one of the members of the ensemble ($\hat{K}_j(\mathbf{x})$). This way some of the members can be emphasized. The main difficult is to find the weights that fit best. The expression is,

$$\hat{K}_{ens}(\mathbf{x}) = sgn\left[\sum_{j=1}^{n} \omega_j \hat{K}_j(\mathbf{x})\right]; \quad \sum_{j=1}^{n} \omega_j = 1. \quad (5)$$

2. Weighted averaging: In order to solve the emphasizing problem of the simple averaging method, different weights (ω_j) can be assigned to each ensemble's member depending on the generalization ability of each one (see (6)). The problem involving this method is the way the weights are computed [20].

$$\hat{y}_{ens}(\mathbf{x}) = \sum_{j=1}^{n} \omega_j \hat{y}_j(\mathbf{x}); \quad \sum_{j=1}^{n} \omega_j = 1. \quad (6)$$

The weighted methods add an extra component of complexity, as the most complex part of building an ensemble is to figure out which weight should be assigned to each model. Despite this added complexity, the created models have more adaptability and yields to better results [11].

For overcoming the added complexity, some authors use heuristics approaches, such as the work of Shen et al. [22], where they used Genetic Algorithms in order to optimize the initial weights (the weights assigned before training) of an ensemble of neural networks. In the present work, another heuristics method, the Particle Swarm Optimization, has been used as it seems to converge faster, it has a bigger continuity (density) of search space, and it presents the ability to reach good solution without local search comparing with other methods [19].

2.3 Particle Swarm Optimization

Swarm intelligence is based on the study of collective behaviour in decentralized and self-organized systems. One of the most spreaded swarm intelligence algorithm is the Particle Swarm Optimization (PSO). This landmark paper triggered waves of publications in the last decade on various successful applications of PSO to solve many difficult optimization problems [19].

This optimization algorithm had been applied in Neural Network Ensembles, as an optimizer for membership function in a Type-2 Fuzzy Neural Network. Hence, it seems a good first approach to use this algorithm for computing the weights of the weighted averaging combining method.

The PSO works by evaluating points of the d-dimensional space over the loss function. These particles change their position iteratively according to the best position found within all the particles (called best global position $\mathbf{p}_g$) and the best position each individual particle had found over the iterations $\mathbf{p}_b$. The algorithm starts with the random initialization of the particles and their velocities in the d-dimensional space. Secondly, each one of the particle is evaluated by the loss function. When all particles have been evaluated, the best global position can be found as well as each individual particle's best position. After that, the particle's position is updated by the formula

$$\mathbf{p}_{i+1} = \mathbf{p}_i + \mathbf{v}_{i+1}, \tag{7}$$

where $\mathbf{p}_i$ represents the particle position at ith iteration and $\mathbf{v}_i$ stands for the velocity, that is computed as

$$\mathbf{v}_{i+1} = \omega \cdot \mathbf{v}_i + c_1 \cdot r_1(\mathbf{p}_i - \mathbf{p}_b) + c_2 \cdot r_2(\mathbf{p}_i - \mathbf{p}_g), \tag{8}$$

where ω, c_1 and c_2 are arbitrary constants known as inertia weight and acceleration coefficients respectively. r_1 and r_2 are random numbers generated at each iteration. When all the positions have been updated, the whole process is repeated until reaching some target criteria or reaching a maximum number of iterations.

3 Method

3.1 ELM Training

For testing the Ensemble of ELMs, the well-known Santa Fe laser data time series benchmark [23] has been used. The Santa Fe data time series (see Fig. 2a) is composed of 10,000 points of the fluctuations in a far-infrared laser, approximately described by three coupled nonlinear ordinary differential equations. For training the ELMs the first 1000 points (10% of the data) have been used, and the remaining 9000 have been used for testing the ELMs in the Ensemble. The inputs of the ELM are the past samples of the signal plus the signal value at time step t and the output will be the forecasted value at $t + 1$. To determine how many past samples have to be taken an autocorrelation plot has been done.

Analyzing the Fig. 2b, it can be concluded that the most significant past samples, that the signal depends on, are between 1 to 10 previous samples, because of the change of the correlation sign and its seasonality.

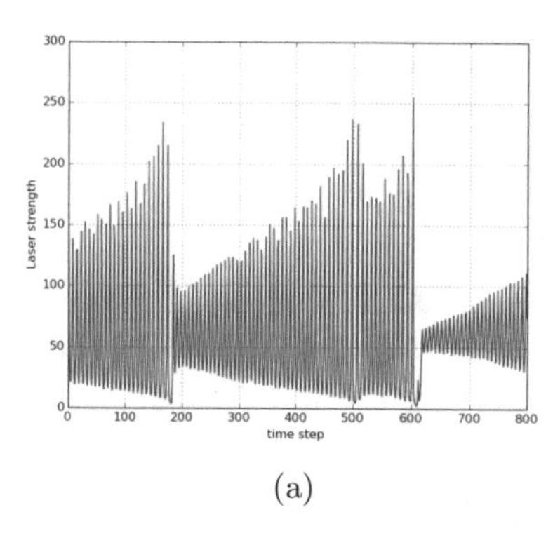

(a)

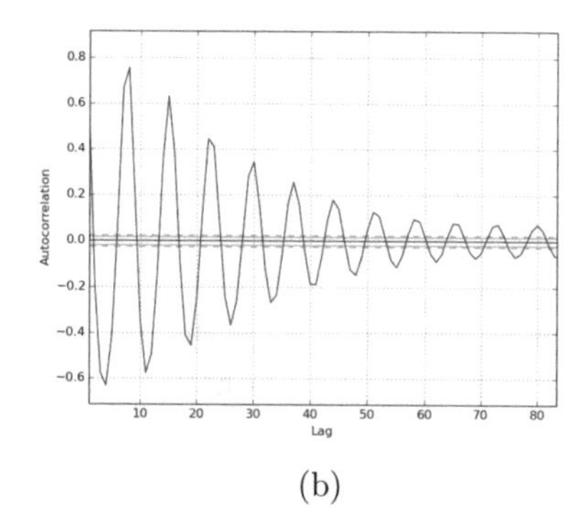

(b)

Fig. 2. (a) First 800 points of Santa Fe's Laser dataset (b) Autocorrelation plot of Santa Fe time series

As the aim of the work is to built up an Ensemble, several different ELM architectures must be trained. As we are constrained to use only past samples of the data, we only can modify the number of past samples introduced to the net and the number of nodes in the hidden layer (since the ELM is a SLFN). Hence, a bunch of ELMs have been trained with a different number of nodes in the hidden layer and different number of inputs. Specifically, the number of nodes oscillates between 150 and 200 (determined heuristically) and the inputs between 2 and 10 past samples, determined by the autocorrelation function. The activation function used in the hidden layer is sigmoid for all cases.

3.2 PSO Tuning and Adaptation

The Loss Function: The PSO by itself is not capable of assuming any constriction of the search space, i.e. is not capable of discriminate the search for certain range of the space. For this purpose the loss-function is introduced. The loss-function is in charge of assessing how good is a particle of the search space. In this case, as we are facing a time-series problem, the suitable loss function that has been used is the Mean Squared Error (MSE) over the forecasted samples, as proposed in [24]. The expression for obtaining the MSE is

$$MSE = \frac{1}{D}\sum_{i=1}^{D}(\hat{y}(\mathbf{x}_i) - y_i)^2, \tag{9}$$

where D is the total amount of forecasted samples, $\hat{y}(\mathbf{x}_i)$ is the prediction of the model before a $\mathbf{x}_i$ input and y_i is the ith real value.

PSO Adaptation: The PSO by definition explores the whole search space without any constrain [25]. In this work an Ensemble of ELMs with a weighted averaging method was used and, as shown by (6), the sum of the weights must be 1. Hence, for reducing the search space to a space where the condition by (6) is fulfilled, an adaptation of the PSO has been done. In this adaptation, the weights have been normalized before being evaluated by the loss function, using the equation

$$\mathbf{w} = \frac{\mathbf{w}}{\sum_{j=1}^{n} \omega_j} \tag{10}$$

where $\mathbf{w}$ represents the vector of weights $\mathbf{w} = [\omega_1, \ldots, \omega_n]$ and ω_j stands for the individual weight of the jth model. This way the condition imposed by the weighted averaging method is always fulfilled.

Parameter Tuning: The PSO has three configurable parameters, the inertia weight ω and the two acceleration coefficients c_1 and c_2, see (8). As suggested in [26] the weight inertia has been set to 0.7298 as it seems to be the best relation between the velocity of the particle and the swarm dependence velocity. Analogously, Eberhart et al. [27] demonstrated that the best acceleration coefficients correspond to $c_1, c_2 = 1.49618$. Hence, in this study, following the others researchers criteria, the parameters of the PSO had been chosen as $\omega = 0.7298$ and $c_1, c_2 = 1.49618$.

3.3 Ensemble Model Building

The Ensemble Model is a bunch of models that are combined somehow in order to improve the performance comparing to an individual model. In the case involving this work, the models are combined using the weighted averaging method. To build the ensemble, the next steps are followed:

1. First, the trained ELMs are sorted by the error they commit in ascending order.
2. Second, the first model in the sorted list is added to the ensemble with a weight of 1.
3. Third, the following respecting model is added to the ensemble, and the PSO is called in order to compute the best weights.
4. The third step is repeated until the maximum number of experts in the EM is reached.

4 Results

Firstly, the MSE error of the trained ELMs over the test set is presented in Table 1. The 'ELM Name' tag represents the ELM's configuration, the number after 'N' stands for the number of nodes in the hidden layer, and the number following the 'M' represents the number of inputs that are introduced to the ELM (see Fig. 1).

Building the ensemble and applying the PSO algorithm to compute the weights, the error committed by the ensemble over the test set can be obtained, as it is shown in Fig. 3. In this figure the evolution of the MSE can be seen as more ELMs are introduced in the ensemble.

Table 1. The best five ELMs found after the batch training, and the corresponding error

ELM name	MSE test set
N-182-M-4	42.30
N-167-M-2	42.60
N-152-M-9	43.55
N-199-M-7	44.03
N-174-M-6	45.02

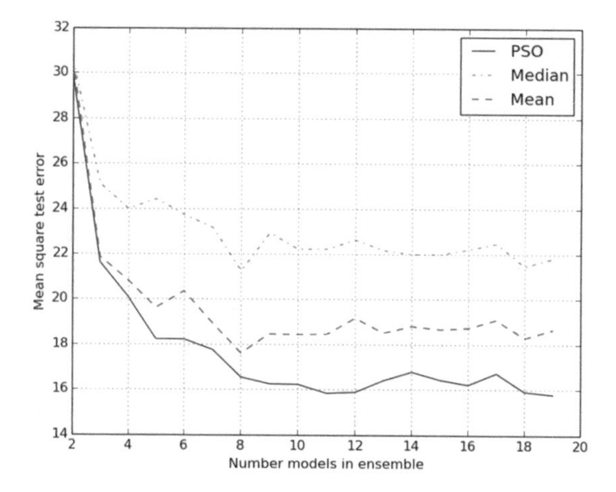

Fig. 3. Mean Squared Error vs number of ELMs in the ensemble

5 Conclusion

In this work, a first approach of the Particle Swarm Optimization algorithm for Ensemble Model optimization is proposed. The selected EM combining method is the weighted average. As the chosen combining method implies that the sum of the weights must be equal to one, an adaptation of the PSO has been proposed in order to overcome this constrain. It has been demonstrated that the PSO is a suitable algorithm to optimize EM's, for gaining both accuracy and performance while reducing the error. In Fig. 3 is shown how the performance of the EM evolves as the number of ELMs in the ensemble grows for three different combining methods, the mean, the median and the PSO boosted weighted average. As seen in the figure, the PSO boosted weighted average performs best, improving the error ratios that commit both the mean and the median.

The best individual model commits a MSE of 42.30 while the EM in its best configuration commits a MSE of 16.22, what is a reduction in the error about the 68.5%. Comparing the error rates with the median and the mean, the PSO improves about 14% the performance of the median and 12% for the mean.

Besides, the error rates reached by the EM allow it to compete with the best models [23].

The main weakness of heuristics algorithm is that they do not guarantee the obtaining of the optimal point. Also, the computational cost of the weight computing is high and increases linearly with the size of the EM, although the computing time is not an important issue in the researched field, but it could be critical in a future application. As future work, a deeper research of the PSO algorithm is proposed that overcome this mayor problems, by finding only one weight in each iteration, instead whole set of weights.

Acknowledgement. This work comes under the framework of the project IT874-13 granted by the Basque Regional Government. The authors would like to thank the company IK4-IDEKO that has supported this work.

References

1. Qazi, A., Fayaz, H., Wadi, A., Raj, R.G., Rahim, N., Khan, W.A.: The artificial neural network for solar radiation prediction and designing solar systems: a systematic literature review. J. Clean. Prod. **104**, 1–12 (2015)
2. Zhang, G.P.: Time series forecasting using a hybrid ARIMA and neural network model. Neurocomputing **50**, 159–175 (2003)
3. Martinetz, T.M., Berkovich, S.G., Schulten, K.J.: 'Neural-gas' network for vector quantization and its application to time-series prediction. IEEE Trans. Neural Netw. **4**(4), 558–569 (1993)
4. Porto, A., Irigoyen, E.: Gas consumption prediction based on artificial neural networks for residential sectors. In: Proceedings of the International Joint Conference SOCO 2017-CISIS 2017-ICEUTE 2017, León, Spain, 6–8 September, pp. 102–111. Springer (2017)
5. Kaastra, I., Boyd, M.: Designing a neural network for forecasting financial and economic time series. Neurocomputing **10**(3), 215–236 (1996)
6. Kurogi, S., Sawa, M., Ueno, T.: Time series prediction of the cats benchmark using fourier bandpass filters and competitive associative nets. Neurocomputing **70**(13), 2354–2362 (2007)
7. Hansen, L.K., Salamon, P.: Neural network ensembles. IEEE Trans. Pattern Anal. Mach. Intell. **12**(10), 993–1001 (1990)
8. Soto, J., Melin, P., Castillo, O.: A new approach for time series prediction using ensembles of ANFIS models with interval type-2 and type-1 fuzzy integrators. In: 2013 IEEE Conference on Computational Intelligence for Financial Engineering Economics (CIFEr), pp. 68–73, April 2013
9. Perrone, M.P., Cooper, L.N.: When networks disagree: ensemble methods for hybrid neural networks. In: How We Learn; How We Remember: Toward An Understanding of Brain and Neural Systems: Selected Papers of Leon N Cooper, pp. 342–358. World Scientific (1995)
10. Wichard, J.D., Ogorzałek, M.: Time series prediction with ensemble models applied to the cats benchmark. Neurocomputing **70**(13), 2371–2378 (2007)
11. Opitz, D.W., Shavlik, J.W.: Actively searching for an effective neural network ensemble. Connection Sci. **8**(3–4), 337–354 (1996)

12. Jafari, S.A., Mashohor, S.: Robust combining methods in committee neural networks. In: 2011 IEEE Symposium on Computers Informatics, pp. 18–22, March 2011
13. Soto, J., Melin, P., Castillo, O.: Particle swarm optimization of the fuzzy integrators for time series prediction using ensemble of IT2FNN architectures, pp. 141–158. Springer International Publishing, Cham (2017)
14. Huang, G.-B., Zhu, Q.-Y., Siew, C.-K.: Extreme learning machine: theory and applications. Neurocomputing **70**, 489–501 (2006)
15. Wang, H., Fan, W., Sun, F., Qian, X.: An adaptive ensemble model of extreme learning machine for time series prediction. In: 2015 12th International Computer Conference on Wavelet Active Media Technology and Information Processing (ICCWAMTIP), pp. 80–85, December 2015
16. Dietterich, T.G.: Ensemble methods in machine learning. In: International Workshop on Multiple Classifier Systems, pp. 1–15. Springer (2000)
17. Claesen, M., De Smet, F., Suykens, J., De Moor, B.: EnsembleSVM: a library for ensemble learning using support vector machines. arXiv preprint arXiv:1403.0745 (2014)
18. Kocev, D., Vens, C., Struyf, J., Džeroski, S.: Ensembles of multi-objective decision trees. In: European Conference on Machine Learning, pp. 624–631. Springer (2007)
19. Kachitvichyanukul, V.: Comparison of three evolutionary algorithms: GA, PSO, and DE. Ind. Eng. Manag. Syst. **11**(3), 215–223 (2012)
20. Wichard, J.D., Ogorzalek, M.: Time series prediction with ensemble models. In: 2004 IEEE International Joint Conference on Neural Networks (IEEE Cat. No. 04CH37541), vol. 2, pp. 1625–1630, July 2004
21. Bishop, C.M.: Pattern recognition and machine learning. Springer (2006)
22. Shen, Z.-Q., Kong, F.-S.: Optimizing weights by genetic algorithm for neural network ensemble. In: Advances in Neural Networks, ISNN 2004, pp. 323–331 (2004)
23. Weigend, A.S., Gershenfeld, N.A.: Results of the time series prediction competition at the Santa Fe Institute. In: IEEE International Conference on Neural Networks, 1993, pp. 1786–1793. IEEE (1993)
24. Lee, T.-H.: Loss functions in time series forecasting. In: International Encyclopedia of the Social Sciences, vol. 9, pp. 495–502 (2008)
25. Eberhart, R.C., Shi, Y., Kennedy, J.: Swarm Intelligence. Elsevier, London (2001)
26. Van Den Bergh, F., Engelbrecht, A.P.: A study of particle swarm optimization particle trajectories. Inf. Sci. **176**(8), 937–971 (2006)
27. Eberhart, R.C., Shi, Y.: Comparing inertia weights and constriction factors in particle swarm optimization. In: Proceedings of the 2000 Congress on Evolutionary Computation, vol. 1, pp. 84–88. IEEE (2000)

Fall Detection Analysis Using a Real Fall Dataset

Samad Barri Khojasteh[1,2], José R. Villar[2(✉)], Enrique de la Cal[2], Víctor M. González[3], and Javier Sedano[4]

[1] Department of Industrial Engineering, Sakarya University, Serdivan, Sakarya, Turkey
samad.khojasteh@ogr.sakarya.edu.tr, samad.khojasteh@gmail.com
[2] Computer Science Department, EIMEM, University of Oviedo, Oviedo, Spain
{villarjose,delacal}@uniovi.es
[3] Control and Automatica Department, EPI, University of Oviedo, Gijón, Spain
vmsuarez@uniovi.es
[4] Instituto Tecnológico de Castilla y León, Pol. Ind. Villalonquejar, 09001 Burgos, Spain
javier.sedano@itcl.es

Abstract. This study focuses on the performance of a fall detection method using data coming from real falls performed by relatively young people and the application of this technique in the case of an elder person. Although the vast majority of studies concerning fall detection place the sensory on the waist, in this research the wearable device must be placed on the wrist because it's usability. A first pre-processing stage is carried out as stated in [1,17]; this stage detects the most relevant points to label. This study analyzes the suitability of different models in solving this classification problem: a feed-forward Neural Network and a rule based system generated with the C5.0 algorithm. A discussion about the results and the deployment issues is included.

1 Introduction

Fall Detection (FD) is a very active research area, with many applications to healthcare, work safety, etc. Even though there are plenty of commercial products, the best rated products only reach a 80% of success [20]. There are basically two types of FD systems: contex-aware systems and wearable devices [14]. FD has been widely studied using context-aware systems, i.e. video systems [28]; nevertheless, the use of wearable devices is crucial because the high percentage of eldel people and their desire to live autonomously in their own house [19].

Wearables-based solutions may combine different sensors, such as a barometer and inertial sensors [22], 3DACC and gyroscope [23], 3DACC and intelligent tiles [7] or a 3DACC and a barometer in a necklace was also reported in [3]. However, 3DACC is by far the most chosen option [4,12,13,27,29], with a variable number of sensors and locations, even some of them proposed the use of the smartphone sensory system. Different solutions have been proposed to perform

M. Graña et al. (Eds.): SOCO'18-CISIS'18-ICEUTE'18, AISC 771, pp. 334–343, 2019.
https://doi.org/10.1007/978-3-319-94120-2_32

the fall event detection, for instance, a feature extraction stage and Support Vector Machines have been applied directly in [27,29], using some transformations and thresholds with very simple rules for classifying an event as a fall [4,13,16]. A comparison of classifiers has been presented in [12], comparing Decision tree, SVM, Nearest neghbor and Discrimenent analysis. Several threshold-based fall detection algorithms were presented in [4,9,10]. The two latter employed three threshold algorithms based to compare with the acceleration magnitude. Igual et al. compared several public datasets for fall detection via support machine vector (SVM) and nearest neighbor (NN) and analyzed results them [15]. The common characteristic in all these solutions is that the wearable devices are placed on the waist or in the chest. This research limits itself to use a single sensor -a marketed smartwatch- placed on the wrist in order to promote its usability.

Interestingly, the previous studies do not focus on the specific dynamics of a falling event: although some of the proposals report good performances, they are just machine learning applied to the focused problem. There are studies concerned with the dynamics in a fall event [2,8], establishing the taxonomy and the time periods for each sequence. Additionally, Abbate et al. proposed the use of these dynamics as the basis of the FD algorithm [1]. A very interesting point of this approach is that the computational constraints are kept moderate, although this solution includes a high number of thresholds to tune. In [17], this solution was analyzed with data gathered from sensors placed on the wrist, using the Abate solution plus a SMOTE balancing stage and a feed-forward Neural Network. In this research, an alternative based on C5.0 rule based systems is proposed.

2 Adapting Fall Detection to a Wrist-Based Solution

Abate et al. [1] proposed the following scheme to detect a candidate event as a fall event (refer to Fig. 1). A time t corresponds to a **peak time** (point 1) if the magnitude of the acceleration a is higher than $th_1 = 3 \times g, g = 9.8\,\mathrm{m/s}$. After a peak time there must be a period of 2500 ms with relatively calm (no other a value higher than th_1). The **impact end** (point 2) denotes the end of the fall event; it is the last time for which the a value is higher than $th_2 = 1.5 \times g$. Finally, the **impact start** (point 3) denotes the starting time of the fall event, computed as the time of the first sequence of an $a <= th_3$ ($th_3 = 0.8 \times g$) followed by a value of $a >= th_2$. The impact start must belong to the interval $[impact\ end - 1200\ \mathrm{ms}, peak\ time]$. If no impact end is found, then it is fixed to peak time plus 1000 ms. If no impact start is found, it is fixed to peak time.

Whenever a peak time is found, the following transformations should be computed:

- Average Absolute Acceleration Magnitude Variation, $AAMV = \sum_{t=is}^{ie} \frac{|a_{t+1} - a_t|}{N}$, with is being the impact start, ie the impact end, and N the number of samples in the interval.

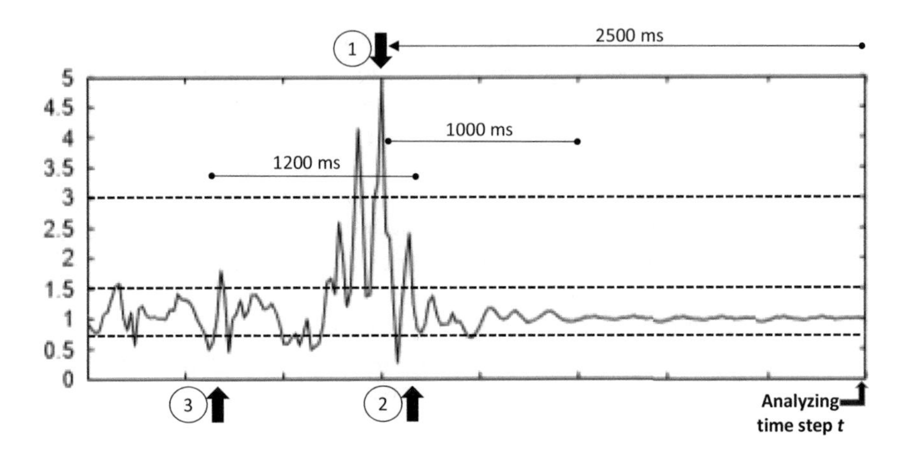

Fig. 1. Evolution of the magnitude of the acceleration -y-axis, extracted from [1].

- Impact Duration Index, $IDI = impact\ end - impact\ start$. Alternatively, it could be computed as the number of samples.
- Maximum Peak Index, $MPI = max_{t \in [is,ie]}(a_t)$.
- Minimum Valley Index, $MVI = min_{t \in [is-500,ie]}(a_t)$.
- Peak Duration Index, $PDI = peak\ end - peak\ start$, with peak start defined as the time of the last magnitude sample below $th_{PDI} = 1.8 \times g$ occurred before peak time, and peak end is defined as the time of the first magnitude sample below $th_{PDI} = 1.8 \times g$ occurred after peak time.
- Activity Ratio Index, ARI, measuring the activity level in an interval of 700 ms centered at the middle time between impact start and impact end. The activity level is calculated as the ratio between the number of samples not in $[th_{ARIlow} 0.85 \times g, th_{ARIhigh} = 1.3 \times g]$ and the total number of samples in the 700 ms interval.
- Free Fall Index, FFI, computed as follows. Firstly, search for an acceleration sample below $th_{FFI} = 0.8 \times g$ occurring up to 200 ms before peak time; if found, the sample time represents the end of the interval, otherwise the end of the interval is set 200 ms before peak time. Secondly, the start of the interval is simply set to 200 ms before its end. FFI is defined as the average acceleration magnitude evaluated within the interval.
- Step Count Index, SCI, measured as the number of peaks in the interval $[peak\ time - 2200, peak\ time]$. SCI is the step count evaluated 2200 ms before peak time. The number of valleys are counted, defining a valley as a region with acceleration magnitude below $th_{SCIlow} = 1 \times g$ for at least 80 ms, followed by a magnitude higher than $th_{SCIhigh} 1.6 \times g$ during the next 200 ms. Some ideas on computing the time between peaks [26] were used when implementing this feature.

Evaluating this approach was proposed as follows. The time series of acceleration magnitude values are analyzed searching for peaks that marks where a fall event candidate appears. When it happens to occur, the *impact end* and the *impact start* are determined, and thus the remaining features. As long as this fall events are detected when walking or running, for instance, a Neural Network (NN) model is obtained to classify the set of features extracted.

In order to train the NN, the authors made use of an Activities of Daily Living (ADL) and FD dataset, where each file contains a Time Series of 3DACC values corresponding to an activity or to a fall event. Therefore, each dataset including a fall event or a similar activity -for instance, running can perform similarly to falling- will generate a set of transformation values. Thus, for a dataset file we will detect something similar to a falling, producing a row of the transformations computed for each of the detected events within the file. If nothing is detected withing the file, no row is produced. With this strategy, the Abbate et al. obtained the training and testing dataset to learn the NN.

2.1 The Modifications on the Algorithm

As stated in [11,25], the solutions to this type of problems must be ergonomic: the users must feel comfortable using them. We considered that placing a device on the waist is not comfortable, for instance, it is not valid for women using dresses. When working with elder people, this issue is of main relevance. Therefore, in this study, we placed the wearable device on the wrist. This is not a simple change: the vast majority of the literature reports solutions for FD using waist based solutions. Moreover, according to [24] the calculations should be performed on the smartwatches to extend the battery life by reducing the communications. Therefore, these calculations should be kept as simple as possible.

A second modification is focused on the training of the NN. The original strategy for the generation of the training and testing dataset produced a highly imbalanced dataset: up to 81% of the obtained samples belong to the class FD, while the remaining belong to the different ADL similar to a fall event.

To solve this problem a normalization stage is applied to the generated imbalanced dataset, followed by a SMOTE balancing stage [6]. This balancing stage will produce a 60%(FALL)–40%(no FALL) dataset, which would allow to avoid the over-fitting of the NN models. As usual, there is a compromise between the balancing of the dataset and the synthetic data samples introduced in the dataset.

These above mentioned changes have already been studied in [17]. In this research we proposed to analyze the performance of rule based systems in this context, which represents more simpler models that can be easily deployed in wearable devices and with a very reduced computational complexity. Therefore, they could represent a very interesting improvement, either if they work similarly to the NN or just similarly to them.

3 Experiments and Results

A ADL and FD dataset is needed to evaluate the adaptation, so it contains time series sample from ADL and for falls. This research made use of the UMA-FALL dataset [5] among the publicly available datasets. This dataset includes data for several participants carrying on with different activities and performing forward, backward and lateral falls. Actually, this falls are not real falls -demonstrative videos have been also published, but they can represent the initial step for evaluating the adapted solution problem. Interestingly, this dataset includes multiple sensors; therefore, the researcher can evaluate the approach using sensors placed on different parts of the body.

The thresholds used in this study are exactly the same as those mentioned in the original paper. All the code was implemented in R [21] and caret [18]. The parameters for SMOTE were perc.over set to 300 and perc.under set to 200 -that is, 3 minority class samples are generated per original sample while keeping 2 samples from the majority class. These parameters produces a balanced dataset that moves from a distribution of 47 samples from the minority class and 200 from the majority class to a 188 minority class versus 282 majority class (40%/60% of balance).

To obtain the parameters for the NN a grid search was performed; the final values were size set to 20, decay set to 10^{-3} and maximum number of iterations 500, the absolute and relative tolerances set to 4×10^{-6} and 10^{-10}, respectively. In this research, we use the C5.0 implementation of the C4.5 that is included in the R package to obtain the rule based systems. The parameters found optimum for this classification problem are cf set to 0.25, bands set to 2, the fuzzy Threshold parameter set to TRUE, the number of trials set to 15, and winnow set to FALSE.

Both 5x2 cross validation (cv) and 10-fold cv were performed to analyzed the robustness of the solution. The latter cv would allow us to compare with existing solutions, while the former shows the performance of the system with an increase in the number of unseen samples. The results are shown in Tables 1 and 2 for 10-fold cv and 5x2 cv, respectively.

3.1 Discussion on the Results

From the tables it can be seen that both modelling techniques perform exceptionally well once the SMOTE is performed and using test folds from 10-fold cv: the models even perform ideally for several folds. And more importantly, the two models are interchangeable with no apparent loss in the performance. Actually, these results are rather similar to those published in the original work [1]. However, when using 5x2 cv the results diverts from those previously mentioned.

With 5x2 cv, the size of the train and test datasets are of similar number of samples, producing a worse training and, what is more interesting, introduces more variability in the test dataset. Therefore, the results are worse. The point is that these results suggest the task is not solved yet as the number of false alarms increased unexpectedly (Fig. 2).

Table 1. 10 fold cv results obtained for the NN (up) and C5.0 rule based system (bottom). From left to right, the main statistical measurements are shown: accuracy (Acc), Kappa factor (Kp, sensitivity (Se), the specificity (Sp), the precision (Pr) and the geometric mean of the Acc and Pr, $G = \sqrt[2]{Pr \times Acc}$.

Fold	Feed forward NN					
	Acc	Kp	Se	Sp	Pr	G
1	0.9362	0.8686	0.9286	0.9474	0.9630	0.9456
2	0.9787	0.9562	0.9643	1.0000	1.0000	0.9820
3	0.9575	0.9131	0.9286	1.0000	1.0000	0.9636
4	0.9583	0.9129	0.9655	0.9474	0.9655	0.9655
5	1.0000	1.0000	1.0000	1.0000	1.0000	1.0000
6	0.9565	0.9069	1.0000	0.8889	0.9333	0.9661
7	0.9167	0.8319	0.8621	1.0000	1.0000	0.9285
8	0.9565	0.9087	0.9643	0.9444	0.9643	0.9643
9	0.8936	0.7846	0.8571	0.9474	0.9600	0.9071
10	1.0000	1.0000	1.0000	1.0000	1.0000	1.0000
mean	0.9554	0.9083	0.9470	0.9675	0.9786	0.9623
median	0.9570	0.9108	0.9643	0.9737	0.9828	0.9649
std	0.0337	0.0682	0.0531	0.0383	0.0243	0.0293
Fold	C5.0 rule based system					
	Acc	Kp	Se	Sp	Pr	G
1	0.9575	0.9117	0.9643	0.9474	0.9643	0.9643
2	0.9787	0.9555	1.0000	0.9474	0.9655	0.9826
3	0.9362	0.8686	0.9286	0.9474	0.9630	0.9456
4	0.8958	0.7719	1.0000	0.7368	0.8529	0.9236
5	0.9787	0.9555	1.0000	0.9474	0.9655	0.9826
6	0.9348	0.8589	1.0000	0.8333	0.9032	0.9504
7	0.9583	0.9144	0.9310	1.0000	1.0000	0.9649
8	0.9783	0.9539	1.0000	0.9444	0.9655	0.9826
9	0.9362	0.8640	1.0000	0.8421	0.9032	0.9504
10	0.9787	0.9562	0.9643	1.0000	1.0000	0.9820
mean	0.9533	0.9011	0.9788	0.9146	0.9483	0.9629
median	0.9579	0.9131	1.0000	0.9474	0.9649	0.9646
std	0.0276	0.0605	0.0297	0.0838	0.0470	0.0203

This problem is important because in this experimentation we used the UMA-Fall dataset [5]. This dataset used was generated with young participants using a very deterministic protocol of activities. The falls were performed with the participants standing still and letting them fall in the forward/backward/lateral

Table 2. 5x2 cv results obtained for the NN (up) and C5.0 rule based system (bottom). From left to right, the main statistical measurements are shown: accuracy (Acc), Kappa factor (Kp, sensitivity (Se), the specificity (Sp), the precision (Pr) and the geometric mean of the Acc and Pr, $G = \sqrt[2]{Pr \times Acc}$.

Fold	Feed forward NN					
	Acc	Kp	Se	Sp	Pr	G
1	0.8936	0.7772	0.9220	0.8511	0.9028	0.9123
2	0.9319	0.8571	0.9575	0.8936	0.9310	0.9442
3	0.9192	0.8359	0.8794	0.9787	0.9841	0.9303
4	0.9532	0.9027	0.9575	0.9468	0.9643	0.9609
5	0.9362	0.8682	0.9291	0.9468	0.9632	0.9460
6	0.9106	0.8148	0.9149	0.9043	0.9348	0.9248
7	0.9362	0.8682	0.9291	0.9468	0.9632	0.9460
8	0.9404	0.8750	0.9645	0.9043	0.9379	0.9511
9	0.9149	0.8252	0.9007	0.9362	0.9549	0.9274
10	0.9021	0.7935	0.9433	0.8404	0.8987	0.9207
mean	0.9238	0.8418	0.9298	0.9149	0.9435	0.9364
median	0.9255	0.8465	0.9291	0.9202	0.9464	0.9372
std	0.0188	0.0394	0.0271	0.0446	0.0277	0.0154
Fold	C5.0 rule based system					
	Acc	Kp	Se	Sp	Pr	G
1	0.9234	0.8387	0.9575	0.8723	0.9184	0.9377
2	0.9362	0.8654	0.9716	0.8830	0.9257	0.9484
3	0.9149	0.8214	0.9433	0.8723	0.9172	0.9302
4	0.9404	0.8746	0.9716	0.8936	0.9320	0.9516
5	0.9319	0.8556	0.9787	0.8617	0.9139	0.9458
6	0.9106	0.8087	0.9787	0.8085	0.8846	0.9305
7	0.8894	0.7679	0.9220	0.8404	0.8966	0.9092
8	0.9362	0.8649	0.9787	0.8723	0.9200	0.9489
9	0.9149	0.8162	1.0000	0.7872	0.8758	0.9358
10	0.9021	0.7898	0.9787	0.7872	0.8734	0.9246
mean	0.9200	0.8303	0.9681	0.8479	0.9058	0.9363
median	0.9192	0.8301	0.9752	0.8670	0.9156	0.9368
std	0.0166	0.0356	0.0220	0.0398	0.0214	0.0132

direction. Therefore, the differences with real falls might be relevant; even if they are not so different, the variability that might be introduced will severely punish the performance of the obtained models.

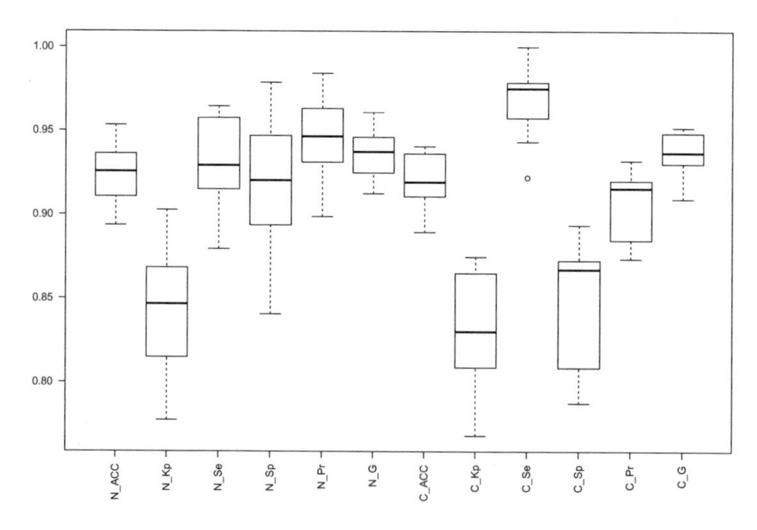

Fig. 2. 5x2 cv Boxplot for the different measurements -Accuracy (Acc), Kappa (Kp), Sensitivity (Se) and Specificity (Sp), Precision (Pr) and the geometric mean of the Acc and Pr, $G = \sqrt[2]{Pr \times Acc}$, both for the feed-forward NN (six boxplots to the left, with the N_ prefix) and C5.0 (six boxplots to the right, with the C_ prefix).

4 Conclusions

This study compares the performances of two classification techniques when tackling the problem fall detection with data gathered from accelerometers located on one wrist. The original proposal detected fall events and performed a feature extraction which was classified with a feed-forward NN. A SMOTE stage is included to balance the transformed dataset previous modelling. Two different techniques are compared: the feed-forward NN and C5.0 rule based systems. A publicly available dataset with falls has been used in evaluating the proposal. Interestingly, the two modelling techniques performed similarly, which suggest that in real world applications with the solution embedded in smartwatches perhaps the rule based systems is more likely to be used.

Although exceptional results have been found using 10 fold cv, the 5x2 cv results suggest that still a high number of false alarms is obtained. Although the percentages are better that those reported for commercial devices, some design aspects must be analyzed in depth: the robustness to the variability in the behaviour of the user, or the tuning of the threshold to fit specific populations like the elderly.

Acknowledgment. This research has been funded by the Spanish Ministry of Science and Innovation, under project MINECO-TIN2014-56967-R and MINECO-TIN2017-84804-R.

References

1. Abbate, S., Avvenuti, M., Bonatesta, F., Cola, G., Corsini, P., Vecchio, A.: A smartphone-based fall detection system. Pervasive Mob. Comput. **8**(6), 883–899 (2012)
2. Abbate, S., Avvenuti, M., Corsini, P., Light, J., Vecchio, A.: Wireless Sensor Networks: Application - Centric Design. In: Monitoring of human movements for fall detection and activities recognition in elderly care using wireless sensor network: a survey, p. 22. Intech (2010)
3. Bianchi, F., Redmond, S.J., Narayanan, M.R., Cerutti, S., Lovell, N.H.: Barometric pressure and triaxial accelerometry-based falls event detection. IEEE Trans. Neural Syst. Rehab. Eng. **18**(6), 619–627 (2010)
4. Bourke, A., O'Brien, J., Lyons, G.: Evaluation of a threshold-based triaxial accelerometer fall detection algorithm. Gait Posture **26**, 194–199 (2007)
5. Casilari, E., Santoyo-Ramón, J.A., Cano-García, J.M.: Umafall: a multisensor dataset for the research on automatic fall detection. Procedia Comput. Sci. **110**(Supplement C), 32–39 (2017). http://www.sciencedirect.com/science/article/pii/S1877050917312899
6. Chawla, N.V., Bowyer, K.W., Hall, L.O., Kegelmeyer, W.P.: Smote: synthetic minority over-sampling technique. J. Artif. Intell. Res. **16**, 321–357 (2002)
7. Daher, M., Diab, A., Najjar, M.E.B.E., Khalil, M.A., Charpillet, F.: Elder tracking and fall detection system using smart tiles. IEEE Sens. J. **17**(2), 469–479 (2017). http://ieeexplore.ieee.org/document/7733127/
8. Delahoz, Y.S., Labrador, M.A.: Survey on fall detection and fall prevention using wearable and external sensors. Sensors **14**(10), 19806–19842 (2014). http://www.mdpi.com/1424-8220/14/10/19806/htm
9. Fang, Y.C., Dzeng, R.J.: A smartphone-based detection of fall portents for construction workers. Procedia Eng. **85**, 147–156 (2014)
10. Fang, Y.C., Dzeng, R.J.: Accelerometer-based fall-portent detection algorithm for construction tiling operation. Autom. Constr. **84**, 214–230 (2017)
11. González, S., Sedano, J., Villar, J.R., Corchado, E., Herrero, Á., Baruque, B.: Features and models for human activity recognition. Neurocomputing **167**, 52–60 (2015)
12. Hakim, A., Huq, M.S., Shanta, S., Ibrahim, B.: Smartphone based data mining for fall detection: Analysis and design. Procedia Comput. Sci. **105**, 46–51 (2017). http://www.sciencedirect.com/science/article/pii/S1877050917302065
13. Huynh, Q.T., Nguyen, U.D., Irazabal, L.B., Ghassemian, N., Tran, B.Q.: Optimization of an accelerometer and gyroscope-based fall detection algorithm. J. Sens. **2015**, 8 (2015)
14. Igual, R., Medrano, C., Plaza, I.: Challenges, issues and trends in fall detection systems. BioMed. Eng. OnLine **12**(1), 66 (2013). http://www.biomedical-engineering-online.com/content/12/1/66
15. Igual, R., Medrano, C., Plaza, I.: A comparison of public datasets for acceleration-based fall detection. Med. Eng. Phys. **37**(9), 870–878 (2015). http://www.sciencedirect.com/science/article/pii/S1350453315001575
16. Kangas, M., Konttila, A., Lindgren, P., Winblad, I., Jämsaä, T.: Comparison of low-complexity fall detection algorithms for body attached accelerometers. Gait Posture **28**, 285–291 (2008)

17. Khojasteh, S.B., Villar, J.R., de la Cal, E., González, V.M., Sedano, J., Yazğan, H.R.: Evaluation of a Wrist-based Wearable Fall Detection Method. In: 13th International Conference on Soft Computing Models in Industrial and Environmental Applications (2018, submitted)
18. Kuhn, M.: The caret package (2017). http://topepo.github.io/caret/index.html. Accessed 15 Jan 2018
19. Kumari, P., Mathew, L., Syal, P.: Increasing trend of wearables and multimodal interface for human activity monitoring: a review. Biosens. Bioelectron. **90**(15), 298–307 (2017)
20. Purch.com: Top ten reviews for fall detection of seniors (2018). http://www.toptenreviews.com/health/senior-care/best-fall-detection-sensors/
21. R Development Core Team: R: A Language and Environment for Statistical Computing. R Foundation for Statistical Computing, Vienna, Austria (2008). http://www.R-project.org. ISBN 3-900051-07-0
22. Sabatini, A.M., Ligorio, G., Mannini, A., Genovese, V., Pinna, L.: Prior-to- and post-impact fall detection using inertial and barometric altimeter measurements. IEEE Trans. Neural Syst. Rehab. Eng. **24**, 774–783 (2016)
23. Sorvala, A., Alasaarela, E., Sorvoja, H., Myllyla, R.: A two-threshold fall detection algorithm for reducing false alarms. In: Proceedings of 2012 6th International Symposium on Medical Information and Communication Technology (ISMICT) (2012)
24. Vergara, P.M., de la Cal, E., Villar, J.R., González, V.M., Sedano, J.: An iot platform for epilepsy monitoring and supervising. J. Sens. **2017**, 18 (2017)
25. Villar, J.R., González, S., Sedano, J., Chira, C., Trejo, J.: Human Activity Recognition and Feature Selection for Stroke Early Diagnosis. In: Pan, J.S., Polycarpou, M.M., Woźniak, M., de Carvalho, A.C.P.L.F., Quintián, H., Corchado, E. (eds.) HAIS 2013. LNCS (LNAI), vol. 8073, pp. 659–668. Springer, Heidelberg (2013). https://doi.org/10.1007/978-3-642-40846-5_66
26. Villar, J.R., González, S., Sedano, J., Chira, C., Trejo-Gabriel-Galán, J.M.: Improving human activity recognition and its application in early stroke diagnosis. Int. J. Neural Syst. **25**(4), 1450036–1450055 (2015)
27. Wu, F., Zhao, H., Zhao, Y., Zhong, H.: Development of a wearable-sensor-based fall detection system. Int. J. Telemedicine Appl. **2015**, 11 (2015). https://www.hindawi.com/journals/ijta/2015/576364/
28. Zhang, S., Wei, Z., Nie, J., Huang, L., Wang, S., Li, Z.: A review on human activity recognition using vision-based method. J. Healthc. Eng. **2017**, 31 (2017)
29. Zhang, T., Wang, J., Xu, L., Liu, P.: Fall detection by wearable sensor and one-class svm algorithm. In: Huang, D.S., Li, K., Irwin, G. (eds.) Intelligent Computing in Signal Processing and Pattern Recognition, Lecture Notes in Control and Information Systems, vol. 345, pp. 858–863. Springer, Berlin Heidelberg (2006)

Implementation of a Non-linear Fuzzy Takagi-Sugeno Controller Applied to a Mobile Inverted Pendulum

C. H. Rodriguez-Garavito[1(✉)], Miguel F. Arevalo-Castiblanco[2], and Alvaro A. Patiño-Forero[1]

[1] Automation Engineering, La Salle University, Carrera 2 10-70, Bogota, Colombia
{cerodriguez,alapatino}@unisalle.edu.co

[2] Master in Industrial Automation Engineering, Nacional University of Colombia, Carrera 30-45, Bogota, Colombia
miarevaloc@unal.edu.co

Abstract. Applications based on inverted pendulum principle, have proliferated in recent years, surveillance system such as two-wheeled vehicles or manipulators attached to robotics legs in humanoids robots are some examples. Even though the control of inverted pendulum has been deeply studied in the academic community for decades, there are few implementations with a Fuzzy Takagi Sugeno model for overlapped membership functions joint to Linear Quadratic Regulator, because just in [11], this kind of generalized T-S modeling was addressed. This paper presents the implementation of a nonlinear Fuzzy Takagi-Sugeno control with overlapped membership functions applied to a mobile inverted pendulum mechanism and its comparison against LQR in terms of state space behavior and robustness.

Keywords: Nonlinear control · Takagi-Sugeno fuzzy control
Mobile inverted pendulum · LQR

1 Introduction

Within the context of non-linear control, one of the most studied plants over the years has been the simple, double and triple inverted pendulum, given its unstable dynamics, its ease of construction and/or commercial availability, and its field of application [1]. Likewise, the inverted pendulum offers an attractive complexity for the implementation of new control strategies, which has caused great interest amongst researchers, seeking to test, evaluate and compare different modern control techniques. In the field of control engineering, working with robust pendulum platforms allows the validation of various controllers according to specific applications [2]. As in the case of robotics and transport systems, where control applications in pendulum systems have appeared [3]. In humanoid robotics the balance of the body can be seen as a pendulum control issue for stable walking trajectory [4]. As well, the new forms of single-person transportation,

M. Graña et al. (Eds.): SOCO'18-CISIS'18-ICEUTE'18, AISC 771, pp. 344–353, 2019.
https://doi.org/10.1007/978-3-319-94120-2_33

which use the inverted pendulum principle, the SegWay used for surveillance and entertainment, is an example, along with the single-wheel bicycle [5–7]. Other applications of more sophisticated use of inverted pendulum are found in [8], where a single-track vehicle which uses the gyroscopic effects of two flywheels as an active stabilizing system is described, and in [9], where a semi-active in-car crib system with the joint application of regular and inverted pendulum mechanisms is applied to reduce the collision shock and risk of injury to an infant.

To address the modeling of these pendulum platforms, for example in the SegWay structure, two strategies have been commonly used, the identification of a model based on input-output data, and the analytical modeling using Newton-Euler-based methods and the Lagrange's classic energy approach [10].

Amongst the most common identification methods, there is the FTS (Fuzzy Takagi-Sugeno) approach. Its objective is to address a set of auto-tuned rules based upon the operating data of the plant, which allows to approximate the dynamics of the system in different points of operation, as many as rules are contained in the knowledge base. Once, each rule is approached and taking into account that its structure, in the consequent of each rule, can be interpreted as a linear system in the state-space representation (time-domain approach), it is possible to propose an independent control design for each sub-state space representation, under any modern control strategy of such as LQR (Linear Quadratic Regulator).

The literature report some hardware applications of a non-linear identification and control methodology, FTS-LQR with overlapped membership functions [11]. At the simulation level, in [12], a Ball and Beam model is addressed. In [13], simulation and experimental results are reported from a real prototype of a rotary chain pendulum. Likewise, applications have been reported in control of chaotic systems [14], control of power quality systems [15] and in the control of a helicopter-like twin rotor [16].

Under this line, the work presented in this article is a contribution in the validation of the FTS-LQR methodology, showing a real implementation and an analysis of experimental results, on a NxtWay type mobile pendulum platform from Lego Mindstorms^{TM}.

The article addresses the two ways described above for the control of an inverted Pendulum SegWay type, in Sect. 2 addresses a linear approach, based upon the linearized analytical model presented in [17,18], for the design of a LQR controller. In Sect. 3, a non-linear strategy for controlling the NxtWay is developed; first the FTS identification of the real model is made, and subsequently, an LQR controller is designed for each operation rule. In Sect. 4, the results and their respective discussion are presented, to finally, in Sect. 5, enunciate the main conclusions of the work.

2 Modeling and Linear Control

The first step to address the control of a dynamic system, is to have a full knowledge of its model according to the physical laws that govern it; As mentioned in the previous section, the platform to be analyzed is the Lego NxtWay™mobile pendulum. Its modeling is taken from [17,18].

In Fig. 1, the platform's base structure can be observed, which defines the positions and speeds that will be used as state variables in the system's model. The position q_1 of the wheels and orientation with respect to z, q_3 is defined, next to the position of the pole q_2.

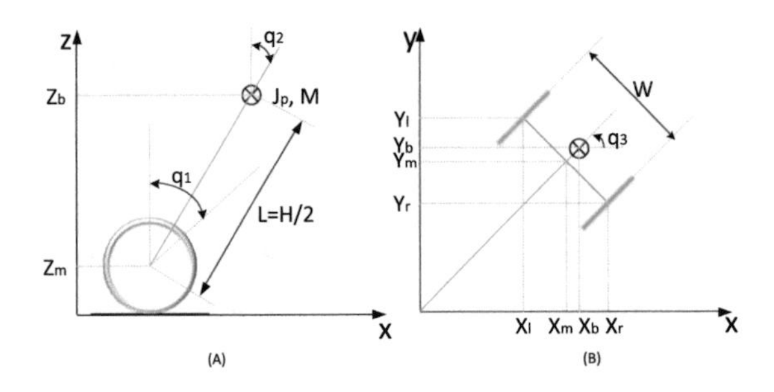

Fig. 1. Schematic of the NxtWay mobile inverted pendulum, figure (A) indicates the front view of the platform and figure (B) its top view. Figure taken from [17]

The definition of the robotics equation [19] or expression of the non-linear model, is presented in Eq. (1). It should be noted that the input of the non-linear model will be the voltage of the actuators, which for that matter, are the motors of each wheel.

$$M_x(q)\ddot{q} + C_x(q, \dot{q})\dot{q} + G_x(q) = V \tag{1}$$

For the design of a linear controller, the linearization of the NxtWay dynamic model in its standard state-space form is required: $\dot{x}(t) = Ax(t) + Bu(t)$ and $y(t) = Cx(t) + Du(t)$. The state vector of the platform $x(t)$ is made-up of the positions shown in Fig. 1 and their corresponding speeds; the values of each matrix are defined in [17], while the values of matrices C and D are defined in a standard way.

With the state matrices, it is possible to perform the calculation of controllers. This model requires two control loops, an LQR control loop for the position of the pole q_2, where part of the state vector $[q_2, \dot{q_1}, \dot{q_2}]$ intervenes, and a second proportional-integrative loop, for the position of the mobile q_1, state variable which results independent from the acceleration of pole $\ddot{q_2}$.

The blocks diagram of the closed loop controlled system is as shown in Fig. 2. In the control loop, a correction of the mobile position can be seen due to the slidings and asynchronisms in the movement that may occur when the energy of the virtual wheel actuation, abstraction of the operation of the two wheels of the model, is derived for each of the two physical actuators, this block is called Steer. Additionally, another saturation block is included over the actuation signal, with a value equal to the maximum voltage under normal operation for the motors.

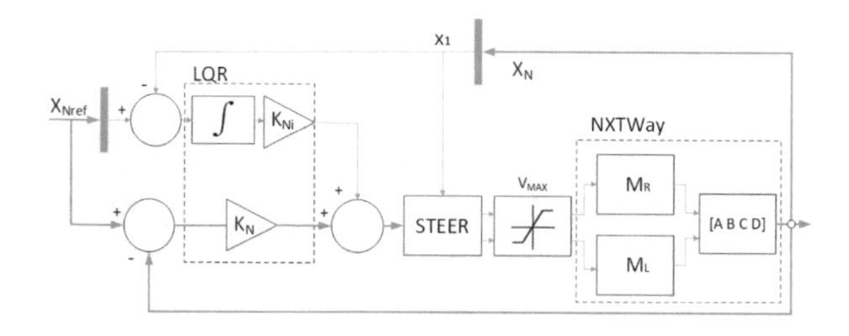

Fig. 2. LQR control blocks diagram of NxtWay Pendulum

A fifth state variable is added to the model, defined by the integral of q_1 to guarantee the position of the platform given its independence from the dynamic model.

The optimum System control feedback constants, according to the LQR design are consolidated in Eq. (2).

$$\begin{aligned} K_N &= [-0.8211 \quad -69.4743 \quad -1.0739 \quad -9.0738] \\ K_{Ni} &= -0.4472 \end{aligned} \tag{2}$$

3 Identification-Control Strategy FTS-LQR

In this section we present the design of the fuzzy Takagi-Sugeno identification-control technique with an optimum LQR controller.

For the validation of the dynamic performance of the platform, the linear LQR control algorithm described in the previous section, is implemented within the RobotC^{TM} development environment. Once the correct operation of the closed loop system has been verified; that is, the equilibrium of the pendulum structure is reached with minimum oscillation, an additive disturbance is added to the output of the controller with random frequency, according to an uniform probability distribution; the response of the controlled pendulum in the presence of external disturbance, only for the position of the pole and actuator voltage, can be seen in Figs. 3a and b.

3.1 FTS Identification

Based on the response of the controlled system, the Takagi-Sugeno identification methodology is proposed in [20] presents a method for estimating a fuzzy model of the type $y = f(x_1, x_2, \ldots x_k), f : \Re^k \rightarrow \Re$, with consequent as a linear input-output relationship in the following way:

$$\begin{aligned} R^n : \quad & If\; x_1\; is\; A_1^n, \ldots, and\; x_k\; is\; A_k^n \\ & then\; y = p_0^n + p_1^n x_1 + \cdots + p_k^1 x_k \end{aligned} \tag{3}$$

The synthesis of the Takagi-Sugeno model is presented in Eq. (4).

$$y = \frac{\sum_{i=1}^{n} \left(A_1^i(x_1) \wedge \cdots \wedge A_k^i(x_k)\right)\left(p_0^i + p_1^i x_1 + \cdots + p_k^i x_k\right)}{\sum_{i=1}^{n} \left(A_1^i(x_1) \wedge \cdots \wedge A_k^i(x_k)\right)} \tag{4}$$

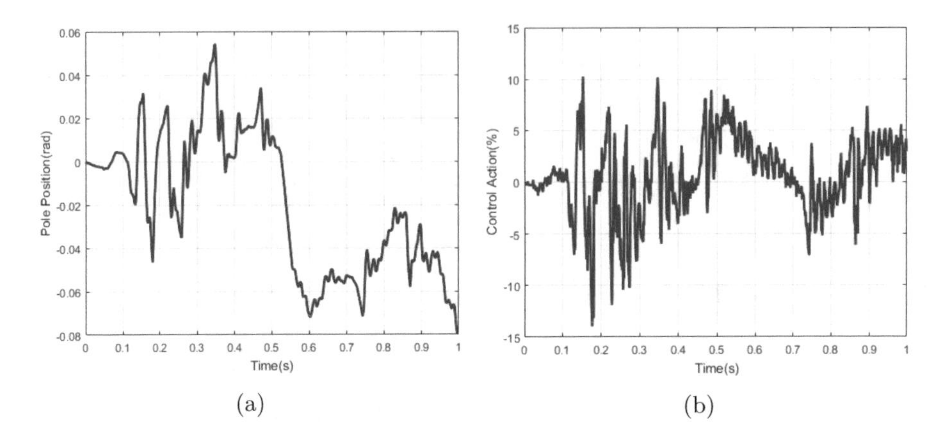

Fig. 3. (a) LQR control response, pole position q_2, in presence of disturbance. (b) LQR control response, control action V, in presence of disturbance

In compact form is:

$$\beta_i = \frac{A_1^i(x_1) \wedge \cdots \wedge A_k^i(x_k)}{\sum_{i=1}^{n}\left(A_1^i(x_1) \wedge \cdots \wedge A_k^i(x_k)\right)} \tag{5}$$

Then, the FTS model can be expressed as:

$$y = \sum_{i=1}^{n}\left(p_0^i\beta_i + p_1^i x_1\beta_i + \cdots + p_k^i x_k\beta_i\right) \tag{6}$$

Now, if an input output set is available, of the form: $\{x_{1j}, x_{2j}, \ldots, x_{kj} \rightarrow y_j, (j = 1, \ldots, m)\}$, then it is possible to obtain the parameters of the consequent: $p_0^i, p_1^i, \ldots, p_k^i\ (i = 1, \ldots, n)$ through the method of least squares minimizing the error between the output generated by the model and the expected values as follows:

$$J = \sum_{j=1}^{m}\left(y_j - \hat{y_j}\right)^2 = \|Y - XP\|^2 \tag{7}$$

X,Y and P are assembled in the following form:

$$X = \begin{bmatrix} \beta_{11}, .\beta_{n1}, \beta_{11}x_{11}, .\beta_{n1}x_{11}, .\beta_{11}x_{k1}, .\beta_{n1}x_{k1} \\ \vdots \\ \beta_{1m}, .\beta_{nm}, \beta_{1m}x_{1m}, .\beta_{nm}x_{1m}, .\beta_{1m}x_{km}, .\beta_{nm}x_{km} \end{bmatrix} \tag{8}$$

For this case the β_{ij} coefficients are defined as follows:

$$\beta_{ij} = \frac{A_1^i(x_{1j}) \wedge \cdots \wedge A_k^i(x_{kj})}{\sum_{i=1}^{n}\left(A_1^i(x_{1j}) \wedge \cdots \wedge A_k^i(x_{kj})\right)} \tag{9}$$

$$Y = \left[y_1, \ldots, y_m\right]^T \tag{10}$$

$$P = \left[p_0^1, \ldots, p_0^n, p_1^1, \ldots, p_1^n, \ldots, p_k^1, \ldots, p_k^n\right]^T \tag{11}$$

The solution to the minimization problem by least squares is as follows:

$$P = \left(X^T X\right)^{-1} X^T Y \tag{12}$$

This solution does not apply to overlapping sets, for which in [11] a variation is proposed by adding a factor in the index to minimize J.

$$\begin{aligned} J &= \textstyle\sum_{j=1}^{m} \left(y_j - \hat{y}_j\right)^2 + \gamma^2 \sum_k P_k^2 \\ &= \left\| \begin{bmatrix} Y \\ 0 \end{bmatrix} - \begin{bmatrix} X \\ \gamma I \end{bmatrix} P \right\|^2 \end{aligned} \tag{13}$$

The matrix X_a is full range and therefore the solution to the problem of minimization by least squares can now be found, as in the original case, for standard sets with overlapping partitions.

Finally, the nonlinear dynamical systems of the form: $x^n = f\left(x, \dot{x}, \ddot{x}, \ldots, x^{n-1}, u\right)$, can be specified through a Takagi-Sugeno model with fuzzy rules of the type:

$$\begin{aligned} R^i : if\ x\, is\, A_i^{j_1}\ and\ \dot{x}\, is\, A_i^{j_2} \ldots, and\ x^{n-1}\, is\, A_i^{j_k} \\ then\ y^i = a_0^i + a_1^i x + a_2^i \dot{x} + \cdots + b^i u \end{aligned} \tag{14}$$

The Fuzzy Takagi-Sugeno identification allows to describe the non-linear behavior of the wheels acceleration $\ddot{q_1}$, Fig. 4a, and the acceleration of the pole $\ddot{q_2}$, Fig. 4b, which equations completely describe the non-linear dynamics of the NxtWay in a linear system fashion (15).

$$\begin{aligned} R^i : if\ \ & q_2\, is\, A_i^{Q_2}\ and\ \dot{q_2}\, is\, A_i^{\dot{Q}_2} and\ \dot{q_1}\, is\, A_i^{\dot{Q}_1} \\ then\ \ & \ddot{q_1} = a_{01}^i + a_{11}^i q_2 + a_{21}^i \dot{q_2} + a_{31}^i \dot{q_1} + b_1^i u\ and \\ & \ddot{q_2} = a_{02}^i + a_{12}^i q_2 + a_{22}^i \dot{q_2} + a_{32}^i \dot{q_1} + b_2^i u \end{aligned} \tag{15}$$

All coefficients $a_{01}^i \ldots a_{32}^i$ and b_1^i and b_2^i can be recovered from (12).

3.2 Design of LQR Controllers for the FTS Model

From the identified system, the generation of the set of constants for the FTS-LQR controller is done, for this design, 3 sets were taken for each of the state variables involved in the system, in the ranges in which they were captured during the previous data collection process. The position of the pole q_2 variable, was handled in a range of (−10,10) degrees, the speed of the pole $\dot{q_2}$ variable, was treated in the range of (−50,50) degrees, and the speed of the wheels $\dot{q_1}$ variable, was marked at (−500,500) degrees.

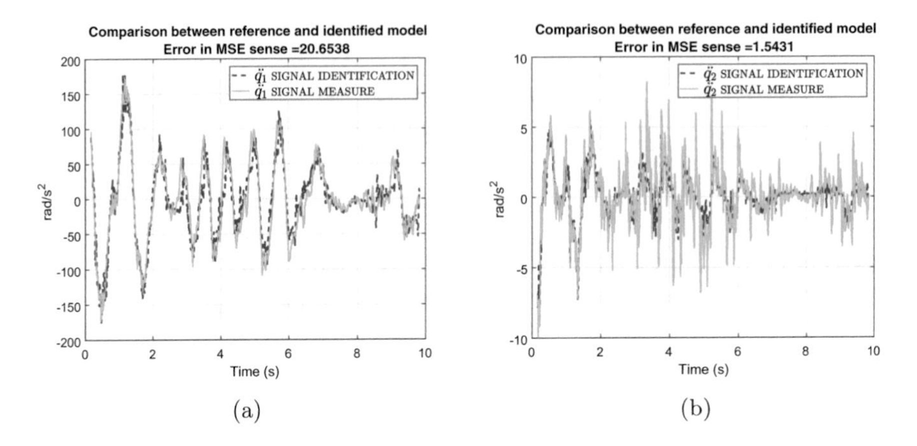

(a) (b)

Fig. 4. (a) Identified wheel's acceleration. (b) Identified pole's acceleration.

With this system, 27 rules are obtained, therefore, having the same number of linear subsystems $[A_i, B_i]$. For every rule, a controller LQR is calculated as shown in (16).

$$A_i = \begin{bmatrix} 0 & 0 & 1 \\ a_{11}^i & a_{21}^i & a_{31}^i \\ a_{12}^i & a_{22}^i & a_{32}^i \end{bmatrix} \; B_i = \begin{bmatrix} 0 \; b_1^i \; b_2^i \end{bmatrix} \; A0_i = \begin{bmatrix} 0 \; a_{01}^i \; a_{01}^i \end{bmatrix}$$
$$K(i) = \begin{bmatrix} -B_i/A0_i \; LQR(A_i, B_i, Q_i, R_i) \end{bmatrix} \tag{16}$$

4 Results and Discussion

Once, the FTS-LQR controller is implemented in the NxtWay platform, its behavior is then compared with that exhibited by the linear LQR controller designed in Sect. 2. The testing protocol includes a comparison of the control bearings in the state-space representation, of the linear and nonlinear LQR controllers in their FTS version, through a disturbance of the system in equilibrium with: an external force, and a gradual change in one parameter of the plant, the diameter of the wheels.

In Figs. 5a and b, the evolution of the state vector of the pendulum is observed, with the LQR and FTS-LQR controllers respectively, starting from its equilibrium position and as a response to an external force produced by the impact of a 10 g mass from a 10° degrees angle, suspended by a 50 cm long rope. There is a similar stabilization time of 2.5 s between LQR and FTS-LQR controller, although in the case of LQR version, with a greater extra impulse in the position of the wheels, inferred from the integral of the signal $\int q_1$. That is, the FTS-LQR controller, for this case, has a softer response, given that the behavior of the other components of the state vector is similar in magnitude.

Finally, in relation to the parametric variation experiment in the NxtWay model, in Figs. 6a and b, the behavior of the controlled system under the two

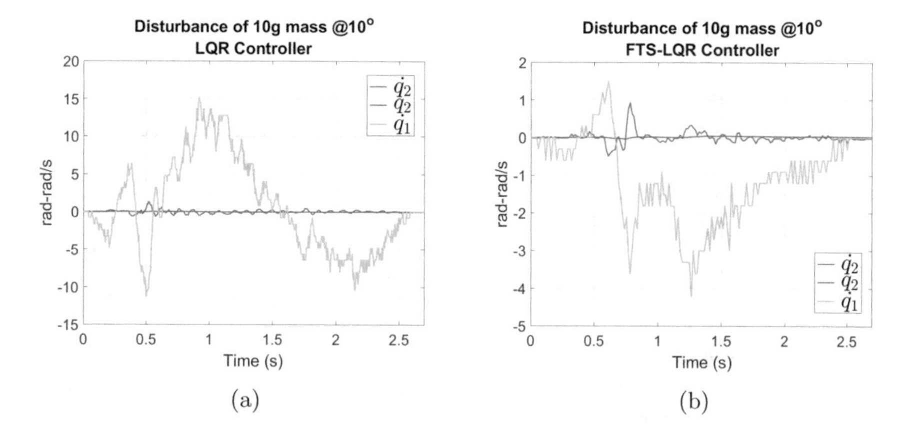

Fig. 5. (a) LQR behaviour. (b) FTS-LQR behaviour

approaches is presented, linear and non-linear, expressed in their corresponding phase diagrams, for different wheel diameters (sum of support and pneumatic): T1 (21.3 mm), T2 (22 mm), T3 (28 mm), T4 (30 mm), T5 (30 mm) and T6 (42.5 mm). In this experiment, stability was also achieved in all the previous types, except in T1 with the smaller diameter wheel type and stop the LQR controller. Again, the state-space representation where trajectories were taken was lower for the FTS-LQR case. These tests made validate the operation of the proposed controllers, showing particular characteristics of operation such as: the oscillation present in each state variable, the levels of extra impulse and inverted energy, as well as the stabilization of the pole.

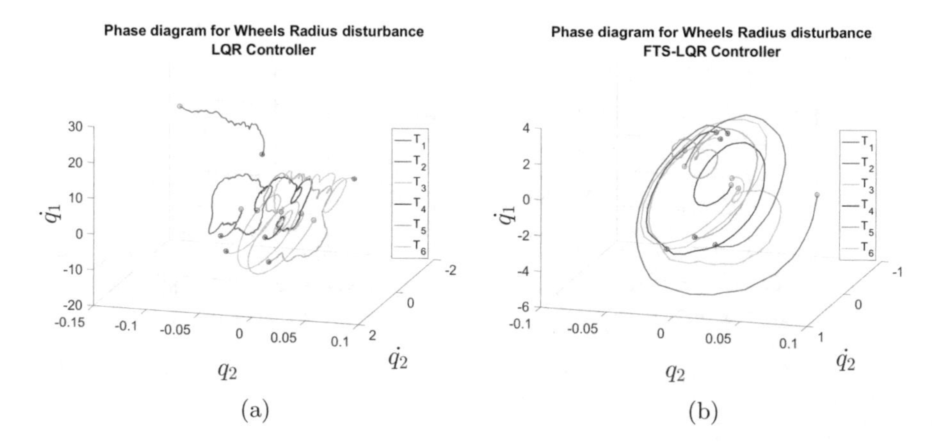

Fig. 6. (a)Phase diagram of the LQR controller. (b) Phase diagram of the FTS-LQR controller.

5 Conclusions

In a real implementation, an important aspect when identifying a highly unstable system given its own non-linearity as in the case of the mobile inverted pendulum, is the capture of input-output data under different operating modes. When the data capture is made from a simulated analytical model, the data is collected from the execution of a simulation cycle under different initial conditions, such as described in [21]. In the case of a real implementation, the data is taken by disturbing the control signal in closed loop, with a square signal at random frequency, defined by a uniform probability function.

The performance of a pendulum platform controlled by an LQR regulator, however, should not depend on the re-feed gain vector, since it is obtained from an optimization process, it requires a heuristic search of the values of the Q and R matrices that in practice, define a fine tuning, to achieve the maximum performance of the controller. In the case of the FTS-LQR controller, the adjustment is applied to the ranges in which the input universes are defined, corresponding to the state variables involved in the model. Likewise, the behavior of the controller is also influenced by the partitions of each universe, number of sets per universe.

The FTS-LQR controlled system, showed in the tests performed with additive disturbance to the model input and variation of the wheel radius, a lower energy expenditure and a lower oscillation in the position of the wheels q_1, additionally, it was evidenced a greater robustness, by controlling the platform in all the experiments.

Acknowledgement. This work was supported by La Salle University, Bogotá-Colombia, through the VRIT project (24326205).

References

1. Quanser^{TM}: Quanser - real-time control experiments for education and research, Feburary 2017. http://www.quanser.com/
2. Hercus, R., Wong, K.-Y., Shee, S.-K., Ho, K.-F.: Control of an inverted pendulum using the neurabase network model. In: Lee, M., Hirose, A., Hou, Z.-G., Kil, R.M. (eds.) Neural Information Processing, pp. 605–615. Springer, Heidelberg (2013)
3. Nawawi, S.W., Ahmad, M.N., Osman, J.H.S.: Control of two-wheels inverted pendulum mobile robot using full order sliding mode control. In: Proceedings of International Conference on Man-Machine Systems, Langkawi, Malaysia, 15–16 September 2006
4. Olcay, T., Özkurt, A.: Design and walking pattern generation of a biped robot. Turk. J. Electr. Eng. Comput. Sci. **25**(2), 761–769 (2017)
5. Back, S., Lee, S.-G.: Modeling and control system design of a ball segway. In: 2016 16th International Conference on Control, Automation and Systems (ICCAS), pp. 481–483. IEEE (2016)
6. Pousti, A., Bodur, M.: Kinematics and dynamics of a wheeled mobile inverted pendulum. IEEE Xplore, 10–12 July 2008

7. Nawawi, S.W., Ahmad, M.N., Osman, J.H.S.: Real-time control of a two-wheeled inverted pendulum mobile robot. Int. J. Electr. Comput. Eng. **39**, 214–220 (2008)
8. Chu, T.-D., Chen, C.-K.: Design and implementation of model predictive control for a gyroscopic inverted pendulum. Appl. Sci. **7**(12), 1272 (2017)
9. Kawashima, T.: Simulation study on an acceleration control system for semi–active in-car crib with joint application of regular and inverted pendulum mechanisms. Mech. Eng. J. **4**(4), 17–27 (2017)
10. Dini, N., Majd, V.J.: Model predictive control of a wheeled inverted pendulum robot. In: 2015 3rd RSI International Conference on Robotics and Mechatronics (ICROM), pp. 152–157. IEEE (2015)
11. Al-Hadithi, B.M., Jimenez, A., Matia, F.: A new approach to fuzzy estimation of Takagi-Sugeno model and its applications to optimal control for nonlinear systems. Appl. Soft Comput. **12**(1), 280–290 (2012)
12. Adanez, J.M., Al-Hadithi, B.M., Jimenez, A., Matia, F.: Optimal control of a ball and beam nonlinear model based on Takagi-Sugeno fuzzy model. In: Advances in Fuzzy Logic and Technology, pp. 1–11. Springer, Cham (2017)
13. Aranda-Escolástico, E., et al.: Control of a chain pendulum: a fuzzy logic approach. Int. J. Comput. Intell. Syst. **9**(2), 281–295 (2016)
14. Wang, D., Han, P.: Controlling chaotic systems using aggregated linear quadratic regulator. Prz. Elektrotech. **88**(7b), 336–340 (2012)
15. Ko, H.-S., Jatskevich, J.: Power quality control of wind-hybrid power generation system using fuzzy-LQR controller. IEEE Trans. Energy Conv. **22**(2), 516–527 (2007)
16. Tao, C.-W., Taur, J.-S., Chen, Y.C.: Design of a parallel distributed fuzzy LQR controller for the twin rotor multi-input multi-output system. Fuzzy Sets Syst. **161**(15), 2081–2103 (2010)
17. Arevalo-Castiblanco, M.F., Rodriguez-Garavito, C., Patiño-Forero, A.A., Salazar-Caceres, J.F.: Controlador lqr y smc aplicado a plataformas pendulares. Rev. Iberoam. Autom. Inform. Ind. **15**(3), (2018)
18. Yamamoto, Y.: NXTway-GS model-based design-control of self-balancing two-wheeled robot built with LEGO mindstorms NXT, Technical report. Cybernet Systems Co., Ltd., Mathworks Inc. (2009)
19. Barrientos, A.: Fundamentos de Robotica. Mcgraw-Hill/Interamericana De Espana, Madrid (2007)
20. Takagi, T., Sugeno, M.: Fuzzy identification of systems and its applications to modeling and control. IEEE Trans. Syst. Man Cybern. **15**(1), 116–132 (1985)
21. Arevalo-Castiblanco, M.F., Rodriguez-Garavito, C.H., Patiño-Forero, A.A.: Identification of a non-linear model type inverted rotary pendulum. In: 2017 IEEE 3rd Colombian Conference on Automatic Control (CCAC), pp. 1–6, November 2017

Special Session: Soft Computing Applications in the Field of Industrial and Environmental Enterprises

SVR-Ensemble Forecasting Approach for Ro-Ro Freight at Port of Algeciras (Spain)

Jose Antonio Moscoso-López(✉), Ignacio J. Turias,
Juan Jesús Ruiz Aguilar, and Francisco Javier Gonzalez-Enrique

Intelligent Modelling of Systems Research Group, Polytechnic School of Engineering (Algeciras), University of Cádiz,
Avda. Ramón Puyol s/n, 11202 Algeciras, Cádiz, Spain
{joseantonio.moscoso, ignacio.turias, juanjesus.ruiz, javier.gonzalezenrique}@uca.es

Abstract. The forecasting of the freight transportation provides a helpful information in the management of ports environment and can be used as a decision-making tool. This work addresses the forecasting of ro-ro (roll-on roll-off) freight flow in a port using a two-stage approach by an ensemble of the best Support Vector Regression (SVR) models. The time series used for forecasting is daily ro-ro freight in the port of Algeciras during the period from 2000 to 2007. Additionally, the time series was preprocessed through an exponential smoothing in order to improve the performance. The experiment results show that the proposed approach is a promising tool in freight forecasting.

Keywords: Time series forecasting · Support vector regression
Freight transport · Ro-Ro ensemble

1 Introduction

Ports play an important role of the entire supply chain where the information knowledge of freight flow, is a key factor in ports operations and management. An accuracy freight forecasting is a valuable decision making tool and can help to the different actors in the planification and management of the ports operations. The capacity to accurately forecast demand in freight could minimize the operation cost and improve the service quality in the ports services, and thereby increase competitiveness.

In European Union (EU), all the maritime import freight from third countries must be checked in ports facilities. The complexity of the import process during the port phase and the variation of the scheduled arrival of the cargo causes problems as delays and congestions that provoke a critical point in the supply chain. These problems imply continuous pre-planning modifications and generate high variability in ports environments [1, 2].

The agility of the import processes depends on the future behavior, in short-terms of the freight flow. Hence, is very important the use of decision making tools to anticipate the planification of the transport system [3]. This fact is more important in the case of perishable freight, because delays or bottlenecks in the process could be supposed losses of the cargo value.

M. Graña et al. (Eds.): SOCO'18-CISIS'18-ICEUTE'18, AISC 771, pp. 357–366, 2019.
https://doi.org/10.1007/978-3-319-94120-2_34

During last few decades, numerous forecasting approaches (a wide variety of techniques) have been used for the problem of traffic flows prediction in ports operations. These problems have been addressed by times series modelling approaches such statistical methods and soft computing. Generally, these methods develop their predictions on historical data analysis and present good performance when the time series vary temporally. Nowadays, statistical methods have been used for comparison purposes whenever a new forecasting approach is proposed [4–7]. In recent times, Support Vector Machines (SVM), have been introduced in the context of traffic flow forecasting and have been extensively adopted to solve many engineering problems in pattern recognition. This method is proposed by Vapnik [8] which is based on statistical learning theory. SVM established the structural risk minimization principle (minimizing the fitting error and reducing the upper bound of the generalization error). Support Vector Regression (SVR) has been developed by incorporating Vapnik´s ε-intensive loss function into SVMs to solve non-linear regression estimation problems. SVR has been applied to solve short-term transport forecasting problems obtained good performance [9–12].

Combining several models can improve the forecasting performance. In turn, ensemble approaches have been proposed in order to overcome the deficiencies of the single models and avoid the limitations of each individual models. Combined methods are proposed for accurate prediction in transport engineering. Many authors employ the combinations of intelligence techniques with statistical methods to improve their forecasting. These combined models based in intelligence techniques always outperform simple statistical models and single pattern models [13]. Peng [14] presented two statistical combination models among six univariate forecasting models proposed with the purpose to predict container throughput in the three major ports in Taiwan. Another example is the work of Tian et al. [15] where Ports container throughput also were forecasted using a multistep model of combined techniques including econometric models. Others works used SVR in hybrid approaches to forecast container throughput. The employment of SVR with statistical methods achieves better forecasting performance than individual approaches [16, 17]. SVR was also successfully by authors employed in combination with SOM to forecast maritime freight traffic at (port) Border Inspection Post (BIPs) [18].

In this work, we propose a two-stage algorithm of prediction approach for short-term freight forecasting. Thus, the proposed approach integrates the advantages of each individual SVR model in an ensemble of experts to develop a robust prediction model.

The remainder of the work is organized as follows, in Sect. 2, the basic concepts of the SVR is briefly reviewed, Sect. 3 presents the database and Sect. 4 the experiment procedure. Section 5 showed the results and the comparison of the performance with simple models. Finally, the concluding remarks are presented.

2 Methodology

2.1 Support Vector Regression (SVR)

SVR is based on statistical learning theory [8] and approximates an unknown function in order to map the input data into a high-dimensional feature space through a nonlinear mapping function, an then a linear regression problem is constructed in this feature space. Given a set of data points $(x_1, y_1), (x_2, y_2), \ldots, (x_m, y_m)$ where $x_1, y_1 \in \Re$ are the input and output vector correspondingly and m is the total number of training samples. In this work, x_i is an input and the number of input depends on the number of the past samples selected. On the other hand, y_i is the output was the forecasted value of the vegetable freight (in Kg) for 7 days of prediction horizon. In the high-dimensional feature space, there theoretically exists a linear function, f, to formulate the non-linear relationship between input and output data. Such a linear regression function, namely SVR function, is as Eq. 1.

$$y = f(x, w) = w^T \varphi(x) + b \tag{1}$$

Where $f(x,w)$ is a function that produces the forecasting values; $\varphi(x)$ denotes the Kernel function which maps the input (x) to a vector in a feature space; the coefficients w (weight vector) and b (bias term) are adjustable and they are estimated by the minimizing the regularized risk function R (Eq. 2):

$$R(w, b) = \frac{1}{2}\|w\|^2 + \frac{C}{N}\sum_{i=1}^{N} |y_i - f(x_i, w)|_\varepsilon \tag{2}$$

where C is the regularization parameter and determines the trade-off between the flatness of the $f(x)$ and the amount up to which deviations larger than ε are tolerated, y-$f(x,w)$ is the ε-insensitive loss function as defined as (Eq. 3)

$$|y - f(x, w)|_\varepsilon = \begin{cases} y - f(x, w) - \varepsilon, & \text{if } |y - f(x, w)| \geq \varepsilon \\ 0, & \text{otherwise} \end{cases} \tag{3}$$

Using Lagrange Multiplier techniques, the minimization of the Eq. 2 leads to the following dual optimization problem.

$$\min_{w,b,\xi,\xi^*} \frac{1}{2} w^T w + C(\xi_i + \xi_i^*) \tag{4}$$

$$\text{subject to} \begin{cases} y_i - (w^T \varphi(x_i) + b \leq \varepsilon + \xi_i \\ (w^T \varphi(x_i) + b) - y_i \leq \varepsilon + \xi_i^* \\ \xi_i, \xi_i^* \geq 0 \end{cases} \tag{5}$$

Where ε denotes the maximum deviation allowing during the training, $C > 0$ decides the trade-off of generalization ability and training error. The slack variables ξ_i and ξ_i^*, correspond to the distance between actual values and the corresponding boundary values of *ε-tube*. This problem can be solved by making use of the Karush-Kuhn-Tucker′s conditions [19]. The performance of SVR is determined by the type of kernel function and the setting of kernel parameters. In this work, is employed the Gaussian Kernel (Radial Basic Function)

$$K(x, z) = exp\left(\frac{-(x-z)^2}{2\delta^2}\right) \tag{6}$$

where δ is the width of Gauss Kernel (positive scalar).

3 The Data

We have used a database provided by the Port Authority of Algeciras Bay where the daily traffic of Ro-Ro transport in the Strait of Gibraltar is include. The strait of Gibraltar is the main gate between North Africa and Europe for the Ro-Ro transport. The original database is composed of over three millions of records among January 2000 to December 2007. We use in this work a daily database of the fresh vegetable freight during the period mentioned above. Vegetable freight is the main Ro-Ro freight in the Strait of Gibraltar. In the import of fresh freight, the full process in port may be the most efficient possible due to the value of the good decrease along the days. Obviously, inefficiencies in the process could be supposed losses of the freight value.

Fresh vegetables represent the 75% of the total freight for human consumption and this kind of trade must be inspected in the Border Inspection Post. This checking process is an important bottleneck and need an accuracy decision-making tool to planning the resources involved. The origin of the ro-ro cargo, unload in Port of Algeciras, is frequently from North of Africa and the point of consumption is Spain (40%) or rest of EU (60%). The trip by ship is covered in only 2 h. The fresh cargo has the special treatment, which need stay in the market as soon as possible because the economic value decrease quickly.

4 Experimental Procedure

We propose a two stage approach with the purpose to improve the prediction of the daily vegetable freight in the logistic node of the Strait of Gibraltar. At the first stage (Stage-I), we developed a procedure to obtain a prediction for 7 days ahead using a SVR models with different parameters (C, ε and γ) and using a resampling procedure with two-fold crossvalidation (Fig. 1). At the second stage (Stage-II), we obtained the best forecasting values for each prediction approach from Stage-I. First, we selected the best model configuration applying several multicomparison tests and finally an ensemble approach determined the best output.

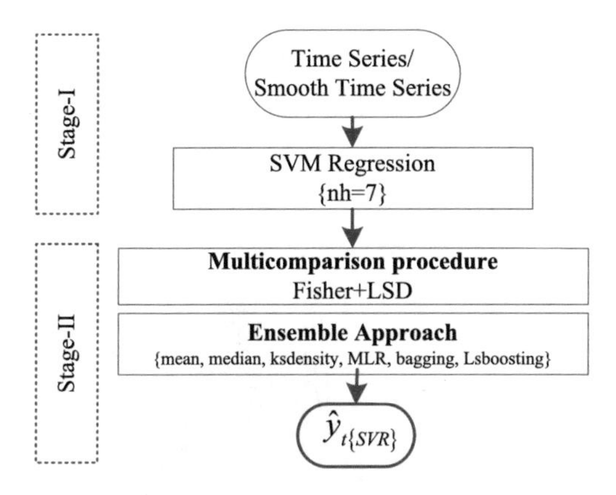

Fig. 1. Procedure for seven days forecasting with SVR and Ensemble.

4.1 Stage-I Simple Forecasting Approach

The forecasting procedure is illustrated in Fig. 2 where the different lags and autoregressive windows are represented.

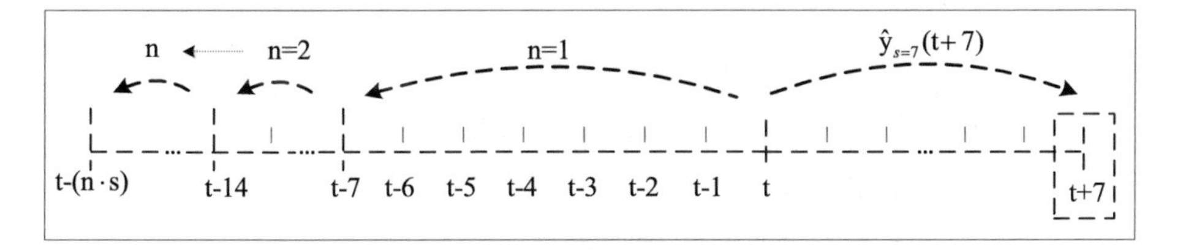

Fig. 2. Forecasting procedure

To address the daily prediction, we analysed the features of the time series in order to perform an adequate selection of forecasting approach. The daily weight fluctuates greatly, and can vary from one day to the next. For this reason, we analysed the results between two approaches and compared the time series with an exponential smooth-pretreatment and without pretreatment (Fig. 3). The exponential smooth-pretreatment is carried out as outlier filter with the purpose to improve the performance in the forecasting method [20].

In order to select the best configuration of the parameters, we used an iterative procedure varying the values of the different input parameters (C, ε and γ) to the models. The parameter's ranges used in this work are collected in Table 1. In this work, 27,648 configurations have been analyzed in Stage-I, (2 time series configurations x 2 lags x 9 autoregressive windows x 768 SVRs configurations).

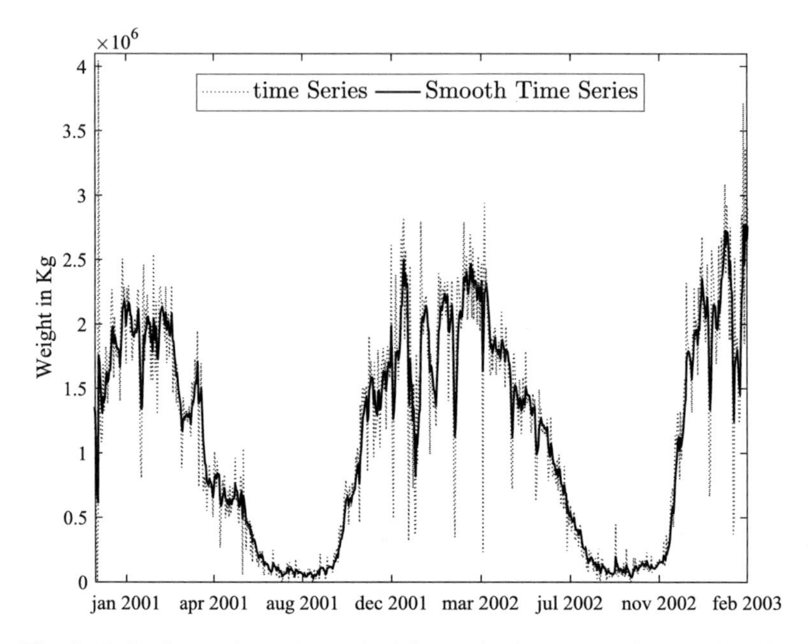

Fig. 3. Daily time series and smoothed time series in a twenty five-month period

Table 1. Parameters used in SVM (Stage-I)

Parameters	SVR
Time horizon	7
Number of lags	1;7
Size of autoregressive window	1:4,7,14,21,28,52
Kernel function	Radial Basic Function
C	0.05;0.5;1:5;10:10:50
Epsilon (ε)	2^{-8}, 2^{-7},..., 2^{-2}
*Gamma (γ)	2^{-8}, 2^{-7},..., 2^{-2}

4.2 Stage-II: Multicomparison and Ensemble Approach

Firstly, Stage-II determines the configuration of the best model (and the models without significant differences) At this stage, we have used a multicomparison test. Secondly, an ensemble mixture of expert is developed using the results of the best configuration and the results of the models not significantly different.

In order to select the configuration of the best performance models a multicomparison approach is developed. A Friedman/LSD-Fisher test [21] compare the accuracy of the models analysed in order to determine if the models have significant difference between them [22]. Friedman test is the non-parametric alternative to the ANOVA test. Friedman test assess the mean and variance of the performance indexes, with a probability lower than 0.05, and determines the existence of significant difference when the null hypothesis is rejected. This test is carried out when the assumptions of the ANOVA test (normality and homogeneity of variance) are not achieved.

Once the null hypothesis is rejected, the post-hoc LSD Fisher test determines which models are not equivalent. LSD Fisher test is a post-hoc method, which computed the error of the difference between two means and compared it with LSD. The LSD test is expressed as:

$$\left|\bar{y}_i - \bar{y}_j\right| \geq t_{\alpha/2}\sqrt{S_w^2\left(\frac{2}{n}\right)} \Rightarrow \text{Reject } \mathrm{H}_0 \quad (7)$$

Where $t_{\alpha/2}$ is the t value for a significance level of $\alpha/2$, in this case $\alpha = 0.05$, n is the number of observation for each model and S_w^2 is the variance of the residual error.

Finally, at Stage-II, the outputs of the best models selected by the multicomparison test, are improved by using an ensemble approach. The accuracy of the results of the best model and the models without significantly differences generally is similar, therefore, establishing which model provides the best overall value is no easy task. With the purpose to determine the best overall model a combination of the best outputs is carried out. An ensemble approach is proposed using as input a combination of the outputs obtained in the previous stage. These outputs are those obtained using the best configurations previously selected through the multicomparison test.

The outputs obtained in the proposed ensemble are based in the follow six operators:

1. Mean. The ensemble output is the mean of the inputs selected by the multicomparison test.
2. Median. The output is obtained by the middle value in the list of the inputs of the ensemble. The median sorts all the values in numerical order and picks the middle value.
3. Kernel Density Function. The output is obtained by a kernel distribution which represent the probability function for a random variable. The selection of the output is given by the probability of the value that appears most often in inputs data.
4. Multiple Linear Regression. In this case, the output (dependent variable) is obtained as a relationship between the inputs (independent variable).
5. Least Squared Boosting (LSB). This method is one of the most commonly used in ensemble task. Boosting method [23, 24] generate a high accuracy hypothesis using a weak learning algorithm and shows good resistance to overfitting. Boosting procedure starts with a sensible estimation of the learner and their improvement is carried out by iterations based on the performance of the training data set. The procedure of the LSB has a simple structure, in the first step (initialization) inputs are fitted using a least square fit. Then compute residuals and fit the real-valued learned to the current residuals by least squares of the generic functional gradient descent and finally increase a new interaction with new inputs weight and compute the residuals again.
6. Bagging. In the same way as boosting, bagging (boostrap aggregation) is a popular ensemble method proposed by Breiman [25]. This method uses boostrapping

(equiprobable selection with replacement) on the training set to generate many varied new subsets. The proposed of the based algorithm is the creation of different samples of the base model outputs for a given input.

4.3 Performance Criteria

In order to evaluate the prediction performance of the proposes models, standard correlation coefficient (R), agreement index (d), and root mean squared error (RMSE) have been used. Authors have previously used this indexes in different works [26, 27].

The correlation coefficient (R) determines the strength of the linear relationship between observed and forecasted values. The index of agreement varies from 0 to 1 with higher index values indicating that the forecasted values F have better agreement with the observations O. On the other hand, RMSE was used to measure the difference between the values predicted and the observed ones.

5 Results

In this section, the prediction performance of the proposed approaches is presented, compared and interpreted. The results represent the forecasting values of the daily vegetable freight by Ro-Ro transport which are imported in the Port of Algeciras Bay. In this work, we estimated 7 days of prediction horizon. For a clear discussion, the results achieved are presented and discussed for each stage. The prediction performance is assessed by d, R, and RMSE and presented in Table 2.

Table 2. Best models comparison throughout the two stages

Stage	Methodology	Prediction approach	d	R	RMSE
I	SVR	$\hat{y}_{t+7}$	0,7708	0,6216	797.870,38
I	Smooth TS+SVR	$\hat{y}_{t+7(smooth)}$	0,8355	0,7273	709.655,49
II	SVR+ENSEMBLE	$\hat{y}_{t+7} + ENSBL$	0,9304	0,8751	491423,49
II	Smooth TS+SVR+EMSEMBLE	$\hat{y}_{t+7(smooth)} + ENSBL$	**0,9344**	**0,8818**	**487.109,26**

Table 2 shows the outcome of the two forecasting approaches proposed in this work. Overall, the time series pretreatment (smoothing) presents good influence in the performance improvement of all the models. Comparing the results achieved it worth mentioning that smoothing time series presented a better performance that the data without preprocessing.

On the other hand, the combination with ensemble approach achieves better results in all indexes. The best performance is achieved with LSB ensemble.

Therefore, the time series pretreatment (smoothing) and the combination of SVR and Ensemble provides an improvement in the forecasting performance in all the cases.

6 Conclusions

The freight flow forecasting is one of the main problem in logistic and ports environment. Obtaining accurate results in short-term forecasting is an important challenge in Ports. As case study we have selected the Port of Algeciras Bay, one of the most relevant ports in Europe.

An ensemble approach of SVR modes have been carried out using the outputs of models not significantly different with the best model in order to achieve a reliable prediction of Ro-Ro freight. Additionally, the time series were preprocessed by exponential smoothing. The obtained results clearly indicate that the performance is improved using pre-processed data. Besides, the results suggest that the two-stage approach enhanced those obtained using single SVR models.

Future work point out to integrate some exogenous information and using some different techniques such as genetic algorithms or deep learning.

Acknowledgments. This work is part of the coordinated research projects TIN2014-58516-C2-1-R and TIN2014-58516-C2-2-R supported by (MICINN Ministerio de Economía y Competi-tividad-Spain). The data have been kindly provided by Port Authority of Algeciras Bay.

References

1. Bilegan, C.I., Crainic, T.G., Gendreau, M.: Forecasting freight demand at intermodal terminals using neural networks–an integrated framework. Eur. J. Oper. Res. **13**, 22–36 (2008)
2. Romero, G., Durán, G., Marenco, J., Weintraub, A.: An approach for efficient ship routing. Int. Trans. Oper. Res. **20**, 767–794 (2013)
3. Vlahogianni, E.I., Golias, J.C., Karlaftis, M.G.: Short-term traffic forecasting: overview of objectives and methods. Transp. Rev. **24**, 533–557 (2004)
4. Al-Deek, H.M.: Which method is better for developing freight planning models at seaports - neural networks or multiple regression? Transp. Res. Rec. **1763**, 90–97 (2001)
5. Murat Celik, H.: Modeling freight distribution using artificial neural networks. J. Transp. Geogr. **12**, 141–148 (2004)
6. Mostafa, M.M.: Forecasting the Suez Canal traffic: a neural network analysis. Marit. Policy Manag. **31**, 139–156 (2004)
7. Ratrout, N.T., Gazder, U.: Factors affecting performance of parametric and non-parametric models for daily traffic forecasting. Procedia Comput. Sci. **32**, 285–292 (2014)
8. Vapnik, V.: Statistical Learning Theory. Wiley, New York (1998)
9. Castro-Neto, M., Jeong, Y., Jeong, M.K., Han, L.D.: AADT prediction using support vector regression with data-dependent parameters. Expert Syst. Appl. **36**, 2979–2986 (2009)
10. Bhattacharya, A., Kumar, S.A., Tiwari, M., Talluri, S.: An intermodal freight transport system for optimal supply chain logistics. Transp. Res. Part C Emerg. Technol. **38**, 73–84 (2014)
11. Hwang, W.-Y., Lee, J.-S.: A new forecasting scheme for evaluating long-term prediction performances in supply chain management. Int. Trans. Oper. Res. **21**, 1045–1060 (2014)
12. Marković, N., Milinković, S., Tikhonov, K.S., Schonfeld, P.: Analyzing passenger train arrival delays with support vector regression. Transp. Res. Part C Emerg. Technol. **56**, 251–262 (2015)

13. Karlaftis, M.G., Vlahogianni, E.I.: Statistical methods versus neural networks in transportation research: Differences, similarities and some insights. Transp. Res. Part C Emerg. Technol. **19**, 387–399 (2011)
14. Peng, W.Y., Chu, C.W.: A comparison of univariate methods for forecasting container throughput volumes. Math. Comput. Model. **50**, 1045–1057 (2009)
15. Tian, X., Liu, L., Lai, K.K., Wang, S.: Analysis and forecasting of port logistics using TEI@I methodology. Transp. Plan. Technol. **36**, 669–684 (2013)
16. Xie, G., Wang, S., Zhao, Y., Lai, K.K.: Hybrid approaches based on LSSVR model for container throughput forecasting: a comparative study. Appl. Soft Comput. **13**, 2232–2241 (2013)
17. Geng, J., Li, M.-W., Dong, Z.-H., Liao, Y.-S.: Port throughput forecasting by MARS-RSVR with chaotic simulated annealing particle swarm optimization algorithm. Neurocomputing. **147**, 239–250 (2015)
18. Ruiz-Aguilar, J.J., Turias, I.J., Jiménez-Come, M.J.: A two-stage procedure for forecasting freight inspections at Border Inspection Posts using SOMs and support vector regression. Int. J. Prod. Res. **53**, 2119–2130 (2015)
19. Schölkopf, B., Smola, A.J.: Learning with Kernels: Support Vector Machines, Regularization, Optimization, and Beyond. MIT Press, Cambridge (2002)
20. Bergmeir, C., Hyndman, R.J., Benítez, J.M.: Bagging exponential smoothing methods using STL decomposition and Box-Cox transformation. Int. J. Forecast. **32**, 303–312 (2016)
21. Fisher, R.A.: Statistical Methods and Scientific Inference. Hafner Publishing Co, Oxford (1956)
22. Kourentzes, N., Barrow, D.K., Crone, S.F.: Neural network ensemble operators for time series forecasting. Expert Syst. Appl. **41**, 4235–4244 (2014)
23. Schapire, R.E.: The strength of weak learnability. Mach. Learn. **5**, 197–227 (1990)
24. Freund, Y., Schapire, R.E.: Experiments with a new boosting algorithm. In: Proceedings of the ICML, pp. 148–156 (1996)
25. Breiman, L.: Bagging predictors. Mach. Learn. **24**, 123–140 (1996)
26. Moscoso-López, J.A., Ruiz-Aguilar, J.J., Turias, I., Cerbán, M., Jiménez-Come, M.J.: A comparison of forecasting methods for ro-ro traffic: a case study in the strait of gibraltar. In: Proceedings of the Ninth International Conference on Dependability and Complex Systems DepCoS-RELCOMEX, 30 June–4 July, 2014, Brunów, Poland, pp. 345–353. Springer (2014)
27. Ruiz-Aguilar, J.J., Turias, I.J., Moscoso-López, J.A., Come, M.J.J., Cerbán, M.M.: Forecasting of short-term flow freight congestion: A study case of Algeciras Bay Port (Spain) (2016). http://www.revistas.unal.edu.co/index.php/dyna/article/view/47027

Current Research Trends in Robot Grasping and Bin Picking

Marcos Alonso[1], Alberto Izaguirre[2], and Manuel Graña[1(✉)]

[1] Computational Intelligence Group, CCIA Department, UPV/EHU, Paseo Manuel de Lardizabal 1, 20018 San Sebastian, Spain
malonso117@ikasle.ehu.es, ccpgrrom@sc.ehu.es

[2] CIS and Electronics Department, University of Mondragon, Loramendi Kalea 5, 20500 Mondragon, Spain
aizagirre@mondragon.edu

Abstract. We provide a view of current research issues in Robotic Grasping and Bin Picking focused on the perception aspects of the problem, mainly related to computer vision algorithms. After recalling the evolution of the topics in the last decades, we focus on the modern use of Deep Learning Algorithms. Two main trends are followed in the approaches to innovative grasping techniques. First, Convolutional Neural Networks are used for grasping perceptual aspects. We discuss the different degrees of success of several published approaches. Second, Deep Reinforcement Learning is being extensively tested in order to develop integrated eye-hand coordination systems not requiring delicate calibration. We provide also a discussion of possible future lines of research.

Keywords: Robot grasping · Bin Picking · Deep Learning
Convolutional Neural Networks · Robot gripper · Robot planning

1 Introduction

Grasping is the act of taking hold of some object by some kind of manipulator, usually the human hand. Robot grasping is the act of grasping by a robotic manipulator, which may have quite different designs, some of them inspired on the human hand, hence we talk of robotic hands, or on other means of taking hold of things. Grasping is often referred as eye-hand coordination, meaning that the perception (eye) drives the motion of the manipulation end effector. Grasping has been a hot topic in robotics and automation for a long time. Central to robot grasping is the task of calculating the contact points of an object to ensure a stable holding or manipulation by a robotic hand. Also, the design and construction of robotic hands is a critical issue in robot because different hands lead to quite different grasping strategies and computational needs.

Bin picking is the action of taking objects from a bin (a box with its top side open) for subsequent manipulation or positioning in a different place and pose.

M. Graña et al. (Eds.): SOCO'18-CISIS'18-ICEUTE'18, AISC 771, pp. 367–376, 2019.
https://doi.org/10.1007/978-3-319-94120-2_35

Bin picking is a central task in many industrial processes, so its automation has a great economical impact and many research efforts have been pursued following diverse directions aiming to find specific or general solutions to the bin picking problem. For example, initial implementations of this system used mechanical vibratory feeders as sensors [10]. Advances in bin picking have come from many diverse technological areas, including computational science and new hardware designs for grasping, manipulation as well as perception and sensing. For instance, advances in computer vision have had strong impact on the automation of bin picking. However, bin picking in the most general situations remains a challenging task. The aim is always to increase system flexibility and robustness.

Most current industrial bin picking systems plan in advance a set of predefined grasp points for a known object CAD-Model and use perception to estimate the pose of those identified objects in order to pick them with a specifically built gripper. Nowadays research deals with unknown parts in clutter environments and usage of learning algorithms to successfully grasp them. Although some bin picking scenarios have been successfully solved, the general problem remains open. Machine Perception defined as the capability of a computer vision system to interpret visual data acquired by specific hardware (e.g. RGB-D Sensors) is a critical issue in the evolution of bin picking and grasping systems. Robust and flexible perception provides means to achieve increasingly general task solutions. Recently, artificial neural networks have been rediscovered in the form of deep learning architectures, specially the convolutional neural networks [17] are being successfully applied to many computer vision tasks achieving results that are significantly above the previous state of the art. The main distinctive feature of deep learning architectures is that they extensively exploit hierarchical structures of weights that can be mapped to a hierarchical structure of representations of increasing levels of abstraction [11].

Subsequent to this introduction, the outline of the paper is as follows: Sect. 2 introduces grasping from a historical point of view. Section 3 introduces the topic of perception. Section 4 presents the current works on Deep Learning applied to grasping. Finally, a discussion of actual work is given in Sect. 5, where we also include some conclusions and ideas for future work.

2 Robot Grasping

The perception tasks for robot grasping involves finding the optimal grasp points of an object to carry out a specific manipulation task. Mason and Salisbury [22] and Asada [2] pioneered the field of robotic grasping using a three finger hand.

Two different grasping approaches can be identified,

- dexterous manipulation, i.e. manipulation with fingers, and
- enveloping grasps, i.e. formed by wrapping both fingers and the palm around the object.

Enveloping grasps have superior stability than dexterous manipulation, thus most industrial grippers are designed to do enveloping grasps, and there are few examples of dexterous manipulation in industrial environments. Bicchi and Kumar [5] provide an excellent survey covering the theoretical conditions for a stable grasp. Most of the literature ignores the kinematics of the fingers or the joints involved in the contact with the object. Trinkle et al. [29] was the first to study the kinematics of two or three fingers with palms. However, the optimal design of robots hands to increase grasp robustness and decrease hardware complexity remains an open problem still.

As the theoretical research work on grasping still lacks the maturity for industrial implementations, the group of Allen carried out extensive experimental research of real grasping tasks, computing the quality of them [23]. This work has progressed into a Grasp Simulator (GraspIt! [24]) that has been made publicly available to other researchers. Figure 1 shows some of the diverse body specification possible in GraspIt!, and the formal definition of the data types available to build simulations.

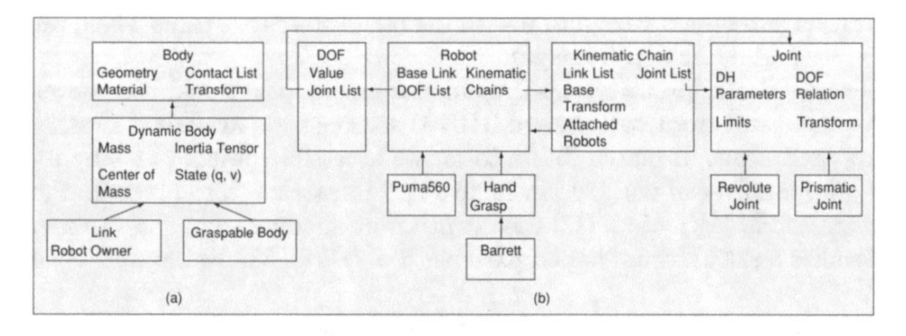

Fig. 1. (a) The body types defined within GraspIt! and their associated data. (b) The robot class definition and its associated data types

Although many research groups have used anthropomorphic type of grippers (i.e. mechanisms for grasping), Bicchi [4] states that other grasping hands can be built for specific applications, increasing the grasp robustness and decreasing the difficulty of their design and control. A survey of multi-finger control grasping and manipulation has been made by Yoshikawa [33], concluding that still many control problems remain unsolved, and the tremendous skill and versatility of human hands will continue to be a great challenge for researchers of new robot hands.

3 Machine Perception in Bin Picking

Historically, visual feedback was introduced in automated bin picking in the 1960s with the works of Roberts [27], extracting edges from 2D monocular camera images with the aim of locating polyhedral parts (Roberts Cross Edge Detector). In order to obtain the 3D pose (position and orientation of a part with respect to a known frame, e.g. robot or camera frame) other techniques were developed during the 1970s. Passive stereo vision allowed 3D reconstruction from two images after matching them and computing the point disparities [3]. Further research works have employed image rectification based on the epipolar line criteria, making the stereo matching a one dimensional problem [6]. Other techniques based on active stereo vision used laser emitters or structured light projectors in order to simplify the stereo matching problem. Faugeras et al. [8] were able to grab and analyze 3D range data from complex parts representing them by linear primitives, such as points, lines, and planes. However, most of those 3D optical range sensors suffered from poor depth of view. Rioux et al. [26] designed a compact laser scanner based on Scheimpflug condition of the lens respect to the camera sensor to achieve a larger depth of view.

Finding the pose of an object from the 3D range data has been extensively researched. Chen et al. [32] developed the Interactive Closest Point algorithm (ICP) that calculates the alignment and best fitting of a cloud of points with respect to a reference model (CAD data). Based on this approach, several commercial implementations exists nowadays in the market that provide mapping, localization and object recognition with the use of RGB-D sensors which provide both color and dense depth images.

The very first approaches to solve bin picking algorithms had been developed by Dessimoz and Horn, well before RGB-D sensors were available. Dessimoz et al. proposed usage of match filters to locate accessible object parts by a XYZ robot with a universal parallel jaw hand Fig. 2 illustrates this approach. Parallel to this research, Horn et al. [13] used depth from shading techniques to segment and localize a set of torus shaped parts with a PUMA 560 Robot and a parallel grip.

Currently most bin picking algorithms rely on segmentation of RGB-D data. 3D object recognition is done matching the 3D data to their known CAD models. At the current state of the art, image segmentation is the main performance bottleneck. Drost et al. [7] proposed a method that creates a global model description using an oriented point pair feature (describes the relative position and orientation of two oriented points, Fig. 3). This model description represents a mapping from the feature space into the model description. In addition to it, a fast voting scheme very similar to the Generalized Hough Transform is used, improving the above mentioned ICP algorithm performance. When bin picking is restricted to only one type of object, these techniques provide quite satisfactory results in practice. However, in the general case, bin picking may involve hundred of classes of objects, increasing the recognition and learning computational time. Wohlkinger et al. [31] introduce a methodology for learning 3D descriptors from synthetic CAD models. Besides, they created 3DNet (3d-net.org) a free open source framework with a large-scale hierarchy of CAD-model databases and increasing number of object classes and scenarios. Their pose estimation algorithms use these descriptors and an ICP algorithm.

4 Deep Learning

Deep learning has historical roots going back decades, LeCunn et al. [17] or Krizhevsky et al. [15] generalized the backpropagation algorithm for training multilayer nets, which was the foundation of the Deep Learning revolution of the 2010s [18]. Deep learning architectures, such as deep neural networks, deep belief networks and recurrent neural networks, have been applied successfully to image characterisation, image explanation and tagging, speech recognition, translation, games, and the list is increasing everyday. Computer vision has been greatly improved in classification and labeling images [15]. Speech and voice recognition systems produce text in several languages better than humans [1]. AlphaZero [28] has outperformed the best Go and Chess players and machines after 24 h of training time, starting from knowledge of the game rules only.

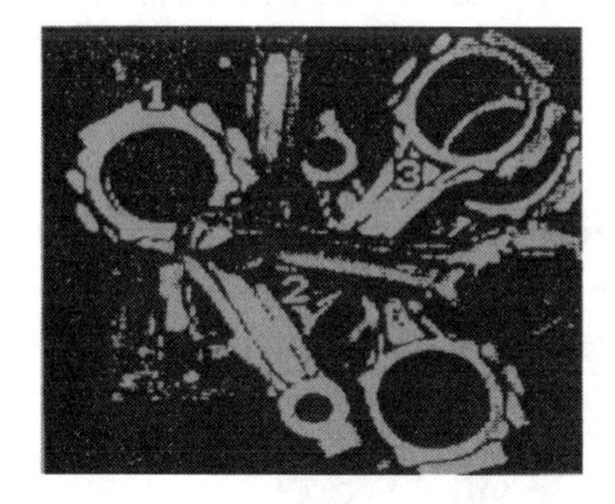

Fig. 2. Input gray level image (left) and parallel jaw gripper (right).

4.1 Deep Learning Methodologies for Bin Picking

As previously stated, traditional grasping methods have shown poor performance except in scenarios, in which the parts and grasping contact points are well defined. Lenz et al. [19] solved detection of robotic grasp points using a two different Deep Neural Networks. The first one evaluates a large number of grasping candidates pruning them in order to filter the best ones, which are fed into the second network. The second network uses a learning algorithm for multi-modal

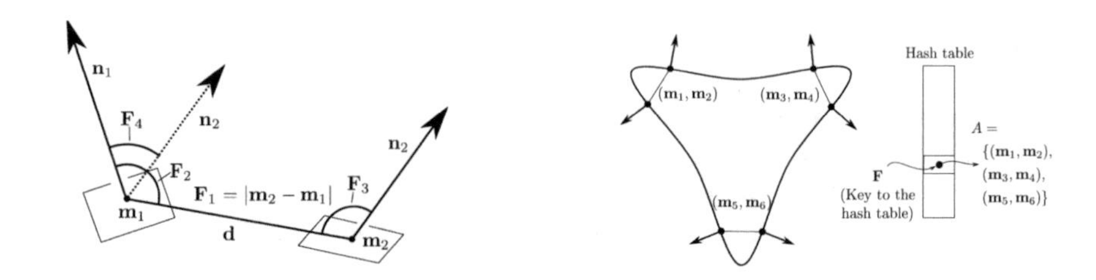

Fig. 3. Point pair feature F of two oriented points.

data based on group regularization. The experiments show that this approach outperform classical grasping methods. Varley et al. [30] presented a deep learning architecture for inducing the robot palm and fingertip stable grasp positions from partial object views obtained using an RGB-D sensor. They used the above mentioned GraspIt! Simulator achieving better results in the Grasp Quality Calculation in cases of incomplete object views. Gualtieri et al. [12] found precision grasp poses in densely cluttered scenarios using a parallel finger gripper tool. They used a Convolutional Neural Networks (CNN) training it in two steps; in the first one idealized CAD-Models are used together with simulated RGB-D data. The results of this pre-training are exploited afterwards in the second step to train the CNN together with real data. The grasp success reaches a 93% rate. Laskey et al. [16] trained a Deep Neural Network in order to grasp a desired object in a cluttered situation where other objects obstruct both visibility and motion. The learning process is based on demonstration by a human supervisor, reaching a successful grasp rate of 90%. Figure 4 illustrates a process of roll-out of clutter object to get access to the target object in a 2D environment.

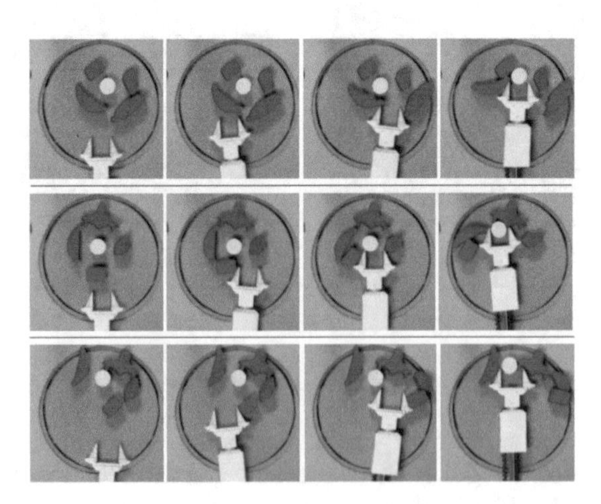

Fig. 4. Zymark 3-DOF Robot used by Laskey et al. [16] roll-out red objects in a cluttered environment to grasp the yellow on

Pinto et al. [25] trained a CNN network, increasing the amount of data reducing the overfitting problems occurred in previous experiments. The learned CNN has good generalization performance for novel objects. Levine et al. [20] employed fourteen robotic manipulators and monocular cameras for collecting over eight hundred grasp attempts per robot to learn a CNN grasp prediction model for a hand eye coordination. The approach required no previous robot to camera calibration (Hand Eye). The method demonstrated that it can effectively grasp a wide range of different objects including novel ones. Finally Mahler et al. [21] trained a Grasp Quality CNN that predicts the probability grasp success from depth images. A synthetic database of 3D Models was used for training giving a 93% success rate. It also had a 99% precision success rate on a set of 40 novel objects.

5 Discussion and Future Works

Advocates of Deep Learning methodologies strongly believe that they are the panacea to solve all problems in Bin Picking, by learning from multiple robot grasping attempts, having as input the RBG-D data. In the presented robot grasping experiments illustrated in Fig. 5 using Deep Learning, the scenarios have consisted mostly of common daily used objects (usually office, home and lab objects) placed on flat surface or bins with wide open walls, using parallel finger grippers.

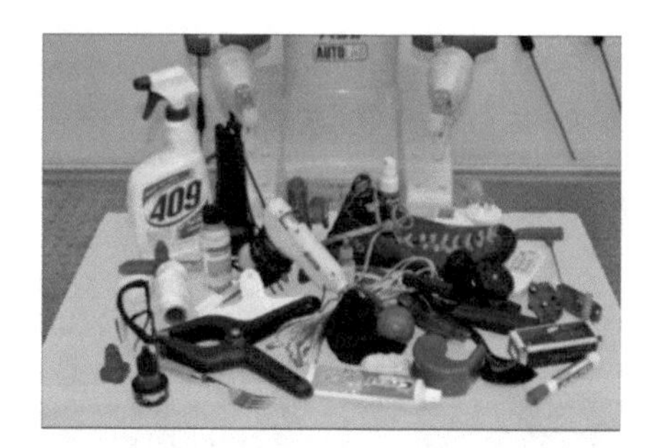

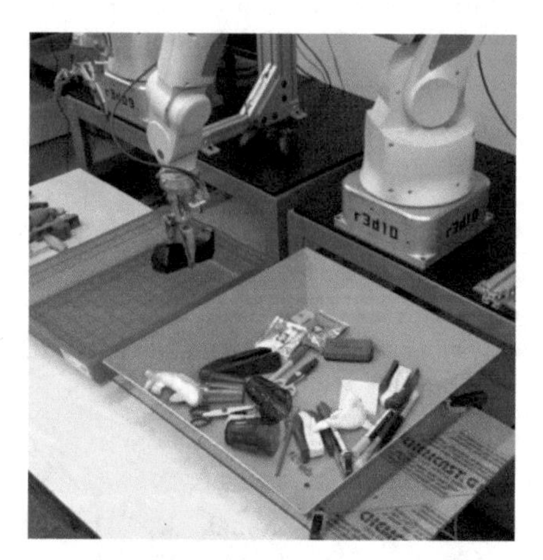

Fig. 5. Dex-Net 2.0 [21] Train and novel objects, and open wall container in Levine's experiments [20]

The experiments have consisted on learning the best contact points of the observed objects in order to find the most stable grasps. However, the manipulation capability (defined as a function of both the task execution waypoints and the object grasping contact points) of these estimated poses have not been taken into account. It may happen that a grasp position could be very stable but not reachable by the gripper or robot arm because it interferes with other parts or the bin walls. Ghalamzan et al. [9] address this problem defining task relevant kinematic manipulability measure defined as follows:

$$m(J) = \sqrt{\det(JJ^T)} = (\lambda_1 \lambda_2 ... \lambda_n)^{\frac{1}{2}} \quad (1)$$

Where, λ_j is j^{th} eigenvalue of JJ^T, and n denotes the dimension of output-space. This function is used to calculate the manipulability along a trajectory, including the joint limits and obstacle penalty functions, e.g bin wall can be avoid using these penalty functions. The work concludes that this approach to computing grasping contact points results in solutions that are both good at grasping parts and generating trajectories with good manipulability.

To our best knowledge, Deep Learning hasn't yet take into account grasping points, manipulability and penalty functions simultaneously. In a classical approach, Khatib et al. [14] proposal forcing strategies for cooperative tasks could be incorporated into Deep Learning loss functions in order to train the network to avoid collisions. Additionally wrist force and torque sensors could be used as additional information data source to reinforce generation of good grasping decisions.

As previously stated by Bicchi [4] grasping hands design can be tailored for specific applications, increasing the grasp robustness and decreasing the complexity of their control. The Deep Learning algorithms should be able to identify the optimal hand among a given set of task specific hands. This way, the combination of grasping contact points, robot trajectories, data from force and torque sensors and chosen hand allows a more robust and stable grasping. It is possible that some parts of the Deep Learning process need human supervision, to ensure a higher success rate.

Finally, it may be possible that the Deep Learning approach classifies objects into classes, for which the same gripper, similar trajectory types and contact points can be determine. This way, a humans user could introduce new objects as belonging to a previously identified specific class.

References

1. Amodei, D., Anubhai, R., Battenberg, E., Case, C., Casper, J., Catanzaro, B., Chen, J., Chrzanowski, M., Coates, A., Diamos, G., Elsen, E., Engel, J., Fan, L., Fougner, C., Han, T., Hannun, A.Y., Jun, B., LeGresley, P., Lin, L., Narang, S., Ng, A.Y., Ozair, S., Prenger, R., Raiman, J., Satheesh, S., Seetapun, D., Sengupta, S., Wang, Y., Wang, Z., Wang, C., Xiao, B., Yogatama, D., Zhan, J., Zhu, Z.: Deep speech 2: end-to-end speech recognition in English and Mandarin. CoRR abs/1512.02595 (2015)
2. Asada, H.: Studies on prehension and handling by robot hands with elastic fingers. Ph.D. thesis, Kyoto University (1979)

3. Barnard, S.T., Thompson, W.B.: Disparity analysis of images. IEEE Trans. Pattern Anal. Mach. Intell. PAMI **2**(4), 333–340 (1980)
4. Bicchi, A.: Hands for dexterous manipulation and robust grasping: a difficult road toward simplicity. IEEE Trans. Robot. Autom. **16**(6), 652–662 (2000)
5. Bicchi, A., Kumar, V.: Robotic grasping and contact: a review. In: Proceedings of the IEEE International Conference on Robotics and Automation, vol. 1, pp. 348–353 (2000)
6. Deriche, R., Zhang, Z., Luong, Q.T., Faugeras, O.: Robust recovery of the epipolar geometry from an uncalibrated stereo rig. In: Proceedings of the European Conference on Computer Vision, vol. 800, pp. 567–576 (1994)
7. Drost, B., Ulrich, M., Navab, N., Ilic, S.: Model globally, match locally: efficient and robust 3D object recognition. In: Proceedings of the IEEE Computer Society Conference on Computer Vision and Pattern Recognition, pp. 998–1005 (2010)
8. Faugeras, O.D., Hebert, M.: The representation, recognition, and locating of 3-D objects. Int. J. Robot. Res. **5**(3), 27–52 (1986)
9. Ghalamzan, A.M., Mavrakis, N., Kopicki, M., Stolkin, R., Leonardis, A., E., A.M.G., Mavrakis, N., Kopicki, M., Stolkin, R., Leonardis, A.: Task-relevant grasp selection: a joint solution to planning grasps and manipulative motion trajectories. In: 2016 IEEE/RSJ International Conference on Intelligent Robots and Systems (IROS), pp. 907–914, October 2016
10. Ghita, O., Whelan, P.: A systems engineering approach to robotic bin picking, November 2008
11. Graña, M., Torrealdea, F.: Hierarchically structured systems. Eur. J. Oper. Res. **25**(1), 20–26 (1986). http://www.sciencedirect.com/science/article/pii/0377221786901104
12. Gualtieri, M., ten Pas, A., Saenko, K., Platt, R.: High precision grasp pose detection in dense clutter. In: 2016 IEEE/RSJ International Conference on Intelligent Robots and Systems (IROS), pp. 598–605, October 2016
13. Horn, B., Ikeuchi, K.: The mechanical manipulation of randomly oriented parts. Sci. Am. **251**(2), 100–111 (1984)
14. Khatib, O., Yokoi, K., Chang, K., Ruspini, D., Holmberg, R., Casal, A., Baader, A.: Force strategies for cooperative tasks in multiple mobile manipulation systems. In: International Symposium on Robotics Research (1995). https://doi.org/10.1007/978-1-4471-0765-1_37
15. Krizhevsky, A., Sutskever, I., Hinton, G.E.: ImageNet classification with deep convolutional neural networks. In: Advances in Neural Information Processing Systems, NIPS 2012, vol. 25, pp. 1–9 (2012)
16. Laskey, M., Lee, J., Chuck, C., Gealy, D., Hsieh, W., Pokorny, F.T., Dragan, A.D., Goldberg, K.: Robot grasping in clutter: using a hierarchy of supervisors for learning from demonstrations. In: 2016 IEEE International Conference on Automation Science and Engineering (CASE), pp. 827–834 (2016)
17. LeCun, Y., Bottou, L., Bengio, Y., Haffner, P.: Gradient-based learning applied to document recognition. Proc. IEEE **86**(11), 2278–2323 (1998)
18. LeCun, Y.A., Bengio, Y., Hinton, G.E.: Deep learning. Nature **521**(7553), 436–444 (2015)
19. Lenz, I., Lee, H., Saxena, A.: Deep learning for detecting robotic grasps. Int. J. Robot. Res. **34**(4–5), 705–724 (2015)
20. Levine, S., Pastor, P., Krizhevsky, A., Quillen, D.: Learning hand-eye coordination for robotic grasping with deep learning and large-scale data collection (2016)

21. Mahler, J., Liang, J., Niyaz, S., Laskey, M., Doan, R., Liu, X., Ojea, J.A., Goldberg, K.: Dex-Net 2.0: deep learning to plan robust grasps with synthetic point clouds and analytic grasp metrics (2017)
22. Mason, M., Salisbury, J.J.: Robot Hands and the Mechanics of Manipulation. The MIT Press, Cambridge (1985)
23. Miller, A.T., Allen, P.K.: Examples of 3d grasp quality computations. In: Proceedings of the 1999 IEEE International Conference on Robotics and Automation (Cat. No.99CH36288C), vol. 2, pp. 1240–1246 (1999)
24. Miller, A.T., Allen, P.K.: Graspit: a versatile simulator for robotic grasping. IEEE Robot. Autom. Mag. **11**(4), 110–122 (2004)
25. Pinto, L., Gupta, A.: Supersizing self-supervision: learning to grasp from 50K tries and 700 robot hours. In: Proceedings of the IEEE International Conference on Robotics and Automation, pp. 3406–3413, June 2016
26. Rioux, M., Bechthold, G., Taylor, D., Duggan, M.: Design of a large depth of view 3-dimensional camera for robot vision. Opt. Eng. **26**(12), 1245–1250 (1987)
27. Roberts, L.G.: Machine perception of three-dimensional solids. Ph.D. thesis (1965)
28. Silver, D., Schrittwieser, J., Simonyan, K., Antonoglou, I., Huang, A., Guez, A., Hubert, T., Baker, L., Lai, M., Bolton, A., Chen, Y., Lillicrap, T., Hui, F., Sifre, L., Van Den Driessche, G., Graepel, T., Hassabis, D.: Mastering the game of Go without human knowledge. Nature **550**(7676), 354–359 (2017)
29. Trinkle, J.C., Abel, J.M., Paul, R.P.: An investigation of frictionless enveloping grasping in the plane. Int. J. Robot. Res. **7**(3), 33–51 (1988)
30. Varley, J., Weisz, J., Weiss, J., Allen, P.: Generating multi-fingered robotic grasps via deep learning. In: IEEE International Conference on Intelligent Robots and Systems, pp. 4415–4420, December 2015
31. Wohlkinger, W., Aldoma, A., Rusu, R.B., Vincze, M.: 3DNet: large-scale object class recognition from CAD models. In: Proceedings of the IEEE International Conference on Robotics and Automation, pp. 5384–5391 (2012)
32. Yang, C., Medioni, G.: Object modelling by registration of multiple range images. Image Vis. Comput. **10**(3), 145–155 (1992)
33. Yoshikawa, T.: Multifingered robot hands: control for grasping and manipulation. Ann. Rev. Control **34**(2), 199–208 (2010)

Visualizing Industrial Development Distance to Better Understand Internationalization of Spanish Companies

Alfredo Jiménez[1] and Alvaro Herrero[2(✉)]

[1] Department of Management, KEDGE Business School, Bordeaux, France
alfredo.jimenez@kedgebs.com

[2] Grupo de Inteligencia Computacional Aplicada (GICAP), Departamento de Ingeniería Civil, Escuela Politécnica Superior, Universidad de Burgos, Av. Cantabria s/n, 09006 Burgos, Spain
ahcosio@ubu.es

Abstract. The analysis of bilateral distance between home and host countries is a key issue in the internationalization strategy of companies. As a multi-faceted concept, distance encompasses multiple dimensions, with psychic distance being one of the most critical ones for the overseas investments of firms. Among all the psychic distance stimuli that have been proposed until now, the present paper focuses on Industrial Development Distance (IDD). Together with data from both the countries and the companies, IDD is analysed by means of neural projection models based on unsupervised learning, to gain deep knowledge about the internationalization strategy of Spanish large companies. Informative projections are obtained from a real-life dataset, leading to useful conclusions and the identification of those destinations attracting large flows of investment but with a particular idiosyncrasy.

Keywords: Artificial neural networks · Unsupervised learning
Exploratory projection · Industrial development distance · Internationalization

1 Introduction and Previous Work

Managing international operations is a critical component of many companies' strategy nowadays. Understanding how the differences in the idiosyncrasies of the host countries compared to the home countries may disrupt the activities of the firms or, on the contrary, be sometimes a source of opportunities, is a critical challenge faced by those companies daring to invest abroad. This pivotal role of distance between countries is so critical that the whole domain of international management has been argued to fundamentally rely on it. As Zaheer and colleagues point out, "*essentially, international management is management of distances*" [1].

Distance has been conceptualized in recent works as a multi-faceted construct [2]. As a result, different frameworks have been proposed to account for the multiple dimensions of distance which may have an impact on firm operations overseas. Thus, [3] proposed the famous CAGE framework in which distance is disentangled in cultural, administrative, geographic and economic. [4] go beyond and propose nine

M. Graña et al. (Eds.): SOCO'18-CISIS'18-ICEUTE'18, AISC 771, pp. 377–386, 2019.
https://doi.org/10.1007/978-3-319-94120-2_36

dimensions to measure cross-national distance, namely: economic, financial, political, administrative, cultural, demographic, knowledge, global connectedness and geographic. In turn, these dimensions of distance have also been dissected into components or dimensions to better understand this complex phenomenon. For instance, the seminal work of [5] proposed four dimensions of cultural distance (power distance, individualisms, uncertainty avoidance, and masculinity), which were later complemented with two more (long-term orientation and indulgence) in subsequent works [6]. This framework, and its operationalization in a single construct by [7] picked up huge popularity as measured by the number of citations these works accumulate.

However, recently, the literature has started to point out that "*cultural dimensions and measures do not fully capture psychic distance, which is really the key parameter affecting many managerial choices in an international business*" [8]. Psychic distance is a broader framework encompassing cultural but also other dimensions of distance [9, 10]. Given the difficulty of assessing a manager's perceptions just at the moment of making a decision, or avoiding the causality problems due to *ex-post* perceptions being biased by the final outcome of the decision, [9] propose six macro-level factors, known as psychic distance stimuli, that shape the context in which perceptions are formed [11]. Specifically, these stimuli measure national differences in education, industrial development, language, democracy, social system, and religion that make the flow of information from and to the host market harder [9, 12]. The real relevance of these stimuli has been demonstrated in empirical studies which have showed a significant impact on performance, online internationalization, market selection, entry mode choice, trade flows, and Foreign Direct Investments (FDI) [13].

Many of these studies, however, continue the tradition of aggregating several stimuli into a single construct, despite various authors arguing that this procedure may lead to erroneously think that all the components are equally [14, 15] significant. Aware of this criticism, in this paper we focus on only one stimulus, namely Industrial Development Distance (IDD), to concentrate on the individual effects irrespective of the potential confounding effects of the other stimuli. While we acknowledge that all stimuli can have an important role, we focus on industrial development as it is arguably one of the most important factors affecting the majority of firms regardless their size, sector and experience as this indicator is calculated based on differences in GDP, the consumption of energy, vehicle ownership, employment in agriculture, the number of telephones, radios and televisions, and percentage of non-agricultural labour. As a consequence, difference in industrial development can have a critical impact on FDIs, especially those following an efficiency-seeking approach, i.e. those looking to increase efficiency and cost reduction [16].

Artificial Neural Networks have been applied to many different fields in a wide variety of systems for quite a long time [17, 18, 19, 20, 21]. In present paper, internationalization data are visualized through three different neural projection methods. Differentiating from previous work, in present paper they are applied to internationalization data for the first time. Additionally, this paper goes one step further, enriching the visualization with IDD information to ease the analysis of the dataset.

The rest of this paper is organized as follows: the applied neural methods are described in Sect. 2, the setup of experiments and the dataset under analysis are described in Sect. 3, together with the results obtained and the conclusions of present study that are stated in Sect. 4.

2 Neural Visualization

This work proposes the application of unsupervised neural models for the visualization of internationalization data. Visualization techniques are considered a viable approach to information seeking, as humans are able to recognize different features and to detect anomalies by means of visual inspection. The underlying operational assumption of the proposed approach is mainly grounded in the ability to render the high-dimensional data in a consistent yet low-dimensional representation.

This problem of identifying patterns that exist across dimensional boundaries in high dimensional datasets can be solved by changing the spatial coordinates of data. However, an a priori decision as to which parameters will reveal most patterns requires prior knowledge of unknown patterns.

Projection methods project high-dimensional data points onto a lower dimensional space in order to identify "interesting" directions in terms of any specific index or projection. Having identified the most interesting projections, the data are then projected onto a lower dimensional subspace plotted in two or three dimensions, which makes it possible to examine the structure with the naked eye.

2.1 Principal Component Analysis

Principal Component Analysis (PCA) is a well-known statistical model, introduced in [22], that describes the variation in a set of multivariate data in terms of a set of uncorrelated variables each, of which is a linear combination of the original variables.

From a geometrical point of view, this goal mainly consists of a rotation of the axes of the original coordinate system to a new set of orthogonal axes that are ordered in terms of the amount of variance of the original data they account for. PCA can be performed by means of neural models such as those described in [23] or [24]. It should be noted that even if we are able to characterize the data with a few variables, it does not follow that an interpretation will ensue.

2.2 Maximum Likelihood Hebbian Learning

Maximum Likelihood Hebbian Learning [25] is based on Exploration Projection Pursuit (EPP). The statistical method of EPP was designed for solving the complex problem of identifying structure in high dimensional data by projecting it onto a lower dimensional subspace in which its structure is searched for by eye. To that end, an "index" must be defined to measure the varying degrees of interest associated with each projection. Subsequently, the data is transformed by maximizing the index and the associated interest.

MLHL is a family of learning rules that is based on maximizing the likelihood of the residual from a negative feedback network whenever such residuals are deemed to come from a distribution in the exponential family. The main advantage of this model is that by maximizing the likelihood of the residual with respect to the actual distribution, we are matching the learning rule to the probability density function of the residual by applying different values of the p parameter specified in the learning rule.

2.3 Cooperative Maximum Likelihood Hebbian Learning

The Cooperative MLHL (CMLHL) model [26] extends the MLHL model, by adding lateral connections between neurons in the output layer of the model. These lateral connections are derived from the Rectified Gaussian Distribution (RGD) [27], that is a modification of the standard Gaussian distribution in which the variables are constrained to be non-negative, enabling the use of non-convex energy functions. In a more precise way, CMLHL includes lateral connections based on the mode of the cooperative distribution that is closely spaced along a nonlinear continuous manifold. By including these lateral connections, the resulting network can find the independent factors of a dataset in a way that captures some type of global ordering in the dataset.

Considering an N-dimensional input vector (x), and an M-dimensional output vector (y), with W_{ij} being the weight (linking input j to output i), then CMLHL can be expressed as defined in Eqs. 1, 2, 3 and 4.

1. Feed-forward step:

$$y_i = \sum_{j=1}^{N} W_{ij} x_j, \forall i \tag{1}$$

2. Lateral activation passing:

$$y_i(t+1) = [y_i(t) + \tau(b - Ay)]^+ \tag{2}$$

3. Feedback step:

$$e_j = x_j - \sum_{i=1}^{M} W_{ij} y_i, \forall j \tag{3}$$

4. Weight change:

$$\Delta W_{ij} = \eta . y_i . sign(e_j) |e_j|^{p-1} \tag{4}$$

Where: η is the learning rate, τ is the "strength" of the lateral connections, b the bias parameter, p a parameter related to the energy function and A a symmetric matrix used to modify the response to the data. The effect of this matrix is based on the relation between the distances separating the output neurons.

3 Experiments and Results

As previously mentioned, some neural visualization models (see Sect. 2) have been applied to visualize IDD on a dataset gathering information about internationalization. Present section introduces the analyzed dataset as well as the main obtained results.

3.1 Dataset

The dataset analysed in present study is based on a sample of all Spanish MNEs registered with the Foreign Trade Institute (ICEX) and from the website www.oficinascomerciales.es, both managed by the Spanish Ministry of Industry, Tourism, and Trade. In order to analyze a representative sample of companies with sufficient autonomy, we restricted the sample to keep only those large and independent enough to conduct and decide their own internationalization strategy. Thus, following a well-established cut-off point in International Business literature, we dropped from the sample those with less than 250 employees. We also dropped those companies with a foreign majority owner controlling more than half of the capital.

It is also important to note the huge impact of the financial crisis on the Spanish economy, which forced many multinational enterprises to sell or postpone international operations in order to focus on the problems of the home market. To avoid distortions in the results due to this exogenous effect, we took the year 2007 as our base year. Overall, the sample consists of 164 companies investing in 119 countries worldwide. Unfortunately, Afghanistan, Andorra, Puerto Rico, and São Tomé and Príncipe are not included in the sample due to a lack of data. In addition, Serbia, Montenegro, and Kosovo are included as a group because at the time of the study they constituted a single country.

For the companies and countries above mentioned, the following data about each one of the cases of international presence were collected (further details about the different features can be found in [13]):

- Company sector: 5 binary features stating the economy sector the company belongs to (manufacturing, food, construction, regulated and others).
- Company product diversification: 3 binary featuring (non-diversified, related or unrelated diversification).
- Other company characteristics: Assets, number of employees, Return on Assets (ROA), ROA growth, age, number of countries where the company operates, leverage and whether or not the company is included in a stock market.
- Host country characteristics: GDP, GDP growth, total inward FDI, population, unemployment, level of corruption and Economic Freedom Index.
- Geographic and psychic distance stimuli between home and hot countries [9]. The education distance stimulus is based on differences on literacy rate and enrolment in second and third-level education. The industrial development stimulus takes into account differences in ten dimensions such as in energy consumption, vehicle ownership, employment in agriculture, number of telephones and televisions, etc. The language stimulus is based on the differences between the dominant languages and the bilateral influence of each country's major language in the other country.

The democracy stimulus includes differences in political rights, civil liberties and POLCON and POLITY IV indices. The political ideology stimulus is based on the ideological leanings of the chief executive's political party and the largest political party in the government. Finally, the religion stimulus is calculated based on the differences between the dominant religions and the bilateral influence of each country's dominant religion in the other country.

As a result, a dataset containing 1456 samples and 33 features was obtained and the obtained projections are presented in the following subsection.

3.2 Results

For comparison purposes, three different projection models have been applied, whose results are shown below. In all of them, the IDD has been added to the projection by means of the glyph metaphor. In order to do that, every sample (international presence) is depicted in a certain way, according to the quartile of the original IDD value: ✳ 1st quartile, ○ 2nd quartile, ▷ 3rd quartile, and + 4th quartile.

PCA Projection

Figure 1 shows the principal component projection, obtained by applying PCA to the previously described data, and combining the two principal components. It can be seen that data are split in several different groups; some of them are labeled (1–4).

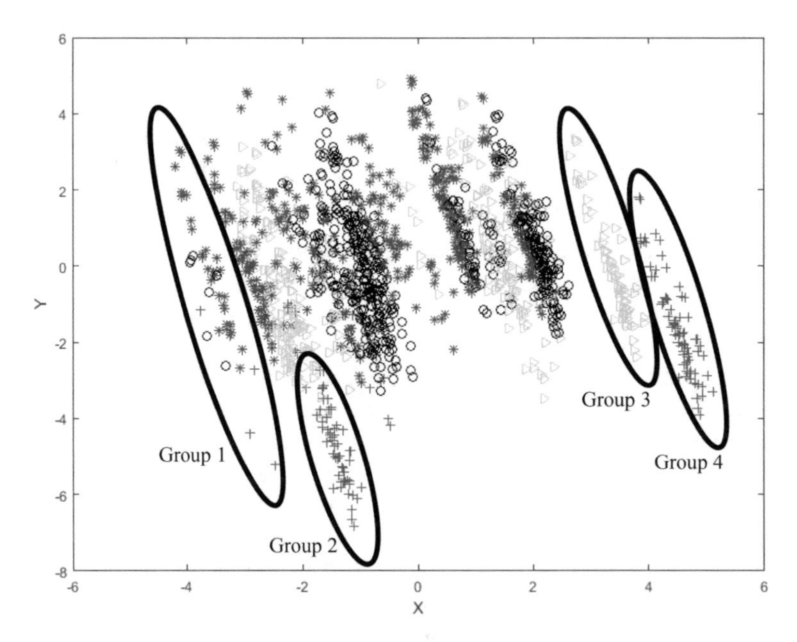

Fig. 1. PCA projection of the dataset, visualizing IDD.

A thorough analysis of the most representative groups of PCA projection has been carried out. As a result, it can be said that most samples with highest values of IDD are placed in groups 2 and 4, at the bottom of the projection. The samples that are closest to these two groups are all of them from the 2nd quartile of IDD. Group 1 comprises investments in under-developed countries such as Venezuela, Bangladesh and Nigeria. Group 2 is entirely made of all investments in China. While this country has become one of the top recipients of FDI in the world due to its emerging economy, its particular idiosyncrasy makes it a distinct group and the large IDD value differentiates it from the other groups. Group 3 is also entirely made of all investments in one country, in this case UK. While being one of the top destinations for Spanish FDI flows, this country is singled out due to its significant differences compared to the rest of main destinations in the EU. Finally, Group 4 comprises all investments in the US. As the largest economy of the word, this country started to be a main destination for Spanish FDI flows since the beginning of the 21st century [28], but, it also exhibits the largest IDD score as it appears to be perceived as much more advanced.

MLHL Projection

Figure 2 shows the IDD-enriched projection on the two main components identified by MLHL from the analyzed data. The parameter values of the MLHL model for the projection shown in Fig. 2 are: number of output dimensions: 3, number of iterations: 3000, learning rate: 0. 08009, p: 0.54.

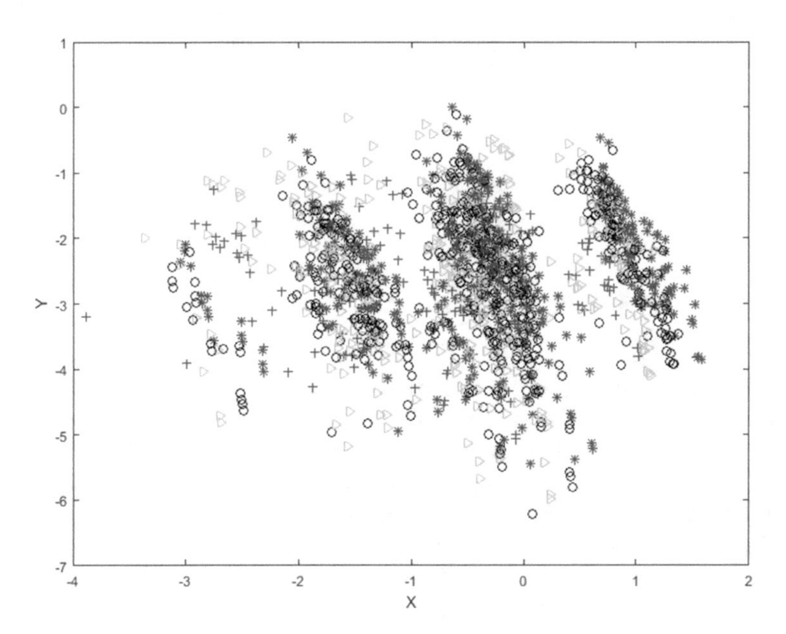

Fig. 2. MLHL projection of the dataset, visualizing IDD.

As the MLHL projection does not reveal a clear organization of samples according to IDD, and for the sake of brevity, this projection is not further studied in present paper.

CMLHL Projection

When applying CMLHL to the analysed dataset, the projection (on the two main components) shown in Fig. 3 has been obtained and IDD information has been added. The parameter values of the CMLHL model for the projections shown in Fig. 3 are; number of output dimensions: 3, number of iterations: 3000, learning rate: 0.000175, p: 1.96, τ: 0.034.

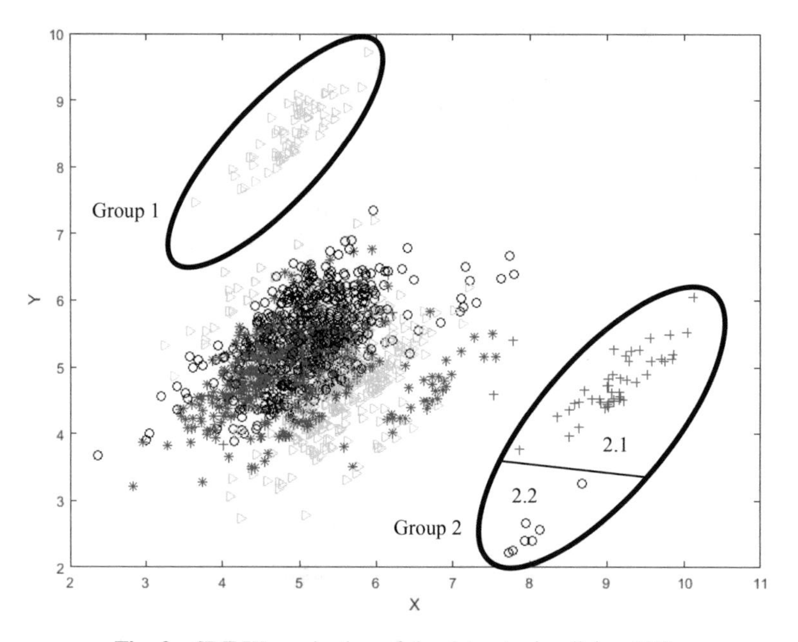

Fig. 3. CMLHL projection of the dataset, visualizing IDD.

In Fig. 3 it is easy to visually identify some groups of data; the most distant ones are labeled as 1 and 2. Group 1 displays all investments in UK, as the previously described Group 3 in the PCA projection. Group 2.1 gathers all investments in China, as the previously described Group 2 in the PCA. Finally, Group 2.2 displays all investments in Serbia, Montenegro, and Kosovo. This group may be unexpected, especially since the companies investing in this country are quite heterogeneous (infrastructures, shoes, clothing, …). While we can speculate maybe it can be due to the specific situation previous to the split of this territory into multiple countries short after, further research is needed to disentangle this finding.

4 Conclusions and Future Work

From the projections in Sect. 3, it can be concluded that neural projection models are an interesting proposal to visually analyse internationalization data in order to better understand it. More specifically, when visualizing IDD, neural projections let us gain deep knowledge about the nature of such dataset.

After the analysis of the projections and the associated allocation of international presence, it can be said that these techniques are useful to identify those destinations attracting a large share of subsidiaries from multinational enterprises but at the same time exhibiting a distinct environment (e.g., China, UK, US) compared to other more homogenous regions where Spanish FDI is also prominent (e.g., UE, Latin America).

This finer-grained analysis would allow firms to obtain a more nuanced understanding on the specific characteristics and idiosyncrasies of the foreign markets where they are currently conducting operations as well as other potential host markets where they might be considering as potential destinations for investment. By signaling those host countries exhibiting some particular differences compared to other alternative destinations, managers will be able to make better location-decision choices and will be more aware of the need to get familiarized with the specificities of the local market.

In future work, some other neural visualization models will be applied to the same dataset to better understand its nature and gaining deep knowledge of the internationalization strategy of Spanish companies.

Acknowledgments. The work was conducted during Álvaro Herrero's research stay at KEDGE Business School in Bordeaux.

References

1. Zaheer, S., Schomaker, M.S., Nachum, L.: Distance without direction: restoring credibility to a much-loved construct. J. Int. Bus. Stud. **43**, 18–27 (2012)
2. Jiménez, A., Benito-Osorio, D., Puck, J., Klopf, P.: The multi-faceted role of experience dealing with policy risk: the impact of intensity and diversity of experiences. Int. Bus. Rev. **27**, 102–112 (2018)
3. Ghemawat, P.: Distance still matters. Harv. Bus. Rev. **79**, 137–147 (2001)
4. Berry, H., Guillén, M.F., Zhou, N.: An institutional approach to cross-national distance. J. Int. Bus. Stud. **41**, 1460–1480 (2010)
5. Hofstede, G.: Culture's consequences: international differences in work-related values. Sage Publications, Beverly Hills (1980)
6. Hofstede, G., Hofstede, G.J., Minkov, M.: Cultures and organizations: software of the mind. McGraw-Hill, New York (2010)
7. Kogut, B., Singh, H.: The effect of national culture on the choice of entry mode. J. Int. Bus. Stud. **19**, 411–432 (1988)
8. Tung, R.L., Verbeke, A.: Beyond Hofstede and globe: improving the quality of cross-cultural research. J. Int. Bus. Stud. **41**, 1259–1274 (2010)
9. Dow, D., Karunaratna, A.: Developing a multidimensional instrument to measure psychic distance stimuli. J. Int. Bus. Stud. **37**, 575–577 (2006)

10. Håkanson, L., Ambos, B.: The antecedents of psychic distance. J. Int. Manag. **16**, 195–210 (2010)
11. Brewer, P.A.: Operationalizing psychic distance: a revised approach. J. Int. Mark. **15**, 44–66 (2007)
12. Johanson, J., Vahlne, J.-E.: The internationalization process of the firm: a model of knowledge development and increasing foreign market commitments. J. Int. Bus. Stud. **8**, 23–32 (1977)
13. Jiménez, A., de la Fuente, D.: Learning from others: the impact of vicarious experience on the psychic distance and FDI relationship. Manag. Int. Rev. **56**, 633–664 (2016)
14. Shenkar, O.: Cultural distance revisited: towards a more rigorous conceptualization and measurement of cultural differences. J. Int. Bus. Stud. **32**, 519–536 (2001)
15. Kirkman, B.L., Lowe, K.B., Gibson, C.B.: A quarter century of culture's consequences: a review of empirical research incorporating Hofstede's cultural values framework. J. Int. Bus. Stud. **37**, 285–320 (2006)
16. Dunning, J.H.: Multinational enterprises and the global economy. Addison-Wesley, Wokinham (1993)
17. Li, E.Y.: Artificial neural networks and their business applications. Inf. Manag. **27**, 303–313 (1994)
18. Hill, T., Marquez, L., O'Connor, M., Remus, W.: Artificial neural network models for forecasting and decision making. Int. J. Forecast. **10**, 5–15 (1994)
19. Zhang, G., Hu, M.Y., Patuwo, B.E., Indro, D.C.: Artificial neural networks in bankruptcy prediction: general framework and cross-validation analysis. Eur. J. Oper. Res. **116**, 16–32 (1999)
20. Garcia, R.F., Rolle, J.L.C., Gomez, M.R., Catoira, A.D.: Expert condition monitoring on hydrostatic self-levitating bearings. Expert Syst. Appl. **40**, 2975–2984 (2013)
21. Herrero, Á., Corchado, E., Jiménez, A.: Unsupervised neural models for country and political risk analysis. Expert Syst. Appl. **38**, 13641–13661 (2011)
22. Pearson, K.: On lines and planes of closest fit to systems of points in space. Phil. Mag. **2**, 559–572 (1901)
23. Oja, E.: Principal components, minor components, and linear neural networks. Neural Netw. **5**, 927–935 (1992)
24. Fyfe, C.: A neural network for PCA and beyond. Neural Process. Lett. **6**, 33–41 (1997)
25. Corchado, E., MacDonald, D., Fyfe, C.: Maximum and minimum likelihood Hebbian learning for exploratory projection pursuit. Data Min. Knowl. Discov. **8**, 203–225 (2004)
26. Corchado, E., Fyfe, C.: Connectionist techniques for the identification and suppression of interfering underlying factors. Int. J. Pattern Recognit. Artif. Intell. **17**, 1447–1466 (2003)
27. Seung, H.S., Socci, N.D., Lee, D.: The rectified gaussian distribution. Adv. Neural. Inf. Process. Syst. **10**, 350–356 (1998)
28. Jiménez, A.: Does political risk affect the scope of the expansion abroad? evidence from Spanish MNEs. Int. Bus. Rev. **19**, 619–633 (2010)

Studying Road Transportation Demand in the Spanish Industrial Sector Through *k*-Means Clustering

Carlos Alonso de Armiño, Miguel Ángel Manzanedo, and Álvaro Herrero(✉)

Departamento de Ingeniería Civil, Escuela Politécnica Superior, Universidad de Burgos, Av. Cantabria s/n, 09006 Burgos, Spain
{caap,mmanz,ahcosio}@ubu.es

Abstract. Transportation is the economic activity that is the most tightly coupled with the other ones. As a result, knowledge about transportation in general, and market demand in particular, is key for an economic analyisis of a sector. In present paper, the official data about the industrial sector, coming from the Ministry of Public Works and Transport in Spain, is analysed. In order to do that, k-means clustering technique is applied to find groupings or patterns in the dataset that contains data from a whole year (2015). Samples allocation to clusters and silhouette values are used to characterize the demand of the industrial transportation. Useful insights into the analysed sector are obtained by means of the clustering technique, that has been applied with 4 different criteria.

Keywords: Clustering · *k*-means · Road transportation · Logistics Industrial sector

1 Introduction

The concept of economic activity is an aggregate of various sector functions that are developed in a system, and that are largely interrelated. A common frame of aggregation of such activities are the geographical territories and the periods of time that we understand as natural years.

Transportation is the economic activity that is the most tightly coupled with the other ones, given that it is in those relationships in which its essence is based. This may have motivated that this activity has historically been controlled and supervised by the governments, for which it has constituted, in a large part of its history. Nowadays, being a logically liberalized activity, transportation continues to be a subject closely supervised and controlled by governments in its different territories. Under the current geopolitical framework, the European Union has been demonstrating for decades its interest in transport activities that unite their territories through their exchanges of people and goods.

Scant attention has been devoted until now to apply clustering techniques in order to analyse transportation data. In [1] authors apply clustering techniques in order to identify areas of high porosity, or high permeability, for pedestrians along the border

M. Graña et al. (Eds.): SOCO'18-CISIS'18-ICEUTE'18, AISC 771, pp. 387–396, 2019.
https://doi.org/10.1007/978-3-319-94120-2_37

region of some countries using terrain, land use, and road data along with geocomputational methods. Obtained results could be potentially useful for decision making processes for tourism development and road transportation management in that region.

Although many previous studies have been focused in vehicle route optimization, few of them applied clustering algorithms. That is the case of [2] where *k*-means is proposed to improve the computational performance of a new algorithm aimed at solving vehicle routing problems. Authors claim that the new extended formulation where the clustering algorithm is applied, captures truckload and travel distance, supporting the development of systems that respond fast, possibly online, to changes in the real problem situations.

The analysis of transport activity is normally developed under two aggregation frameworks: (1) the one that distinguishes between travellers and merchandise, and (2) the one that distinguishes the so-called different modes of transport (rail, highway, and navigable roads). Focusing on the transport of goods, and analyzing the different modes in the European Union, the road presents an overwhelming supremacy over the rest of the modes, representing 75.4% of the total amount of tons per kilometres transported within and between EU member states in 2016 (latest available data) [3]. For all the above, the European Union has imposed on its member states control mechanisms of transport activity, especially the transport of goods by road, and has designed a homogenized data collection framework for its member states [4]. In the case of Spain, these processes have been integrated into the National Statistical Plan, and materialized in the issuance and collection of data from the Permanent Survey of Transport of Goods by Road (PSTGR), that is issued, collected and supervised by the General Directorate of Transport integrated in the Spanish Ministry of Public Works and Transport.

The data collected in PSTGR are designed for a regional statistical representativeness, with the observation unit being the vehicle-week, analyzing the transport activity performed by the vehicle sample in that period. It gathers data about the characteristics of the vehicle, the transported goods, the origin, destination and distance of the operation and, when appropriate, the price of the service. Not all these data are publicly available; the prices of services are specially protected data from this agency.

According to the situation above described, present paper focuses on analysing the demand on the transportation of industrial goods by analysing some data from the PSTGR. In order to gain deep knowledge of such real-life dataset, a standard clustering technique (*k*-means) is applied for descriptive tasks.

The rest of this paper is organized as follows: the k-means clustering technique and associated measurements are described in Sect. 2, the setup of experiments and the dataset under analysis are described in Sect. 3. Finally, the obtained results and the conclusions of present study are stated in Sect. 4.

2 Clustering

Cluster analysis [5, 6] consist in the organization of a collection of data items or patterns (usually represented as a vector of measurements, or a point in a multidimensional space) into clusters based on similarity. Hence, patterns within a valid cluster are more similar to each other than they are to a pattern belonging to a different cluster.

Pattern proximity is usually measured by a distance function defined on pairs of patterns. A variety of distance measures are in use in various communities [7, 8]. There are different approaches to data clustering [5], but in general terms, there are two main types of clustering techniques: hierarchical and partitional approaches. Hierarchical methods produce a nested series of partitions (illustrated on a dendrogram which is a tree diagram) based on a similarity for merging or splitting clusters, while partitional methods identify the partition that optimizes (usually locally) a clustering criterion. Hence, obtaining a hierarchy of clusters can provide more flexibility than other methods. A partition of the data can be obtained from a hierarchy by cutting the tree of clusters at certain level. A representative and well-known technique for partitional clustering is applied in present study, namely *k*-means [9], that is described in Subsect. 2.1.

As similarity is fundamental to the definition of a cluster, a measure of the similarity is essential to most clustering methods and it must be carefully chosen. Present study applies well-known distance criteria used for examples whose features are all continuous when applying the *k*-means algorithm.

Additionally, different techniques have been applied in present study in order to estimate the optimal value of *k* for the *k*-means algorithm. Such techniques evaluate the goodness of clustering results [10] by taking into account a certain criterion. The following criteria have been applied in present work, for comparison purposes:

- Calinski-Harabasz Index [11]: it evaluates the cluster validity based on the average between-and within-cluster sum of squares. Index measures separation based on the maximum distance between cluster centers, and measures compactness based on the sum of distances between objects and their cluster center.
- Davies-Bouldin Index [12]: it computes the sum of the maximum ratios of the intra-cluster distances to the inter-cluster distances for each cluster.
- Gap criterion [13]: it uses the output of any clustering algorithm, comparing the change in within-cluster dispersion with that expected under an appropriate reference null distribution. This index is especially useful on well-separated clusters and when used with a uniform reference distribution in the principal component orientation.
- Silhouette Index [14]: it validates the clustering performance based on the pairwise difference of between and within cluster distances. In addition, the optimal cluster number is determined by maximizing the value of this index.

2.1 *k*-Means Clustering Technique

The well-known *k*-means [9] is a partitional clustering technique for grouping data into a given number of clusters. Its application requires two input parameters: the number of clusters (*k*) and their initial centroids, which can be chosen by the user or obtained through some pre-processing. Each data element is assigned to the nearest group centroid, thereby obtaining the initial composition of the groups. Once these groups are obtained, the centroids are recalculated and a further reallocation is made. The process is repeated until there are no further changes in the centroids. Given the heavy reliance of this method on initial parameters, a good measure of the goodness of the grouping is simply the sum of the proximity Sums of Squared Error (SSE) that it attempts to minimize, Where *p()* is the proximity function, *k* is the number of the groups, c_j are the centroids, and *n* the number of rows:

$$SSE = \sum_{j=1}^{k} \sum_{x \in G_j} \frac{p(x_i, c_j)}{n} \tag{1}$$

In the case of Euclidean distance [15], the expression is equivalent to the global mean square error.

K-means technique takes distance into account to cluster the data. Different distance criteria were defined and the distance measures applied in the study are described in this subsection.

An *mx*-by-*n* data matrix *X*, which is treated as *mx* (1-by-*n*) row vectors $x_1, x_2, \ldots, x_{mx}$, and *my*-by-*n* data matrix *Y*, which is treated as *my* (1-by-*n*) row vectors $y_1, y_2, \ldots, y_{my}$.. are given. Various distances between the vector x_s and y_t are defined as follows:

Seuclidean distance

In Squared Euclidean metrics (Seuclidean), each coordinate difference between rows in *X* is scaled, by dividing it by the corresponding element of the standard deviation:

$$d_{st}^2 = (x_s - y_t)V^{-1}(x_s - y_t)' \tag{2}$$

Where *V* is the *n*-by-*n* diagonal matrix the *j*th diagonal element of which is $S(j)^2$, where *S* is the vector of standard deviations.

Cityblock distance

In this case, each centroid is the component-wise median of the points in that cluster.

$$d_{st} = \sum_{j=1}^{n} |x_{sj} - y_{tj}| \tag{3}$$

Cosine Distance

This distance is defined as one minus the cosine of the included angle between points (treated as vectors). Each centroid is the mean of the points in that cluster, after normalizing those points to unitary Euclidean lengths:

$$d_{st} = 1 - \frac{x_s y_t'}{\sqrt{(x_s x_s')(y_t y_t')}} \tag{4}$$

Correlation Distance

In this case, each centroid is the component-wise mean of the points in that cluster, after centering and normalizing those points to a zero mean and a unit standard deviation.

$$d_{st} = 1 - \frac{(x_s - \bar{x}_s)(y_t - \bar{y}_t)'}{\sqrt{(x_s - \bar{x}_s)(x_s - \bar{x}_s)'}\sqrt{(y_t - \bar{y}_t)(y_t - \bar{y}_t)'}} \quad (5)$$

3 Real-Life Case Study on Industrial Road Transportation

As previously stated, in present study data about Road Transportation in the Industrial sector have been analysed through *k*-means. Data were obtained from the 2015 PSTGR, picking those transportation services related to the industrial sector (raw material for industry, mainly), according to the NST 2007 standard goods classification for transport statistics [16]. Among all the services in 2015 PSTGR, only those associated to the following types of goods were included in the dataset:

- 01 - Products of agriculture, hunting, and forestry; fish and other fishing products.
- 02 - Coal and lignite; crude petroleum and natural gas.
- 03 - Metal ores and other mining and quarrying products; peat; uranium and thorium.
- 06 - Wood and products of wood and cork (except furniture); articles of straw and plaiting materials; pulp, paper and paper products; printed matter & recorded media.
- 08 - Chemicals, chemical products, and man-made fibers; rubber and plastic products; nuclear fuel.
- 09 - Other non-metallic mineral products.
- 10 - Basic metals; fabricated metal products, except machinery and equipment
- 11 - Machinery and equipment n.e.c.; office machinery and computers; electrical machinery and apparatus n.e.c.; radio, television and communication equipment and apparatus; medical, precision and optical instruments; watches and clocks.
- 12 - Transport equipment.
- 16 - Equipment and material utilized in the transport of goods.

For the included services, the following features (from 3 different areas) were gathered:

- Time:
  - Day of the week: the day of the week in which the goods were transported (from Monday to Sunday).
- Vehicle:
  - Combination: vehicle combination used in the operation. It takes one of the following values: Lorry, Cab + Body, Cab or Body.
  - Number of axes: it takes values from 1 to 8.
  - Load capacity (in hundreds of kilos).

- Permissible Maximum Weight (in hundreds of kilos): it includes the vehicle and the load.
- Service:
  - Distance of the operation (in kilometers).
  - Type of goods. It takes one of the values from [16] above listed.
  - Type of packaging. It takes one of the following values: liquid in bulk, solid in bulk, big container, other containers, freight package, pre-slinged, motor vehicles and live animals, other mobile units, other kinds of cargos, no load.
  - Type of service: It takes one of the following values: normal, delivery and/or pickup, repetitive, normal with no load.
  - Stratum: whether the service is public/private and load capacity.
  - Contracting party. It takes one of the following values: 1 (contract) or 2 (on its own).
  - Geographical range. It takes one of the following values: 0 (within a city), 1 (from one city to another, in the same region), 2 (from one region to another one), 3 (import), 4 (export) or 5 (between other countries, what is called cabotage).
  - Weight (in tons): transported weight multiplied by the coefficient of elevation of the statistical representativeness of the sample element. It is an estimation of the real quantity of merchandise transported by the whole market represented by this statistical item.
  - Number of operations: number of operations multiplied by the coefficient of elevation of the statistical representativeness of the sample element. It is an estimation of the real quantity of merchandise transported by the whole market represented by this statistical item.

All in all, 4586 samples were gathered, which extrapolates to 3086000 realized transportation services in that year based on their statistical representativeness.

4 Clustering Results

The technique and parameters described in Sect. 2 were applied to the case study presented in Sect. 3 and the results are discussed below. Table 1 shows the information on the k estimation for the whole normalized dataset, performed by applying the four different measures. In this table, column 'Estimated Optimal *k*' represents the optimum number of clusters calculated by each one of the measures from the initial range (from 1 to 10).

Once the cluster evaluation was performed, values of 2 and 4 for *k* paramenter were selected for subsequent experiments.

Table 2 shows the results obtained for the *k*-means, with k-means++ [17] algorithm for cluster center initialization and different distance criteria and values of *k* equal to 2 and 4 (as stated before). In this table, '*k*' is the value of such parameter, 'Distance' is the applied distance criteria, 'Sum Distance' is the within-cluster sums of point-to-centroid distances and 'Mean Silhouette' is the mean of silhouette values for all the points [−1, 1] according to the same distance criteria used by *k*-means.

Table 1. k-estimation for the whole dataset.

Measure	*Estimated Optimal k*
Calinski-Harabasz	2
Davies-Bouldin	4
Gap	9
Silhouette	4

Table 2. *k*-means clustering results from different experiments.

#	*k*	Distance	Sum Distances	Mean Silhouette
1	2	Seuclidean	[3.9146 3.5818] $\cdot$ 10^4	0.3149
2	2	Cityblock	[2.2486 1.7686] $\cdot$ 10^4	0.3423
3	2	Cosine	[1.2920 1.1683] $\cdot$ 10^3	**0.3436**
4	2	Correlation	[1.3448 1.2621] $\cdot$ 10^3	0.3030
5	4	Seuclidean	[3.2812 3.3197 0.1074 0.2121] $\cdot$ 10^4	**0.3314**
6	4	Cityblock	[0.7196 1.3503 1.0884 0.3393] $\cdot$ 10^4	0.2943
7	4	Cosine	[509.5208 382.8418 204.9995 855.7570]	0.2879
8	4	Correlation	[819.4047 585.7896 475.0979 195.2167]	0.2521

From the results in Table 2, the following experiments were selected for subsequent analysis:

- #3: it is selected as the representative experiments with *k* parameter equal to 2, as it obtained the highest Mean Silhouette value. Samples assigned to each one of the clusters in this experiment: 50.8% to cluster 1 and 49.2% to cluster 2.
- #5: it is selected as the representative experiments with *k* parameter equal to 4, as it obtained the highest Mean Silhouette value. Samples assigned to each one of the clusters in this experiment: 46.1% to cluster, 49.3% to cluster 2, 1.9% to cluster 3, and 2.7% to cluster 4.

In order to visually check the goodness of the clustering, the silhouette graphs of such experiments are shown in Figs. 1 and 2. These graphs depict the silhouette index [14] (horizontal axis) for each one of the data assigned to each one of the clusters (vertical axis).

From Fig. 1 it can be clearly seen that some of the samples assigned to clusters 1 and 2 (those at the top of each one of the clusters in the graph) are clearly different to samples from the other cluster. Although some of the samples from both clusters have a low silhouette value, there is not any sample with a negative silhouette value.

From a thorough analysis of the samples assigned to each one of the two clusters, it has been noticed that samples in cluster 1 are mainly those from lorry combination and low distance of the operation (mean value within the cluster: 53.31 km). Additionally, the average transported weight (multiplied by the coefficient of elevation) of this cluster is 4.040 tons. On the other hand, samples allocated in cluster 2 are mainly those from cab + body combination and high distance of the operation (mean value within the

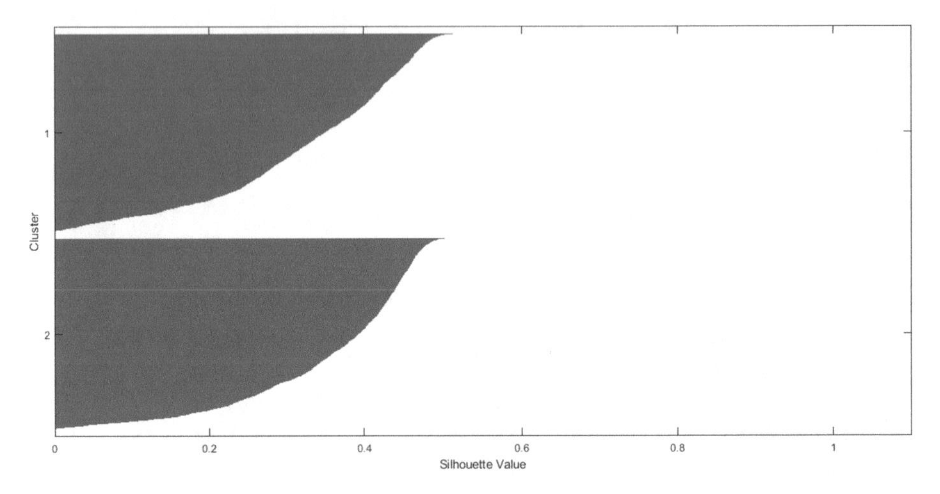

Fig. 1. Silhouette graph generated from clustering results of experiment #3.

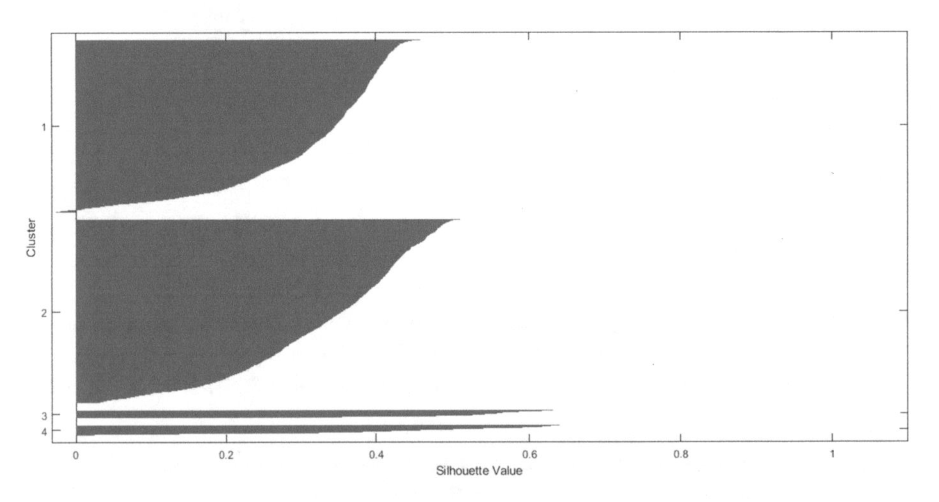

Fig. 2. Silhouette graph generated from clustering results of experiment #5.

cluster: 271.37 km). The average transported weight (multiplied by the coefficient of elevation) of this cluster is 12.946 tons.

From the silhouette graph based on results from experiment#5, it can be said that only very few samples (all of them from cluster 1) have negative silhouette, while most of them have positive values. More precisely, a reduced number of samples were assigned to clusters 3 and 4 but those have the highest values (greater than 0.6 in some case). This graph confirms that 4 is an appropriate number of clusters as higher silhouette values were obtained when compared with those in experiment #3.

As in the case of experiment #3, a thorough analysis of the samples assigned to each one of the two clusters, let us characterize them in the following way:

- Cluster 1: it gathers samples that could be defined as transportation from one province to the other one with heavy loads. Most of them are cab + body combination and medium/low distance of the operation (mean value within the cluster: 183.98 km). The accumulated transported weight (multiplied by the coefficient of elevation) of this cluster is 29,696,675 tons; that is 76.89% of the total transported weight (whole dataset).
- Cluster 2: it gathers samples that could be defined as transportation within the same province with medium loads. Most of them are lorry combination and low distance of the operation (mean value within the cluster: 51.45 km). The accumulated transported weight (multiplied by the coefficient of elevation) of this cluster is 7,718,800 tons; that is 19.99% of the total transported weight (whole dataset).
- Cluster 3: it gathers samples that could be defined as non-freelance. Most of them are body combination and medium/low distance of the operation (mean value within the cluster: 209.31 km). The accumulated transported weight (multiplied by the coefficient of elevation) of this cluster is 258,597 tons; that is 0,67% of the total transported weight (whole dataset).
- Cluster 4: it gathers samples that could be defined as international. Most of them are cab + body combination and very high distance of the operation (mean value within the cluster: 1,688.52 km). The accumulated transported weight (multiplied by the coefficient of elevation) of this cluster is 945,883 tons; that is 2,45% of the total transported weight (whole dataset).

Once results from the two experiments are compared, it can be concluded that clustering from experiment #5 ($k = 4$) could be more interesting for the transportation analysis. Clusters 3 and 4 from this experiment gather samples with higher silhouette values than any other samples from the two experiments.

5 Conclusions and Future Work

Main conclusions derived from obtained results are that clustering is a useful tool to gain deep knowledge of a previously unknown dataset. By a standard clustering technique such as k-means, clear patterns are identified in a road transportation dataset.

Best results have been obtained when applying k-means with the following parameters: initialization = kmeans++, $k = 4$, distance = Seuclidean.

More precisely, certain trends have been identified in the demand of transportation services in the Spanish industrial sector during 2015. The majority of the transportation services are associated to short distances (within the same province of from one province to another one), while a small amount of them are associated to an international context.

Conclusions obtained from the analysis of the datasets are useful for companies within transportation sector as knowledge about demand is obtained. Non-evident aggregations are obtained, what could also be interesting for States as well in order to gain abstract knowledge about such an important sector, to identify subjacent typologies of services, and to find emerging services increasingly demanded.

Future work will consist of extending proposed analysis to a wider time period, adding additional features ton the dataset and applying some other clustering techniques for a wider comparison.

References

1. Hisakawa, N., Jankowski, P., Paulus, G.: Mapping the porosity of international border to pedestrian traffic: a comparative data classification approach to a study of the border region in Austria, Italy, and Slovenia. Cartogr, Geog. Inf. Sci. **40**, 18–27 (2013)
2. Cinar, D., Gakis, K., Pardalos, P.M.: A 2-phase constructive algorithm for cumulative vehicle routing problems with limited duration. Expert Syst. Appl. **56**, 48–58 (2016)
3. Report from the Commission to the European Parliament and the Council: Fifth Report on Monitoring Development of the Rail Market. European Comission (2016)
4. Council Regulation (EC) No 1172/98 of 25 May 1998 on Statistical Returns in Respect of the Carriage of Goods by Road. Council of the European Union (2000)
5. Jain, A.K., Murty, M.N., Flynn, P.J.: Data clustering: a review. ACM Comput. Surv. **31**, 264–323 (1999)
6. Xu, R., Wunsch, D.C.: Clustering. Wiley (2009
7. Andreopoulos, B., An, A., Wang, X., Schroeder, M.: A roadmap of clustering algorithms: finding a match for a biomedical application. Brief. Bioinf. **10**, 297–314 (2009)
8. Zhuang, W.W., Ye, Y.F., Chen, Y., Li, T.: Ensemble clustering for internet security applications. IEEE Trans. Syst. Man Cybern. Part C Appl. Rev. **42**, 1784–1796 (2012)
9. Ding, C., He, X.: K-means clustering via principal component analysis. In: Proceedings of the 21st International Conference on Machine learning, p. 29. ACM (2004)
10. Liu, Y., Li, Z., Xiong, H., Gao, X., Wu, J.: Understanding of internal clustering validation measures. In: IEEE 10th International Conference on Data Mining (ICDM), pp. 911–916. IEEE (2010)
11. Caliński, T., Harabasz, J.: A dendrite method for cluster analysis. Commun. Stat. Theory Methods **3**, 1–27 (1974)
12. Davies, D.L., Bouldin, D.W.: A cluster separation measure. IEEE Trans. Pattern Anal. Mach. Intell. **1**, 224–227 (1979)
13. Tibshirani, R., Walther, G., Hastie, T.: Estimating the number of clusters in a data set via the gap statistic. J. R. Stat. Soc. Ser. B (Stat. Methodol.) **63**, 411–423 (2001)
14. Rousseeuw, P.: Silhouettes: a graphical aid to the interpretation and validation of cluster analysis. J. Comput. Appl. Math. **20**, 53–65 (1987)
15. Danielsson, P.-E.: Euclidean distance mapping. Comput. Graph. Image Process. **14**, 227–248 (1980)
16. Standard goods classification for transport statistics (NST 2007). http://ec.europa.eu/eurostat/statistics-explained/index.php/Glossary:Standard_goods_classification_for_transport_statistics_(NST). Accessed 09 Mar 2018
17. Arthur, D., Vassilvitskii, S.: k-means++: the advantages of careful seeding. In: Proceedings of the 18th Annual ACM-SIAM Symposium on Discrete Algorithms, Society for Industrial and Applied Mathematics, pp. 1027–1035 (2007)

LiDAR Applications for Energy Industry

Leyre Torre-Tojal[1(✉)], Jose Manuel Lopez-Guede[2,4], and Manuel Graña[3,4]

[1] Department of Mining and Metallurgical Engineering and Materials Science, Faculty of Engineering, University of the Basque Country (UPV/EHU), Nieves Cano 12, 01006 Vitoria-Gasteiz, Spain
leyre.torre@ehu.es

[2] Department of Systems Engineering and Automatic Control, Faculty of Engineering, University of the Basque Country (UPV/EHU), Nieves Cano 12, 01006 Vitoria-Gasteiz, Spain

[3] Department of Computer Science and Artificial Intelligence, Faculty of Computer Science, University of the Basque Country (UPV/EHU), Paseo Manuel de Lardizabal, 1, 20018 Donostia-San Sebastian, Spain

[4] Computational Intelligence Group, University of the Basque Country (UPV/EHU), Leioa, Spain

Abstract. The first step for an optimum energy consumption reducing planning supposes the accurate estimation of the primary sources. For this purpose, Light Detection and Ranging (LiDAR) remote sensing technique is being widely applied because its ability to collect huge amounts of data with good accuracy. This study focuses on the application of this technology to the improvement of the assessment of wind, solar and biomass energies. In the case of the biomass, a proof of concept of the estimation for the Pinus Radiata specie in the Arratia-Nervión region (Spain) has been explained. Due to the promising results obtained with this technique, LiDAR has stand out as a powerful and versatile tool for energy consumption reduction in the industrial sector.

Keywords: LiDAR · Energy industry · Remote sensing · Biomass

1 Introduction

In the context of global economic and rapid population growth, the US Energy Information Administration estimates that world energy consumption will grow by 56% between 2010 and 2040 [1]. Developing these resources and providing them to the country's industrial enterprises and population imply an interaction between the public and private sectors that affects the structure and the functioning of domestic and international markets of these resources.

Industry accounts for approximately one-third of global final energy use, more than any other end-use sector. The industrial sector comprises a diverse set of industries, including manufacturing (food, paper, chemicals, refining, iron and steel, nonferrous metals, nonmetallic minerals and others) and nonmanufacturing (agriculture, mining and construction). The mix and intensity of fuels consumed in the industrial sector

M. Graña et al. (Eds.): SOCO'18-CISIS'18-ICEUTE'18, AISC 771, pp. 397–406, 2019.
https://doi.org/10.1007/978-3-319-94120-2_38

varies across regions and countries, depending on the level and mix of economic activity and technological development, among other factors.

In order to reduce the energy consumption of the industrial sector on a global scale, new policies must be implemented, and unprecedented investment in best practices and new technologies is required. If best available recent technologies were deployed globally, industrial energy use could be reduced by 20% to 30%. There is an urgent need for a major development in technologies that have the potential to significantly change industrial energy use [2]. Energy efficiency in industry has improved significantly in the last decade, but additional improvement is still possible through the implementation of new available technologies.

Future generations of industrial machinery, plants, and decision support systems may be expected to carry out round-the-clock operation, with minimal human intervention in manufacturing products or providing services. This will require the appropriate integration of such devices as sensors, actuators, and controllers appropriately distributed throughout the systems [3]. Light Detection and Ranging (LiDAR) has now become an industrial standard tool for collecting accurate and dense topographic data at very high speed. Laser scanning is a remote sensing tool with the ability to retrieve surface elevations at high spatial resolutions, in rough terrain and in heavily forested areas [4]. LiDAR applications for different purposes are continuously being developed with very good results [5].

In this paper focus on LiDAR applications for the three main renewable energy sources, i.e., wind, solar and biomass energy, showing an example for this last scope.

The harnessing of energy from wind and turning it into renewable electricity has many advantages. The technology has never been better or more cost effective than it is right now. An accurate assessment of wind energy potential is essential to increase the effectiveness of this resource. The utilization of solar energy in buildings plays an increasingly important role in a sustainable urban development. The calculation of the solar irradiance of a site is the first step to be able to exploit this energy source optimally. If biomass is processed efficiently, either chemically or biologically, by extracting the energy stored in the chemical bonds and the subsequent energy product combined with oxygen, the carbon is oxidized to produce CO2 and water [6]. However, as it happens with other energy sources, a previous estimation of the resource is crucial.

The structure of the paper is as follows. The paper begins presenting the LiDAR systems basic principles in Sect. 2. Section 3 reviews a number of LiDAR applications in the scope of energy industry, while Sect. 4 gives a brief example of its utilization for the biomass estimation. Finally, the last section presents our main conclusions.

2 LiDAR Background

Light Detection and Ranging (LiDAR) is an active remote sensing technology that is able to measure positions on a massive scale. It is based on a laser sensor (working in the infrared spectral region) emitting pulses that are returned to the sensor once they have impacted against any object or surface. Once the round-trip time of each emitted pulse is measured and taking into account that the light speed is known, the distance to an object is easily calculable. Besides x, y and z coordinates for every impact, LiDAR system provides information about the intensity of the reflected signal. The LiDAR system can be mounted in a satellite, in an aircraft or helicopter, drones, or in the case of terrestrial LiDAR, it can be static (in a tripod) or dynamic (in ground based vehicles). Despite of the basic principle of the technology is the same, there are some differences in data capture and processing steps, applications of data, etc. For more information about the basic physical principles behind the technology, please see [7]. When the sensor is mounted in a dynamic platform (aircraft, car, etc.) the system needs to integrate additional positioning systems, a Global Positioning System (GPS) and an Inertial Movement Unit (IMU) to reach an accurate positioning of airborne lasers, as shown in Fig. 1.

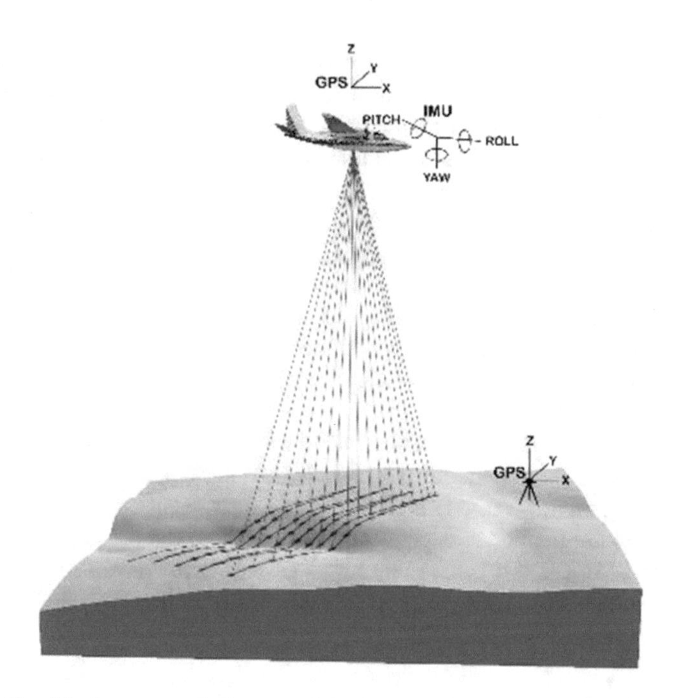

Fig. 1. Simplified conceptual diagram of the LiDAR system (source: SRM Consulting)

Major part of the last two decades has been dominated by single wavelength single pulse Linear Mode LiDAR (LML) with single or multi-pulse ability. However, recently several new types of LiDAR sensors have appeared enabling new applications in earth sciences [8].

In addition, LiDAR is a key technology in urban planning because of its great value in applications such as building information modelling (BIM), assess management, pollution monitoring and assistance in the creation of emergency response strategies for natural disasters amongst others.

3 LiDAR Applications

As have been mentioned, LiDAR applications for three of the main renewable energy resources are going to be analyzed. Firstly, in Subsect. 3.1, the integration of this technology for wind energy assessment is analyzed. In Subsect. 3.2, how this sensor can contribute to a more accurate estimation of solar energy is shown. Finally, in Subsect. 3.3, biomass estimation using LiDAR data is introduced.

3.1 Wind Energy

In European countries such as Spain and Denmark, wind energy already contributes supplying the 22.8% and 37.6% respectively of the annual electricity demand according to the 2016 Spanish Electricity System preliminary report [9].

Despite of being significant advantages using wind energy, there are also disadvantages. When planning to build a wind farm, it needs to be built at locations with high wind speed in order to get high-energy output performance. Seeking a suitable location for a wind farm is the first and most important step, but it requires a detailed analysis of the local wind conditions.

A number of researches have evaluated wind power potential in many countries in the last few decades [10, 11], requiring large amounts of data from meteorological stations, which are not available at locations far away from meteorological stations. Traditionally, tall and costly meteorological masts are installed with cup and sonic anemometers at the proposed site. Another disadvantage of the traditional method is that the meteorological mast can only measure wind speed up to 80 m of height due to the limit of typical mast height.

However, LiDAR is a ground-based remote sensing instrument with many advantages: it is cheaper than building a metrological mast, it can measure wind data up to 200 m above ground level, it is a ground based instrument and it is easy to remove and it implies minimum human dependence since most of the processes are automatic. The development of LiDAR technology offers easier, more convenient, more accurate and continuous operation instruments to measure wind profiles near the ground (up to 200 m above ground level) [12, 13].

3.2 Solar Energy

Currently, solar photovoltaic technology, despite its low efficiency (between 10% and 20%) depending on the type of technology, can reach grid parity in specific geographic areas [14].

As the rate of urbanization continues accelerating, the utilization of solar energy in buildings plays an increasingly important role in sustainable urban development. Different studies have applied LiDAR technology to arrange solar irradiation maps of the buildings of a city [15, 16]. Solar irradiation models are useful for regional planning of energy systems, distributed and real-time photovoltaic forecasting analysis, validating photovoltaic upscaling methodologies and for accurate photovoltaic modelling for photovoltaic grid integration studies.

Different LiDAR points densities have been compared in order to analyse the influence of this parameter on the solar irradiation estimation, ranging from low-resolution LiDAR data set (0.5–1 pts/m^2) to high-resolution LiDAR data set (6–8 pts/m^2) [15]. Although the high-resolution LiDAR data is more accurate than the low-resolution LiDAR data, due the limited coverage of this kind of data, it has been demonstrated that using low-resolution LiDAR data it is possible to perform competitively for solar resource assessment.

3.3 Biomass

One important aspect about the utilization of biomass as energy resource is its accurate estimation in order to apply optimal energy policies.

The estimation of the forest biomass has been traditionally carried out using two main methodologies: destructive and non-destructive methods. The first one requires the destruction of the sample trees, and later, when the material is dry, it is weighted [17]. The indirect methods use allometric equations, where the normal diameter and the height of the trees measured in field are the most used variables in such equations [18]. The ability of the LiDAR technique to gather huge amount of points allows a more efficient and cheaper tool for forest biomass management [19].

4 Roadmap to Biomass Estimation

In this section we present a roadmap to Above Ground Biomass estimation of the Pinus Radiata specie using LiDAR data in the Arratia-Nervión region (40,000 ha) located in Biscay (Spain), as shown in Fig. 2.

Two main data sources were used in this study, both are public data repositories with periodical actualization:

- The Fourth National Forestal Inventory (NFI 4), where the dendometric data (such as tree height and diameter) were gathered during the summer of 2011. This project is oriented to obtain the maximum information about the Spanish forests, focusing in the situation, land ownership and the productive capacity, amongst others.
- The LiDAR data were obtained from the LiDAR flight of the Autonomous Community of the Basque Country carried out during the summer of 2012 by the Basque Government [19].

From a total of 118 plots placed in the area of interest, 63 were chosen, having more than 80% of occupation of Pinus Radiata. Allometric equations were applied to estimate the biomass for every tree, depending on the diameter and the tree height [20], as indicated by Eq. (1):

$$W_{tree} = 0.009892\left(d^2h\right)^{1.023} - 0.00434d^2h + 61.57 - 6.978d + 0.346d^2 \quad (1)$$

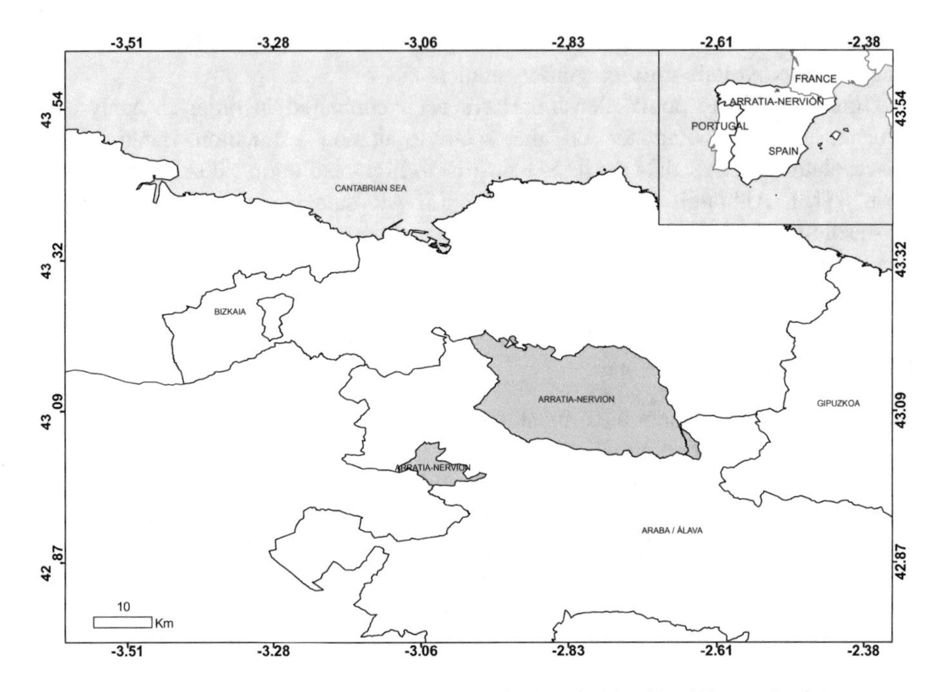

Fig. 2. Location of the study area, namely Arratia-Nervión (Biscay, Spain).

Once the reference values of the biomass for every tree were calculated, these values were expanded to a hectare extension, using expansion factors. Then, LiDAR data were processed to obtain various results for the area, i.e., Digital Terrain Models (DTM) and Digital Surface Models (DSM). The DTM is a continuous function that represents the elevation of every planimetric position, z = f(x,y), while the DSM includes only the points that are in the highest part of the area. The Canopy Height Model (CHM) was later computed by the subtraction of the previous two models. From the 63 plots included in the analysis, Fig. 3 shows a subplot of the plot number 443, once the subtraction of the two models have been done, resulting the normalized point cloud. As can be appreciated the CHM represents the real height of the element, in the case of forest, we will get tree heights directly.

On the one hand, the conventional metrics given by commercial software were calculated, such as height percentiles, maximum, minimum and average height and canopy density metrics amongst others. An algorithm was implemented to obtain specific density metrics from the cloud that has been successfully used in similar studies [21, 22]. The main steps of that algorithm are the following:

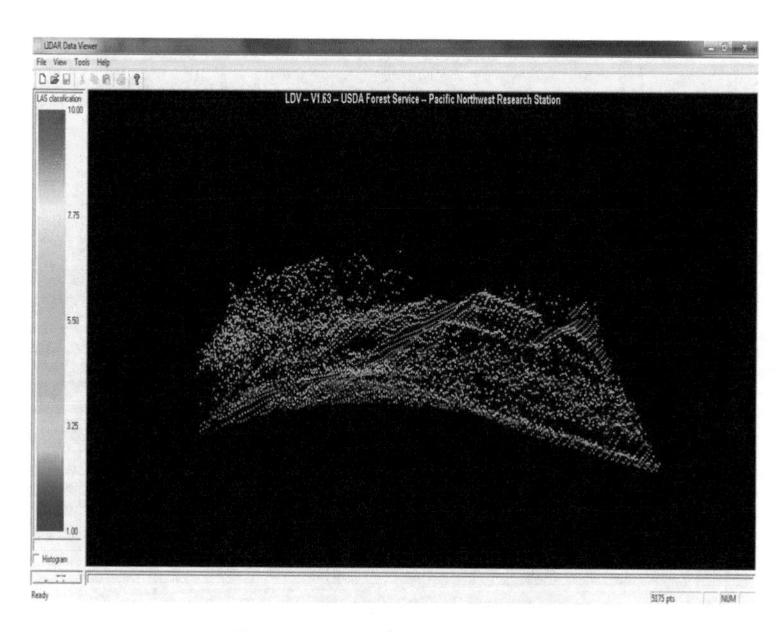

Fig. 3. CHM of a subset of a plot

1. Firstly, the correlation matrix between the calculated field biomass and the obtained LiDAR metrics is computed to check which variables were more correlated. We have decided to use the logarithm of the biomass because of possible problems of homoscedasticity in the model [23].
2. Secondly, these linear relationships between the biomass (and the logarithm of the biomass) measured in field and the LiDAR metrics were verified graphically, creating dispersion diagrams between the measured field biomass and it´s logarithm and each obtained LiDAR metric, as shown in Fig. 3.
3. Then, the models were generated using combinations of two and three variables. The main criteria to choose the appropriate model is the determination coefficient (R^2), but several statistical tests were carried out to evaluate the level of fulfilment of the hypothesis of the linear regression technique.

The best results with $R^2 = 0.76$ and a Root Mean Square Error (RMSE) of 0.26 in logarithmic units involve high percentiles of the LiDAR heights and point density metrics. The values for the applied tests show that almost all the models satisfy the hypothesis of the multiple linear regression technique, as shown in Fig. 4.

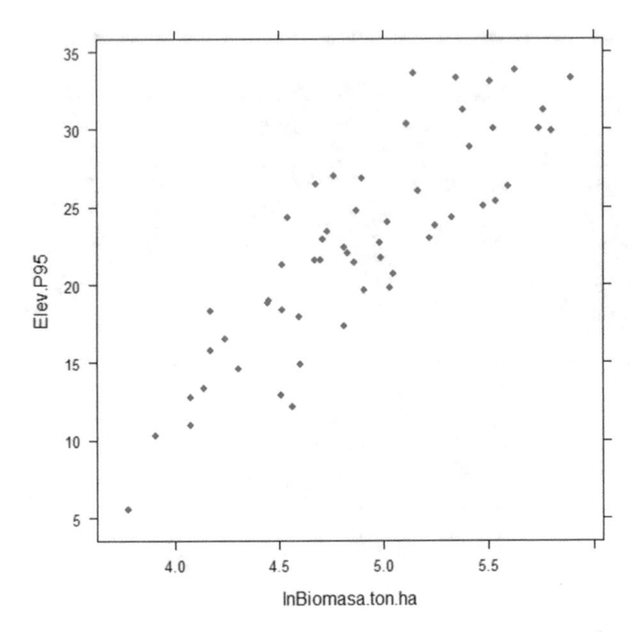

Fig. 4. Dispersion diagram between the logarithm of the biomass and the 95th percentiles of the LiDAR height

5 Conclusions and Future Work

Different applications of the LiDAR remote sensing technique related to energy industry have been discussed, among them, wind resource assessment and solar irradiation assessment and mapping. The rapidly increasing potential of the information that can be extracted from LiDAR has been presented for three important energy sources.

The potential implementation of LiDAR system for biomass estimation has been evaluated obtaining promising results. This study has shown the application of an automatic and straightforward process to estimate biomass of Pinus Radiata using totally public and periodical LiDAR data from a low point density flight (0.5 points/m^2). The developed models got a coefficient of determination (R^2) of 0.76 with a RMSE of 0.26 in logarithmic units.

The accuracy and increasing availability of the huge among of information provided by the LiDAR, make this technology a very capable tool to estimate energy resources. Although LiDAR is an instrumentation technology in continuous development, very powerful for energy quantification, the limited availability of appropriate datasets and the enormous size of the LiDAR datasets are some of the challenges that must be overcomed in order to profit from them, obtaining efficient visualization, management and analysis.

References

1. U.S. Energy Information Administration (EIA): International Energy Outlook 2013 (2013)
2. International Energy Agency: Energy Technology Transitions for Industry. Strategies for the Next Industrial Revolution (2016)
3. Karray F.O., De Silva C.: Soft Computing and Intelligent Systems Design. Pearson Education Limited (2004)
4. Reutebuch, S.E., McGaughey, R.J., Andersen, H., Carson, W.W.: Accuracy of a high-resolution lidar terrain model under a conifer forest canopy. Can. J. Remote Sens. **29** (5), 527–535 (2003)
5. Eitel, J.U.H., Höfle, B., Vierling, L.A., Abellán, A., Asner, G.P., Deems, J.S., Glennie, C.L., Joerg, P.C., LeWinter, A.L., Maney, T.S., Morton, D.C.: Vierling, Kerri T.: Beyond 3-D: The new spectrum of lidar applications for earth and ecological sciences. Remote Sens. Environ. **186**, 372–392 (2016)
6. McKendry, P.: Energy production from biomass (Part 1): overview of biomass. Biores. Technol. **83**(1), 37–46 (2002)
7. Baltsavias, E.P.: Airborne laser scanning: basic relations and formulas. ISPRS J. Photogramm. Remote Sens. **54**, 199–214 (1999)
8. Lohani, B., Ghosh, S.: Airborne LiDAR technology: a review of data collection and processing systems. Proc. Natl. Acad. Sci. India Sect. A Phys. Sci. **87**, 567–579 (2017)
9. Red Eléctrica de España: The Spanish Electricity System, Preliminary Report 2016 (2016)
10. Khahro, S.F., Tabbassum, K., Soomro, A.M.: Evaluation of wind power production prospective and Weibull parameter estimation methods for Babaurband, Sindh Pakistan. Energy Convers. Manag. **78**, 956–967 (2014)
11. Nouri, A., Babram, M.A., Elwarraki, E., Enzili, M.: Moroccan wind farm potential feasibility. Case Study. Energy Convers. Manag. **122**, 39–51 (2016)
12. Kim, D., Kim, T., Oh, G., Huh, J., Ko, K.: A comparison of ground-based LiDAR and met mast wind measurements for wind resource assessment over various terrain conditions. J. Wind Eng. Ind. Aerodyn. **158**, 109–121 (2016)
13. Li, J., Yu, X.: LiDAR technology for wind energy potential assessment: Demonstration and validation at a site around Lake Erie. Energy Convers. Manag. **144**, 252–261 (2017)
14. Martínez-Rubio, A., Sanz-Adan, F., Santamaría-Peña, J., Martínez, A.: Evaluating solar irradiance over facades in high building cities, based on LiDAR technology. Appl. Energy **183**, 133–147 (2016)
15. Lingfors, D., Bright, J.M., Engerer, N.A., Ahlberg, J., Killinger, S., Widén, J.: Comparing the capability of low- and high-resolution LiDAR data with application to solar resource assessment, roof type classification and shading analysis. Appl. Energy **205**, 1216–1230 (2017)
16. Liang, C., Hao, X., Shuyi, L., Yanming, C., Fangli, Z., Manchun, L.: Use of LiDAR for calculating solar irradiance on roofs and façades of buildings at city scale: Methodology, validation, and analysis. ISPRS J. Photogram. Remote Sens. **138**, 12–29 (2018)
17. Parresol, B.R.: Assessing tree and stand biomass: a review with examples and critical comparisons. For. Sci. **45**(4), 573–593 (1999)
18. Moore, J.R.: Allometric equations to predict the total above-ground biomass of radiata pine trees. Ann. Sci. **67**(8), 806 (2010)
19. Shao, G., Shao, G., Gallion, J., Saunders, M.R., Frankenberger, J.R., Songlin, F.: Improving Lidar-based aboveground biomass estimation of temperate hardwood forests with varying site productivity. Remote Sens. Environ. **204**, 872–882 (2018)

20. Spatial Data Infrastructure of the Basque Government. ftp://ftp.geo.euskadi.eus/lidar/LIDAR_2012_ETRS89/LAS/. Accessed 12 March 2018
21. Canga, E., Dieguez-Aranda, I., Afif-Khouri, E., Cámara-Obregón, A.: Above-ground biomass equations for Pinus radiata D. Don in Asturias. For. Syst. (INIA) **22**(3), 408–415 (2013)
22. Næsset, E., Gobakken, T., Solberg, S., Gregoire, T.G., Nelson, R., Stähl, G., Weydahl, D., et al.: Model-assisted regional forest biomass estimation using LiDAR and InSAR as auxiliary data: a case study from a boreal forest area. Remote Sens. Environ. **115**, 3599–3614 (2011)
23. Næsset, E., Gobakken, T.: Estimation of above- and below-ground biomass across regions of the boreal forest zone using airborne laser. Remote Sens. Environ. **112**(6), 3079–3090 (2008)
24. Valbuena, M.A.: Determinación de variables forestales de masa y de árboles individuales mediante delineación de copas a partir de datos LiDAR aerotransportado (2014)

Waste Management as a Smart Cognitive System: The Wasman Case

Evelyne Lombardo[1(✉)], Pierre-Michel Riccio[2], and Serge Agostinelli[3]

[1] Kedge Business School, Researcher LSIS, CNRS, Toulon, France
eve_lombardo@hotmail.com
[2] LGI2P, IMT Mines Ales - Univ Montpellier, Ales, France
[3] LAMIA, Université des Antilles, Pointe-à-Pitre, France

Abstract. The Wasman (Waste Management as Policy Tools for Corporate Governance) project is a program comprising six southern European countries that share best practices to achieve collective and intelligent decisions regarding waste management. The partnership among the actors is envisioned as a social structure in a socially distributed cognitive system that is the product of action and the conditional element of future action. The authors call this a system of "intelligent cognitive waste management."

Keywords: Waste management · Intelligent cognitive system

1 Introduction

For most countries, waste management is an important topic, especially in terms of increasing political and environmental requirements. Although solving waste management problems requires technical, economic, and environmental solutions, the scope of the problem is also fundamentally social. It therefore becomes a task of making all the actors responsible for reducing environmental waste, from the marketing of products until their elimination or revalorization.

Waste management is an open-ended enterprise that requires multiple strategies with different stakeholders (waste producers, institutions, associations, etc.). It can be broken down into two distinct missions: collecting and processing.

The aim of the Wasman Project is to set up a methodology to improve waste management in southern Europe through several actions: raising awareness of stakeholders, minimizing waste upstream, improving collection and treatment of waste, and enhancing the system of territorial governance through the involvement of public and private actors. The project brought together partners from local governments, development agencies, and research organizations in six countries in southern Europe: Spain, Italy, Slovenia, Greece, Cyprus, and France. The project lasted three years and had a budget of 1,616,961 euros.

In France, local authorities are considered to be the most competent in coordinating waste management activities. This service can be managed by the municipalities or by a public institution of inter-municipal cooperation. Waste management sectors are set

M. Graña et al. (Eds.): SOCO'18-CISIS'18-ICEUTE'18, AISC 771, pp. 407–414, 2019.
https://doi.org/10.1007/978-3-319-94120-2_39

up according to the socioeconomic and environmental criteria imposed by a national regulatory framework enacted in a 13 July 1992 law, requiring departmental household waste plans. A departmental plan for waste prevention and management has three objectives: reduction, recovery, and optimization of waste. This national regulatory framework is itself subject to greater European strategies for sustainable development.

The French Environment and Energy Management Agency (ADEME) is proposing to create a portal for environmental action aimed at all French stakeholders. This site offers help in understanding the regulatory framework and also supports action for financial assistance for territorial projects and research and provision of methodological tools for communities, businesses, and citizens.

The objective of the French part of the Wasman Project, acting as a scientific partner specializing in collective decision-making, is to elaborate, with the help of the other partners, a model of decision support for the management of household waste. The ambition is to make this model transferable to the whole of Mediterranean Europe to reinforce the synergies for better area governance enhancing sustainable development.

The area studied in France included all municipalities in the Nîmes metropolis in the Languedoc-Roussillon region. Indeed, this area has obtained 93% results in terms of recovery of household waste compared to much lower rates in other country participants. In addition, the cost of waste treatment in the French area is also lower than that of its European partners.

2 Problem

The question this article attempts to answer is as follows: "How can a sustainable development approach be built and implemented for a Mediterranean model that could help the stakeholders make best practice decisions about the management of waste in specific locations?" The authors call this an "intelligent cognitive system." The next section describes the article's theoretical framework and identifies different theories to clarify the concept. The authors then focus on defining what an "intelligent cognitive system" is through clarifying the theoretical terms and concepts that underpin this notion. Next the article describes the specific case of the Wasman Project and its methodology. Finally the results of the study and an analysis are provided.

3 Theoretical Framework

3.1 Community of Social and Cultural Practice

Territorial development and integrated waste management can connect at the intersection of use and practice. Use establishes a relation between people and technology (here, communication devices), whereas practice associates actors and actions related to a particular field of activity (here, the management of waste). Use provides the operative efficiency of the devices and generates a symbolic representation, allowing the implementation of procedures. For example, citizens use a town hall's web page to consult schedules and find practical local information about waste collection. Practice, however, provides operational efficiency of activity, which is associated with a mental schema based on experience and associates operations with expected ends. This research focuses on citizen interaction as a social and cultural practice. It describes a shared and distributed knowledge through which waste management practices can be improved. The approach offered in this article proposes an alternative to a planning model of analysis, which refers to the "good actions" in term of management of waste from a framework and how these actions could be carried out [1]. Similar to sustainable development, the management of household waste is everyone's business and is a matter of practice.

The Giddens [5] model is used to analyze the communities of these social waste practices. For Giddens [5], there are two ways of considering action in situations that combine human interaction based on mutual knowledge of activities and rules: the resources produced and reproduced in social interaction over time and the actors involved. Common knowledge, or socially shared cognition [3], is what the actors have to share.

In other words, the partnership envisioned as a social structure is a socially distributed cognitive system that is the product of the action and the conditional element of future action. Therefore, the set of rules that allows for interaction is primarily a cognitive network in which partners derive knowledge of their activities in a situation and the interactions fostered by that knowledge produce rules and resources that in turn enable future actions.

For Garfinkel [4] the model of action is inseparable from cognition, which allows for a global vision of people and the world in which they live. The link between action and cognition takes effect in the principle of identity, which postulates that the steps that enable one to perform an activity are the same as the steps for understanding and describing it.

On the one hand, the means of action are identical to cognitive processes; on the other hand, this principle enables the synthesis of different aspects of an object's appearance and links it to a general form inseparable from its meaning. Also, actions and the attributions of meaning are associated with a single practice of accomplishment, which enables the explanation of the actions. Here, knowledge or practical skills are presuppositions of the action studied as different modes of cognitive activity that people can sift through as a stock of socially available knowledge [9]. This stock of knowledge provides a foundation of prior knowledge (foreknowledge) for the principle of identity to interpret and understand the practices of social life. Daily skills and knowledge are therefore the temporal result of local work in situ; they form the content of common sense [9] related to social life.

The stock of socially available knowledge through common sense highlights the collective aspect of knowledge and proposes a vision of how it is structured and distributed. The establishment of a cognitive intelligent waste system can be seen through the prism of a local structure, which requires at least partial synchronization of the actors and their visions of the world. An intelligent cognitive system refers to a set of devices (actors, systems, and rule norms). The next section describes the Wasman Project context.

3.2 Background of the Wasman Project

The Wasman Project aims to build a concept model for waste management to promote waste prevention and minimization through source reduction, encourage responsible action by waste generators, and improve citizen awareness of waste (applicable ideas, knowledge, methods, participation processes, local pilot actions, and innovative technologies for efficient and environmentally sound waste management). In addition, the project promotes synergies and disseminates transnational best practices among Mediterranean countries for waste management governance, taking into account the specific problems of each partner region.

The project also hopes to assist municipal governments in developing a more cost-effective waste management system (zero waste strategy) and to strengthen local economies by promoting economical production chain initiatives and identifying niche markets for local businesses. This project was co-financed by the European National Fund, ERDF.

The objectives of the project are as follows:

- Provide guidance to decision-makers on how to allocate resources and develop sustainable achievable strategies to reduce solid waste.
- Set up a local waste management model for the Mediterranean region.
- Develop more-efficient and cost-effective waste management for solid waste collection, treatment, and disposal in small towns and less populated areas.
- Improve the waste management system in various territories, involving public and private institutions.
- Adapt the waste management model to increase added value for local economies and thus create new development opportunities for local businesses.
- Increase citizen awareness and commitment to waste collection and recycling processes and processes.
- Set up eight waste management partnerships.
- Organize technical seminars and workshops to sensitize the actors to waste sorting and dissemination campaigns (good practices).
- Develop multicriteria analysis and methodology training, networking, and transfer of best practice models.
- Develop a community model program based on exemplary municipal waste management plans.
- Organize and implement pilot and innovative actions in waste sorting.

In this context, the researchers wanted to develop a decision support tool (an "intelligent cognitive system") to choose best practices in terms of waste management for stakeholders in different communities of practice in the six countries of the southern Mediterranean.

4 Methodology

This research follows a case study approach similar that advocated by Yin [10], who says the case study is "an empirical investigation that studies a contemporary phenomenon in its real-life context, where the boundaries between the phenomenon and the context are not clearly evident, and in which multiple sources of information are used." This analysis seems particularly adapted to a field in which the actors and the interactions among the actors are numerous so as to account for the complexity and richness of these interactions in real life. The links among the actors during the three years of the study seemed appropriate for this approach [10]. This research also fulfills the criteria of a multiple case study [10], because of the recurrent phenomena (good practices of waste management) among a large number of situations (in six southern Europe countries) with a large number of actors interacting.

The research uses a deductive approach to verify the theoretical representation (good waste management practices) in the field by checking if the theoretical elaboration actually corresponds to the phenomena in its natural context, taking into account the different actors. The initial idea was to build an ideal model by consolidating the different practices, assess the situation of each country, then recommend a path of progression for each country. A multicriteria analysis methodology was created that consisted of the design of a prototype decision-making system based on a waste management model in the Mediterranean area.

Multicriteria techniques were proposed to evaluate the performance of waste management solutions in each area to select the most relevant and best practices or consider new practices to implement in other territories. The criteria were linked to specific objectives of the area waste management model: reduction of the quantity of waste, improvement of the cost-effectiveness ratio, reuse of the raw material, recycling for sustainable economic growth, reduction of the changes climatic conditions, and so on.

The model was based on a hierarchical decomposition of area waste management scenarios in elementary processes (incineration, composting, etc.). Evaluation of these process performances was used, in a second step, to assign a single synthesis score to each scenario. The aggregation model allowed for analytical formalization that facilitated scenario comparisons, diagnosed the strengths and weaknesses of waste management policies, and proposed improved performance through multicriterion optimization.

Information was also collected from different sources to use Yin's [10] method of triangulation to cover the object of analysis from several angles: documents, archives, and files that concerned waste management practices in the different countries were read, and also interviews and observations were conducted of the different communities of actors on their practices (local authorities, development agencies, and research organizations in Spain, Italy, Slovenia, Greece, Cyprus, and France). The analysis of the interviews was then processed by identifying the recurrences of the different themes as advocated by Huberman and Miles [7].

5 Results

Satisfactory results in terms of valuation and processing costs hide several important elements in the particular situation of each actor.

In France, we noted that progress needs to be made in sorting both from a technological and organizational point of view. Moreover, the analysis of Nîmes's 93% performance in terms of recovery of household waste shows that this high rate is linked largely to the choice of the main treatment of household waste: incineration to generate energy reinjected into the common electricity supply network. This type of waste treatment has been challenged by many environmental protection associations and is complex to set up. Although the incinerator's releases are constantly monitored, this solution is limited in its processing capacity.

Additionally, the other European partners of the Wasman Project have chosen different waste management practices, which must be evaluated in context. Spreading is one of those practices, which will not be continued because it does not meet the objectives of sustainable development. To implement a management system that would replicate the energy recovery achieved in Nîmes represents a heavy investment for most communities. The cost of this change makes the operation difficult or impossible in some cases.

In this context, imposing on these communities an ideal unified model based on French practices is problematic. The situation cannot be judged based on simple performance indicators. The analysis of the waste management problem has therefore been enriched by a study extended to the particular contexts of the various project leaders. The attention to the context led the researchers to ask questions specific to the idea of an integrated waste management. The issue of waste rarely presents itself as a lively theme of discussion for citizens and members of the industry. Bringing awareness to this issue is therefore paramount.

In France, the cost of waste treatment and the amount of the contribution paid through the waste tax are relatively low. Yet this tax is not necessarily legitimate to the citizens. The citizens make efforts to sort their waste and feel aggrieved by a system that generates this tax, because they think they contribute to the waste management system, although their efforts need improvement. More generally, there is still a lot of work to be done to raise awareness of the need to optimize waste management throughout society. As such, manufacturers must realize that the reduction of packaging will ultimately be necessary to preserve the environment. There is also a need to highlight the value of organic materials, optimization of collection routes, and glass recovery, which is well below that found across Europe. In sum, communication with citizens should be revised to be integrated with the various public and private actors.

Waste management begins with recognizing the meaning of routine practices and related vocabulary and evolves through the rites of interaction [6]. It manifests itself in genres of discourse [2] to give meaning to reality, to authorize communication, and to provide a given cultural context. Waste management discourse can happen at the public and macrosocial levels, so that activities can be organized and managed on a daily basis. The aim of this article is to describe a system of principles, strategies, and values that guides the creation and interpretation of a local meaning of practices in a given cultural and social setting. The results of this research call for the integrated management of household waste as an embodied cognitive system, that is to say, capable of perception and action as well as intelligence.

- At the periphery of this development is a global system of waste management in which knowledge will be specified. The communication device diffuses the mastered and recognized knowledge that makes decisions in complex cases possible.
- In the middle a relational system is set up in which information can be organized in a sustainable way and provided independent of the perceptive context.
- Finally, at the heart of this relational system are people's thoughts and actions about waste management

Waste management depends on the organization of the knowledge base available to each individual and the relationship between this knowledge base and the know-how related to the action. Individuals have their own knowledge and know-how, even if this information does not correspond to recognized practices. Citizen action results from this relational system, in which organized information is associated with personal and particular know-how.

6 Conclusions

This study highlights a wide understanding of ideal cognitive waste treatment. For example, although Nîmes maximized the value of waste (using incineration), the ideal in Andalusia produced individual raw material units (fuel ball bags) for free use by citizens in their heating or lighting systems. This encouraged a much more complex model (in which many progression paths were possible) to provide for the possibility that each territory could set up an approach (by giving specific weightings to the different elements of the model).

The study also showed that the common sense shared in a problem of this type is different according to location. The advantage of this study is that the communities studied were geographically distant enough that cultural differences were productively important. This explains the need to build a more complex model, but it also justifies the process of more closely considering varying practices (understood as the set of competences and motivations of specific actors) and uses (understood as all functions made available to stakeholders by stakeholders).

Smart recycling of waste has become a necessity. It is noteworthy that following this study a new ultramodern sorting center has just been inaugurated in Nîmes, which indicates the in situ usefulness of pursuing such scientific research capable of changing attitudes and practices within the communities involved in this type of sustainable development.

References

1. Camhis, M.: Planning Theory and Philosophy. Tavistock Publications, London (1979)
2. Clo, Y., Faïta, D.: Genres et styles en analyse du travail: concepts et méthodes. Travailler, Revue internationale de psychopathologie et de psychodynamique du travail 4, 7–42 (2000)
3. Cole, M.: Conclusion. In: Resnick, L., Levine, J., Teasley, S. (eds.) Perspectives on socially shared cognition, pp. 398–417. American Psychological Association, Washington, DC (1991)
4. Garfinkel, H.: Recherches en ethnométhodologie, (1° éd. 1967). PUF, Paris (2007)
5. Giddens, A.: La constitution de la société: eléments d'une théorie de la structuration. PUF, Paris (1987)
6. Goffman, E.: L'ordre de l'interaction. In: Winkin, Y. (ed.) (dir) Les moments et leurs hommes, pp. 186–230. Le Seuil/minuit, Paris (1988)
7. Huberman, A.M., Miles, M.B.: Analyse des données qualitatives. Traduit de l'anglais par Catherine De Backer et Vivian Lamongie. Éditions du Renouveau pédagogique, Montréal, De Boeck-Wesmael, Bruxelles (1991)
8. Hutchins, E., Klausen, T.: Distributed cognition in an airline cockpit. In: Middleton, D., Engeström, Y. (eds.) Communication and cognition at work, pp. 15–35. Cambridge University Press, Cambridge (1996)
9. Schütz, A.: Le chercheur et le quotidien, p. 286. Méridiens/Klincksieck, Paris (1987)
10. Yin, R.K.: Case Study Research: Design and Methods. Sage Publications, London (1984)

A New Approach for System Malfunctioning over an Industrial System Control Loop Based on Unsupervised Techniques

Esteban Jove[1(✉)], José-Luis Casteleiro-Roca[1], Héctor Quintián[1], Juan Albino Méndez-Pérez[2], and José Luis Calvo-Rolle[1]

[1] Departamento de Ingeniería Industrial, University of A Coruña, Avda. 19 de febrero s/n, 15495 Ferrol, A Coruña, Spain
esteban.jove@udc.es

[2] Department of Computer Science and System Engineering, Universidad de La Laguna, Avda. Astrof. Francisco Sánchez s/n, 38200 S/C de Tenerife, Spain
jamendez@ull.edu.es

Abstract. Systems optimization is one of the great challenges to improve the industry plants performance. From an economical point of view, a proper optimization means, among others, energy, material and maintenance savings. Furthermore, the quality of the final product is improved. So fault detection techniques development plays a very important role to achieve the system optimization. Under this topic, the present research shows the developed work over a real common system, the level control. A new proposal based on unsupervised techniques were used to detect the system malfunction states, taking into account a dataset collected during the right operation. The proposal is validated with ad-hoc created faults for the different system operation points. The performance is very satisfactory in general terms.

Keywords: Fault detection · Anomaly detection · Control system Unsupervised techniques

1 Introduction

The vast majority of industrial companies present expensive and complex processes whose operation can be optimized [1,2]. The optimal operation of industrial systems leads to an increase in energy efficiency, resources saving, product quality and, therefore, this let the companies achieve more economic benefits and be more competitive [3]. Hence, system optimization plays a key role in industrial activities.

A proper optimization requires the right operation of the different system components, such as sensors, actuators and so on. Abnormal operation or anomalies can have multiple sources [4]; mechanical faults, changes in system behavior,

M. Graña et al. (Eds.): SOCO'18-CISIS'18-ICEUTE'18, AISC 771, pp. 415–425, 2019.
https://doi.org/10.1007/978-3-319-94120-2_40

sensor error and human mistakes. Hence, anomalies that occur during plant operation represent an important and frequent problem that is mandatory to be addressed as soon as it happens.

From a technical point of view, anomalies are data patterns that do not conform the expected behavior in a specific application [5]. Figure 1 shows an example with two normal data groups T_1 and T_2 and three abnormal data that are outside of the expected normal function a_1, a_2 and a_3. The main difficulty related to anomalies lies in selecting a limit between anomaly and normal data.

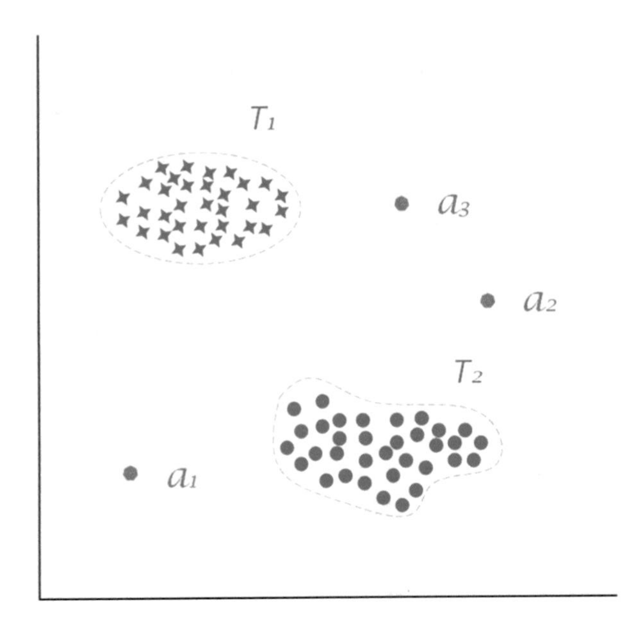

Fig. 1. Anomaly example in a two dimension data set

1.1 State of the Art

Depending on the information of the dataset, three main cases of anomaly detection are contemplated [4]:

- Case 1: a model is obtained only with normal cases or with a few abnormalities. In this type, the model is taught with normal data and an anomaly is identified when abnormal data arrives. In [6], a virtual sensor for failure detection is implemented in the aircraft folding/unfolding wings system. This aim is achieved by modeling the system dynamics. When an abnormal data is measured, the model is able to detect an anomaly.
- Case 2: the anomaly detection has to be determined without previous knowledge of the data. The detection works as an unsupervised clustering, distributing the data in groups. New input data can be classified by comparing it with the data acquired during system operation. This detection assumes

that normal data is well separated from the outliers. It offers successfully results once the system has a large dataset with good coverage.
- Case 3: initially, normal and abnormal data are available. The data is pre-labelled between right and wrong data before the classifier is implemented. In [7] an outliers detection applied to medicine is performed labelling artificially generated anomaly examples.

In some applications, specially the critical ones, only right operation data from the plant is available and failure data is not statistically representative or even unknown. In the cases where most of abnormal functioning situations have not occurred yet, a One Class Classification (OCC) is commonly used [8]. The well-known Support Vector Machine (SVM) algorithm employed in many different applications [9–11] is frequently used to solve OCC problem. To solve anomalies issues in different parts of industrial plants, the use of virtual sensors or missing data imputation techniques are very common [12–16].

This work presents a Case 1 identification problem with a significant modification. This modification consists on using Dimensionality Reduction Techniques (DTRs) to identify the different data boundaries in a two dimensional map. Instead of identifying automatically the data groups using clustering algorithms, the limits are defined by the user.

The approach proposed was tested in a didactic real plant used to control the water level of a tank. The speed of a pump is controlled in order to maintain a constant water level while emptying through an output valve. Different abnormalities were induced during the plant operation and the model proposed is able to recognize this situation.

The outline of this paper is as follows. Section 2 describes briefly the case of study. Then the model approach is presented. After Sect. 3, the techniques applied to validate the proposed model are shown. Section 5 explains experiments and results and finally, conclusions and future works are exposed in Sect. 6.

2 Case of Study

The proposed fault detection technique was developed and verified over a laboratory control liquid level system. The physical system is described as follows.

2.1 Laboratory of Liquid Level Control

This maquette was built with industrial equipment with the aim to achieve a very reliable performance, to export the implementations easily to the real cases. The scheme of the system is shown in Fig. 2. The level of the tank is controlled through varying the turning speed of the water three-phase pump.

The real plant used for these experiments consists of a tank, fed at the top by the liquid from the pump; the pump speed will be controlled to maintain a constant water level while emptying through the output valve.

The control system is a virtual controller that takes signals from the plant through a data acquisition card. As a set point signal, the plant receives the

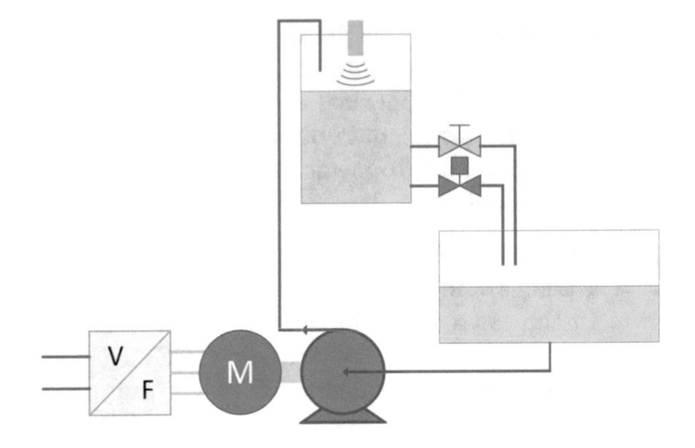

Fig. 2. Laboratory control liquid level scheme

required water level and set the speed of centrifugal pump to set the input flow into the tank.

2.2 The Control Implementation

The method was run in the Matlab environment. For operations at the plant, a National Instruments data acquisition card (model USB-6008 12-bit 10 KS/s Multifunction I/O) was chosen. The diagram of the process is implemented in Matlab with the control Scheme shown in Fig. 3.

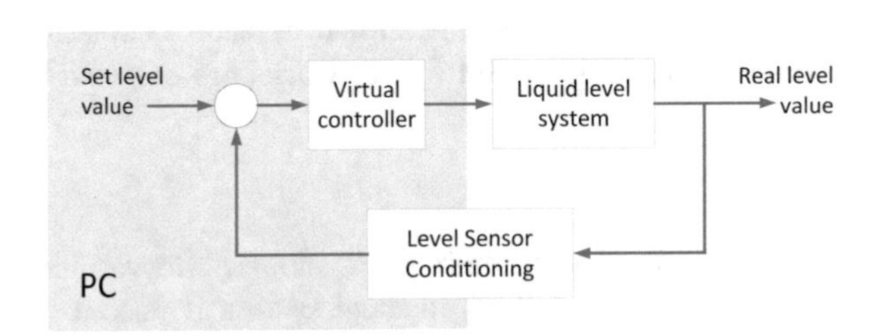

Fig. 3. Laboratory of liquid level control scheme

3 Model Approach

The approach for this research is the creation of a model to perform fault detection in real industrial plants. The good performance of the model would involve a significant decrement in the fault detection devices used on the plant.

Another important point is to avoid the use of labeled data; thus the non-supervised techniques should be used.

The unsupervised techniques selected are used in order to allow the visualization of the operation point. Some of the variables at the plant are used for this objective, and the techniques employed allow to represent an operation point in a two dimension graph regardless the number of variables. As the industrial plants usually work in one or a few working point, all the visualized data in an operation point should be near to each other. Hence, data from different working points displayed in the two dimension graph must be clearly separated.

The data represented in the two dimension graph can belong to different operation points, and the user is able to define a contour in the data. With the defined contour, the algorithm used detects if the working point is out of the working area defined.

The creation of a working contour is not an automatic task, it allows to identify areas, for example, a fault situation in the system. Making this task automatically, it could be possible that the contour do not detect failures that the operator of the plant could take into account with the manual contour definition.

3.1 The Dataset Obtaining and Description

The used dataset in this research is obtained performing several tests at the laboratory plant. In each test, before a fault situation is created, the plant was working during, at least, 10 min, with a sample time of one second. After that, a failure was created and then, the test was stopped.

The different experiments are based on an adaptive PID (*Proportional, Integral, Derivative*) algorithm to control the water level. For the PID implementation, the RLS (*Recursive Least Squared*) algorithm was used to identify the parameters of the plant transfer function. The transfer function weights allow to calculate the controller parameters according to the actual plant operating point. The adaptive PID algorithm helps to control non-linear systems, adjusting the plant transfer function each time.

In this research, the dataset was created with the three parameters of the PID controller along the time. These parameters vary depending of the transfer function variables. A total of 4871 samples were obtained, where only 21 are failures.

4 Techniques Applied to Validate the Proposed Model

In order to provide a visual representation of the abnormal behavior of the tested system, this work proposes the application of several Dimensionality Reduction Techniques to detect anomalies by means of visual inspection [17]. The problem of identifying patterns of anomalies that exist across dimensional boundaries in high dimensional datasets, can be solved by using projection methods. These methods project high dimensional data points onto a lower dimensional space in order to identify "interesting" directions in terms of any specific index or projection.

In this work it has been applied Principal Component Analysis (PCA), MLHL (Maximum Likelihood Hebbian Learning) and Beta Hebbian Learning Algorithm (BHL DTR) techniques to a real control liquid level system to validate our approach.

4.1 Principal Component Analysis

Principal Component Analysis (PCA) is a well-known statistical model, introduced in [18], that describes the variation in a set of multivariate data in terms of a set of uncorrelated variables each, of which is a linear combination of the original variables. From a geometrical point of view, this goal mainly consists of a rotation of the axes of the original coordinate system to a new set of orthogonal axes that are ordered in terms of the amount of variance of the original data they account for.

PCA can be performed by means of neural models such as those described in [19] or [20]. It should be noted that even if it is possible to characterize the data with a few variables, it does not follow that an interpretation will ensue.

4.2 Maximum Likelihood Hebbian Learning

Maximum Likelihood Hebbian Learning (MLHL) [21] which is based on Exploration Projection Pursuit (EPP). The statistical method of EPP [22] was designed for solving the complex problem of identifying structure in high dimensional data by projecting it onto a lower dimensional subspace in which its structure is searched visually. To that end, an "index" must be defined to measure the varying degrees of interest associated with each projection. Subsequently, the data is transformed by maximizing the index and the associated interest. From a statistical point of view the most interesting directions are those that are as non-Gaussian as possible.

4.3 Beta Hebbian Learning Algorithm

Beta Hebbian Learning algorithm (BHL) [23], belongs to a novel family of learning rules derived from the Probability Density Function (PDF) of the residual based on Beta distribution.

In general, the minimization of the cost function associated with this network, may be thought to make the probability of the residuals more dependent on the PDF of the residuals. Thus, if the probability density function of the residuals is known, this knowledge could be used to determine the optimal cost function. So, the residual (e) is draw from the Beta distribution, $B(\alpha, \beta)$, with the following probability density function (Eq. 1):

$$p(e) = e^{\alpha-1}(1-e)^{\beta-1} = (x - Wy)^{\alpha-1}(1 - x + Wy)^{\beta-1} \tag{1}$$

Where α and β are the parameters that determine the shape of the PDF curve of the Beta distribution, x is the input to the network, W is the weight vector associated with network neurons and y is the output to the network.

Then, to maximize the likelihood of the data with respect to the weights, the gradient descent is performed using Eq. 2:

$$\frac{\partial p}{\partial W} = (e_j^{\alpha-2}(1-e_j)^{\beta-2}(-(\alpha-1)(1-e_j)+e_j(\beta-1))) = \\ (e_j^{\alpha-2}(1-e_j)^{\beta-2}(1-\alpha+e_j(\alpha+\beta-2))) \tag{2}$$

In the case of the BHL, by maximizing the likelihood of the residual with respect to the actual distribution, the learning rule is matched to the PDF of the residual. The BHL may also be linked to the standard statistical method of Exploratory Projection Pursuit, as the nature and quantification of the interestingness is in terms of how likely the residuals are under a particular model of the PDF of the residuals.

Therefore, the new neural architecture for BHL is defined as follows:

$$Feedforward : y_i = \sum_{j=1}^{N} W_{ij} x_j, \forall i \tag{3}$$

$$Feedback : e_j = x_j - \sum_{i=1}^{M} W_{ij} y_i \tag{4}$$

$$Weightsupdate : \Delta W_{ij} = \eta(e_j^{\alpha-2}(1-e_j)^{\beta-2}(1-\alpha+e_j(\alpha+\beta-2)))y_i \tag{5}$$

Where α and β are the parameters that determine the shape of the PDF curve of the Beta distribution, x is the input to the network, W is the weight vector associated with the network neurons, e is the residual and y is the output to the network.

5 Experiments and Results

This section presents the results obtained in the validation process of the proposed system.

Both linear and nonlinear models such as PCA, MLHL and BHL have been applied to the previously described dataset (Sect. 3.1), in order to identify the system malfunction states.

5.1 Results

For comparison purposes, three different projection models have been applied, whose results are shown below.

In Fig. 4, it is presented the best projection of each algorithm (PCA, MLHL and BHL) by using the parameter presented in Table 1. In each figure, anomalies are displayed using red diamonds shape ($\diamond$) and normal samples with green dots ($\cdot$).

Results obtained by PCA shows normal samples (green dots - Fig. 4) very sparse over the graph so it is difficult to set the boundaries with the anomalies.

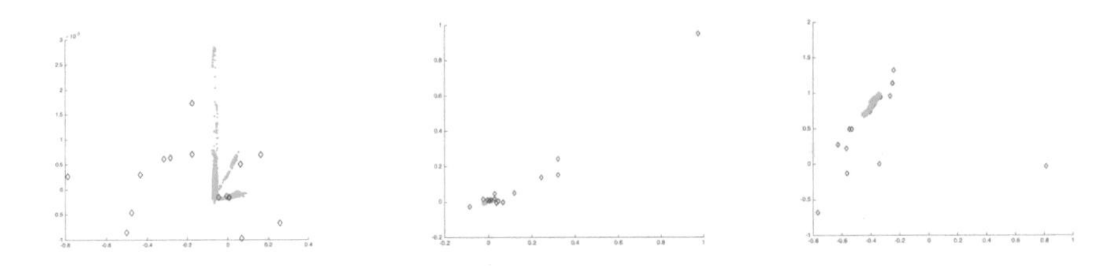

Fig. 4. PCA, MLHL and BHL projections

Table 1. PCA, MLHL and BHL parameters

PCA	–
MLHL	iters = 1000, lrate = 0.01, p = 0.5
BHL	iters = 5000, lrate = 0.01, $\alpha = 3$, $\beta = 4$

However, in case of MLHL, its results present the normal samples in a very compact group, but several anomalies are over or very near to this group.

In spite of none of the 3 methods are able to separate the 100% of anomalies results of BHL clearly overcome the projections of PCA and MLHL, providing a clear visualization of samples which represent anomalies. BHL is able to present in a compact group (G1, see Fig. 5) samples belonging to the right system operation and presenting the anomalies as samples separated from this compact group (G1).

The Exploratory Projection Pursuit (EPP) [24] is the statistical method aimed at solving the difficult problem of identifying structure in high dimensional data. It does this by projecting the data onto a low dimensional subspace in

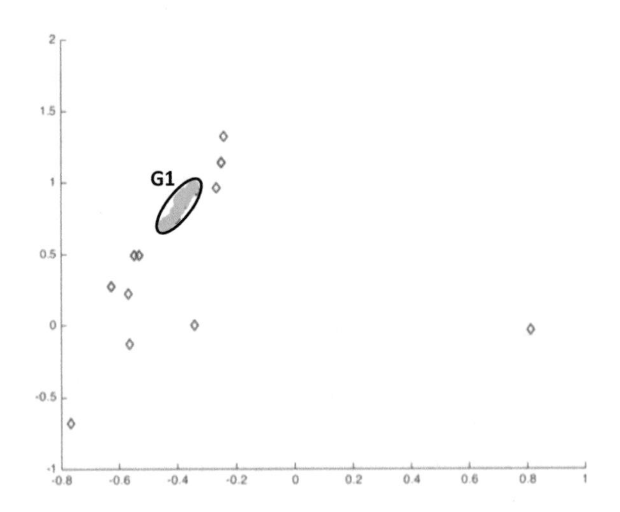

Fig. 5. BHL projection of the dataset

which we search for structures by visual inspection (by eye). Therefore, the visual presentation is the standard measure widely accepted by the EPP community.

Hence, the capability to present these anomalies in a visual mode becomes to BHL in a powerful tool to monitor and supervise the right operation of industrial processes.

6 Conclusions and Future Works

In this work it is proposed a method to accomplish fault, anomaly or malfunction detection with unsupervised learning techniques. The novel approach has been validated over a real laboratory plant where a level control loop is implemented. The achieved results have been very satisfactory in general terms, obviously depending on the used technique.

In contrast with typical fault detection techniques, after the results interpretation, this work allows a skilled operator to define the contour detection limit next to the unsupervised techniques application. With this contribution, it is possible to include expert knowledge of the user in the fault detection process and, consequently, a better performance. Also, it allows to visually monitor the status of the industrial process.

In general terms the contribution allows an optimal operation of the systems, giving advantages and improvements in terms of efficiency, resources saving, profit increasing, quality, and so on. If it is possible to prevent not desired situations, the approach will contributes to it.

The optimal operation of industrial systems leads to an increase in energy efficiency, raw material saving, product quality and, therefore, this let the companies achieve more economic benefits and be more competitive [3].

Based on obtained results the use of unsupervised techniques applied to anomaly detection is a powerful tool to monitor and supervise the right operation of industrial processes, specially when complexity of the systems are high and number of variables to be monitor are high dimensional.

As future works, the capability of defining new contours in real time as new data arrives to the system could be contemplated. Furthermore, the possibility of validating additional fault situations using a bigger dataset from the control level plant will be accomplished. It could also include the study of other fault detection techniques to make a comparison with the results obtained in the present work. Also, some kind of unsupervised approaches could be used to the present aim.

References

1. Jove, E., Alaiz-Moretón, H., García-Rodríguez, I., Benavides-Cuellar, C., Casteleiro-Roca, J., Calvo-Rolle, J.L.: PID-ITS: an intelligent tutoring system for PID tuning learning process. In: International Joint Conference SOCO17-CISIS17-ICEUTE17, León, Spain, 6–8 September 2017, Proceedings, pp. 726–735. Springer, Cham (2018)

2. Alaiz Moretón, H., Calvo Rolle, J., García, I., Alonso Alvarez, A.: Formalization and practical implementation of a conceptual model for PID controller tuning. Asian J. Control **13**(6), 773–784 (2011)
3. Parmee, I., Hajela, P.: Optimization in Industry. Springer, Heidelberg (2012)
4. Hodge, V., Austin, J.: A survey of outlier detection methodologies. Artif. Intell. Rev. **22**(2), 85–126 (2004)
5. Chandola, V., Banerjee, A., Kumar, V.: Anomaly detection: a survey. ACM Comput. Surv. (CSUR) **41**(3), 15 (2009)
6. Heredia, G., Ollero, A.: Virtual sensor for failure detection, identification and recovery in the transition phase of a morphing aircraft. Sensors **10**(3), 2188–2201 (2010)
7. Abe, N., Zadrozny, B., Langford, J.: Outlier detection by active learning. In: Proceedings of the 12th ACM SIGKDD International Conference on Knowledge Discovery and Data Mining, pp. 504–509. ACM (2006)
8. Khan, S.S., Madden, M.G.: A survey of recent trends in one class classification. In: Irish Conference on Artificial Intelligence and Cognitive Science, pp. 188–197. Springer, Heidelberg (2009)
9. Calvo-Rolle, J.L., Quintian-Pardo, H., Corchado, E., del Carmen Meizoso-López, M., García, R.F.: Simplified method based on an intelligent model to obtain the extinction angle of the current for a single-phase half wave controlled rectifier with resistive and inductive load. J. Appl. Log. **13**(1), 37–47 (2015)
10. Casteleiro-Roca, J.L., Pérez, J.A.M., Piñón-Pazos, A.J., Calvo-Rolle, J.L., Corchado, E.: Modeling the electromyogram (EMG) of patients undergoing anesthesia during surgery. In: 10th International Conference on Soft Computing Models in Industrial and Environmental Applications, pp. 273–283. Springer, Cham (2015)
11. Jove, E., Gonzalez-Cava, J.M., Casteleiro-Roca, J.L., Pérez, J.A.M., Calvo-Rolle, J.L., de Cos Juez, F.J.: An intelligent model to predict ANI in patients undergoing general anesthesia. In: International Joint Conference SOCO17-CISIS17-ICEUTE17, León, Spain, 6–8 September 2017, pp. 492–501. Springer, Cham (2018)
12. Casteleiro-Roca, J.L., Jove, E., Sánchez-Lasheras, F., Méndez-Pérez, J.A., Calvo-Rolle, J.L., de Cos Juez, F.J.: Power cell SOC modelling for intelligent virtual sensor implementation. J. Sens. **2017**, 10 (2017)
13. Fernández-Serantes, L.A., Vázquez, R.E., Casteleiro-Roca, J.L., Calvo-Rolle, J.L., Corchado, E.: Hybrid intelligent model to predict the SOC of a LFP power cell type. In: International Conference on Hybrid Artificial Intelligence Systems, pp. 561–572. Springer, Cham (2014)
14. Jove, E., Blanco-Rodríguez, P., Casteleiro-Roca, J.L., Moreno-Arboleda, J., López-Vázquez, J.A., de Cos Juez, F.J., Calvo-Rolle, J.L.: Attempts prediction by missing data imputation in engineering degree. In: International Joint Conference SOCO17-CISIS17-ICEUTE17, León, Spain, 6–8 September 2017, pp. 167–176. Springer, Cham (2018)
15. Casteleiro-Roca, J.L., Quintián, H., Calvo-Rolle, J.L., Corchado, E., del Carmen Meizoso-López, M., Piñón-Pazos, A.: An intelligent fault detection system for a heat pump installation based on a geothermal heat exchanger. J. Appl. Log. **17**, 36–47 (2016)
16. Gonzalez-Cava, J.M., Reboso, J.A., Casteleiro-Roca, J.L., Calvo-Rolle, J.L., Méndez Pérez, J.A.: A novel fuzzy algorithm to introduce new variables in the drug supply decision-making process in medicine. Complexity **2018**, 15 (2018)
17. Sánchez, R., Herrero, Á., Corchado, E.: Clustering extension of MOVICAB-IDS to distinguish intrusions in flow-based data. Log. J. IGPL **25**(1), 83–102 (2017)
18. Karl Pearson, F.R.S.: LIII. On lines and planes of closest fit to systems of points in space. Lond. Edinb. Dublin Philos. Mag. J. Sci. **2**(11), 559–572 (1901)

19. Oja, E.: Principal components, minor components, and linear neural networks. Neural Netw. **5**(6), 927–935 (1992)
20. Fyfe, C.: A neural network for PCA and beyond. Neural Process. Lett. **6**(1), 33–41 (1997)
21. Corchado, E., MacDonald, D., Fyfe, C.: Maximum and minimum likelihood Hebbian learning for exploratory projection pursuit. Data Min. Knowl. Discov. **8**(3), 203–225 (2004)
22. Hou, S., Wentzell, P.D.: Re-centered kurtosis as a projection pursuit index for multivariate data analysis. J. Chemom. **28**(5), 370–384 (2014). CEM-13-0108.R1
23. Quintián, H., Corchado, E.: Beta Hebbian learning as a new method for exploratory projection pursuit. Int. J. Neural Syst. **27**(6), 1–16 (2017)
24. Friedman, J.H., Tukey, J.W.: A projection pursuit algorithm for exploratory data analysis. IEEE Trans. Comput. **100**(9), 881–890 (1974)

Swarm Intelligence Methods on Inventory Management

Dragan Simić[1(✉)], Vladimir Ilin[1], Svetislav D. Simić[1], and Svetlana Simić[2]

[1] Faculty of Technical Sciences, University of Novi Sad, Trg Dositeja Obradovića 6, 21000 Novi Sad, Serbia
dsimic@eunet.rs, {dsimic, v.ilin, simicsvetislav}@uns.ac.rs

[2] Faculty of Medicine, University of Novi Sad, Hajduk Veljkova 1–9, 21000 Novi Sad, Serbia
svetlana.simic@mf.uns.ac.rs

Abstract. Inventory control is the science-based art of controlling the amount of inventory (or stock) held, in various forms. Inventory control techniques are very important components and the most organizations can substantially reduce their costs associated with the flow of materials. This paper presents biological swarm intelligence in general, and in particularly two models: particle swarm optimization and firefly algorithm for modelling on inventory control in production system. The aim of this research is to create models to minimize production cost according to price of items and inventory keeping cost.

Keywords: Inventory control · Particle swarm optimization · Firefly algorithm

1 Introduction

Inventory control is the science-based art of controlling the amount of inventory (or stock) in various forms, within an organization, to meet the economic demand placed upon that business. Interest in the problems of optimal stock management at a scientific level goes back to the start of the 20th century [1].

The importance of inventory control in business increased dramatically with the increasing interest rates of the 70s. It was the rule of the hour to release surplus operating capital tied up in excessive inventories and to use the resulting liquidity to finance new investments. To control the amount of inventory it is necessary to forecast the level of future demand, where such demand can be regarded as essentially either independent or dependent. Most organizations can substantially reduce their costs associated with the flow of materials [2].

This paper presents biological swarm intelligence in general, and two models particularly: particle swarm optimization (PSO) and firefly algorithm (FFA) for modelling inventory control in production system. The aim of this research is to create models to minimize *production cost*, when production is given on demand while taking into account the price of items and inventory keeping cost. This research continuous the

M. Graña et al. (Eds.): SOCO'18-CISIS'18-ICEUTE'18, AISC 771, pp. 426–435, 2019.
https://doi.org/10.1007/978-3-319-94120-2_41

authors' previous researches in supply chain management, and inventory management system presented in [3–7].

The rest of the paper is organized in the following way: Sect. 2 overviews the inventory management and selection related work. Section 3 shows two models (1) particle swarm optimization (PSO) and (2) firefly algorithm (FFA) implemented in inventory management system. This section also describes used dataset. Experimental results and discussion are presented in Sect. 4 and finally, Sect. 5 gives concluding remarks.

2 Inventory Management and Related Work

Inventory is money bound in a nonmonetary form and unless it can be sold, excess inventory can be a huge drain on profits. This includes the storage cost and the lower value of perishable or dated goods. A unit of money has more value today than a year later, even if the inflation rate is zero. The reason lies in the fact that the use of money brings about a return, which can be paid as interest. Discounting is used when interest goes back in time [8].

A fuzzy logic-based inventory control model is proposed in [9]. It maintains the inventory at a desired level despite fluctuations in the demand, taking into account the dynamics of the production system. The concept of fuzzy set theory had been applied to inventory control models considering the fuzziness of inputs and the dynamics of the systems.

An intelligent system implemented in a web-based environment to integrate multiple stores for providing an effective coordination of all the stores, intelligently determining the different reorder points of all the disparate stores in the systems and communicating the information back to the centralized store, is proposed in [10]. The paper presents optimization of the performance of inventory management integrating multiple systems and providing an efficient coordination and monitoring moving away from a single store into distribution system relating real time status of supplies at the different stores.

The paper [11] presents the effective inquiry of artificial neural network modelling and forecasting in the issues of inventory management, especially in the lot-sizing problem. Several types of neural networks are created and tested within the research, searching for the most efficient neural network architecture.

A major challenge in supply chain inventory optimization is handling, design, analysis, and optimization system under uncertainty. The concept of linguistic variables, fuzzy set, as an alternative approach to modeling human thinking in an approximate manner, serves to summarize information and express it, is presented in [12].

3 Models in Inventory Management

Manufacturing environments - where demand at lower levels of the production process are clearly dependent on demand at higher levels - from a stock holding point of view it may be more sensible to ignore this self-evident dependency and assume trends identified in past data. Two models for modelling on inventory control in production system based on biological swarm intelligence are introduced in this research. They are used to create models to minimize productions cost, when production is given on demand, but taking into account price of items and inventory keeping cost. The first proposed model is based on particle swarm optimization and the second model is based on firefly algorithm.

3.1 Particle Swarm Optimization Algorithm

The original Particle Swarm Optimization (PSO) algorithm was inspired by the social behavior of biological organisms, specifically the ability of groups of some species of animals to work, as one, in locating desirable positions in a given area, e.g. birds flocking to a food source.

Algorithm 1. *The* ***Particle Swarm Optimization*** *algorithm*

Begin

Step 1: ***Initialization.*** A random population of individuals $\{x_i\}$, $i \in [1,N]$
Each individual's n-element velocity vector $\{v_i\}$, $i \in [1,N]$
The best-so-far position of each individuals: $b_i \leftarrow x_i$, $i \in [1,N]$
Neighbourhood size $\sigma < N$; $nPop = 150$; Pop. Size; Max. velocity, v_{max}; $MaxIt = 1000$; Max. Num. of Iter. $w = 1$; Inertia weight
c1 = 2; Personal Learning Coeff.; c2 = 2; Global Learning Coeff.

Step 2: ***While*** (the termination criteria is not satisfied) **or** ($MaxIt = 1000$)

Step 3: ***for each*** individual x_i, $i \in [1,N]$

$H_i \leftarrow \{\sigma$ nearest neighbors of $x_i\}$
$h_i \leftarrow \arg\min_x \{f(x) : x \in H_i\}$
Generate a random matrix $r_p(k) \sim U <0, 1>$ for $k \in [1, N]$
Generate a random matrix $r_g(k) \sim U <0, 1>$ for $k \in [1, N]$
$v_i(k) \leftarrow \omega\, v_i(k) + c_1\, r_p\, (b_i - x_i(k) + c_2\, r_g\, (h_i(k) - x_i(k))$
if $|v_i(k)| > v_{max}$ ***then***
$v_i(k) \leftarrow v_i(k)\, v_{max} / |v_i(k)|$
end if
$x_i(k) \leftarrow x_i(k) + v_i(k)$
$b_i \leftarrow \arg\min\, \{f(x_i(k)),\, f(b_i)\}$

end for *i* Next individual

Step 4: ***end while*** Next generation

Step 5: *Post-processing the results and visualization;*

End.

The PSO is originally due to Kennedy, Eberhart, and Shi [13, 14]. An individual particle i is composed of three vectors: its position in the N-dimensional search space of solutions $\vec{x_i} = (x_{i1}, x_{i2}, \ldots, x_{iN})$, the best position that it has individually found $\vec{p_i} = (p_{i1}, p_{i2}, \ldots, p_{iN})$, and its velocity $\vec{v_i} = (v_{i1}, v_{i2}, \ldots, v_{iN})$. Particles were originally initialized in a uniform random manner throughout the search space; velocity is also randomly initialized. These are also weighted matrixes r_p, r_g dimensions - *search space* x *time horizon* - which items are random numbers of unimodal distribution in the range U <0, 1>. These particles then move throughout the search space. The Euclidian neighborhood model was abandoned in favour of less computationally intensive models for mathematical optimization. The *Algorithm 1* presents PSO used algorithm where the entire swarm is updated at each time step by updating the velocity and position of each particle in every dimension when the termination criteria is not satisfied or maximal number of iteration is not satisfied.

3.2 Mainframe of Firefly Algorithm

The firefly algorithm (FFA) is a relatively new swarm intelligence optimization method, in which the search algorithm is inspired by social behavior of fireflies and the phenomenon of bioluminescent communication. There are two critical issues in the firefly algorithm that represent a variation of light intensity which presents *cost value* and formulation of attractiveness.

Algorithm 2. *The algorithm of* ***firefly algorithm***

Begin

Step 1: ***Initialization.*** *Set the generation counter* **G** *= 1; Initialize the population of n fireflies* **P** *randomly and each firefly corresponding to a potential solution to the given problem; define light absorption coefficient* γ*; set controlling the step size* $\boldsymbol{\alpha}$ *and the initial attractiveness* $\boldsymbol{\beta_0}$ *at* **r** *= 0.*

Step 2: *Evaluate the cost function* ***I*** *for each candidate in* **P** *determined by* ***f(x)***

Step 3: ***While*** the termination criteria is not satisfied ***or*** **G** < MaxGeneration ***do***

    for *i=1:n (all n candidate solution)* ***do***

        for *j=1:n (n candidate solution)* ***do***

            if $(I_j < I_i)$,

                move candidate i towards j;

            end if

            Vary attractiveness with distance r via exp$[-\gamma^{r^2}]$;

            Evaluate new solutions and update cost function;

        end for *j*

    end for *i*

    G = G+1;

Step 4: ***end while***

Step 5: *Post-processing the results and visualization;*

End.

Fireflies communicate, search for pray and find mates using bioluminescence with varied flashing patterns. Attractiveness is proportional to the brightness which decreases with increasing distance between fireflies. If there are no brighter fireflies than one particular candidate solution, will move at random in the space.

The brightness of a firefly is influenced or determined by the objective function. For a *maximization/minimization* problem, brightness can simply be *proportional/inversely proportional* to the value of the cost function. The basic steps of the FA are summarized by the pseudo code shown in Algorithm 2.

3.3 Data Set

The data set is taken from [15]: (1) *Demand for products*; (2) *Item price*; (3) *Inventory keeping cost*; and are presented from Tables 1, 2 and 3. The *maximum production capacity* is 1000 units.

Table 1. Demand for products

Product	Day										Total
	1	*2*	*3*	*4*	*5*	*6*	*7*	*8*	*9*	*10*	
1	59	34	84	69	28	25	55	65	74	36	529
2	84	34	31	46	78	36	80	64	22	40	515
3	57	56	50	79	40	54	22	90	41	37	526

Table 2. Item price

Product	Day										Average
	1	*2*	*3*	*4*	*5*	*6*	*7*	*8*	*9*	*10*	
1	188	138	176	104	153	133	200	189	163	181	162.5
2	149	129	117	181	196	173	182	117	188	183	161.5
3	147	173	188	182	185	103	120	171	200	154	162.3

Table 3. Inventory keeping cost

Product	Day										Average
	1	*2*	*3*	*4*	*5*	*6*	*7*	*8*	*9*	*10*	
1	3	6	10	2	4	7	3	7	3	10	5.5
2	8	10	4	8	5	5	5	4	7	6	6.2
3	3	9	5	7	8	8	8	4	3	5	6.1

4 Experimental Results and Discussion

According to the nature of PSO and FFA algorithm, the experiment is repeated 100 times with the data set from [15] and experimental results for minimizing productions cost are summarized and presented in Table 4. The number for iteration for every method was 1000. The best three results for both methods are presented here.

First it is important to notice that the experimental results show that *Total numbers of products are* satisfied for every product. Production cost is different for every experiment, it is shown that the best experimental results for *Production cost* is 232269 monetary units which belongs to PSO experiment. But, the second-best result belongs to firefly optimization algorithm where *Production cost* is equal to 234400 monetary units. Third and fourth experimental results of *Production cost* amounts to 236766 monetary units and 238004 monetary units belong to PSO methods.

Table 4. Experimental results, Method (*PSO, FFA*), Number of days (*Day*), Total production per each day (*Total*), and Production cost

Method	Best	Pro	Day										Total	Prod. Cost
			1	*2*	*3*	*4*	*5*	*6*	*7*	*8*	*9*	*10*		
	.	1	89	90	0	129	10	149	0	1	54	7	529	
P	1	2	87	41	109	11	41	46	55	125	0	0	515	**232269**
	.	3	128	0	48	67	39	91	85	50	0	18	526	
	.	1	72	99	9	168	45	65	22	23	16	10	529	
S	2	2	84	54	70	27	50	29	78	100	10	13	515	**236766**
	.	3	146	46	11	53	37	69	106	21	0	37	526	
	.	1	62	118	5	114	39	58	12	68	22	34	529	
O	3	2	87	73	65	32	42	50	55	95	8	8	515	**238004**
	.	3	189	2	52	0	43	93	112	0	4	31	526	
	.	1	67	31	79	171	73	30	0	0	78	0	529	
F	1	2	94	52	127	2	7	85	22	126	0	0	515	**234400**
	.	3	85	87	0	71	39	60	74	75	0	35	526	
	.	1	59	76	42	148	47	49	0	15	93	0	529	
F	2	2	179	5	30	2	106	21	46	126	0	0	515	**239332**
	.	3	94	64	24	99	2	67	141	21	0	15	526	
	.	1	126	41	10	222	0	0	37	0	73	20	529	
A	3	2	103	45	104	8	15	91	33	72	27	17	515	**240413**
	.	3	72	77	32	87	14	81	94	1	51	17	526	

Experimental results for both methods PSO and FFA, number of planning days, Total production per each day, and Production cost are presented in Table 4. The experimental results for PSO: (a) Used capacity, Inventory and Order account in Time; (b) *Best Production Cost* on number of iteration are presented on Fig. 1. The experimental results for FFA: (a) Used capacity, Inventory and Order account in Time; (b) *Best Production Cost* on number of iteration are presented on Fig. 2. On figures Fig. 1(b) and Fig. 2(b) present only 250 iterations of *Best Production Cost* because the differences in amount for *Best Production cost* after 250 iterations are very small. Average value for *Best Production Cost* for PSO for 100 experiments is 239102.7 monetary units, and average value for *Best Production Cost* for FFA for 100 experiments is 241895.4 monetary units, which presents difference of 1.15%.

4.1 Experimental Results PSO and FFA – New Data

In order to test and verify previous experimental results *new data set* is used for demand of products and it is presented in Table 5. The experiments are repeated 100 times, by using new data set for *Demand for products*, and *Item price*, *Inventory keeping cost* which are presented in Tables 2 and 3 respectively.

First it is important to notice that the experimental results show that *Total numbers of products are* satisfied for every product. Production cost is different for every

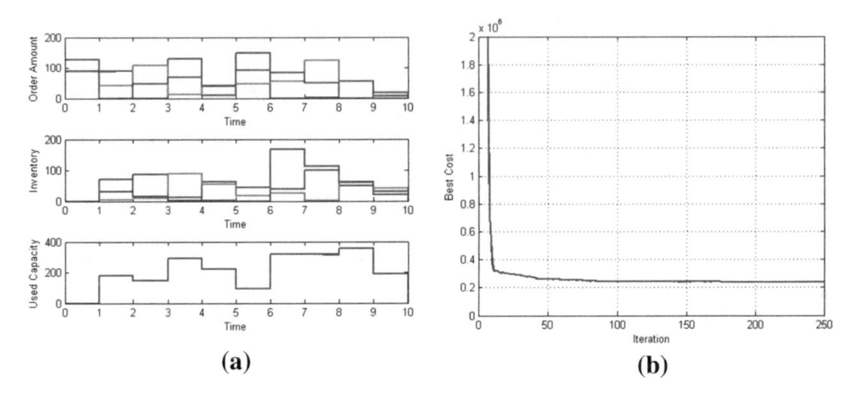

Fig. 1. Experimental results for *particle swarm optimization* method: (a) Used capacity, Inventory and Order account in Time; (b) Best cost on number of iteration

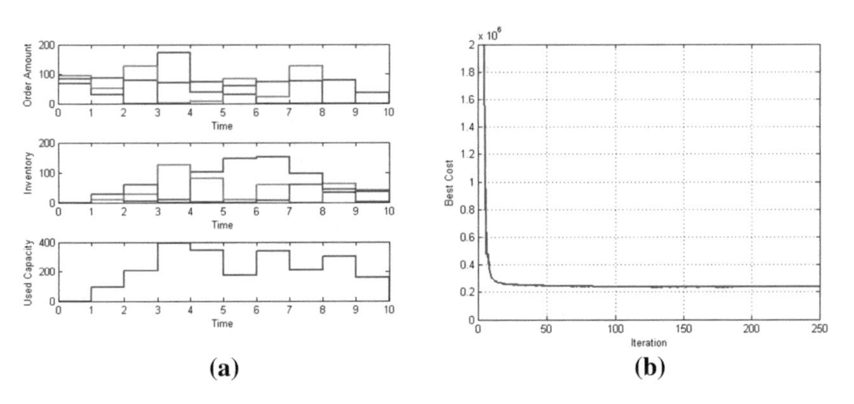

Fig. 2. Experimental results for *firefly optimization algorithm*: (a) Used capacity, Inventory and Order account in Time; (b) Best cost on number of iteration

Table 5. Demand for products – new data set

Product	Day										Total
	1	*2*	*3*	*4*	*5*	*6*	*7*	*8*	*9*	*10*	
1	59	34	104	69	28	75	55	65	74	86	649
2	84	84	31	96	78	36	80	84	62	40	675
3	87	56	50	79	90	54	62	90	41	77	686

experiment, it is shown that the best experimental results for *Production cost* is 299889 monetary units which belongs to firefly optimization algorithm (Fig. 3).

The fourth-best FFA experimental result is not presented in Table 6, only three best results of both optimization methods are presented there. And finally, the fifth-best experimental result for *Production cost* equals 309405 monetary units and belongs to particle swarm optimization method.

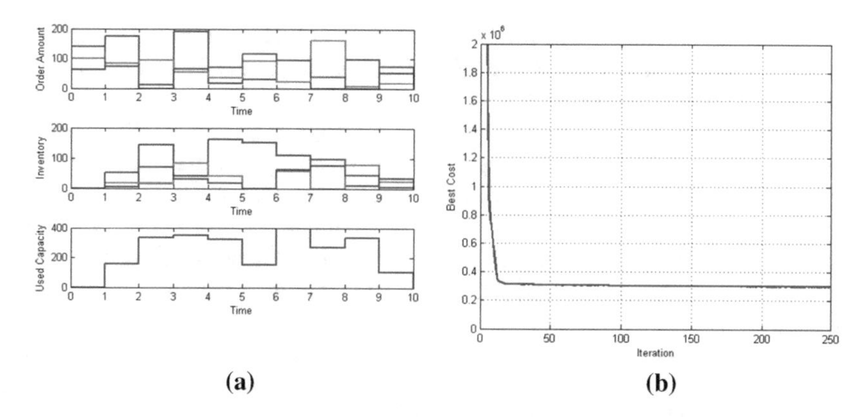

Fig. 3. Experimental results with *new data set* - for *firefly optimization algorithm*: (a) Used capacity, Inventory and Order account in Time; (b) Best cost on number of iteration

Table 6. Experimental results for testing and verifying with *new data set*, Method (*PSO*, *FFA*), Number of days (*Day*), Total production per each day (*Total*), and Production cost

Method	Best	Pro	Day										Total	Prod. Cost
			1	*2*	*3*	*4*	*5*	*6*	*7*	*8*	*9*	*10*		
	.	1	71	104	53	111	72	69	1	25	62	81	649	
P	1	2	134	50	97	66	27	99	23	145	8	26	675	309405
	.	3	111	73	40	87	52	82	152	34	32	23	686	
	.	1	63	116	20	181	63	41	66	2	25	73	649	
S	2	2	126	51	120	38	78	39	46	107	45	27	675	310453
	.	3	142	72	28	67	65	84	106	65	0	57	686	
	.	1	70	129	0	136	78	57	0	42	56	81	649	
O	3	2	142	61	101	59	34	32	87	114	13	32	675	311126
	.	3	111	56	67	63	73	111	46	98	5	56	686	
	.	1	62	175	0	191	18	31	22	0	97	53	649	
F	1	2	100	85	96	55	37	92	24	162	6	18	675	299889
	.	3	139	73	12	66	72	116	96	38	0	74	686	
	.	1	59	138	0	212	3	163	7	0	4	63	649	
F	2	2	84	123	50	56	80	18	78	126	22	38	675	302309
	.	3	96	99	3	129	35	148	43	46	10	77	686	
	.	1	101	85	11	282	2	7	17	18	60	66	649	
A	3	2	99	125	36	90	28	75	36	97	89	0	675	307503
	.	3	159	28	95	1	79	96	142	15	44	27	686	

4.2 Discussion

Experimental results show that both implemented algorithms work well and it is not possible to make final decision which of them is better for inventory management.

This research on minimizing *Production Cost* taking into account only item prices and inventory keeping cost is only a "tip of an iceberg" in a very large and complexly inventory-production management.

First, in some cases it is possible to improve the applied algorithms. PSO algorithm needs some improvements and tuning of its behavioural parameters to perform well when applied on the problem at hand. PSO variants are continually being devised to overcome this decency [16, 17] for a few recent additions. These PSO variants greatly increase the complexity of the original method but they demonstrate satisfactory performance which can be achieved with the basic PSO if only its parameters are properly tuned [18].

Second, in inventory-production management typical dependent demand environment, given a commitment to a future production plan of a specified number of finished products at specified times (*Master Production Schedule*) and also given that information is available such that the number of sub-assemblies, components and raw materials that make up each finished product can be defined (*Bill of Material*), combined with information as to the current stock holding situation of these items (*Status of all Materials*). The future demand for sub-assemblies, components and raw materials in terms of their required quantities and timing can theoretically be forecasted with a high degree of certainty and accuracy is referred to by the generic term *Material Requirements Planning* (MRP) which is essentially a large, and often very complicated, database manipulation exercise which requires large amounts of information regarding the relationships between finished products and their constituent sub-assemblies, components and raw materials.

5 Conclusion and Future Work

This paper presents biological swarm intelligence on the two models: particle swarm optimization and firefly algorithm for modelling inventory control in production system. The proposed models minimize production costs, when production is given on demand, but taking into account item prices and inventory keeping cost.

Experimental results encourage further research. The optimization method, particle swarm optimization has several parameters that determine its behaviour and efficacy. The future work could focus on extending research on good choice of parameters for various optimization scenarios which should help the authors to achieve better results with little effort. Then this model will be tested with original real-world dataset obtained from existing manufacturing company.

References

1. Lewis, C.: Demand Forecasting and Inventory Control: A Computer Aided Learning Approach. Wiley, New York (1998)
2. Bartmann, D., Beckmann, M.J.: Inventory Control: Models and Methods. Springer, Heidelberg (1992)

3. Simić, D., Simić, S.: Evolutionary approach in inventory routing problem. Lecture Notes in Computer Science, vol. 7903, pp. 395–403. Springer (2013)
4. Simić, D., Simić, S.: Hybrid artificial intelligence approaches on vehicle routing problem in logistics distribution. In: Hybrid Artificial Intelligence Systems. LNCS, vol. 7208, pp. 208–220. Springer, Heidelberg (2012)
5. Simić, D., Svirčević, V., Simić, S.: A hybrid evolutionary model for supplier assessment and selection in inbound logistics. J. Appl. Log. **13**(2, Part A), 138–147 (2015)
6. Simić, D., Kovačević, I., Svirčević, V., Simić, S.: 50 years of fuzzy set theory and models for supplier assessment and selection: a literature review. J. Appl. Log. **24**(part A), 85–96 (2017)
7. Simić, D., Kovačević, I., Svirčević, V., Simić, S.: Hybrid firefly model in routing heterogeneous fleet of vehicles in logistics distribution. Log. J. IGPL **23**(3), 521–532 (2015)
8. Keynes, J.M.: The General Theory of Employment, Interest, and Money (reprint edition). Macmillan and Co., London (1949)
9. Samanta, B., Al-Araimi, S.A.: An inventory control model using fuzzy logic. Int. J. Prod. Econ. **73**(3), 217–226 (2001)
10. Madamidola, O.A, Daramola, O.A Akintola, K.G.: Web – based intelligent inventory management system. Int. J. Trend Sci. Res. Dev. **1**(4), 164–173 (2017)
11. Šustrová, T.: A suitable artificial intelligence model for inventory level optimization. Trends Econ. Manag. **25**(1), 48–55 (2016)
12. Zhivitskaya, H., Safronava, T.: Fuzzy model for inventory control under uncertainty. Central Eur. Researchers J. **1**(2), 10–13 (2015)
13. Kennedy, J., Eberhart, R.: Particle swarm optimization. In: IEEE International Conference on Neural Networks, vol. 4, pp. 1942–1948 (1995)
14. Shi, Y., Eberhart, R.: A modified particle swarm optimizer. In: IEEE International Conference on Evolutionary Computation, pp. 69–73 (1998)
15. https://www.mathworks.com/matlabcentral/fileexchange/53142-inventory-control-using-pso-in-matlab. Accessed 8 Mar 2018
16. Niknam, T., Amiri, B.: An efficient hybrid approach based on PSO, ACO and k-means for cluster analysis. Appl. Soft Comput. **10**(1), 183–197 (2010)
17. Mahadevana, K., Kannan, P.S.: Comprehensive learning particle swarm optimization for reactive power dispatch. Appl. Soft Comput. **10**(2), 641–652 (2010)
18. Pedersen, E.M.H., Chipperfeld, A.J.: Simplifying particle swarm optimization. Appl. Soft Comput. **10**(2), 618–628 (2010)

CISIS 2018

Movement Detection Algorithm for Patients with Hip Surgery

Cesar Guevara[1(✉)], Matilde Santos[2], and Janio Jadán[1]

[1] Technological University Indoamérica in the Mechatronics and Interactive Systems (MIST) research center, Quito, Ecuador
{cesarguevara, janiojadan}@uti.edu.ec

[2] University Complutense of Madrid, Madrid, Spain
msantos@ucm.es

Abstract. This work proposes a model of movement detection in patients with hip surgery rehabilitation. Using the Microsoft Xbox One Kinect motion capture device, information is acquired from 25 body points -with their respective coordinate axes- of patients while doing rehabilitation exercises. Bayesian networks and sUpervised Classification System (UCS) techniques have been jointly applied to identify correct and incorrect movements. The proposed system generates a multivalent logical model, which allows the simultaneous representation of the exercises performed by patients with good precision. It can be a helpful tool to guide rehabilitation.

Keywords: Movement detection · Kinect · Bayesian networks sUpervised Classification System · Rehabilitation

1 Introduction

Currently, the use of online assistance systems is having a great impact on different fields. In particular, in the health sector, assisted rehabilitation has greatly helped to improve the physical condition of patients and to alleviate possible problems arising from too slow evolution.

But rehabilitation usually requires the presence of an expert, a physiotherapist, who must be continuously monitoring the execution of the exercises. This is expensive and it is not always possible. Indeed, the incorrect performance of some physical efforts can be counterproductive for the patient.

This work proposes a system that is able to detect correct and incorrect movements in the execution of rehabilitation exercises. It is based on the application of some artificial intelligence techniques, particularly, Bayesian networks and sUpervised Classification System (UCS). The proposed model uses the Kinect 2 camera to collect information from patients who have undergone hip surgery.

The main goal is to identify if the patient is adequately performing each of the rehabilitation exercises and to guide them to do it correctly. This has a direct impact on a better recovery of the mobility.

The document is organized as follows. Section 2 summarizes some related works. In Sect. 3 materials and methods are presented (database, pre-processing, and the

M. Graña et al. (Eds.): SOCO'18-CISIS'18-ICEUTE'18, AISC 771, pp. 439–448, 2019.
https://doi.org/10.1007/978-3-319-94120-2_42

artificial intelligence techniques selected). Section 4 describes the development of the proposed movement detection system, using Bayesian networks and UCS algorithms, and the final results are discussed. Conclusions and future lines end the paper.

2 Related Works

In general, robotics as an aid to rehabilitation has been studied for several decades. In [1], a recent survey can be found.

The approach proposed in this work is specifically focused on intelligent systems that can help patients to perform therapeutic rehabilitation exercises.

In this line, the work of Akdogan [2] presents the design of Physiotherabot, a therapeutic robot of three degrees of freedom, which is applied to the lower extremities of patients with spinal cord injury (SCI). This system learns specific movements and exercises by a control structure based on rules. The experiments carried out in healthy subjects shown that the robot can perform the necessary movements, as well as imitate the manual exercises.

Schmidt [3] proposes HapticWalker, a multi-center robot for motor rehabilitation. It uses information from 155 patients with non-ambulatory cerebrovascular problems (DEGAS). The proposal reveals that the experimental group has superior ability to march and in other basic activities of life than the traditional rehabilitation group.

The work by Duschau-Wicke [4] presents the immediate effects of training march assisted by cooperative robots, in patients with incomplete spinal cord injury (iSCI). It is shown that assisted training can improve mobility. In this way, patients showed greater temporal and spatial kinematic variability, a greater increase in heart rate and more muscle activity.

In the paper by Kazerooni [5], an improved control model called BLEEX "hybrid", which has an effective robustness when changing the payload, is described. The gait cycle is divided into posture control and swing control phases. The position control is used for leg posture. A sensitivity amplification controller is used for the leg in oscillation.

Jovanov [6] presents a multi-level telemedicine system for applications of physical rehabilitation, assisted by computer and ambulatory monitoring. The system performs real-time analysis of sensor data, and provides guidance and feedback to the user. The system can generate warnings depending on the user's status, activity level, and environmental conditions. In addition, all recorded information can be transferred to medical servers through the Internet and integrated into the databases of medical records and users histories. This system applies artificial intelligence techniques such as fuzzy logic and self-organized Kohonen maps, obtaining good results.

Our work focuses on the rehabilitation of the lower limbs of patients with hip surgery, applying a synergy of some artificial intelligence techniques to the identification of the correctness of some rehabilitation movements.

3 Materials and Methods

3.1 Description of the Dataset

To carry out this study, a dataset of four patients who have undergone hip surgery has been generated. For a period of 4 weeks, these patients have received help from conventional physiotherapy and, at the same time, from the proposed rehabilitation tool.

The capture of the movements is done with a low cost device [7–9], the Kinect v2 camera, while the patients are performing the rehabilitation exercises. The Kinect sensor is essentially a camera that can detect body movement information in depth. When a person is within its field of vision, it internally determines the basic skeleton. That is, it creates a 3D skeleton model that represents the extremities and how they are interconnected. The skeleton of Kinect v2 consists of 25 articulations with their respective coordinates (x, y, z), distributed throughout the body (Fig. 1).

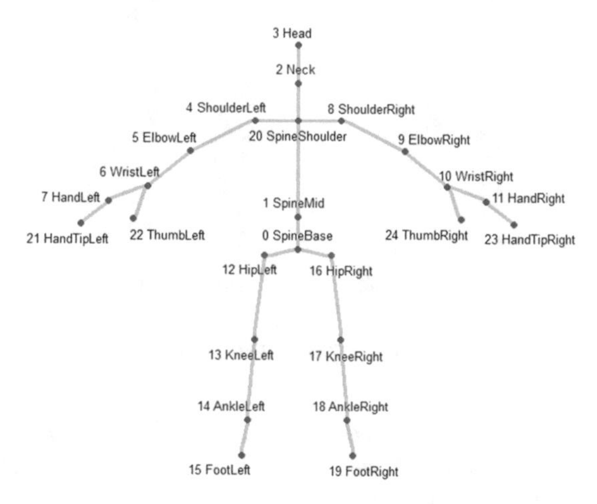

Fig. 1. Body points detected by the Kinect 2 for Xbox One (MSDN Microsoft).

The data set consists of 3292 samples with 75 attributes, divided into 4 users as shown in Table 1.

Table 1. Distribution of data by user

User	Side step		Knee elevation	
	Correct	Incorrect	Correct	Incorrect
1	364	459	459	364
2	486	345	345	486
3	301	408	408	301
4	461	468	468	461
Total	**1612**	**1680**	**1680**	**1612**

There is a wide range of exercises that the patient must perform, i.e. side step, front step, knee lift, leg lift, rotating hip movement, etc. In this first approach, we have selected only two of them, that are considered key exercises for patient's functional movement recovery. The model will be further developed with more exercises once the methodology here proposed will be validated.

These two physiotherapeutic exercises are: side step and knee elevation with one and another leg (Fig. 2). The objective of our system is to measure the movement characteristics during the execution of the exercise: elevation or distance from the starting point. For the present case, it is not necessary to take into account the speed, strength or intensity of the execution of the exercise, because the rehabilitation focuses on movements that help to give more elasticity to the muscles surrounding the hip.

Fig. 2. Rehabilitation exercises: "Side step" (left) and "Knee lift" (right).

The Kinect camera captures the body movements efficiently at a distance of 2 m, generating the skeleton images shown in Fig. 3. The sensor provides detailed information of the movements, angles, etc.

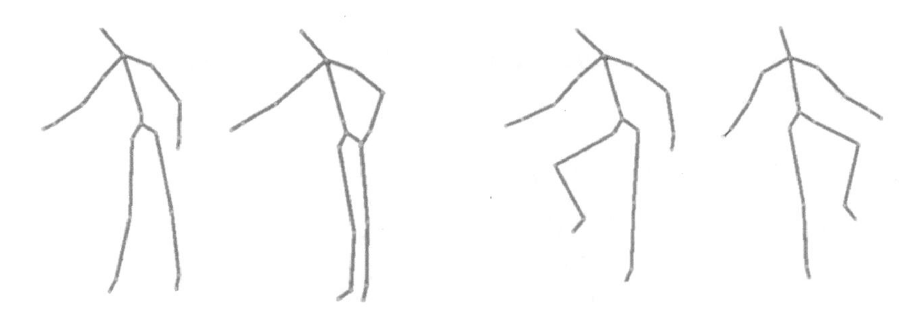

Fig. 3. Skeleton of points captured by the Kinect, exercises of "Side step" (left) and "Knee elevation" (right)

3.2 Attribute Selection

To facilitate the detection of the movements, it was necessary to apply a pre-processing to the data consisting of selecting the most relevant attributes, normalizing the information, and eliminating noise and inconsistent values in the measurements.

The selection of attributes is done using the Greedy Stepwise algorithm, which performs an exhaustive forward or backward search through the space of attribute subsets [10]. The algorithm could start with none or all of the attributes, or from an arbitrary point in the space. The algorithm stops when the addition or deletion of any remaining attribute results in a decrease in the evaluation (1). It can also produce a classified list of attributes by crossing the space from one side to the other and it records the order in which the attributes are selected [11].

$$Goodness\_attribute\_subset = \frac{\sum_{all\ attributes\ x} C(x, class)}{\sqrt{\sum_{all\ attributes\ x} \sum_{all\ attributes\ y} C(x, y)}} \quad (1)$$

As a result of the application of this algorithm, 18 relevant characteristics are selected (in our case only exercises centered on the hip and on the lower extremities, legs and feet, are considered). Specifically, these attributes are 0y, 6x, 6y, 10x, 10y, 12x, 12y, 13x, 13y, 14x, 14y, 15x, 15y, 16x, 16y, 17x, 17y, 19x, where the number corresponds to the sensor (Fig. 1). That is, these attributes identify the coordinates (x, y) of the movement of the lower limbs that will allow the detection of the correctness of movements.

3.3 Classification Techniques

Four different artificial intelligence techniques were tested to select the most suitable for this application.

3.3.1 Bayesian Networks

It is a probabilistic graphical model that represents a set of random variables and their conditional dependencies through a directed acyclic graph. The nodes represent random variables, which can be observable quantities, latent variables, unknown parameters or hypotheses. The edges describe the conditional dependencies, where each node has an associated probability function x, which takes as input a particular set of values of the parent variables of the node, and returns the probability of the variable represented by the node [12]. The joint probability density function can be written as a product of the individual density functions, which are conditioned to the variables of their parents (2), where X is a Bayesian network with respect to graph G.

$$(x) = \prod_{v \in V} p\left(X_v | X_{pa(v)}\right) \quad (2)$$

In our case, a Bayesian network has been used to obtain the conditioned probability of the movement of each point of the left and right legs while performing the lateral step exercise and knee elevation. It will be compared with the pattern to identify if the movement is correct depending on a threshold.

3.3.2 sUpervised Classifier System (UCS)

The sUpervised Classifier System (UCS) is an automatic learning method based on rules derived from XCS [13]. It uses a genetic algorithm (discovery component) to perform learning tasks. This technique seeks to identify a set of context-dependent rules, which are stored and applied in a distributed manner, to establish behavior models, classification, data extraction, regression, approximation function and game strategy. This approach allows the division of complex solution spaces into smaller and simpler parts.

UCS develops from an initial population of individuals, *P*. Each initial individual is a classifier that consists of a rule, with a set of parameters that estimate the quality of that rule. A rule has the structure *condition* → *class*, where the condition specifies the set of examples that will be classified with the class learned in the rule. The equation to select by votes each classifier of the UCS algorithm from a population *P* is given by (3).

$$\mathrm{dv} = \begin{cases} \mathrm{ns}\frac{\bar{\mathrm{F}}}{\mathrm{F}} & \text{if } \exp > \theta_{\mathrm{del}} \text{ and } \mathrm{F} < \delta\bar{\mathrm{F}} \\ \mathrm{ns} & \text{otherwise} \end{cases} \tag{3}$$

Where ns is the initial dataset, F the individual fitness function, $\bar{\mathrm{F}}$ is the fitness average of the whole population *P*, and θ_{del} and δ are parameters set by the user [14].

In our case, the mobility ranges for each of the selected attributes (defined as correct or incorrect) were considered as the initial population. Subsequently, the algorithm was trained so that it learned to identify the execution of the movements.

3.3.3 Decision Tree C4.5

Decision trees are one of the most widely used classification methods in machine learning, especially on unbalanced datasets. They have some similarity with the flow diagrams. Its operation consists in continuously forming partitions of the data set based on a criterion until the final partitions contain elements of a single class [15].

3.3.4 Fuzzy Unordered Rule Induction Algorithm (FURIA)

FURIA (Fuzzy Unordered Rule Induction Algorithm) is a novel classification method based on rules and developed from the well-known RIPPER algorithm [16, 17]. Based on fuzzy logic, FURIA uses simple and interpretable rules, as well as a series of modifications and extensions. The FURIA algorithm learns fuzzy rules instead of conventional rules, and unordered rule sets instead of rule lists, as presented in the following pseudo-code (Algorithm 1):

Algorithm 1. Furia Algorithm pseudo-code

```
1: Let A be the set of numeric antecedents of r
2: while A ≠ ∅ do
3:   a_max ← null {a_max denotes the antecedent with the highest purity}
4:   pur_max ← 0 {pur_max is the highest purity value, so far}
5:   for i ← 1 to size(A) do
6:     compute the best fuzzification of A[i] in terms of purity
7:     pur_A[i] ← be the purity of this best fuzzification
8:     if pur_A[i] > pur_max then
9:       pur_max ← pur_A[i]
10:      a_max ← A[i]
11:    end if
12:  end for
13:  A ← A \ a_max
14:  Update r with a_max
15: end while
```

This algorithm was applied to the movement rules for each of the points given by the sensors (coordinates), using the maximum and minimum limits.

3.4 Selection of Classification Techniques

Each of these intelligent techniques has been applied to the dataset of Iris [18]. The results are shown in Table 2, with the number of correctly (CC%) and incorrectly classified (IC%). The entire dataset of each patient was used for the training and testing, applying the k-fold cross validation technique with k = 10.

Table 2. Detection of movements with different classification algorithms.

Algorithm	Training		Tests	
	CC%	IC%	CC%	IC%
Bayesian network	100%	0%	99.99%	0.01%
UCS	99.98%	0.02%	98.17%	1.83%
C4.5	99.64%	0.36%	96.47%	3.53%
FURIA	99.87%	0.13%	98.78%	0.12%

As it can be seen in Table 1, Bayesian Networks algorithm gives the best results, followed by the UCS algorithm. Thus, both of them will be applied and the other two, C4.5 and FURIA, are discarded due to their higher error ratio, even if the results are not so bad.

4 Movement Detection Model and Discussion of the Results

The proposed model uses the two selected classification techniques, Bayesian Networks and UCS, in parallel (Fig. 4), for both the "Side Step" and "Knee Lift" exercises.

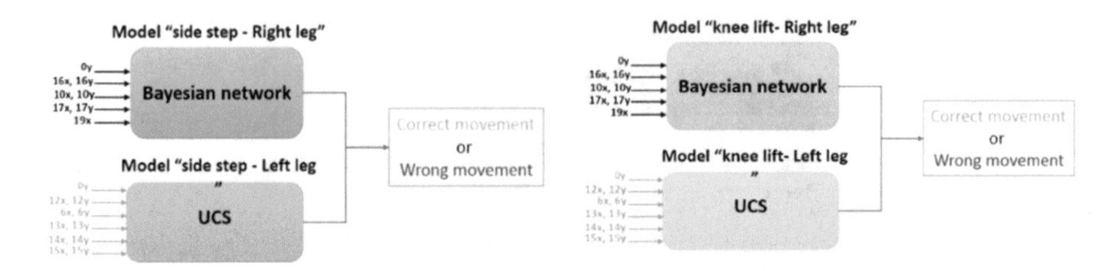

Fig. 4. Detection model for "Side step" and "Knee lift" rehabilitation exercises

Bayesian Networks are applied to the right leg and the UCS algorithm to the left leg. The 0y point (center of gravity of the patient) belongs to both limbs. The response of the proposed model is unique for each exercise, since the output is correct or incorrect whatever the leg or the exercise. These results are described in detailed and discussed.

4.1 "Side Step" Model

The correct identification of the side step exercise for the right leg (Bayesian networks) reaches 98.90% of hits, true positive rate (TP) of 98.90% and false positive (FP) rate 0.90%. For the left leg (UCS), 99.95% samples are correctly classified; TP = 99.90% and FP = 0.24%. Table 3 presents the confusion matrices.

Table 3. Confusion matrix "side step" exercise.

	Bayesian		UCS	
	Correct	Incorrect	Correct	Incorrect
Correct	363	1	363	1
Incorrect	8	451	1	458

4.2 "Knee Lift" Model

The correct identification of the knee lift exercise for the right leg (Bayesian networks) reaches 98.66% of hits, TP = 98.70% and FP = 1.20%. For the left leg (UCS), 99.99% samples are correctly classified; TP = 99.90% and FP = 0.12%. Table 4 presents the confusion matrices.

Table 4. Confusion matrix "knee lift" exercise.

	Bayesian		UCS	
	Correct	Incorrect	Correct	Incorrect
Correct	451	8	459	0
Incorrect	3	361	1	363

The results obtained in the testing phase gave an average of 99.34% accuracy and an average percentage of false positives of 0.61%. These results show that the proposed model gives slightly better results than the work by Rybarczyk [7], where the model accuracy in the capture of movements, using Dynamic Time Warping (DTW), was 98%.

5 Conclusions and Future Work

In this work we have developed a model for the detection of correct and incorrect movements in the execution of rehabilitation exercises in patients with hip surgery. The work uses information from several patients in rehabilitation phase, executing two exercises: "Side step" and "Knee lift".

Movements were captured by a Kinect sensor and different artificial intelligence techniques were used to identify the correctness. The combination of Bayesian networks and UCS has been proved very effective.

The proposed model was developed for specific exercises with the idea of generalizing it to other movements. Another future works includes the automatic processing of the movements.

Acknowledgments. We want to thank the Ecuadorian Corporation for the Development of Research and the Academy (CEDIA) for the support given to this research.

References

1. Barrios, L.J., Hornero, R., Perez-Turiel, J., Pons, J.L., Vidal, J., Azorin, J.M.: State of the art in neurotechnologies for assistance and rehabilitation in spain: fundamental technologies. Revista Iberoamericana De Automatica E Informatica Industrial **14**(4), 346–354 (2017)
2. Akdoğan, E., Adli, M.A.: The design and control of a therapeutic exercise robot for lower limb rehabilitation: Physiotherabot. Mechatronics **21**(3), 509–522 (2011)
3. Schmidt, H., Werner, C., Bernhardt, R., Hesse, S., Krüger, J.: Gait rehabilitation machines based on programmable footplates. J. Neuroeng. Rehab. **4**(1), 2 (2007)
4. Duschau-Wicke, A., Caprez, A., Riener, R.: Patient-cooperative control increases active participation of individuals with SCI during robot-aided gait training. J. Neuroeng. Rehab. **7**(1), 43 (2010)
5. Kazerooni, H., Steger, R., Huang, L.: Hybrid control of the Berkeley lower extremity exoskeleton (BLEEX). Int. J. Robot. Res. **25**(5–6), 561–573 (2006)
6. Jovanov, E., Milenkovic, A., Otto, C., De Groen, P.C.: A wireless body area network of intelligent motion sensors for computer assisted physical rehabilitation. J. NeuroEng. Rehab. **2**(1), 6 (2005)
7. msdn.microsoft.com/en-us/library/microsoft.kinect.kinect.cameraintrinsics.aspx
8. Rybarczyk, Y., Deters, J.K., Gonzalo, A.A., Esparza, D., Gonzalez, M., Villarreal, S., Nunes, I.L.: Recognition of physiotherapeutic exercises through DTW and low-cost vision-based motion capture. In: International Conference on Applied Human Factors and Ergonomics, pp. 348–360. Springer, Cham, July, 2017

9. Ayed, I., Moyà-Alcover, B., Martínez-Bueso, P., Varona, J., Ghazel, A., Jaume-i-Capó, A.: Validación de dispositivos RGBD para medir terapéuticamente el equilibrio: el test de alcance funcional con Microsoft Kinect. Revista Iberoamericana de Automática e Informática Industrial RIAI **14**(1), 115–120 (2017)
10. Gevrey, M., Dimopoulos, I., Lek, S.: Review and comparison of methods to study the contribution of variables in artificial neural network models. Ecological. Model. **160**(3), 249–264 (2003)
11. Xue, B., Zhang, M., Browne, W.N.: Particle swarm optimization for feature selection in classification: a multi-objective approach. IEEE Trans. Cybern. **43**(6), 1656–1671 (2013)
12. Bielza, C., Larrañaga, P.: Discrete Bayesian network classifiers: a survey. ACM Computing. Surv. **47**(1), 5 (2014)
13. Wilson, S.W.: Classifier fitness based on accuracy. Evol. Comput. **3**(2), 149–175 (1995)
14. Bernadó-Mansilla, E., Garrell-Guiu, J.M.: Accuracy-based learning classifier systems: models, analysis and applications to classification tasks. Evol. Comput. **11**(3), 209–238 (2003)
15. Quinlan, J.R.: C4.5: Programming for Machine Learning. Morgan Kauffmann, San Francisco (1993)
16. Hühn, J., Hüllermeier, E.: FURIA: an algorithm for unordered fuzzy rule induction. Data Mining Knowl. Discov. **19**(3), 293–319 (2009)
17. Cohen, W.W.: Fast effective rule induction. In: 1995 Machine Learning Proceedings, pp. 115–123 (1995)
18. https://archive.ics.uci.edu/ml/datasets/Iris

A Study of Combined Lossy Compression and Person Identification on EEG Signals

Binh Nguyen, Wanli Ma, and Dat Tran(✉)

Faculty of Science and Technology, University of Canberra, Canberra, Australia
{Binh.Nguyen,Wanli.Ma,Dat.Tran}@canberra.edu.au

Abstract. Biometric information extracted from electroencephalogram (EEG) signals is being used increasingly in person identification systems thanks to several advantages, compared to traditional ones such as fingerprint, face and voice. However, one of the major challenges is that a huge amount of EEG data needs to be processed, transmitted and stored. The use of EEG compression is therefore becoming necessary. Although the lossy compression technique gives a higher Compression Ratio (CR) than lossless ones, they introduce the loss of information in recovered signals, which may affect to the performance of EEG-based person identification systems. In this paper, we investigate the impact of lossy compression on EEG data used in EEG-based person identification systems. Experimental results demonstrate that in the best case, CR could achieve up to 70 with minimal loss of person identification performance, and using EEG lossy compression is feasible compared to using lossless one.

Keywords: Biometric information · EEG lossy compression
SPIHT · DWT-AAC · EEG-based person identification

1 Introduction

Electroencephalogram (EEG) has been widely used in measuring the electrical activities of the brain and diagnosing some brain-related diseases such as dementia, seizure, psychiatric disorders and sleep apnea [11]. Recently, EEG has also been used as a new type of biometric for person recognition [2,4,15,19,21]. Using EEG-based biometric has some advantages, compared to traditional ones such as fingerprint, face, voice and hand which can be subject to physical damages, so they are not universal. Moreover, face, fingerprint and iris can be photographed while voice can be recorded, introducing of fake biometric samples. In contrast, every living and functional person has a recordable EEG signal, and brain damage is something that rarely occurs, so the EEG feature is universal. In addition, it is very hard to fake an EEG signature or to attack an EEG biometric system [16]. Beside applying to criminal identification and police work, EEG-based person identification has also applications in many civilian purposes, such as access control and financial security systems in which the high level of accuracy

M. Graña et al. (Eds.): SOCO'18-CISIS'18-ICEUTE'18, AISC 771, pp. 449–458, 2019.
https://doi.org/10.1007/978-3-319-94120-2_43

and security is essential [21]. Therefore, it is said that using EEG in person identification is a promising research direction.

However, storing and circulating a huge amount of EEG data are the most challenges in using EEG signals. Actually, the number of channels, the amount of time, sampling frequency and bits resolution that signals being used, captured, sampled and digitised will affect EEG data size. The number of channels can exceed 256 while the time can be several hours or even several months [12]. Additionally, the more EEG signals are captured and used, the better accuracy and reliability that EEG-based applications could obtain. Hence, reducing the size of EEG data using compression techniques whilst still keeping the maximum of the fidelity of reconstructed signals is necessary. Lossless and lossy compressions are two kinds of EEG compression methods. Lossless compression decreases the size of the compressed EEG data without any loss of information in the reconstructed data, while lossy one introduces a loss in reconstructed data. However, the compression ratio (CR) of lossy techniques is much higher than that of the lossless ones. Consequently, EEG data compression studies mostly focus on the lossy techniques.

The use of EEG compression techniques in order to reduce the data size is necessary, so the impact of these techniques on EEG-based applications should be studied. With regard to EEG-based seizure detection systems, Higgins *et al.* and Daou *et al.* investigated the effect of EEG compression to the accuracy of these systems [7,10]. With respect to EEG-based user authentication system, Nguyen *et al.* stated that applying lossy compression on authentication system is feasible and the system accuracy is unchanged if the information loss is not greater than 11% [13]. In contrast, the relationship between lossy compression and EEG-based person identification systems has not been investigated.

To this end, this paper focuses on finding solutions to the following research questions: (1) What is the impact of lossy compression on EEG-based person identification system? (2) Is it feasible for EEG-based person identification systems to use reconstructed signals processed by lossy compression? and if it is, how much information loss can be tolerated?

2 EEG Lossy Compression Techniques

Currently, EEG lossy compression techniques can be classified in to four groups which are Wavelet-based, Filter-based, Predictor and Other [6]. Dao *et al.* pointed out that most of lossy techniques are Wavelet-based, and SPIHT and its extended versions are among the best lossy compression techniques and have been widely used [6]. This section provides a brief information of two lossy techniques, namely DWT-SPIHT and DWT-AAC, and performance measures.

2.1 DWT-SPIHT

SPIHT has been proposed by Said and Pearlman [17] for image compression. Its core principles are based on the embedded zerotree wavelet (EZW) introduced by Shapiro [18]. This algorithm exploits the relationship between wavelet coefficients in different scales to improve the performance. The principles used by the SPIHT algorithm include (1) partial ordering by magnitude, (2) partitioning by significance of magnitudes with regard to a sequence of octavely decreasing thresholds, ordered bit plane transmission, and (3) self-similarity across different scales in wavelet transform.

2.2 DWT-AAC

DWT-AAC has been proposed by Nguyen *et al.* [14]. In this method, the original EEG signals are undergoing a DWT operation, followed by Quantisation and Thresholding, before being coded by an Adaptive Arithmetic Coder (AAC). Putting the thresholding component after quantisation helps to control the impact of using static threshold values on fidelity of signals. Furthermore, using lossless compression technique, AAC, for the binary significance map and indices instead of using Arithmetic Coder or Huffman also helps to improve CR of this algorithm. Because ACC uses an adaptive mechanism in which updating probability of symbols occurs continuously while processing the data. Hence, AAC does not require probability distribution function of the message to be transmitted to the decoder. It is reported in [14] that DWT-AAC achieves good results compared to some recent compression techniques.

2.3 Performance Measures

The widely-used metrics to measure the performance of compression algorithms are Compression Ratio (CR) and Percentage Root-mean-square Difference (PRD). In addition, the standard deviations (STD) of CRs and PRDs are also used.

CR is defined as the ratio between the uncompressed and compressed sizes:

$$CR = \frac{L_0}{L_c} \tag{1}$$

where L_0 and L_c denote the number of bits of original EEG signal and the number of bits of compressed EEG signal, respectively.

PRD is a standard metric to evaluate the distortion between the original and reconstructed signals. It is defined as:

$$PRD = \sqrt{\frac{\sum_{i=1}^{N}(x[i] - x'[i])^2}{\sum_{i=1}^{N}(x[i])^2}} \times 100\% \tag{2}$$

where $x[i]$ represents the original EEG signal, $x'[i]$ represents the compressed signal, and N is the number of samples.

3 EEG-Based Person Identification System

Recently, feature extraction and modelling techniques have been proposed for EEG-based person identification systems. In particular, Autoregressive (AR), Power Spectral Density (PSD), paralinguistic feature and Discrete Fourier Transform (DFT) have been used for feature extraction. Furthermore, Support Vector Machine (SVM), k-Nearest Neighbours, Naïve Bayes and Neural Network are some common techniques for classifiers that have been used in [2,4,15,19].

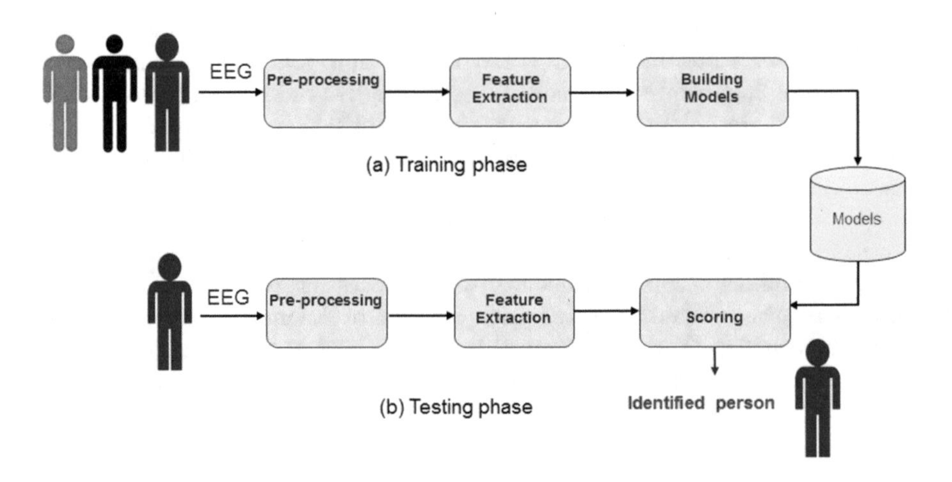

Fig. 1. A typical EEG-based person identification system

In this present work, we employ an EEG-based person identification system proposed in [15]. There are two phases, namely training phase and testing phase, which uses paralinguistic feature and SVM for building person models. Figure 1 illustrates the typical architecture of the person identification system based on EEG signals.

We examine the impact of lossy compression on the person identification performance by studying the identification rate without and with compression at different levels of CR.

4 Experiments and Results

4.1 Test Conditions

Two public datasets were used for this research, which are CHB MIT Scalp EEG Database [9], and Alcoholism Large EEG [3]. CHB MIT Scalp EEG Database contains 10 records of EEG signals from 10 patients with epilepsy while Alcoholism Large EEG has 20 subjects (10 alcoholics and 10 controls).

SPIHT was used and set up to run with 5 levels of biorthogonal 4.4 2-D DWT (DWT-SPIHT) because this mode has already achieved wide acceptance for use in compression algorithms [7]. Beside SPIHT, DWT-AAC was also set up to run using 5 levels of biorthogonal 4.4 2-D DWT and 6 bits uniform quantisation due to its high performance compared to recent techniques [14].

The openSMILE feature extraction toolkit was used to extract EEG features [8]. The feature was extracted individually from each EEG channel, and then features from all channels are merged together. The 8 channels chosen are C3, C4, Cz, P3, P4, Pz, O1 and O2 for all of the two datasets.

We used 2/3 for cross validation training and 1/3 for testing applied to all datasets. Linear SVM classifiers [5] were trained in 3-fold cross validation scheme with parameter C ranging from 1 to 1000 in 5 steps. Features for training and testing were randomly selected ten times, and the final experimental result is the average of these ten results.

4.2 Compression Performance

Figures 2a and 3a illustrate the plots of average PRDs versus average CRs for DWT-SPIHT and DWT-AAC on CHB MIT Scalp EEG and Alcoholism Large EEG, respectively. Generally, DWT-AAC obtains better results than DWT-SPIHT. Particularly, on CHB MIT Scalp EEG, with CR values are smaller than 8, DWT-SPIHT has slightly better compression performance than DWT-AAC whilst there is a reverse trend when CRs are greater than 8. At PRD of 7%, for example, CRs of DWT-SPIHT and DWT-AAC are 7 and 5.5 correspondingly. Conversely, at PRD of 15%, the latter obtains a CR of 15, compared to a CR of 13 for the former. For Alcoholism Large EEG, it can be seen that the compression performance of DWT-AAC is much better than that of DWT-SPIHT. For instance, at CR of 10, PRD of the former is 16% while that of latter is 32%.

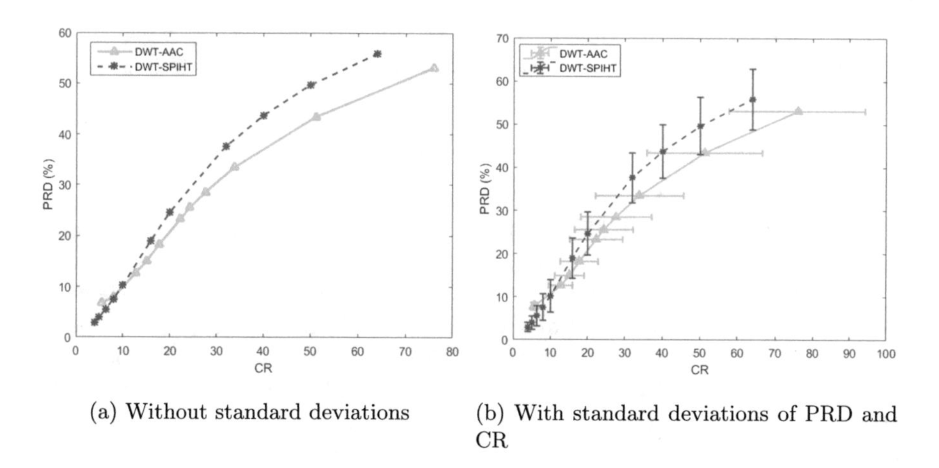

(a) Without standard deviations

(b) With standard deviations of PRD and CR

Fig. 2. Average PRDs versus average CRs for DWT-SPIHT and DWT-AAC on CHB MIT Scalp EEG

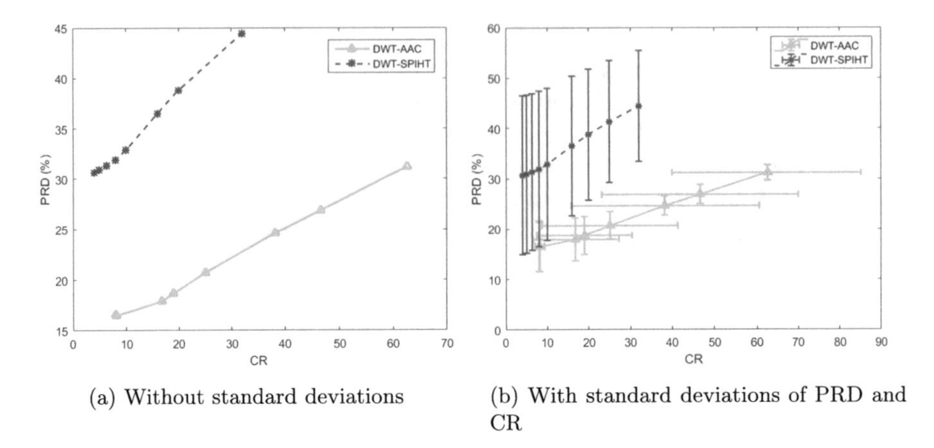

(a) Without standard deviations

(b) With standard deviations of PRD and CR

Fig. 3. Average PRDs versus average CRs for DWT-SPIHT and DWT-AAC on Alcoholism Large EEG

Beside this, the standard deviations (STD) calculated from the CRs and PRDs of subjects' EEG data processed by DWT-SPIHT and DWT-AAC on CHB MIT Scalp EEG and Alcoholism Large EEG are shown on Figs. 2b and 3b, correspondingly. The STDs of CRs of DWT-SPIHT are zero because the size of compressed data is determined exactly based on bit rate before compression, while those of PRDs are quite large. Conversely, the STDs of CRs are much higher than those of PRDs for DWT-AAC. Additionally, the STDs conducted on Alcoholism Large EEG are significantly higher than those performed on CHB MIT Scalp EEG. The reason could be that the differences of EEG signals between alcoholic and normal subjects in the former dataset are bigger than those between epilepsy subjects in the letter. These introduce larger gaps of CRs and PRDs between subjects in Alcoholism Large EEG, creating higher STDs.

4.3 Person Identification Performance with Increasing Compression

Figures 4 and 5 show the plots of identification rates versus corresponding CRs for both DWT-SPIHT and DWT-AAC on CHB MIT Scalp EEG and Alcoholism Large EEG, respectively. As expected, when CR increases, the identification performance for both compression algorithms declines gradually. This is because, when CR augments, PRD will also escalate, which makes more and more the fidelity of the recovered EEG signals being lost. Moreover, identification rates for DWT-AAC are better than those for DWT-SPIHT on both datasets. For CHB MIT Scalp EEG, when CRs are less than 18, identification rates are quite similar for both techniques. However, as CRs are greater than 18, there is a slight reduce trend of identification rates for DWT-AAC, compared to a significant drop one for DWT-SPIHT. Similar on Alcoholism Large EEG, identification performances for DWT-AAC do not fall down until CR reaches 38 whilst those for DWT-SPIHT tumble when CRs are greater than 10.

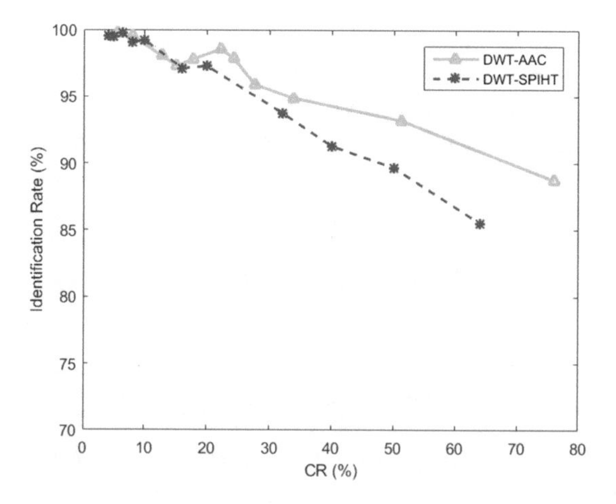

Fig. 4. Identification rate versus CR for DWT-SPIHT and DWT-AAC on CHB MIT Scalp EEG

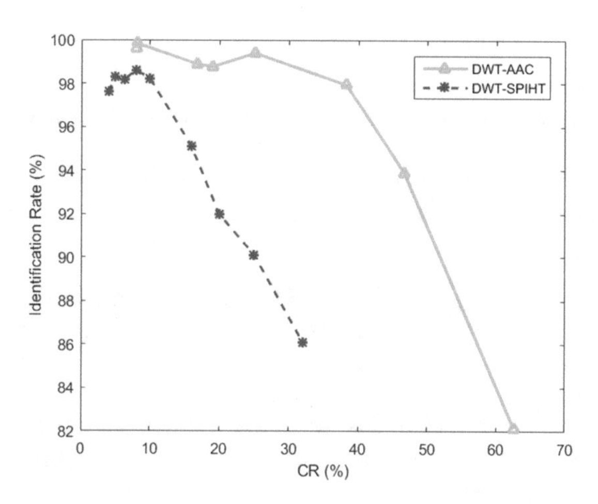

Fig. 5. Identification rate versus CR for DWT-SPIHT and DWT-AAC on Alcoholism Large EEG

Table 1. Performance of person identification using original EEG signals

Dataset	Identification rate(%)
CHB MIT Scalp EEG	99.28
Alcoholism Large EEG	92.61

Table 1 indicates performances of the person identification system using uncompressed (original) EEG data on two datasets. Comparing Table 1, and Figs. 4 and 5, it is said that for both compression algorithms, the system using reconstructed signals at low CRs obtains higher performance than that using original ones. For instance on CHB MIT Scalp EEG dataset, identification rates are 99.52% and 99.72% for DWT-AAC at a CR of 5.8 with a PRD of 7.9% and for DWT-SPIHT at a CR of 6.4 with a PRD of 5.6% respectively, which is higher than that of using the uncompressed signals (99.28%). Similarly, on Alcoholism Large EEG dataset, identification rates are 99.35% and 96.45% for DWT-AAC at a CR of 8.14 with a PRD of 16.48% and for DWT-SPIHT at a CR of 8 with a PRD of 31.95% correspondingly, compared to 92.61% when using the uncompressed data.

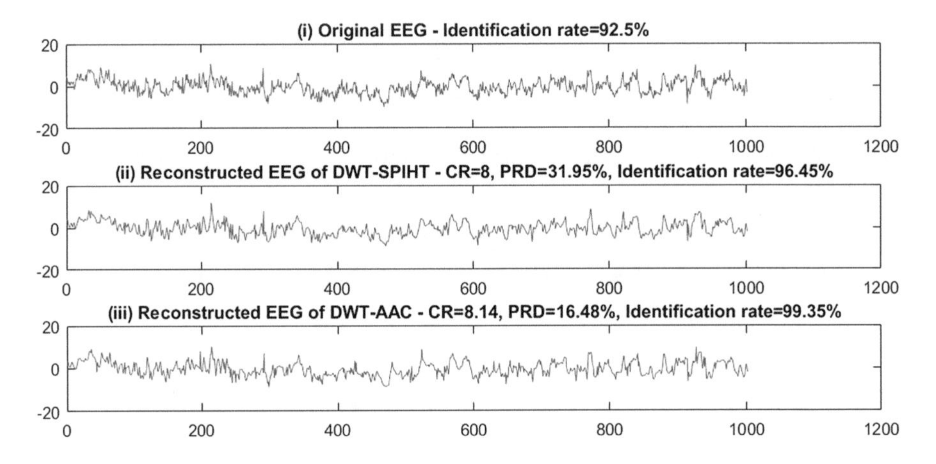

Fig. 6. EEG signals from subject co2a0000365, channel C3, Alcoholism Large. (i) Original signals, reconstructed signals with (ii) DWT-SPIHT and (iii) DWT-SPIHT.

To explain this, some features of both algorithms should be examined. For DWT-AAC, after quantisation, quantised wavelet coefficients are threshold, which results in insignificant coefficients that those values less than the threshold are discarded. Most of insignificant coefficients are the highpass 'detail' coefficients. Based on bit rate, the size of compressed data in DWT-SPIHT will be determined exactly. After quantisation, quantised wavelet coefficients are ordering by magnitude and significance of magnitudes will be coded first during the size of compressed data is still available. Similar to DWT-AAC, unimportant coefficients are removed in DWT-SPIHT. Therefore, the most probable explanation is that applying both compression algorithms at low CRs may actually filter out some irrelevant noise such as signal artifacts present in the EEG signals. This makes recovered signals smoother whilst still preserving the biometric information well, improving the performance of the person identification system. Figure 6 demonstrates a plot of the original signal versus the reconstructed signals for both algorithms at a CR around 8 on Alcoholism Large EEG dataset. Although three signals are nearly similar, two recovered signals have a bit smoother compared to the original ones thanks to removing some irrelevant noise.

4.4 Maximising Compression and Information Loss

One of the purposes of this paper is to maximise the compression rate and information loss without affecting the ability to EEG-based person identification system. For seizure detection, studies in [10] indicated that a percentage greater than 90% is considered very good performing classifier. Therefore, a percentage of more than 90% is also considered very good performance for person identification. In this paper, 90% Identification rate is used as a threshold limit for compression.

Referring back to Fig. 4, it can be seen that at an identification rate of 90% on CHB MIT Scalp EEG, CRs of DWT-SPIHT and DWT-AAC are 50 and 70, respectively. From Fig. 2a, the equivalent PRDs are 49% and 50% for the former and latter correspondingly. From Fig. 5 and 3a on Alcoholism Large EEG at a 90% identification rate, DWT-SPIHT achieves a CR of 25 with a PRD of 41.3% whilst DWT-AAC obtains a higher CR of 52 corresponding to a PRD of 27%.

5 Conclusion

Our experimental results have answered the above-mentioned two research questions as follows (1) Lossy compression affects the EEG-based person identification system by reducing identification performance with increasing compression. However, at low CRs, lossy compression improves identification performance thanks to removing the noise and making reconstructed signals smoother. (2) When 90% identification rate is determined as a threshold, the compression can be tolerated at CR up to 70 (PRD of 50%) for the best case and CR up to 25 (PRD of 41.3%) for the worst case. Beside this, studies in [1,20] indicated that CR of lossless techniques of EEG signals achieves only between 2 and 3. Therefore, it is feasible to apply lossy techniques to EEG-based person identification systems, compared to using the lossless ones. Furthermore, DWT-AAC gives a better result than DWT-SPIHT.

For further investigation, other lossy compression techniques as well as EEG-based person identification systems based on other methods will be conducted on a larger scale of EEG datasets to verify the findings found in this work.

References

1. Antoniol, G., Tonella, P.: EEG data compression techniques. IEEE Trans. Biomed. Eng. **44**(2), 105–114 (1997)
2. Bao, X., Wang, J., Hu, J.: Method of individual identification based on electroencephalogram analysis. In: International Conference on New Trends in Information and Service Science, NISS 2009, pp. 390–393. IEEE (2009)
3. Begleiter, H.: EEG database. http://kdd.ics.uci.edu/databases/eeg/eeg.data.html
4. Brigham, K., Kumar, B.V.: Subject identification from electroencephalogram (EEG) signals during imagined speech. In: 2010 Fourth IEEE International Conference on Biometrics: Theory Applications and Systems (BTAS), pp. 1–8. IEEE (2010)

5. Chang, C.C., Lin, C.J.: LIBSVM: a library for support vector machines. ACM Trans. Intell. Syst. Technol. (TIST) **2**(3), 27 (2011)
6. Dao, P.T., Li, X.J., Do, H.N.: Lossy compression techniques for EEG signals. In: 2015 International Conference on Advanced Technologies for Communications (ATC), pp. 154–159. IEEE (2015)
7. Daou, H., Labeau, F.: Performance analysis of a 2-D EEG compression algorithm using an automatic seizure detection system. In: 2012 Conference Record of the Forty Sixth Asilomar Conference on Signals, Systems and Computers (ASILOMAR), pp. 1632–1636. IEEE (2012)
8. Eyben, F., Wöllmer, M., Schuller, B.: OpenSMILE: the munich versatile and fast open-source audio feature extractor. In: Proceedings of the 18th ACM International Conference on Multimedia, pp. 1459–1462. ACM (2010)
9. Goldberger, A.L., Amaral, L.A., Glass, L., Hausdorff, J.M., Ivanov, P.C., Mark, R.G., Mietus, J.E., Moody, G.B., Peng, C.K., Stanley, H.E.: Physiobank, physiotoolkit, and physionet components of a new research resource for complex physiologic signals. Circulation **101**(23), e215–e220 (2000)
10. Higgins, G., McGinley, B., Faul, S., McEvoy, R.P., Glavin, M., Marnane, W.P., Jones, E.: The effects of lossy compression on diagnostically relevant seizure information in EEG signals. IEEE J. Biomed. Health Inform. **17**(1), 121–127 (2013)
11. Hill, D.: Value of the EEG in diagnosis of epilepsy. Br. Med. J. **1**(5072), 663 (1958)
12. Marsan, C.A., Zivin, L.: Factors related to the occurrence of typical paroxysmal abnormalities in the EEG records of epileptic patients. Epilepsia **11**(4), 361–381 (1970)
13. Nguyen, B., Nguyen, D., Ma, W., Tran, D.: Investigating the possibility of applying EEG lossy compression to EEG-based user authentication. In: 2017 International Joint Conference on Neural Networks (IJCNN), pp. 79–85. IEEE (2017)
14. Nguyen, B., Nguyen, D., Ma, W., Tran, D.: Wavelet transform and adaptive arithmetic coding techniques for EEG lossy compression. In: 2017 International Joint Conference on Neural Networks (IJCNN), pp. 3153–3160. IEEE (2017)
15. Nguyen, P., Tran, D., Huang, X., Sharma, D.: A proposed feature extraction method for EEG-based person identification. In: International Conference on Artificial Intelligence (2012)
16. Riera, A., Soria-Frisch, A., Caparrini, M., Grau, C., Ruffini, G.: Unobtrusive biometric system based on electroencephalogram analysis. EURASIP J. Adv. Signal Process. **2008**, 18 (2008)
17. Said, A., Pearlman, W.A.: A new, fast, and efficient image codec based on set partitioning in hierarchical trees. IEEE Trans. Circuits Syst. Video Technol. **6**(3), 243–250 (1996)
18. Shapiro, J.M.: Embedded image coding using zerotrees of wavelet coefficients. IEEE Trans. Signal Process. **41**(12), 3445–3462 (1993)
19. Shedeed, H.A.: A new method for person identification in a biometric security system based on brain EEG signal processing. In: 2011 World Congress on Information and Communication Technologies (WICT), pp. 1205–1210. IEEE (2011)
20. Srinivasan, K., Reddy, M.R.: Efficient preprocessing technique for real-time lossless EEG compression. Electron. Lett. **46**(1), 26–27 (2010)
21. Sun, S.: Multitask learning for EEG-based biometrics. In: 19th International Conference on Pattern Recognition, ICPR 2008, pp. 1–4. IEEE (2008)

Accelerating DNA Biometrics in Criminal Investigations Through GPU-Based Pattern Matching

Ciprian Pungila(✉) and Viorel Negru

Faculty of Mathematics and Informatics, West University of Timisoara, Timisoara, Romania
{ciprian.pungila,viorel.negru}@e-uvt.ro

Abstract. With the ever-increasing capabilities of modern hardware and breakthroughs in the DNA biometrics field, we are presenting a new, scalable and innovative method to accelerate the DNA analysis process used in criminal investigations, by building an improved methodology for using large-scale GPU-based automata for performing high-throughput pattern-matching. Our approach focuses on all important stages of preparing for the pattern-matching process, tackling with all major steps, from creation, to preprocessing, to the runtime performance. Finally, we experiment using real-world DNA sequences and apply the process to the human DNA genome, for an evaluation of our implementation.

Keywords: DNA · Biometrics · Pattern matching · GPGPU · CUDA

1 Introduction

While the 80s and the 90s have been dominated by major hardware breakthroughs, where Moore's law was not just respected, but even left behind by technological progress, nowadays it has been clear for a good while now that this is no longer the case. Hardware struggles to push the physical limit to the maximum allowed extent, while the next logical step of using multiple computational units in parallel has been proven to be the way to go for the coming decade, if not more. With that in mind, it is commonly agreed today that the most impressive hardware performance breakthroughs come not from CPUs any longer, but from the dominance of SSDs over classic mechanical drives, as well as from heterogeneous systems and their massive parallel computation capabilities (with graphics cards no longer dominating the gaming area, but also having complex roles in high-performance computation fields such as cryptography, biomedical research, digital forensics, medical breakthroughs and criminal investigations).

In this paper, we are focusing on all critical computational aspects of the DNA analysis process, as is commonly used in DNA biometrics in criminal

M. Graña et al. (Eds.): SOCO'18-CISIS'18-ICEUTE'18, AISC 771, pp. 459–468, 2019.
https://doi.org/10.1007/978-3-319-94120-2_44

investigations, discussing the most frequently encountered use-case in the field: that of DNA sequence matching, with a particular emphasis on multiple pattern matching of DNA nucleotide sequences. DNA biometrics are widely used nowadays in various fields of technology: from criminal investigations (where gene identification, parentage, genealogy and body characteristics are used as evidence to incriminate or discriminate the convict in a court of trial) to biomedical research (where DNA sequencing is used for cancer treatment, fighting diseases and pharmacological research of better therapies on various cases). With technology improving through better parallel models (such as those existing in GPGPU-capable graphics cards, using dedicated development frameworks such as CUDA [1] or OpenCL [2]), yet memory retaining an expensive price-tag in such hardware (as opposed to classic RAM memory, which is cheaper), the need to improve the process becomes more evident.

2 Related Work

In this research, we analyze an algorithm based on the most common approach to multiple pattern matching, as used in DNA biometrics nowadays: using finite state machines for exact matching, in particular the widely-spread Aho-Corasick algorithm. While the algorithm itself has been discussed many times and applied in various hardware environments, it is the authors' opinion that the most often overlooked element of all, that of creating the automaton for using it in a real-world environment, is both lacking from literature and is not considered to be as important as, for example, its run-time performance. We prove that the creation step, as well as the run-time step, can both be significantly improved (by significant margins) through a better memory model for storing the automaton, as well as an improved constructional model for performing all necessary pre-computations for building the final automaton, and then we apply the resulting automaton to a real-world test, where we prove that there is no performance impact at run-time. With our findings, proving how such an automaton can be constructed in a massively parallel manner through a heterogeneous CPU-GPU architecture, we speed-up not only the process of preparing for the DNA analysis process, but also open the door to extremely big automata, with hundreds of millions of nodes, to be successfully used in DNA biometrics even in low-memory GPGPU devices.

2.1 The Aho-Corasick Algorithm

The Aho-Corasick algorithm [3] has emerged in 1975, and has established itself as the main approach to performing multiple pattern-matching with linear-performance, with very little to almost no dependence on the number of patterns being used. The algorithm builds a trie tree from all the patterns that are to be searched for, after which it transforms the trie into an automaton by adding a mismatch edge (through a failure function), which is being computed, in simple terms, as the longest suffix of the word at the current node which also exists in the trie (if no such suffix is present in the trie, the mismatch edge points to the root of the automaton). If at a current node, there is a valid edge/transition for the current element, the automaton is traversed to that edge, otherwise the mismatch edge is being used and the input character is being applied to that node where the failure function points to.

For DNA analysis, the primary dictionary symbols are A, C, T and G, corresponding to the four major nucleic acid codes. We are presenting a sample automaton based on these four symbol types, in Fig. 1.

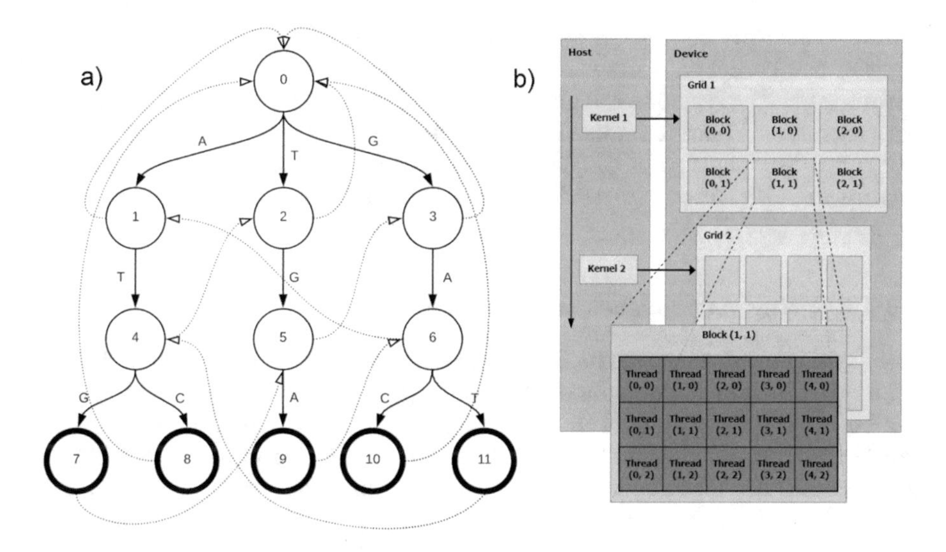

Fig. 1. (a) The Aho-Corasick multiple pattern matching automaton for the keywords *atg, atc, tga, gac, gat*. Mismatch edges are depicted through dashed lines. (b) The CUDA architecture.

In computational terms, its complexity is $O(N+M+K)$, where N is the size of the input data, M is the size of the longest pattern in the dictionary of words to be searched for, and K is the average number of occurrences of matches in the input data. In practice, especially in a field such as DNA biometrics, where sequences are formed by thousands and even millions of nucleotides, both M and K are very small, while N is huge (with billions of nucleotides to be searched).

Additionally, one important issue is that the Aho-Corasick automaton's memory footprint is directly dependent on the number of patterns and their sizes. For a limited alphabet as the one used by the DNA patterns, that poses a significant problem, especially when dealing with huge automata of with hundreds of thousands of patterns.

2.2 The CUDA Architecture

The CUDA framework has been developed by the nVIDIA Corporation, as an attempt to allow custom code execution on their GPUs. Ever since its introduction back in 2007, CUDA has become a widely adopted framework for implementing massively parallel algorithms, both for research and for various computationally-intensive tasks (e.g. video encoding, cryptography, etc.). A warp in CUDA is a group of 32 threads, which is the minimum size of the data processed in SIMD fashion by a CUDA multiprocessor. The CUDA architecture works with blocks that can contain 64 to 512 threads. Blocks are organized into grids. Parallel portions of an application are executed on the device (GPU) as kernels, one kernel at a time, while many threads execute each kernel. Taking advantage of this type of parallelism is essential in order to achieve high throughputs therefore. Figure 1 shows the CUDA architecture used in the implementation. One disadvantage in the CUDA architecture is that all data that is being used in the GPU needs to reside in video memory, therefore requiring moving from RAM to V-RAM.

2.3 Existing Approaches to DNA Analysis

The existing approaches to the DNA analysis process vary significantly, depending on the use-case and domain of application [4]. GPU-accelerated DNA analysis has been used, for example, in cancer detection [5], as well as human identification through polymorphism [6]. A CPU-based SIMD approach (just like the GPUs under the CUDA framework) to accelerating the pattern-matching process with the Intel Xeon Phi has been presented in processor [7]. A particular emphasis for our paper lies in tracking down criminal activity through DNA analysis by pattern-matching, such as the system introduced by the European Union through the Prum system in [8], with more countries looking to exchange such information for preventing crime [9].

Although there are numerous papers discussing GPU-based pattern-matching approaches [7,10–12], the main focus is always the run-time performance, not the stages that the automaton needs to go through, in order to become usable at run-time. For example, there is no mentioning of the implementation details for most of the papers that cover GPU-accelerated pattern matching, or on how much time this requires. Additionally, the amount of preprocessing required for building a large-scale automaton directly involves the ability to use such big data structures for more applications and with more patterns. In one simple experiment we have performed, constructing an automaton with 2 million nodes, and moving it from RAM to video RAM took more than 2 h to complete! In fact, recent research papers (e.g. [7]) are restrained to using simple (and even very simple!) automata, with only a few keywords, for testing the pattern-matching process, which not only reduces the experimental value of the results, but also restricts the applicability of the approach to very small automata of such kind.

3 Implementation

The trie creation on its own is not a severely computationally intensive process. Even with more than 110,000 patterns and a resulting number of more than 8 million nodes, as we have obtained in our experimental results, the creation stage is still very fast when executed on the CPU (host) (totalling 858 ms for DB1, 936 ms for DB2 and 1638 ms for DB3).

3.1 Memory-Compact Automaton

In a classic trie implementation on the CPU, allocating memory through a simple call such as *malloc()* would create a significantly sparse distribution of node pointers in RAM memory, which is detrimental to moving such large trees to video memory, as well as to cache locality. A simple layout of how such memory would look like after a tree memory allocation routines is detailed in Fig. 2, also outlining the ideal memory-compact layout which we are presenting as follows.

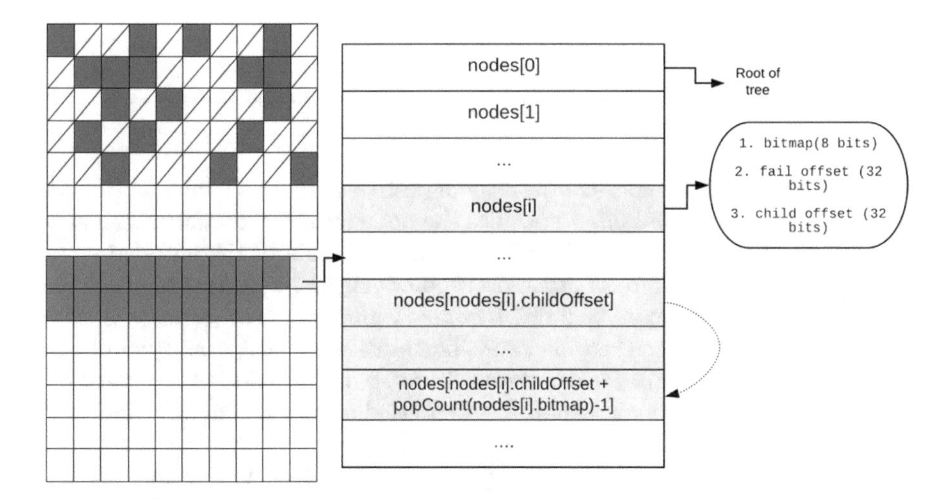

Fig. 2. *Left-side:* Classic tree memory allocation routines layout at top (hashed cells are empty, wasted/unused memory areas), our memory-compact tree layout at the bottom. *Right-side:* depiction of a GPU node structure.

The memory-compact model we are employing focuses on storing all nodes in the automaton as a single, linear list of nodes, modifying data structures accordingly. One important requirement here is that of all children of a node to be stored consecutively in this list. With our layout, transferring the automaton from RAM to V-RAM is done in a single burst, at full PCI-Express bandwidth speeds, paving the way to real-time heterogeneous processing of large automata. In a similar manner to our approach in [13], the construction algorithm for the automaton is presented in Algorithm 1 below.

Algorithm 1. Our proposed automaton memory-compacting model.

```
Data: compressNode(hostNode(hostNode (from RAM), index)
Result: stack of deviceNode (in V-RAM)
deviceNode[index].offset ← top;
deviceNode[index].data ← hostNode.data;
old ← top;
i ← 0;
for i < alphabetSize do
    if hostNode.hasChild(i) then
        setBit(deviceNode[index].bitmap, i);
    end
    i ← i+1;
end
top ← top+popCount(deviceNode[index].bitmap);
i ← 0;
for i < alphabetSize do
    if hostNode.hasChild(i) then
        compressNode(hostNode.child(i), old+i);
    end
    i ← i+1;
end
```

Each node in the memory-compact model uses a bitmapped node, a fail offset and a child offset. The fail offset specifies the node offset for the mismatch edge, while the child offset specifies the offset of the first child of the automaton (all other children offsets can be computed automatically using the *popcount* of that bitmap, up to the index of the child node of interest). The bitmap is being used to indicate whether there is a specific transition in case the symbol with the index of the bitmap is encountered at run-time (a value of 1 indicates a transition, otherwise there is none). In practice, since our alphabet is limited to just 4 symbols, our data structure requires only 9 bytes (1 byte for the bitmap + leaf bit indicator, and 2 integer offsets) for a single node in our automaton.

The memory-compact model construction we proposed took 78 ms (DB1), 140 ms (DB2) and 218 ms (DB3) to build from the classic automaton, making it extremely fast in practice, even for very-large automata.

3.2 Run-Time Benchmarking

For the run-time performance analysis, we used a data-parallel implementation of the Aho-Corasick pruned automaton. Both the classic (CPU) and the memory-compact (GPU) models were employed, for comparing the performance of the two algorithms. The CPU implementation is using a single-core to perform the parsing of the input DNA sequence, to ensure maximum memory bandwidth.

Figure 3 shows a depiction of how the parallel processing is handled in the GPU: a thread here receives a chunk of data, which is being processed and if

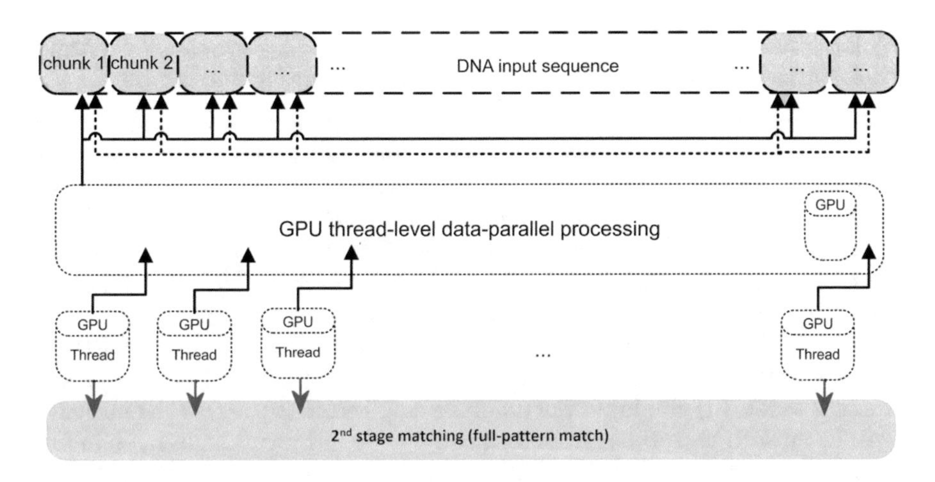

Fig. 3. The data-parallel implementation of the Aho-Corasick automaton in the GPU.

a match is found, a second-stage full-pattern search is employed. To make sure no patterns are lost over boundaries, we extend the boundary of each thread by *depth-1*, where *depth* is the maximum automaton depth (in our case, 100).

4 Experimental Results

For our implementation, we have used DNA sequences of various genomes from the FASTA open database [14], and applied the pattern-matching search on the human (homo sapiens) genome GRCh38 from GenBank [15]. We used three different databases for our testing purposes: the first (DB1) containing 55,687 DNA sequences, the second (DB2) containing 60,776 sequences and a combination of the two (DB3) containing a total of 116,463 sequences.

Many of the sequences of these databases have multiple million symbols, which would create a severely sparsed density of the resulting trie after a certain depth has been reached, so we performed a simple experiment to see how a pruned automaton (to a depth of 100 symbols) would affect the overall matching process - the results have shown that if the first 100 characters of the sequence matched, there is a 99.997% probability that the entire remaining sequence also matches. We therefore pruned the automaton to a depth of 100, and have obtained, for the three databases specified, a total of 3,15 million nodes (DB1), 4,9 million nodes (DB2) and 8,04 million nodes (DB3). In case a leaf is reached when traversing the automaton from the input text, we employ a separate simple search of the remaining pattern of the sequences involved, to ensure accurate matching.

For these experiments, we used an i7 6700HQ CPU, with 32 GB of DDR4 2133 MHz RAM and a 1 TB Samsung EVO SATA SSD, and an nVIDIA Quadro M1000M GPU with 512 CUDA cores and 2 GB DDR5 V-RAM. The i7 has a memory bandwidth of 34.1 GB/s (for a single core), while the Quadro has

80 GB/s, meaning that a maximum theoretical speed-up for the GPU implementation vs CPU should see at most 2.34× higher throughput for the GPU.

Creation and Preprocessing. The first aspect we have focused on when it came to performance evaluation was the time it took us to create and preprocess (compute the failure functions for all nodes) the Aho-Corasick automaton, before actually using traversing it. We handled the test using the 3 database mentioned earlier, both on the CPU and on the GPU. The CPU testing was straightforward: we tested both preprocessing of the classic Aho-Corasick implementation, as well as the same for the memory-efficient version of our proposed model.

For the GPU implementation we first (a) created the trie on the CPU, (b) created our memory-compact automaton on the CPU, (c) copied it to V-RAM in one single burst, (d) employed the preprocessing kernel to prepare the mismatch edges. Each GPU thread has been assigned a node in the automaton, as to fully benefit of the high-throughput SIMD execution. The results (see Fig. 4) have shown that, even when ran on the CPU, our memory-efficient model handles the preprocessing stage faster (4× faster than the classic implementation!), very likely due to the increased cache locality offered by the memory-compact layout we proposed. When executed on the GPU, the preprocessing was 2.15× faster on average than their CPU counterpart, in full consistency with the memory bandwidths of the two devices involved. Storage-wise, our improved model reduced the required storage by a factor of 5.17× as compared to the classic, gap-prone RAM implementation.

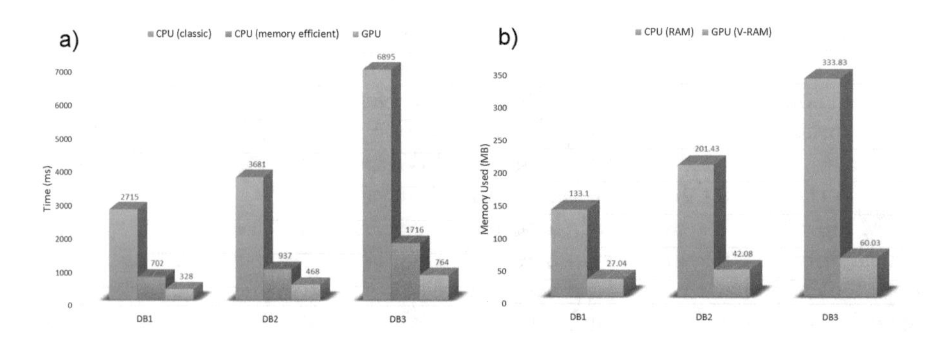

Fig. 4. (a) The experimental results for the preprocessing stage of the pattern-matching automaton. (b) Amount of memory used by a classic CPU implementation vs our GPU-focused, memory-compact model.

Run-Time Performance. The second aspect we have focused on was the run-time performance of our model, compared to the classic CPU implementation. For this purpose, we employed the Aho-Corasick pattern matching both in the CPU and in the GPU, and used the 3 databases above on the human genome GRCh38 (3.2 GB in size). The results are perfectly consistent with the throughput capabilities of each of the devices used, with the speed-up gained from the GPU implementation being 2.6× as compared to the CPU counterpart (Fig. 5).

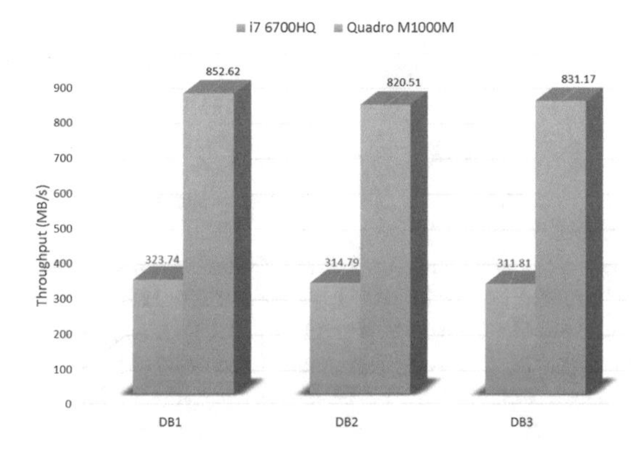

Fig. 5. CPU vs GPU results in run-time performance benchmark.

5 Conclusion

We have presented, implemented and benchmarked a novel, massively-parallel and scalable approach to accelerating GPU-based DNA biometrics, a vital component of criminal investigations, to a degree where we have shown that the entire process can be significantly improved, with up to 5× less memory on average in our tests and significantly reduced construction times and run-times when executed on the GPU, as compared to the CPU implementation counterpart. Unlike other previous models, our own takes into account all stages of the automaton construction and run-time performance, and improves its performance on all fronts. The fact that we used a mid-range graphics card and have obtain massive improvements on all constructional and running aspects of the pattern-matching process, only proves that the design works very well even with low-powered mobile platforms, such as laptops or even smartphones.

Acknowledgement. This work was partially supported by the InnoHPC Interreg - Danube Transnational Programme grant. The views expressed in this paper do not necessarily reflect those of the corresponding project consortium members.

References

1. NVIDIA: NVIDIA CUDA Compute Unified Device Architecture Programming Guide, version 4.1. http://developer.download.nvidia.com/compute/DevZone/docs/html/C/doc/CUDA_C_Programming_Guide.pdf
2. OpenCL: The open standard for parallel programming of heterogeneous systems. https://www.khronos.org/opencl/
3. Aho, A.V., Corasick, M.J.: Efficient string matching: An aid to bibliographic search. Comm. ACM **18**(6), 333–340 (1975)
4. Yin, Z., Lan, H., Tan, G., et al.: Computing platforms for big biological data analytics: perspectives and challenges. Comput. Struct. Biotechnol. J. **15**, 403–411 (2017)

5. Adey, S.: GPU Accelerated Pattern Matching Algorithm for DNA Sequences to Detect Cancer using CUDA, Master's Dissertation, Master of Technology in Computer Engineering, Department of Computer Engineering and Information Technology, College of Engineering, Pune (2013)
6. Memeti, S., Pllana, S.: GSNP: a DNA single-nucleotide polymorphism detection system with GPU acceleration. In: International Conference on Parallel Processing (ICPP), pp. 592–601 (2011)
7. Memeti, S., Pllana, S.: Accelerating DNA sequence analysis using intel xeon Phi. In: Trustcom/BigDataSE/ISPA, vol. 3, pp. 222–227. IEEE (2015)
8. Santos, F., Machado, H.: Patterns of exchange of forensic DNA data in the European Union through the Prüm system. Sci. Justice **57**(4), 307–313 (2017)
9. Taverne, M.D., Broeders, A.P.A.: Cross-border patterns in DNA matches between the Netherlands and Belgium. Sci. Justice **57**(1), 28–34 (2017)
10. Thambawita, D., Ragel, R., Elkaduwe, D.: To use or not to use: Graphics processing units (GPUs) for pattern matching algorithms. In: 7th International Conference on Information and Automation for Sustainability (ICIAfS) (2014)
11. Nehemia, R.: Implementation of multipattern string matching accelerated with GPU for intrusion detection system. In: IAES International Conference on Electrical Engineering, Computer Science and Informatics (2017)
12. Lin, C.H., Liu, C.H., Chien, L.S., Chang, S.C.: Accelerating pattern matching using a novel parallel algorithm on GPUs. IEEE Trans. Comput. **62**(10), 1906–1916 (2012)
13. Pungila, C., Negru, V.: A highly-efficient memory-compression approach for GPU-accelerated virus signature matching. In: Information Security Conference (ISC) (2012)
14. The FASTA Project. https://en.wikipedia.org/wiki/FASTA_format, https://www.ebi.ac.uk/Tools/sss/fasta/
15. GenBank: The NIH genetic sequence database. https://www.ncbi.nlm.nih.gov/genbank/

Towards Secure Transportation Based on Intelligent Transport Systems. Novel Approach and Concepts

Camelia-M. Pintea[1(✉)], Gloria Cerasela Crişan[2], and Petrica Pop[1]

[1] Technical University Cluj-Napoca, North University Center at Baia-Mare, Baia Mare, Romania
dr.camelia.pintea@ieee.org, petrica.pop@cunbm.utcluj.ro
[2] Vasile Alecsandri University of Bacău, Bacău, Romania
ceraselacrisan@ub.ro

Abstract. The security of transportation is nowadays a challenge. The *Intelligent Transport Systems (ITS)* are included to have the newest finding over transportation features. The current work propose a new problem inspired by *ITS* over the *Traveling Salesman Problem (TSP)* highlighting the security constraints. The problem it is called the *Secure Intelligent Transport Systems within the Traveling Salesman Problem (SITS-TSP)*. The optimization problem should satisfy the security constraints. The mathematical model is also included.

1 Introduction

Current developments of human society and complex networks produced by globalization generate new theoretical concepts and practical problems in the field of international, national and individual security. At theoretical level, critical infrastructure-related definitions and methodologies were put in place by international or national organisms. At practical level, each component or sub-system of a network may be assessed and in case of criticality, it may be accordingly protected.

Previous security models are using the well-known academic Traveling Salesman Problem (TSP). For example, in [26], a heuristic method is proposed for efficiently deploy a satellite network for surveillance. A terrestrial patrolling problem is treated in [5]. In [13], a secure protocol including multiple participants to solve in collaboration the *Traveling Salesman Problem* without sharing private information is proposed.

Security in Intelligent Transportation is addressed based on TSP. In [6], a dynamic TSP version is approached as a Markov process in order to minimize the expected total travel time, when only local travel conditions are known. Secure protocols for sharing vehicle-to-vehicle information to optimize fleet-wide fuel efficiency and minimize congestion are presented in [16].

Adding connections with infrastructure sensors from traffic lights and intersections dramatically shortens the time when cars are stopped [3].

M. Graña et al. (Eds.): SOCO'18-CISIS'18-ICEUTE'18, AISC 771, pp. 469–477, 2019.
https://doi.org/10.1007/978-3-319-94120-2_45

A transportation model when the driver receives messages about non-recoverable blockages in his neighborhood is presented in [32]. Treated using a TSP variant, this model allows avoiding some edges. Waze is a real-life app that can be used as additional information provider.

Prerequisites follows in the next section. In Sect. 3 it is introduced the *Secure Intelligent Transport Systems over the Traveling Salesman Problem (SITS-TSP)* with formulation and security constraint. In Sect. 4 it is discussed the way Swarm Intelligence could lead to good solution when using a large number of constraints of transportation risks. The paper concludes with future work proposals.

2 Prerequisites

2.1 Traveling Salesman Problem (TSP)

The *Traveling Salesman Problem (TSP)* is a well known and investigated Combinatorial Optimization Problems [1,2]. Preoccupations in connection to *TSP* are mentioned as early as the last eighteenth century. Its formal definition based on Graph Theory follows.

Definition 1. *To solve the Traveling Salesman Problem (TSP) means to find a minimum weight Hamiltonian cycle in a complete weighted graph* $G = (V, E, w)$.

Usually $V = \{1, 2, ...n\}$ is the set of n vertices and the weight function is specified by the cost matrix $C = (c_{ij})_{1 \leq i,j \leq n}$ (the matrix of strictly positive weights associated with E). The main diagonal of the matrix C is often ignored when the data are provided to solvers. As TSP seeks for Hamiltonian cycles, the loops are excluded.

An equivalent *TSP* formulation seeks for values for $\frac{n(n-1)}{2}$ integer variables described by an integer program which considers:

- the cost matrix C;
- the function $\delta : 2^V \rightarrow 2^E$, where $\delta(S)$ is a set of edges with one from their incident vertices in S;
- the binary 0/1 set of variables $(x_{ij})_{(i,j) \in E}$ taking the value 1 if and only if their corresponding edge belongs to the solution.

Using these notations, *TSP* means solving:

$$\min \sum_{(i,j) \in E}^{n} c_{ij} x_{ij} \text{ when } x_{ij} \in \{0, 1\}, \forall (i,j) \in E \tag{1}$$

$$\text{s.t.} \sum_{(j,k) \in \delta(\{i\})}^{n} x_{jk} = 2, \forall 1 \leq i \leq n \tag{2}$$

$$\sum_{(j,k) \in \delta(S)}^{n} x_{jk} \geq 2, \forall S \subset V, 2 \leq |S| \leq \frac{|V|}{2} \tag{3}$$

The objective function (1) returns the total cost of a feasible solution, the constraints (2) show that the degree of each vertex is 2 (the salesperson comes and goes); the constraints (3) prevent early cycle closings [10]. The constraints (3) are exponential in number; their testing is the most expensive part of *TSP*.

TSP is a *NP*-hard problem from the computational complexity point of view [17]; a polynomial time exact algorithm for all instances is not currently known and is unlikely to be ever found. Large *TSP* instances are difficult to solve due to the computational complexity [2]. This is the reason why, besides the exact solving methods, the researchers also focus on non-exact methods for *TSP* and related complex problems: the approximate and heuristic approaches [7,23,24].

The *Symmetric TSP (STSP)* is a restricted *TSP* which has a symmetric cost matrix. The *Metric TSP (MTSP)* is a more restricted *TSP* with costs forming a metric (so, a *MTSP* is also a *STSP*). The *Euclidean TSP (ETSP)* is a *MTSP* with *Euclidean* norm. The *Asymmetric TSP (ATSP)* is the *TSP* which is not symmetric.

Modern variants consider real-life models. The case of incomplete road network and allowing multiple visits is treated using a memetic algorithm in [25]. The dynamic case of this already complex version, when the weights can change constantly is treated in [20].

2.2 Intelligent Transportation System (ITS)

Intelligent Transportation Systems (ITS) address contemporary transportation problems. They are designed to use information and automation in order to move efficiently people and goods. These systems integrate control, computation and communication, utilizing algorithms, databases, models and human interfaces.

An *Intelligent Transportation System* has to operate with large sets of concepts, requests, platforms and resources, and is supposed to be flexible, reliable and secure. These systems are classified into intelligent infrastructure systems, intelligent vehicle systems and hybrid systems. Data collected from heterogeneous sensors are interpreted online and correct decisions must be made without human interventions. Such examples are: automated openings of new lanes when congestion or accidents appear, automated emergency call systems installed on cars, etc. Other approach is to use mobile applications able to receive instant information about sensitive events and to assist the driver in finding alternative ways of moving.

3 Motivation: Critical Transportation Infrastructure in Europe

During the first years of this century, *European Union* was preoccupied by the impact the natural disasters or negative humans activity may have. The optimization of the Traveling Salesman Problem is reflected in real life, in this particular case, to a better transportation network structure.

The *Council Directive 2008/114/EC* focuses on identify and designate European critical infrastructures and assesses the needs in order to improve the humans protection. It defines as *critical infrastructure* an "*...asset, system or part thereof located in Member States which is essential for the maintenance of vital societal functions, health, safety, security, economic or social well-being of people, and the disruption or destruction of which would have a significant impact in a Member State as a result of the failure to maintain those functions*" [9].

An *European critical infrastructure (ECI)* is considered an infrastructure, a critical one, with a major impact on at least two states, members of the *EU* [9].

The *European Programme for Critical Infrastructure Protection (EPCIP)* aims to integrate measures in order to improve protections of "critical infrastructures" in the *EU*. *EU Joint Research Centre* is the coordinating institution of the *European Reference Network for Critical Infrastructure Protection (ERNCIP)*. From the *ERNCIP* laboratories and specialists, tools and knowledge for protection of "critical infrastructures" against threats are provided [11].

A successful example of an EU-funded project is *Critical Infrastructures Preparedness* and *Resilience Research Network (CIPRNet)* [8].

At national level, the member states were expected to release articulated regulations in order to seamlessly manage all the threats and to easily cooperate. In *Spain*, the *Law 8/2011* and its implementing regulations (*Royal Decree 704/2011*) describe the settings and procedures for good support and effective coordination between all the stakeholders in order to achieve a better security in total. In *France*, the regulations follow the doctrine of defense and national security approach. The *Secteurs dActivité d'Importance Vitale (SAIV)* are described in the Code of Defense, the legislative basis of the national security system. In *UK*, the *Centre for the Protection of National Infrastructure (CPNI)* is the structure with the role of reducing vulnerability of infrastructures from different threats. In *Germany*, the *Internet platform on Critical Infrastructure Protection* includes *Federal Office of Civil Protection and Disaster Assistance (BBK)* and the *Federal Office for Information Security (BSI)*. In *Romania*, the *Government Emergency Ordinance 98/2010* and the *Country's Supreme Defense Council Decision 62/2006* represent the framework for defining and implementing the security national strategy.

4 A Novel Approach of Secure Transportation System Based on Intelligent Transport Systems

The particular new approach of *Secure Transportation System* based on *Intelligent Transport Systems* is inspired by the transportation-related problem of *Traveling Saleman Problem (TSP)* highlighting the security constraints.

Follows the new parameters and their descriptions. The particular *Secure Intelligent Transport Systems within the Traveling Salesman Problem (SITS-TSP)* it is an optimization problem with security constraints. The objective of the *SITS-TSP* problem is to optimize the transportation cost, based on Eq. (1) while satisfying all possible security risk constraints.

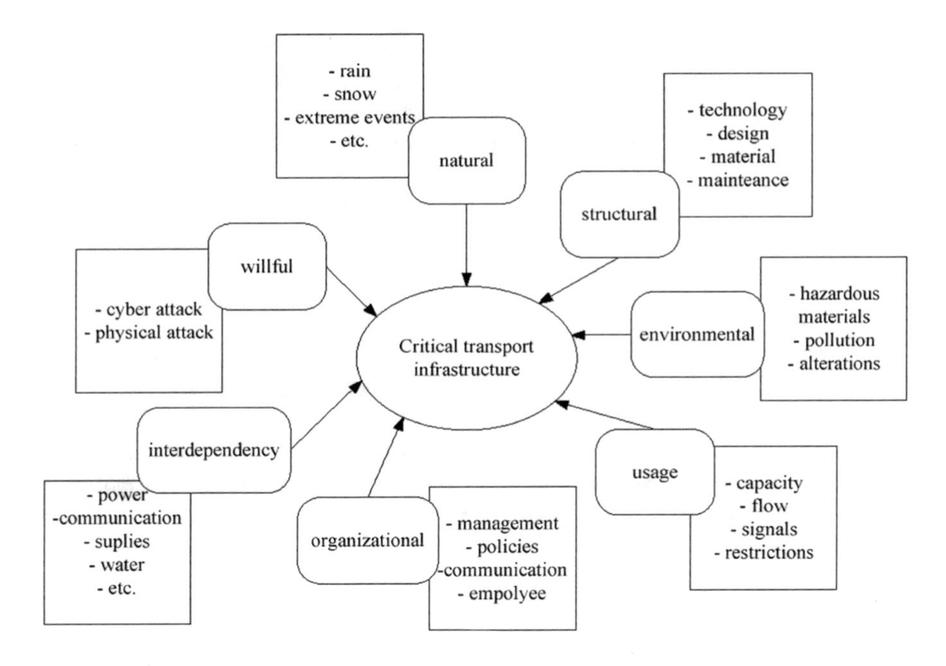

Fig. 1. The main risks influencing the transport infrastructure based on [12].

Each path, whatever it is road, rail or aerial is enhanced with a specific constraint set, a security constraint set. The Set of Constraints (SC) includes security demand *sd* and security rank *sr*:

$$SC = \{sd, sr\}.$$

As a new approach, based on the real life risks as shown in Fig. 1 and based on [14,15,19,30] for both roads and rails, the security demand and security rank are computed based on the following set of parameters, SP:

$$SP = \{nat, wful, idep, org, usg, env, str\}.$$

Parameters Description

There will be included parameters based on Critical transport infrastructure. The introduced parameters, based on the main risks influencing the transport infrastructure [12] from Fig. 1 are:

- *nat* denotes the **Natural** risk.
- *wful* denotes the **Willful** risk.
- *idep* denotes the **Interdependency** risk.
- *org* denotes the **Organizational** risk.
- *usg* denotes the **Usage** risk.
- *env* denotes the **Environmental** risk.
- *str* denotes the **Structural** risk.

In [21] was introduced a *Secure Two-stage Supply Chain Network*, a *Transportation Cost Approach* inspired by the constraint model of security used for data-intensive jobs running on distributed computing [18,28,29]. Furthermore we will use this experience to enhance *Traveling Salesman Problem* with security features.

A secure transportation route between two nodes i and j, is feasible if the secure parameters are in acceptable limits and the constraint from Eq. (4) is met.

$$sd \leq sr. \tag{4}$$

For each country should be particular security intervals values for each type of *Natural, Willful, Interdependency, Organizational, Usage, Environmental* and *Structural* transportation related risks.

In general each risk parameter should have a constraint interval, with inferior (*inf*) and superior (*sup*) values as follows in Eq. (5) for each *transportation risk parameter* (*SP* set): $SP = \{nat, wful, idep, org, usg, env, str\}$. The inferior ($inf$) and superior (*sup*) values differ for each type of parameters and are related to the region or country features and laws.

$$SP_{inf}(i) \leq SP(i) \leq SP_{sup}(i). \forall i \in [1,7]. \tag{5}$$

Security Modes are furthermore considered based on the previous works [18, 21,28,29].

- The *δ-risky* mode includes a probability to measure the risk. The transportation on a *TSP* network edge between two nodes i and j is possible at the most δ-risk. It is considered that the transportation should satisfy the considered security requirements for node j, including the inequalities (4) and (5) for each type of parameter $SP \in \{nat, wful, idep, org, usg, env, str\}$.
- The *Secure* mode it is a particular *δ-risky* level for the 0 value of δ. It is considered the fully secured level. Here the transportation between two nodes in the *TSP* transportation network should satisfy all considered security requirements, including the constraints (4); each type of parameter from the set $SP = \{nat, wful, idep, org, usg, env, str\}$ should have the relation (6) satisfied.

$$SP(i) \leq SP_{inf}(i), \forall i \in [1,7]. \tag{6}$$

- The *Risky* mode it is the opposite level of the *secure* level, with $\delta = 1$ value of the general *δ-risky* level. Here the risky level considers all possible risks transportation between the *TSP* network's nodes. Each type of parameter $SP \in \{nat, wful, idep, org, usg, env, str\}$ should have the relation (7) satisfied.

$$SP(i) \geq SP_{sup}(i), \forall i \in [1,7]. \tag{7}$$

Normalization with Decimal Scaling. All the values of the parameters used: sd, sr, nat, $wful$, $idep$, org, usg, env, str including the related inf and sup values are Normalized with Decimal Scaling, $nds\_value = \frac{value}{10^{no}}$ where no is the considered number of decimals. This normalization leads to values in the $[0, 1]$ interval.

In order to compute the **security rank** sr could be considered different variants depending on the perspective. We propose a linear perspective of computing the sr:

$$sr = \sum_{i=1}^{7} \alpha_i \cdot SP(i), \tag{8}$$

where SP is the set $\{nat, wful, idep, org, usg, env, str\}$ and α_i, $i \in [1, 7]$, is a natural number; in particular could be considered the unit value.

Security Levels in distributed environments are included in terms of a qualitative scale and based on [27]. There are five security levels considered, from very high denoted vh, high h and medium med, to low, l and very low, vl [27].

- If $sd \leq sr$ then the transportation is completely safe.
- If $0 < sd - sr \leq vl$ then it is a possible transport.
- If $vl < sd - sr \leq l$ then the transport is delayed, but before the deadline [31].
- If $l < sd - sr \leq vh$ then the transport is not feasible.

Risk Probability is furthermore introduced as in [18], Eq. (9). The sr_j is the security risk of the j-th TSP node and m is considered the dimension of the TSP network.

$$P(risk_j) = \begin{cases} 0, & sd - sr_j \leq 0 \\ 1 - e^{-\frac{1}{2}sd - sr_j}, & 0 < sd - sr_j \leq vl \\ 1 - e^{-\frac{3}{2}sd - sr_j}, & vl < sd - sr_j \leq l \\ 1, & l < sd - sr_j \leq vh \end{cases} \quad \forall j = 1, \dots m. \tag{9}$$

5 Conclusion, Discussions and Further Work

In order to solve the *Secure Intelligent Transport Systems within the Traveling Salesman Problem (SITS-TSP)* several specified constraints are used. The complexity of the problem is increased because of the security constraints. There are included several examples of *Intelligent Transport Systems* according to their roles and goals on both roads and rails. These examples highlight the security problems on operational transportation. The threats including security ones, could block the transportation operations [4].

The nodes from the *TSP* with the highest probability to be risky are not chosen in the solution set. So, the final solution of the *TSP* problem will include just the secured nodes for transportation purpose. The main risks influencing the transport infrastructure are quantified as mathematical parameters. A δ-risky mode with particular values will be available to a node of the *TSP*, as well as different levels of security in terms of a qualitative scale in distributed environments: from the very high (vh) to the very low (vl) level of security.
In the future study cases and implementation of the newly secure model will be done. The study case will include particular values for each of the included

parameters $SP = \{nat, wful, idep, org, usg, env, str\}$. A model related to secure and green two-stage *Supply Chain Networks* is already made [22].

Acknowledgments. The study was conducted under the auspices of the IEEE-CIS Interdisciplinary Emergent Technologies TF.

References

1. Applegate, D.L., Bixby, R.E., Chvátal, V., Cook, W.J.: The Traveling Salesman Problem: A Computational Study. Princeton University Press, Princeton (2011)
2. Applegate, D.L., Bixby, R.E., Chvátal, V., Cook, W.J.: Finding tours in the TSP, Technical report 99885, University of Bonn (1999)
3. Bento, L.C., Parafita, R., Nunes, U.: Intelligent traffic management at intersections supported by V2V and V2I communications. In: Proceedings 15th IEEE Conference Intelligent Transportation Systems (ITSC), pp. 1495–1502 (2012)
4. Calinescu, A., et al.: Applying and assessing two methods for measuring complexity in manufacturing. J. Oper. Res. Soc. **49**, 723–733 (1998). https://doi.org/10.1057/palgrave.jors.2600554
5. Calvo, R.W., Cordone, R.: A heuristic approach to the overnight security service problem. Comput. Oper. Res. **30**, 1269–1287 (2003)
6. Cheong, T., White, C.C.: Dynamic traveling salesman problem: value of real-time traffic information. IEEE Trans. Intell. Transp. Syst. **13**(2), 619–630 (2012)
7. Chira, C., et al.: Learning sensitive stigmergic agents for solving complex problems. Comput. Inf. **29**(3), 337–356 (2010)
8. CIPRNet Project. https://www.ciprnet.eu/home.html
9. Council Directive 2008/114/EC of 8 December 2008 on the identification and designation of European critical infrastructures and the assessment of the need to improve their protection (2008)
10. Dantzig, G.B., Fulkerson, R., Johnson, S.M.: Solution of a large-scale traveling salesman problem. Oper. Res. **2**, 393–410 (1954)
11. ERNCIP Project platform. https://erncip-project.jrc.ec.europa.eu
12. Fuchs, P.: Transportation infrastructure as an element of critical infrastructure (Dopravn infrastruktura jako prvek kritickej infrastruktury státu). Vysoká škola bezpečnostného manažérstva v Košiciach, 122 p. (2011)
13. Hong, Y., Vaidya, J., Lu, H., Wang, L.: Collaboratively solving the traveling salesman problem with limited disclosure. In: Data and Applications Security and Privacy XXVIII, DBSec 2014. LNCS, vol 8566. Springer, Heidelberg (2014)
14. The intelligent transport society for UK (2012). http://www.itsuk.org.uk/filelibrary/file/Rail%20ITS%20final.pdf
15. James, T.: The importance of protecting transport infrastructure. Eng. Technol. Mag. **7**(12) (2012). https://eandt.theiet.org/content/articles/2012/12/the-importance-of-protecting-transport-infrastructure/
16. Jeong, H., Lee, J.W.W., Jeong, J.P., Lee, E.: DSRC based self adaptive navigation system: aiming spread out the vehicles for alleviating traffic congestion. In: Frontier and Innovation in Future Computing and Communications, pp. 739–749. Springer, Dordrecht (2014)
17. Karp, R.M.: Reducibility among combinatorial problems. In: Miller, R.E., Thatcher, J.W. (eds.) Complexity of Computer Computations. The IBM Research Symposia, pp. 85–103. Plenum Press, New York (1972)

18. Liu, H., Abraham, A., Snášel, V., McLoone, S.: Swarm scheduling approaches for work-flow applications with security constraints in distributed data-intensive computing environments. Inf. Sci. **192**, 228–243 (2012)
19. Martinez, F., Toh, Ch., Cano, J.: Emergency services in future intelligent transportation systems based on vehicular communication network. Intell. Transp. Mag. **2**, 6–20 (2010)
20. Miller, J., Kim, S.I., Menard, T.: Intelligent transportation systems traveling salesman problem (ITS-TSP) – a specialized TSP with dynamic edge weights and intermediate cities. In: 13th International IEEE Conference on Intelligent Transportation Systems, pp. 992–997 (2010)
21. Pintea, C.-M., Calinescu, A., Pop, P., Sabo, C.: Towards a secure two-stage supply chain network: a transportation-cost approach. In: CISIS 2016. Advances in Intelligent Systems and Computing, vol. 527, pp. 547–554 (2016)
22. Pintea, C.-M., Calinescu, A., Pop Sitar, C., Pop, P.C.: Towards secure & green two-stage supply chain networks. Logic J. IGPL (2018, accepted)
23. Pintea, C.-M., et al.: A sensitive metaheuristic for solving a large optimization problem. LNCS, vol. 4910, pp. 551–559 (2008)
24. Pintea, C.-M., Crişan, G.-C., Chira, C.: Hybrid ant models with a transition policy for solving a complex problem. Logic J. IGPL **20**(3), 560–569 (2012)
25. Piwońska, A., Koszelew, J.: A memetic algorithm for a tour planning in the selective travelling salesman problem on a road network. In: Kryszkiewicz, M., Rybinski, H., Skowron, A., Raś, Z.W. (eds.) Foundations of Intelligent Systems. ISMIS 2011. LNCS, vol. 6804. Springer, Heidelberg (2011)
26. Saleh, H.A., Chelouah, R.: The design of the global navigation satellite system surveying networks using genetic algorithms. Eng. Appl. Artif. Intell. **17**, 111–122 (2004)
27. Song, S., Hwang, K., Zhou, R., Kwok, Y.: Trusted P2P transactions with fuzzy reputation aggregation. IEEE Internet Comput. **9**(6), 24–34 (2005)
28. Song, S., Kwok, Y., Hwang, K.: Security-driven heuristics and a fast genetic algorithm for trusted grid job scheduling. In: International Parallel and Distributed Processing Symposium, vol. 65, p. 65a. IEEE Computer Society (2005). https://ieeexplore.ieee.org/xpls/icp.jsp?arnumber=1419889
29. Song, S., Hwang, K., Kwok, Y., et al.: Risk-resilient heuristics and genetic algorithms for security-assured grid job scheduling. IEEE Trans. Comp. **55**(6), 703 (2006)
30. Traintic – Rail industry intelligent transport systems, San Sebastian (2014). http://www.railway-technology.com/contractors/computer/traintic
31. Venugopal, S., Buyya, R.: A deadline and budget constrained scheduling algorithm for eScience applications on data grids. LNCS, vol. 3719, pp. 60–72. Springer, Heidelberg (2005)
32. Zhang, H., Tong, W., Xu, Y., Lin, G.: The Steiner traveling salesman problem with online advanced edge blockages. Comput. Oper. Res. **70**, 26–38 (2016)

Fuzzy-Based Forest Fire Prevention and Detection by Wireless Sensor Networks

Josué Toledo-Castro[1](✉), Iván Santos-González[1], Pino Caballero-Gil[1], Candelaria Hernández-Goya[1], Nayra Rodríguez-Pérez[1], and Ricardo Aguasca-Colomo[2]

[1] University of La Laguna, San Cristóbal de La Laguna, Spain
{jtoledoc,jsantosg,pcaballe,mchgoya,mrodripe}@ull.edu.es
[2] IUSIANI - ULPGC, Las Palmas de Gran Canaria, Spain
ricardo.aguasca@ulpgc.es

Abstract. Forest fires may cause considerable damages both in ecosystems and lives. This proposal describes the application of Internet of Things and wireless sensor networks jointly with multi-hop routing through a real time and dynamic monitoring system for forest fire prevention. It is based on gathering and analyzing information related to meteorological conditions, concentrations of polluting gases and oxygen level around particular interesting forest areas. Unusual measurements of these environmental variables may help to prevent wildfire incidents and make their detection more efficient. A forest fire risk controller based on fuzzy logic has been implemented in order to activate environmental risk alerts through a Web service and a mobile application. For this purpose, security mechanisms have been proposed for ensuring integrity and confidentiality in the transmission of measured environmental information. Lamport's signature and a block cipher algorithm are used to achieve this objective.

Keywords: Forest fires · Internet of Things · Lamport's signature
Fuzzy logic · Real time

1 Introduction

Huge losses may be caused as consequence of emergency situations and natural disasters such as forest fires implicating serious threats to the ecosystems and people's health. The lack of real time systems that allow managing forest resources and activating wildfire incident alerts may cause difficulties in forest fires prevention, detection and fighting. Some environmental events and variables considered as dynamic forest fire risk factors, such as meteorological variables, polluting gases or oxygen, can be used in order to estimate an environmental risk index of wildfire incidents. Their real time monitoring over different forest areas may favor efficiency on the response time of emergency bodies, helping to prevent forest fires, notifying their detection or tracking their progress.

M. Graña et al. (Eds.): SOCO'18-CISIS'18-ICEUTE'18, AISC 771, pp. 478–488, 2019.
https://doi.org/10.1007/978-3-319-94120-2_46

To this end, the use of Internet of Things (IoT) devices and sensors can be relevant for developing Wireless Sensor Networks (WSN) [1] for environmental protection. Regarding their deployment in outdoor forest areas, some challenges such as communication among distributed nodes, possible areas out of network coverage or the implementation of security mechanisms should be considered.

Thus, this proposal is devoted to implementing a real time monitoring system for different environmental variables through a wireless IoT sensor network distributed over different forest areas aimed to prevention, detection and tracking of wildfire incidents. For this purpose, a forest fire risk controller based on fuzzy logic [2] has been integrated. It is responsible for interpreting and analysing whether measurements of meteorological variables, polluting gases and oxygen registered by IoT sensors may evidence forest fire risks. In order to activate and manage environmental alerts and information, a Web service and a mobile application are included. Particular attention has been paid to the system security. Secure communications among the IoT devices and the Web service and mobile devices, based on multi-hop routing [3], has been developed. Lamport's signature scheme [4], a block cipher algorithm and a secure authentication scheme have been implemented as well.

This work is organized as follows. Section 2 deals with the description of some related works. Then, the method applied to evaluate the existence of forest fire risks is detailed in Sect. 3 and the proposed system is outlined in Sect. 4. Finally, Sect. 5 explains the implemented security mechanisms and Sect. 6 provides conclusions, research works in progress and future research lines.

2 Related Works

Currently, many proposals have been implemented with the aim of offering environmental monitoring solutions based on fuzzy logic. In this sense, the work [5] proposes a smart system to analyse carbon dioxide emissions in the same way that the paper [6], both combine fuzzy logic and decision-making trial to improve efficiency in emergency management. To this respect, the paper [7] suggests a long-term forest risk estimation for determining the existence of high or low risk in Mediterranean basin countries through applying an integrated fuzzy model together with machine learning.

Another relevant aspect regarding WSN, IoT and multi-hop routing is the approach to consider communication security as a main goal. In this sense, the implementation of key predistribution schemes should be considered. The paper [8] provides three different mechanisms for key establishment through pre-distributing a random set of keys to each node. Others security issues and challenges need to be also considered, the work [9] identifies security threats.

The system proposed here presents the use of wireless sensor networks, fuzzy logic, multi-hop routing and security mechanisms for developing a secure environmental measurement interface, able to measure dynamic risk factors that make forest fire prevention, detection and tracking more efficient. It also provides a forest fire risks controller for evaluating recent wildfire incidents. It is composed by a

Web service, a mobile application and a particular IoT device containing environmental sensors.

3 Method

A wireless sensor network based on IoT devices and environmental sensors is deployed through forest areas. The objective is to perform a continuous environmental monitoring of dynamic risk factors for forest fires such as meteorological variables (temperature, relative humidity, wind speed and rainfall), polluting gases (such as carbon dioxide and monoxide) and oxygen level. These variables are measured on every forest area and the information gathered is transmitted to a Web service in charge of analysing the possible existence of forest fire risks as well as any evidence of recent wildfire incidents.

Estimating forest fire risk factors is not simple and cannot be executed with complete accuracy. For this purpose, fuzzy logic and Mamdani Inference [10] have been used here in order to obtain a forest fire environmental risk controller. These methods allow to include the uncertainty of environmental data, the imprecision related to the variation in the parameters and the behavior of every monitored variable (see Fig. 1).

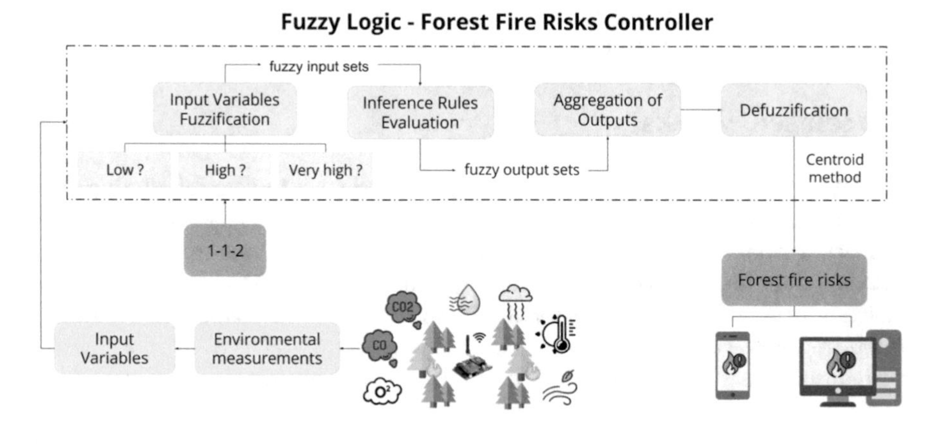

Fig. 1. Fuzzy logic - forest fire risks controller

3.1 Linguistic Variables

The environmental variables aforementioned are considered as the input variables of the proposed fuzzy logic system.

Meteorological variables set includes temperature, relative humidity, wind speed and rainfall as linguistic variables. Their fuzzy sets have been mainly proposed taking into account the "Rule of 30" [11] as forest fire prevention model. It considers temperature measurements above 30 °C, humidity percentages below

30%, wind speed values above 30 km/h and rainfall measurements below 30 mm as relevant environmental conditions that may favor the occurrence of wildfire incidents. As consequence of increasing temperature and wind speed measurements and decreasing humidity percentages or rainfall measure above or below these thresholds, the severity of forest fire risks progressively increases from low to extreme risks. Every meteorological variable has as discourse of universe with its corresponding unit of measurement (Celsius degrees, percentages, kilometers per hour, etc.).

Regarding polluting gases, carbon dioxide and monoxide are also considered as input linguistic variables. These gases are produced in high levels during forest fires, so monitoring atypical increments in the carbon dioxide level (currently measured between 200 – 400 ppm) and the carbon monoxide (concentrations around 1 ppm can be commonly measured at outdoor areas) have to be considered as evidences of environmental measurements at indoor places or consequence of performing industrial activities instead of outdoor forest areas [12]. Taking into account these thresholds, fuzzy sets have been proposed (see Fig. 2). In both cases, particles per million (ppm) are considered as discourse of universe.

Finally, the oxygen level is also considered as input linguistic variable and shows an opposite behavior when comparing with polluting gases behavior. It means that instead of performing unusual increasing, oxygen level decreases progressively as consequence of being progressively consumed by fire. Hence, this variable may also be cataloged as a useful indicator. Despite the changes of oxygen level occurred over last millions of years, the current level may be defined as around 21%. Fire needs at least 16% of oxygen level to occur. In contrast to polluting gases, percentages are used as discourse of universe. In this sense, all the fuzzy sets for every input variable have been proposed according to the knowledge of the experts that have been consulted.

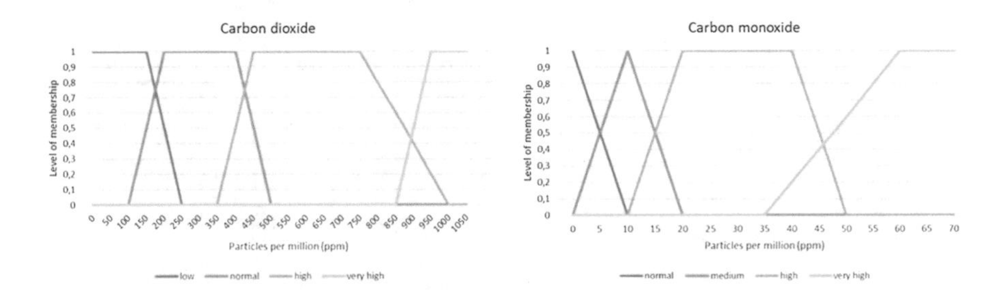

Fig. 2. Example of fuzzy sets proposed for polluting gases

In addition to this, the averages of environmental variables (temperature, humidity, carbon dioxide, etc.) are also used as input linguistic variables. Average calculation depends on prior detected forest fire risks for the current day in every monitored forest area. Taking into account every input linguistic variable, the corresponding average is calculated through all the previously registered measurements (when forest fire risks were not detected until that moment) and, on the other hand, the 15, 10 and 5 latest registered measurements (when low, high or extreme forest fire risk was previously detected in the same area). These averages are compared with the corresponding last registered measurement while evaluating inference rules. Likewise, reducing the measurements number from 15 to 5 for average calculation is also considered when external risk declarations by the emergency services are performed.

3.2 Inference Rules Evaluation

The inference rules proposed are based on evaluating the severity of forest fire risks (nonexistent, low, high or extreme) according to unusual variations between the last registered measurement and the average of every monitored environmental variable. In this way, detecting variations according to the proposed fuzzy sets of a particular variable between its last registered measurement and the average, may help to detect environmental conditions that have recently worsened or evidences of wildfire incidents occurrence. Every proposed inference rule evaluates these fuzzified input values with the aim of determining the fuzzified output result, that is represented on the proposed output variable membership function taking into account a percentage between 0% and 100% of detected forest fire risks.

For this purpose, Fuzzy Associative Memory (FAM) [13] is used as representation tool of the inference rules for evaluating the severity of forest fire risks with respect to the last registered measurement of every monitored environmental variable and its corresponding calculated average. Table 1 shows the sample FAM proposed for the linguistic variable of carbon dioxide:

Table 1. Inference rules evaluation for carbon dioxide variable

Last CO2 measurement/CO2 average	Normal	Medium	High	Very high
Normal	NFR	NFR	NFR	NFR
Medium	LFR	LFR	HFR	EFR
High	HFR	HFR	HFR	EFR
Very High	EFR	EFR	EFR	EFR

In this sense, the comparison between carbon dioxide average and its last registered measurement may return nonexistent forest fire risks (NFR), low risks (LFR), high risks (HFR) or extreme forest fire risks (EFR) in the forest area where measurements were registered. This average is calculated depending on the forest fire risks previously declared. Likewise, this format based on FAM is used to represent the evaluation process of the inference rules for the rest of the linguistic variables (temperature, humidity, wind speed, rainfall, carbon monoxide and oxygen).

3.3 Output Variable

Finally, the severity of detected fire risks in a particular forest area is defined as the output linguistic variable. To this respect, the percentage has been considered as its discourse of universe.

Taking into account the evaluation between averages and the last measurements of every input variable, the inference rules evaluation returns several outputs always ranged from 0% to 100% and fuzzified into the output membership function. Before obtaining a discrete percentage of forest fire risks, the results of rules must be added together to generate the output set (Aggregation of outputs step) and defuzzified through Centroid method [14]. According to the expert knowledge consulted, the triggers of the rules on the variables are previously set through appropriate overlaps of the fuzzy sets of input variables.

For example, the Fig. 3 shows the aggregation of the output results of inference rules evaluation on temperature and carbon dioxide input values into the same output set (previously fuzzified through their corresponding membership functions) in order to be defuzzified and express a discrete percentage for forest fire risks. The results of inference rules evaluated on the rest of input linguistic variables are also aggregated into the same final output set.

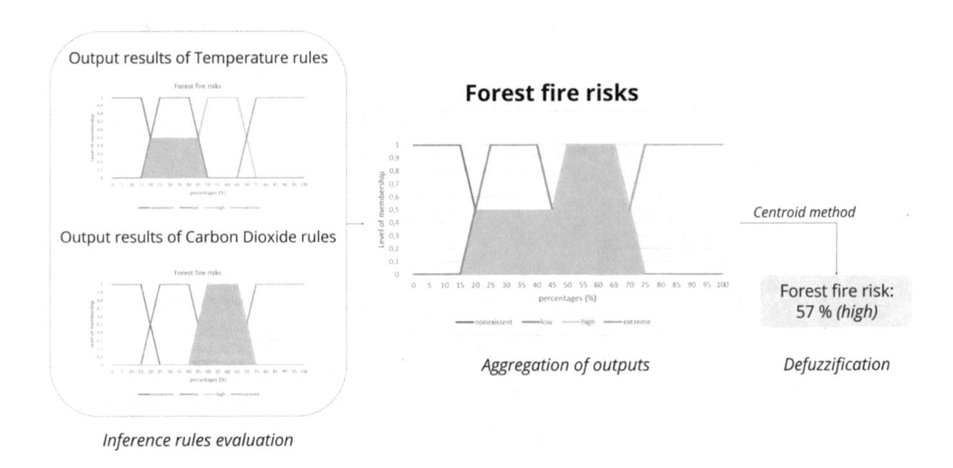

Fig. 3. Aggregation of outputs and defuzzification steps

If forest fire environmental risks are detected by the proposed fuzzy system, the Web service is in charge of updating the environmental information and activating new alerts in order to notify emergency corps members through the mobile application.

4 Proposed System

The system developed is in charge of performing a real time monitoring of dynamic risk factors such as meteorological variables, polluting gases and oxygen level measures through a wireless sensor network based on a particular IoT device prototype together with a Web service and a mobile application.

In this sense, the Web service manages all environmental measurements registered by the IoT devices (through Firebase Realtime Database) and the controller of forest fire risks based on fuzzy logic. This Web service is responsible for updating data (environmental measurements, IoT devices, forest fire alerts, etc.) into different interactive elements disposed on its interface. In this way, the interpretation of environmental conditions is made more efficient through linear and bar graphs, heat maps, gauges, environmental alert information panels and interactive maps. The alerts are sent through notifications to the mobile application which also implements a real time operational module for coordinating emergency tasks between emergency corps headquarters and distributed members around the affected forest areas.

4.1 IoT Devices

Each node in the WSN is able to measure important environmental variables for forest fire prevention, detection and fighting. Arduino platform has been used together with six particular sensors for measuring temperature and humidity, wind speed, rainfall, carbon dioxide, carbon monoxide and oxygen, and a 4G and a Wifi modules have also been assembled. Hence, communication between every sensor node and the Web service is possible. Apart from that, communication among sensor nodes is also available. GPS service has also been implemented to achieve IoT devices locations (see Fig. 4).

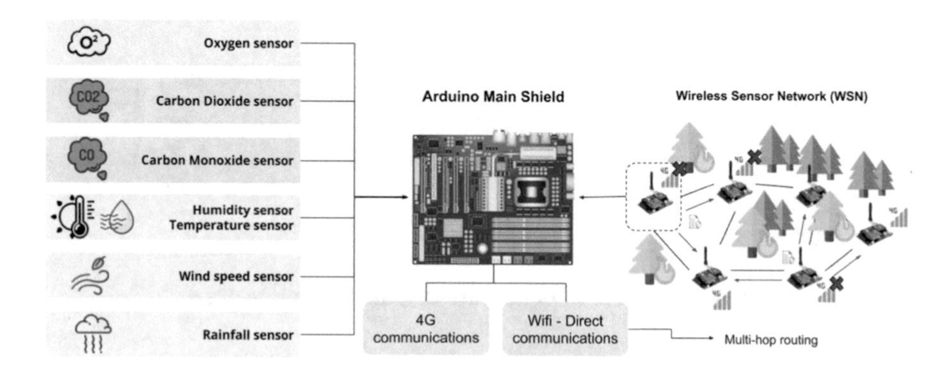

Fig. 4. Wireless sensor network architecture

Once IoT device is deployed, environmental measurement cycle starts. Firstly, a value of each monitored environmental variable is measured through integrated sensors. New environmental data is processed together with specific IoT device parameters such as location (latitude and longitude), battery level and the International Mobile Station Equipment Identity (IMEI). This last private parameter is used to uniquely identify the corresponding device. After data is formatted, it is signed through Lamport's signature scheme and encrypted by using AES encryption algorithm together with Cipher Block Chaining (CBC) mode. After new measured environmental information is sent, the Web service allows to tune up the frequency to receive information from each IoT devices. The default time among every measurement cycle has been considered by default as 5 min but it can be tunned depending on the severity of forest fire risks.

Wifi module is assembled to the main Arduino shield allowing Wifi-Direct [15] communications among IoT nodes. Wifi-Direct makes bidirectional connections between devices easier and allows data transmission through ranges up to 200 meters without needing any access point. Depending on some static factors such as the frequency of wildfire incidents occurrence or the orography, the proposed wireless sensor network has to be designed. Other factor to take into account is human activities nearby forest areas that may affect the ecosystem state.

Multi-hop Routing. Since it is possible that some forest areas may be located out of network coverage, multi-hop routing has been implemented in order to ensure that new environmental information may always reach the Web service through other nodes used as relays when the specific location of the forest area may be temporarily out of the network coverage.

Peer to peer communication is performed through Wifi-direct. When an IoT device is not able to send the recent measured environmental information through 4G module, it is necessary to establish secure communications with other close available sensor nodes in order to transmit this information to the Web service. Every time a sensor node is not able to communicate with the Web service as result of being out of network coverage, multi-hop routing is executed taking into account these steps:

- IoT node out of network coverage connects to a close neighbour node and transmits the recent registered environmental information through Wifi-direct.
- Destination node receives environmental information and checks if network coverage is currently available in its area. Every IoT node keeps a list of environmental information packages previously received. All data packages to be sent to the Web service are signed by its node owner through Lamport's signature scheme.
- Whether a data package reaches more than one time the same neighbour node, retransmission process stops to avoid congestion in the WSN.
- On the other hand, if a data package is received for the first time, it is sent to the Web service, if network coverage is currently available, or retransmitted to other nodes through Wifi-direct, otherwise.

5 System Security

This proposal includes different security mechanisms to provide with secure communications among IoT devices, the Web service and the mobile application.

In this sense, relevant security requirements for IoT deployment such as data privacy, confidentiality and integrity together with authentication have been considered in detail and implemented [16].

One of the challenges to be considered is the authentication of environmental measurements registered by IoT devices in order to avoid node spoofing and ensuring their integrity. For this purpose, Lamport's signature has been used. This type of signature is based on one-way functions [17] and a cryptographic hash function (SHA256). A key pair composed by a private key and a public one are generated. Both contains 2*n elements, but the public one is obtained by applying a hash function on each private key element. After the key pair generation, a hash function is also used on to the content data (measured environmental information together with other device parameters) in order to obtain a specific bits sequence that will be used to select the exact number of bits from the public key and perform Lamport's signature (see Fig. 5).

Hash or Merkle trees are used together with this signature for preventing its one-time use by each key pair generated when one single message is signed, which means many keys should have to be published whether many messages are signed.

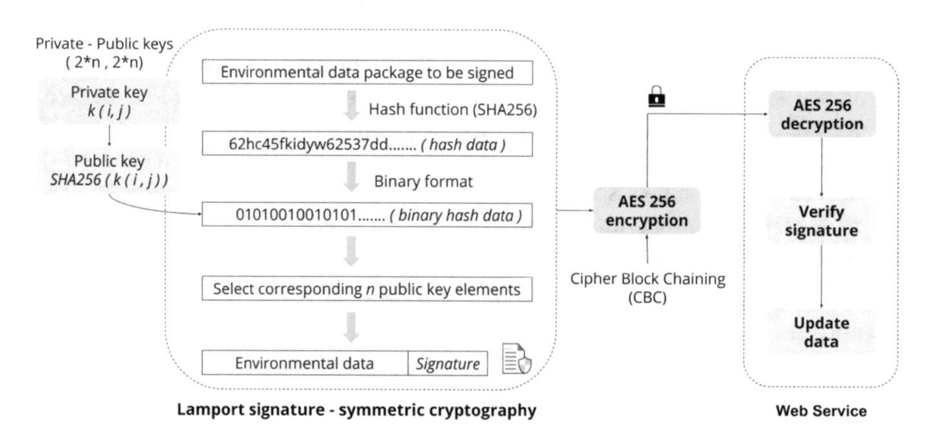

Fig. 5. Lamport's signature and AES 256 CBC mode encryption

A block cipher algorithm has been implemented for encrypting data packages that will be sent to the Web service from the IoT devices. AES encryption algorithm based on a key of 256 bits together with Cipher Block Chaining (CBC) mode (considered as the current standard encryption) and Zero padding have been implemented without involving an additional cost of time in the environmental measurements management. A key predistribution process has been considered to provide particular keys necessaries to perform this encryption process.

Robust authentication has been proposed taking into account the Open Web Application Security Project (OWASP) guidelines. Authentication tokens and secure data transmission protocol through HTTPS are implemented.

Regarding passwords management, hash functions have been used and secret keys (such as API keys or encryption algorithm private keys) are also encrypted. A private administration interface with privileged access has been defined in order to manage system resources or relevant metadata registered.

6 Conclusions

The main objective of this proposal is to contribute to solve some environmental challenges related to forest fires prevention, detection and tracking through the deployment of wireless sensor networks. It makes possible to monitor dynamic forest fire risk factors in real time. In this sense, a measurement interface through novel IoT technologies, devices and sensors together with multi-hop routing, a forest fire risks controller based on fuzzy logic, a Web service and a mobile application have been implemented. All these elements allow to analyze wildfire incidents and to generate forest fire risk alerts providing an enhancement in response time of emergency corps. On the other hand, security mechanisms are proposed ensuring integrity and confidentiality of information through an encrypting process based on AES 256 and Lamport's signature.

Multiple research lines remain open, such as improving the fuzzy-based controller of forest fire environmental risks through introducing machine learning and new interesting environmental sensors. Regarding the use of machine learning, historical weather data may be used for initial generation of training sets since they include the input variables included in the model presented here. The possibility of implementing a secure wireless sensor network including Blockchain technology deserves to be studied. Finally, improving wireless sensor network distribution through interesting forest areas according to static risk factors such as the orography or the widlfire incidents frequency should also be considered.

Acknowledgements. Research supported by Binter-Sistemas grant and the Spanish Ministry of Economy and Competitiveness, the European FEDER Fund, and the CajaCanarias Foundation, under Projects TEC2014-54110-R, RTC-2014-1648-8, MTM2015- 69138-REDT,TESIS- 2015010106 and DIG02-INSITU.

References

1. Kumar, V., Jain, A., Barwal, P.: Wireless sensor networks: security issues, challenges and solutions. Int. J. Inf. Comput. Technol. (IJICT) **4**(8), 859–868 (2014)
2. Zadeh, L.A.: Fuzzy logic-a personal perspective. Fuzzy Sets Syst. **281**, 4–20 (2015)
3. Rani, S., Ahmed, S.H.: Multi-hop routing in wireless sensor networks: an overview, taxonomy, and research challenges. Springer, Singapore (2015)
4. Merkle, R.C.: A certified digital signature. In: Conference on the Theory and Application of Cryptology, pp. 218–238. Springer, New York (1989)
5. Rajkumar, D.M.N., Sruthi, M., Kumar, D.V.V.: IoT based smart system for controlling Co2 emission. Int. J. Sci. Res. Comput. Sci. Eng. Inf. Technol. **2**(2), 284 (2017)

6. Zhou, Q., Huang, W., Zhang, Y.: Identifying critical success factors in emergency management using a fuzzy dematel method. Saf. Sci. **49**(2), 243–252 (2011)
7. Iliadis, L.S.: A decision support system applying an integrated fuzzy model for long-term forest fire risk estimation. Environ. Model. Softw. **20**(5), 613–621 (2005)
8. Chan, H., Perrig, A., Song, D.: Random key predistribution schemes for sensor networks. In: 2003 Symposium on Security and Privacy, IEEE, pp. 197–213 (2003)
9. Pathan, A.S.K., Lee, H.W., Hong, C.S.: Security in wireless sensor networks: issues and challenges. In: The 8th International Conference on Advanced Communication Technology, ICACT 2006, vol. 2, p. 6. IEEE (2006)
10. Mamdani, E.H.: Application of fuzzy logic to approximate reasoning using linguistic synthesis. In: Proceedings of the sixth international symposium on Multiple-valued logic. IEEE Computer Society Press, pp. 196–202 (1976)
11. Lecina-Diaz, J., Alvarez, A., Retana, J.: Extreme fire severity patterns in topographic, convective and wind-driven historical wildfires of mediterranean pine forests. PloS one **9**(1), e85127 (2014)
12. New record for CO2 concentration in the atmosphere (2017). Accessed 15 Jul 2017. https://elpais.com/elpais/2017/10/30/ciencia/1509359304_347557.html
13. Kosko, B.: Fuzzy associative memories (1991)
14. Runkler, T.A.: Selection of appropriate defuzzification methods using application specific properties. IEEE Trans. Fuzzy Syst. **5**(1), 72–79 (1997)
15. Funai, C., Tapparello, C., Heinzelman, W.: Supporting multi-hop device-to-device networks through wifi direct multi-group networking (2015). arXiv preprint. arXiv:1601.00028
16. Khan, M.A., Salah, K.: IoT security: Review, blockchain solutions, and open challenges. In: Future Generation Computer Systems (2017)
17. Naor, M., Yung, M.: Universal one-way hash functions and their cryptographic applications. In: Proceedings of the twenty-first annual ACM symposium on Theory of computing. ACM, pp. 33–43 (1989)

A Greedy Biogeography-Based Optimization Algorithm for Job Shop Scheduling Problem with Time Lags

Madiha Harrabi[1,4(✉)], Olfa Belkahla Driss[2,4], and Khaled Ghedira[3,4]

[1] Ecole Nationale des Sciences de l'Informatique, Université de Manouba, Manouba, Tunisia
madiha.harrabi@gmail.com
[2] Ecole Supérieure de Commerce de Tunis, Université de Manouba, Manouba, Tunisia
olfa.belkahla@isg.rnu.tn
[3] Institut Supérieur de Gestion, Université de Tunis, Tunis, Tunisia
khaled.ghedira@isg.rnu.tn
[4] SOIE-COSMOS Laboratory, Université de Manouba, Manouba, Tunisia

Abstract. This paper deals with the Job shop Scheduling problem with Time Lags (JSTL). JSTL is an extension of the job shop scheduling problem, where minimum and maximum time lags are introduced between successive operations of the same job. We propose a combination between Biogeography-Based Optimization (BBO) algorithm and Greedy heuristic for solving the JSTL problem with makespan minimization. Biogeography-Based optimization is an evolutionary algorithm which is inspired by the migration of species between habitats. BBO has successfully solved optimization problems in many different domains and has reached a relatively mature state using two main steps: migration and mutation. Good performances of the proposed combination between Greedy and BBO algorithms are shown through different comparisons on benchmarks of Fisher and Thompson, Lawrence and Carlier for JSTL problem.

Keywords: Greedy · BBO · Job shop · Optimization · Scheduling Metaheuristic · Time lags

1 Introduction

Scheduling problems consist in allocation of a number of jobs to machines taking into consideration a set of constraints. They are widely studied by many researches since the early 1950s, which aims to find the optimal schedule under various objectives, different machine environments and characteristics of the jobs [24]. The job shop scheduling problem is defined by a set of n jobs which have to be processed on m machines. The dimension of the problem is known as n*m [25]. The job shop problem with time lags is a special case of the classical job shop problem. This problem can be defined as a job shop problem with minimum and maximum delays existing between starting times of

M. Graña et al. (Eds.): SOCO'18-CISIS'18-ICEUTE'18, AISC 771, pp. 489–499, 2019.
https://doi.org/10.1007/978-3-319-94120-2_47

operations of different jobs [16]. Time Lags constraints could be used in several fields of industrial applications of job shop.

Talking about job shop scheduling problem including time lag constraints encompasses two kinds of time lag constraints: either time lags between two successive operations of the same job, it is the Job shop Problem with Time Lags denoted as JSTL or generic time lags between whatever pairs of operations denoted as JSGTL.

Job shop scheduling problem including minimum and maximum time lag constraints is NP-hard [16]. Different methods were proposed for solving the job shop scheduling problem with time lags in the literature.

Caumond et al. [4] introduced a genetic algorithm based on operation insertion heuristic. Caumond et al. [5] proposed a tabu search metaheuristic based on disjunctive graph model of the problem. Caumond et al. [6] proposed a constructive heuristic based on Giffer and Thompson's heuristic. Deppner [8] proposed heuristics for a general scheduling problem which includes the job shop problem with minimal and maximal time lags between every pair of operations. Caumond et al. [7] introduced memetic based approach taking advantages of initial solutions generation by a heuristic which adds each operation iteratively, a powerful local search based on the critical path analysis and on a neighboring generation system. Their approach is validated solving both flow shop, job shop, no-wait and instances with time lags. Karoui et al. [15] investigated a Climbing Discrepancy Search method. Artigues et al. [2] proposed a job insertion heuristic for generating initial solutions and generalized resource constraint propagation mechanisms, which are embedded in a branch and bound algorithm. González et al. [10] proposed a scatter search procedure combining the path relinking and the tabu search metaheuristic. Afsar et al. [1] proposed a disjunctive graph model for the job shop problem with time lags and transport. Recently the multi agent systems are widely used for the resolution of job shop problems. Harrabi and Belkahla Driss [11] proposed a tabu search metaheuristic in a multi agent model. Harrabi et al. [12] proposed parallel tabu searches in a multi agent system composed of competitive agents. Some other methods were proposed for solving the job shop scheduling problem with generic time lags. Lacomme et al. [16] proposed some dedicated constraint propagation using the Bierwirth's vector for presenting a solution, Lacomme et al. [17] proposed a greedy randomized priority rules. Harrabi et al. [13] proposed a combination of genetic algorithm and tabu search metaheuristic. Harrabi et al. [14] proposed a multi-agent model based on hybrid genetic algorithm.

In this paper, we propose a combination between greedy heuristic and Biogeography-Based Optimization algorithm for the job shop scheduling problem with time lags between successive operations of the same job with makespan minimization. The reminder of the paper is organized as follows. In Sect. 2, we explain the job shop scheduling problem with time lags. In Sect. 3, we present the BBO algorithm concept. In Sect. 4, we present the adaptation of Greedy BBO algorithm for our problem. Section 5 analyzes the performance results of GBBO when applied to solve instances of benchmark problems in literature. At last, we come to our conclusion and some possible future directions.

2 Problem Formulation

The problem linear formulation of Job Shop with Time Lags is an extension of the Manne's formalization [20] of classical job shop with additional minimum and maximum time lags constraints.

The Job Shop Scheduling problem with Time Lags is formulated as follows.

- M: a set of machines
- O: a set of operations to schedule
- Z: the completion time of the last operation
- E_k: a set of operations processed on machine k
- t_i: starting time of operation *i*
- p_i: the processing time of operation i
- S_i: the next operation of operation i depending on job
- TL_i^{min}: the minimal time-lags between i and s_i
- TL_i^{max}: the maximal time-lags between i and si
- $a_{ij} = \begin{cases} 1 & \text{if operation j is processed before i} \\ 0 & \text{othewise} \end{cases}$
- H: a positive large number

$$\text{Minimize}\, Z, \tag{1}$$

Subject to

$$\left(H + p_j\right) a_{ij} + \left(t_j - t_i\right) \geq p_i \quad \forall (i,j) \in E_k, \forall k \in M \tag{2}$$

$$\left(H + p_j\right)\left(1 - a_{ij}\right) + \left(t_i - t_j\right) \geq p_j \quad \forall (i,j) \in E_k, \forall k \in M \tag{3}$$

$$t_i + p_i + TL_i^{min} \leq t_{si} \quad \forall i \in O \tag{4}$$

$$t_{si} \leq t_i + p_i + TL_i^{max} \quad \forall i \in O \tag{5}$$

$$Z \geq t_i + p_i \quad \forall i \in O \tag{6}$$

$$t_i \geq 0 \quad \forall i \in O \tag{7}$$

$$a_{ij} \in \{0; 1\} \quad \forall (i,j) \in E_k, \forall k \in M \tag{8}$$

3 Biogeography-Based Optimization

BBO algorithm, proposed by Simon in 2008 [21], is inspired by the mathematics of biogeography and mainly from the work of MacArthur and Wilson [19]. Later, a large amount of theoretical, methodological, and practical studies on BBO have come into being. The two main concepts of BBO are Habitat Suitability Index (HSI) and Suitability Index Variables (SIVs). Features that correlate with HSI include rainfall, diversity of topographic features, land area, and temperature. And SIVs are considered as the independent variables of the habitat. Geographical areas that are well suited for species are said to own a high HSI. Considering the optimization algorithm, a population of candidate solutions can be represented as vectors. Each integer in the solution vector is considered to be an SIV. After assessing performance of the solutions, good solutions are considered to be habitats with a high HSI, and poor ones are considered to be habitats with a low HSI. Therefore, HSI is analogous to fitness in other population based optimization algorithms.

The two main operators of BBO are migration and mutation. Migration operator, including immigration and emigration, bridges the communication of habitats in an ecosystem. The emigration and immigration rates of each solution are used to probabilistically share information between habitats.

Deciding whether or not a habitat performs emigration or immigration is up to its HSI. A habitat with high HSI means that more species has more opportunities to emigrate to neighboring habitats. A habitat with low HSI means sparse species need immigration to increase the diversity of its species. Therefore, habitats with high HSI have larger emigration rates and smaller immigration rates while ones with low HSI have larger immigration rates and smaller emigration rates [23].

4 Greedy Biogeography-Based Optimization for Job Shop Scheduling Problem with Time Lags

4.1 Representing Habitat

For job shop scheduling problems, we can find two strategies for representation of solution, a direct representation, in which a solution is directly encoded into the genotype (i.e., the schedule itself is encoded) and an indirect representation which encodes the instructions to a schedule builder. In this work, we use the direct representation. In this representation, we give the total number of operations of different jobs by order of processing in machines. O_{ij} is the operation j of the job i. See Fig. 1.

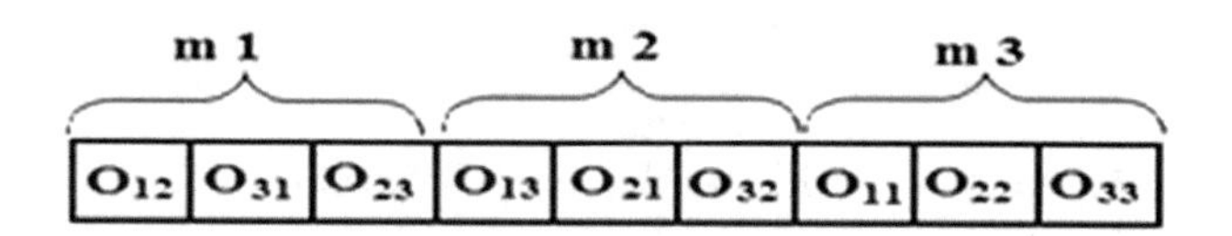

Fig. 1. Habitat representation

4.2 Selection Strategies

This step is one of the distinctive steps of BBO with other algorithms, which is executed through two different strategies, one for migration and one for mutation.

To explain the selection strategies for migration, we should firstly define the immigration rate λ_i and the emigration rate μ_j. During the migration process, we face two types of selection. Firstly, we should determine whether a special habitat H_i should be immigrated or not. To do so, a simple comparison of λ_i with a random number is done. Secondly, we should select habitat H_j for emigrating to H_i.

During the mutation process, a habitat H_i selected to be mutated according to a simple comparison of the mutation probability with a random number.

4.3 Initialization of BBO

The starting step of BBO is the generation of population habitat using a chosen parameter population size PS. The initial population can be generated randomly or using specific constructive heuristics. In this work, we chose to generate the initial population using the greedy constructive heuristic. This choice is based on high performance of the greedy algorithm which can produce solutions with good quality and its use often leads to better quality local optima. The greedy algorithm starts building the solution from one operation to another. After insertion the operation to a defined position of the current solution, the different constraints were checked. If all constraints are satisfied, we proceed to the next operation. Else, this operation was deleted in the current position and added to another position that respects the different constraints.

4.4 Migration Operator

Migration is a probabilistic operator that is used for modifying each solution H_i by sharing features among different solutions. The idea of a migration operator is based on the migration in biogeography which shows the movement of species among different habitats. Solution H_i is selected as immigrating habitat with respect to its immigration rate λ_i, and solution H_j is selected as emigrating habitat with respect to its emigration rate μ_j. It means that a solution is selected for immigrating or emigrating depends on its immigration rate λ_i, or emigration rate μ_j; the migration process can be shown as:

$$H_i(SIV) \leftarrow H_j(SIV)$$

After calculating the HSI for each solution H_i, the immigration rate λ_i and the emigration rate μ_j can be evaluated as follows:

$$\lambda_i = \mathrm{I}\left(1 - \frac{K_i}{\mathrm{n}}\right) \tag{9}$$

$$\mu_j = \mathrm{E}\left(\frac{K_i}{\mathrm{n}}\right) \tag{10}$$

In (9) and (10), k_i represents the rank of the ith habitat after sorting all habitats according to their HSIs. It is clear that since more HSI represents a better solution, more k_i represents the better solution. Therefore, the 1st solution is the worst and the nth solution is the best. I is the maximum immigration rate and E the maximum emigration rate which are both usually set to 1, n is the number of habitats in the population. The two rates, λ_i and μ_j are the functions of fitness or HSI of the solution. Since, according to the biogeography, the SIVs of a high-HSI solution tend to emigrate to low-HSI solutions, a high-HSI solution has a relatively high μ_j and low λ_i, while in a poor solution, a relatively low μ_j and a high λ_i are expected. Figure 2 illustrates an example of migration operator of BBO for our problem.

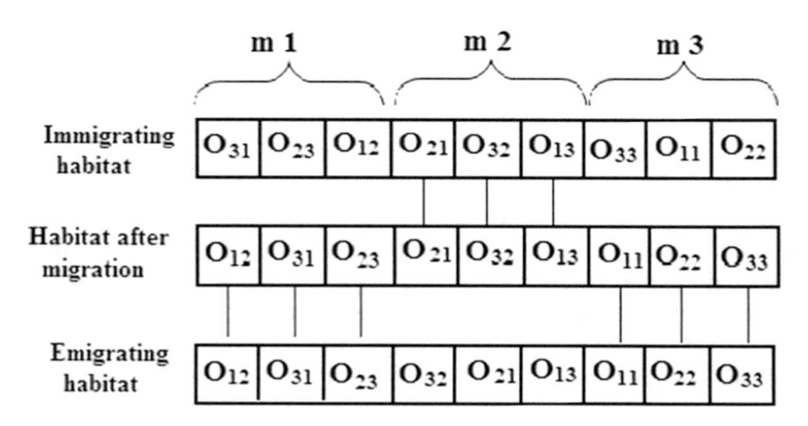

Fig. 2. Migration operator of BBO algorithm

As mentioned earlier, the SIVs from a good habitat tend to migrate into a poor habitat. This migration operator is performed probabilistically based on immigration and emigration rates. In this example we will explain how the migration is implemented in our BBO algorithm. Consider dealing with an instance of job shop scheduling problem with time lags with 3 jobs, 3 operations for each job and 3 machines. Suppose, based on immigration and emigration rates, that an immigrating habitat, H_i = ($O_{31,}$ O_{23}, O_{12}, O_{21}, O_{32}, O_{13}, O_{33}, O_{11}, O_{22}) and an emigrating habitat H_e = ($O_{12,}$ O_{31}, O_{23}, O_{32}, O_{21}, O_{13}, O_{11},O_{22}, O_{13}). The migration process is: H_e (SIVs) $\leftarrow$ H_i (SIVs).

SIVs of H_i will be randomly selected and replace randomly selected SIVs of H_e. Assuming SIVs of H_i (O_{21}, O_{32}, O_{13}), are selected to replace SIVs of H_e (O_{32} O_{21} O_{13}). Therefore, the migration process consists in:

(1) SIVs of machine 2 from H_i (O_{21}, O_{32}, O_{13}) migrate into H_e to replace SIVs of H_e (O_{32} O_{21} O_{13}).
(2) SIVs (O_{21}, O_{23}, O_{32}) replace SIVs (O_{32} O_{23} O_{11}).
(3) SIVs ($O_{12,}$ O_{31}, O_{23}) and (O_{11}, O_{22}, O_{13}) of machine 1 and machine 3 from H_e remain at original places.
(4) Therefore, the new habitat, H_n = ($O_{12,}$ O_{31}, $O_{23,}$ O_{21}, O_{32}, O_{13}, O_{11}, O_{22}, O_{13}) is produced.

4.5 Mutation Operator

Mutation is a probabilistic operator that randomly modifies a solution's SIV based on its priori probability of existence. Mutation is used to enhance the diversity of the population, which helps to decrease the chances of getting trapped in local optima. Solutions with very high HSI and very low HSI are both equally improbable, while medium HSI solutions are relatively probable to mutate. Namely, a randomly generated SIV replaces a selected SIV in the solution H_i according to a mutation probability. Note that an elitism approach is employed to save the features of the habitat that has the best solution in BBO process. The habitat with the best solution has a mutation rate of 0. Figure 3 illustrates an example of mutation operator of BBO.

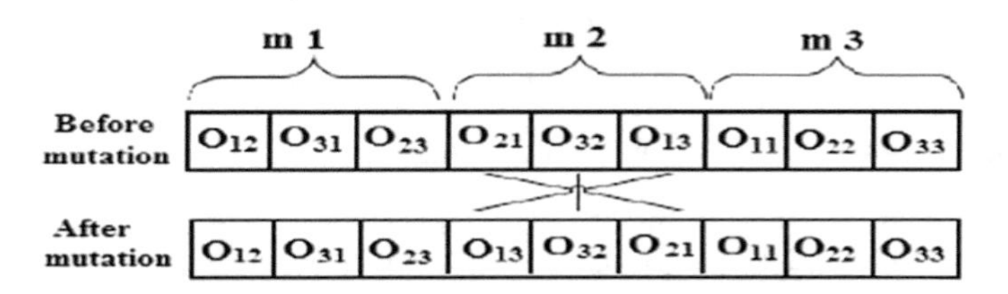

Fig. 3. Mutation operator of BBO algorithm

As mentioned earlier, mutation is performed by simply replacing a selected SIV of a habitat with a randomly generated SIV. In our example the mutation process consists in:

(1) SIV O_{21} is chosen to mutate.
(2) Assume that the new SIV which is randomly generated is $O_{13.}$ SIV O_{21} is replaces with O_{13}.
(3) SIV O_{13} takes the place of O_{21}.
(4) The resulting mutated habitat is produced.

5 Experimental Results

In order to evaluate the performance of the proposed Greedy Biogeography-Based Optimization algorithm for the job shop scheduling problem with minimum and maximum time lags constraints, we give in this section the experimental results.

We use the instances of Fisher and Thompson [9], Lawrence [18] and Carlier [3].
For all instances, $TL^{min}_{ij,\ ij+1} = 0$
For each instace, $TL^{max}_{ij,\ ij+1} = \{0, 0.5, 1, 2\}$.

Instances are designed with Name_ $TL^{min}_{ij,\ ij+1}$_ $TL^{max}_{ij,\ ij+1}$, for example ft06_0_0.5 is the instance of Fisher and Thompson 6 with $TL^{min}_{ij,\ ij+1} = 0$ and $TL^{max}_{ij,\ ij+1} = 0.5$.

5.1 Comparison Results of Fisher and Thompson Instances

Results in Table 1, show that the proposed GBBO algorithm gives better results than the Job Insertion heuristic method [2] and the Operation Insertion heuristic [5] in 75% of instances. Compared with the Tabu Search metaheuristic [4], results show that GBBO gives better results in 100% of instances.

Table 1. Results of JI/OI/TS/GBBO for Fisher and Thompson instances

Instances	n*m	JI	OI	TS	GBBO
ft06_0_0	6*6	96	83	94	83
ft06_0_0.5	6*6	72	109	81	73
ft06_0_1	6*6	72	58	61	61
ft06_0_2	6*6	70	55	55	55

5.2 Comparison Results of Lawrence Instances

For Lawrence instances with 10 jobs and 5 machines, results in Table 2 show that GBBO algorithm gives better results than the Job Insertion heuristic [2] in 60% of instances and better results than Operation Insertion heuristic [5] in 90% of instances. Compared with tabu search method [4], GBBO gives better results in 70% of instances.

Table 2. Results of JI/OI/TS/GBBO for Lawrence instances

Instances	n*m	JI	OI	TS	GBBO
la01_0_0	10*5	1258	1504	1473	1133
la01_0_0.5	10*5	1063	1474	758	959
la01_0_1	10*5	928	1114	916	934
la01_0_2	10*5	967	948	732	889
la02_0_0	10*5	1082	1416	1436	1103
la02_0_0.5	10*5	1011	1207	1153	1019
la02_0_1	10*5	935	1136	900	874
la02_0_2	10*5	928	895	681	801
la03_0_0	10*5	1081	1192	1108	1059
la03_0_0.5	10*5	930	1085	1052	966
la03_0_1	10*5	886	931	847	837
la03_0_2	10*5	808	787	671	651
la04_0_0	10*5	1207	1346	1275	1250
la04_0_0.5	10*5	780	1156	1106	906
la04_0_1	10*5	1010	857	950	867
la04_0_2	10*5	982	838	642	804
la05_0_0	10*5	1080	1224	1128	1049
la05_0_0.5	10*5	935	1208	957	976
la05_0_1	10*5	814	964	761	736
la05_0_2	10*5	749	683	600	700

5.3 Comparison Results of Carlier Instances

For Carlier's instances, results in Table 3 show that GBBO algorithm gives makespan value better than tabu search [4] in 81% of instances and better results than the memetic algorithm [7] in 50% of instances.

Table 3. Results of TS/Mem/GBBO for Carlier instances

Instances	n*m	TS	Mem	GBBO
Car5_0_0	10*6	111495	7821	9406
Car 5_0_0.5	10*6	10293	7821	8854
Car 5_0_1	10*6	8910	7821	7805
Car 5_0_2	10*6	8281	7720	7894
Car 6_0_0	8*9	11243	8313	8453
Car 6_0_0.5	8*9	9100	8330	8312
Car 6_0_1	8*9	9248	8313	8313
Car 6_0_2	8*9	8467	8505	8417
Car 7_0_0	7*7	7704	6558	7528
Car 7_0_0.5	7*7	6953	6558	6958
Car 7_0_1	7*7	6590	6558	6609
Car 7_0_2	7*7	6573	6558	6572
Car 8_0_0	8*8	10144	8407	8423
Car 8_0_0.5	8*8	8856	8407	8459
Car 8_0_1	8*8	8833	8264	8304
Car 8_0_2	8*8	8586	8279	8285

6 Conclusion

Biogeography-Based Optimization (BBO), a relatively new optimization technique based on the biogeography concept, uses the idea of migration strategy of species to derive algorithm for solving optimization problems. In this work, we propose an application of BBO algorithm combined with the greedy constructive heuristic in the initialization step for solving the job shop scheduling problem with time lags between successive operations of the same job which is an NP-hard problem. Our experiments based on benchmark instances of Fisher and Thompson, Lawrence and Carlier for JSTL problem, show that GBBO provides better results than other proposed algorithm in the literature for most instances in term of makespan. Results are promising and different perspectives are possible. We can use the same algorithm to solve other extension of our problem; we can develop the hybridization of BBO with other algorithms in order to solve the same problem.

References

1. Afsar, H.M., Lacomme, P., Ren, L., Prodhon, C., Vigo, D.: Resolution of a job-shop problem with transportation constraints, a master/slave approach. In: IFAC Conference on Manufacturing Modelling, Management and Control (2016)
2. Artigues, C., Huguet, M., Lopez, P.: Generalized disjunctive constraint propagation for solving the job shop problem with time lags. Eng. Appl. Artif. Intell. **24**, 220–231 (2011)
3. Carlier, J.: Ordonnancements a contraintes disjunctives. RAIRO Rech. operationelle/Oper. Res. **12**, 333–351 (1978)
4. Caumond, A., Lacomme, A,P., Tchernev, N.: Proposition d'un algorithme génétique pour le job-shop avec time-lags. In: ROADEF 2005, pp. 183–200 (2005)
5. Caumond, A., Gourgand, M., Lacomme, P., Tchernev, N.: Métaheuristiques pour le problème de job shop avec time lags", Jm|li,s j(i)|Cmax. In: 5ème confèrence Francophone de Modélisation et SIMulation (MOSIM 2004). Modélisation et simulation pour l'analyse et l'optimisation des systèmes industriels et logistiques, pp. 939–946. Nantes, France (2004)
6. Caumond, A., Lacomme, P., Tchernev, N.: Feasible schedule generation with extension of the Giffler and Thompson's heuristic for the job shop problem with time lags. In: International Conference of Industrial Engineering and Systems Management, pp. 489–499 (2005)
7. Caumond, A., Lacomme, P., Tchernev, N.: A memetic algorithm for the job-shop with time-lags. Comput. Oper. Res. **35**, 2331–2356 (2008)
8. Deppner, F.: Ordonnancement d'atelier avec contraintes temporelles entre opérations, Ph.D. thesis, Institut National Polytechnique de Lorraine (2004)
9. Fisher, H., Thompson, G.L.: Probabilistic Learing Combination of Local Job Shop Scheduling Rules, pp. 225–251. Prentice Hall, Industrial Scheduling (1963)
10. González, M.A., Oddi, A., Rasconi, R., Varela, R.: Scatter search with path relinking for the job shop with time lags and set up times. Comput. Oper. Res. **60**, 37–54 (2015)
11. Harrabi, M., Belkahla Driss, O.: MATS-JSTL: a multi-agent model based on tabu search for the job shop problem with time lags. In: International Computational Collective Intelligence Technologies and Applications ICCCI, pp. 39–46 (2015)
12. Harrabi, M., Belkahla Driss, O., Ghedira, K.: Competitive agents implementing parallel tabu searches for job shop scheduling problem with time lags. In: IASTED International Conference on Modelling, Identification and Control, pp. 848–052. MIC (2017)
13. Harrabi, M., Belkahla Driss, O., Ghedira, K.: Combining genetic algorithm and tabu search for job shop scheduling problem with time lags. In: IEEE International Conference on Engineering & MIS, pp. 1–8 (2017)
14. Harrabi, M., Belkahla Driss, O., Ghedira, K.: A multi-agent model based on hybrid genetic algorithm for job shop scheduling problem with generic time lags. In: ACS/IEEE International Conference on Computer Systems and Applications AICCSA, pp. 995–1002 (2017)
15. Karoui, W., Huguet, M.-J., Lopez, P., Haouari, M.: Méthode de recherche à divergence limitée pour les problèmes d'ordonnancement avec contraintes de délais" [Limited discrepancy search for scheduling problems with time-lags]. In: 8ème ENIM IFAC Conférence Internationale de Modélisation et Simulation, Hammamet, Tunisie, pp. 10–12 (2000)
16. Lacomme, P., Huguet, M.J., Tchernev, N.: Dedicated constraint propagation for job-shop problem with generic time-lags. In: 16th IEEE Conference on Emerging Technologies and Factory Automation IEEE Catalog Number: CFP11ETF-USB, Toulouse, France (2011)

17. Lacomme, P., Tchernev, N.: Job-shop with generic time lags: a heuristic based approach. In: 9th International Conference of Modeling, Optimization and Simulation – MOSIM (2012)
18. Lawrence, S.: Supplement to Resource Constrained Project Scheduling: An experimental investigation of Heuristic Scheduling Techniques. Graduate School of Industrial Administration, Carnegie Mellon University (1984)
19. MacArthur, R., Wilson, E.: The Theory of Biogeography. Princeton University Press, Princeton (1967)
20. Manne, S.: On the job-shop scheduling problem. Oper. Res. **8**, 219–223 (1960)
21. Simon, D.: Biogeography-based optimization. IEEE Trans. Evol. Comput. **12**, 702–713 (2008)
22. Wikum, E.D., Lewellynn, D.C., Nemhauser, G.L.: One-machine generalized precedence constrained scheduling problems. Oper. Res. Lett. **16**, 87–99 (1994)
23. Yang, Y.: A Modified Biogeography-Based Optimization for the Flexible Job Shop Scheduling Problem, Mathematical Problems in Engineering (2015)
24. Brucker, P.: Scheduling and constraint propagation. Discrete Appl. Math. **123**, 227–256 (2002)
25. Huang, K.L., Liao, C.J.: Ant colony optimization combined with taboo search for the job shop scheduling problem. Comput. Oper. Res. **35**, 1030–1046 (2008)

PlagZap: A Textual Plagiarism Detection System for Student Assignments Built with Open-Source Software

Elena Băutu[1] and Andrei Băutu[2](✉)

[1] Department of Mathematics and Informatics, "Ovidius" University of Constanta, Constanța, Romania
ebautu@univ-ovidius.ro
[2] Department of Navigation and Naval Transport, "Mircea cel Batran" Naval Academy, Constanța, Romania
andrei.bautu@anmb.ro

Abstract. Plagiarism among university students is an important issue that affects their preparation and undermines the universities' efforts to prepare skilled graduates. Universities try to fight-back this problem with strict ethics policies, but they require the proper plagiarism detection tools, at affordable costs, to implement these policies. In this paper, we present PlagZap, a cost-efficient, high-volume, and high-speed plagiarism detection system built using open-source software and designed to be used on textual student assignments (essays, theses, homework). We discuss the advantages of this design with respect to speed, precision and costs.

Keywords: Plagiarism detection · Open-source software
Information retrieval

1 Introduction

Plagiarism in academic works, from research papers to assignments of students, is regarded as a violation of academic ethics. Books, research papers and theses published by researchers and Ph.D. students are often checked for plagiarism (either through automatic tools or by peer reviewing) to ensure the quality and originality of the research. Publishers lacking such basic quality control mechanisms are usually avoided by honest researchers and "blacklisted" by libraries and the research community [1, 2].

If we move our focus towards the younger members of the academic community (master and bachelor students), originality checking of their work happens at a much lower rate compared to the previous category [3]. Plagiarism in course assignments raises questions about the competencies that students acquired during studies and the validity of their diploma. Most universities have (in theory) strict policies for the academic ethics of students (including plagiarism). In many universities, poor implementation of these policies, which are part of basic academic standards, has many reasons.

M. Graña et al. (Eds.): SOCO'18-CISIS'18-ICEUTE'18, AISC 771, pp. 500–508, 2019.
https://doi.org/10.1007/978-3-319-94120-2_48

A major reason is the economic cost of plagiarism detection services available today. Major players on the market (e.g. TurnItIn, NoPlag) offer various types of subscriptions, both for individual users and for universities. In the first case, teachers split the costs of implementing a university policy among themselves, usually as a voluntary action, in an attempt to optimize their own workflow (i.e. weed out plagiarized assignments from their grading process right from the start). Since this is voluntary, it does not apply to all subjects in the university and students will know soon what subjects are exempt from this type of control. With university subscriptions, all teachers have access to the plagiarism detection service, but it requires a large volume subscription, which is expensive. Either way, when added up, the subscription costs for using commercial plagiarism services on all the teaching levels within a university can be significantly high.

Another reason why universities are reluctant to use commercial plagiarism detection services is related to compliance with privacy and proprietary laws, and the retention policies for the documents submitted for checking. The terms and conditions of such services state that the service provider is granted rights to permanently store copies of the documents for other uses besides the actual checking of the document (e.g. Viper, QueText), or that the teacher/university is responsible for following all legislation on privacy and proprietary (e.g. TurnItIn), or that the teacher/university must own the copyrights on the content they are scanning.

Due to the reasons like the ones mentioned above, many universities lack a proper system for checking the originality of students' assignments, focusing only on final years and/or diploma theses, which is ironic since the junior year is the best period for growing an academic ethics culture with students. In fact, many authors that researched this issue concluded that plagiarism detection solutions are just one piece of the solution, with enforcement of ethics policies, education on publication ethics, and active involvement of readers (teachers, reviewers, etc.) being other parts [3–7].

In order to provide a cost-efficient, high-volume, and high-speed plagiarism detection system for the assignments of students at the "Mircea cel Batran" Naval Academy, we designed PlagZap, a plagiarism detection system that leverages existing industry-proved open-source software packages to perform plagiarism scanning against a large database of textbooks and past student assignments and graduation theses.

2 Open-Source Technologies

Open-source software and technologies have evolved in the past decades to provide stable, high-performance, high-quality solutions for many domains, from agriculture [8, 9] to manufacturing [10], from chemistry [11] and physics [12] to medicine [13, 14], from local transportation [15] to space exploration [16, 17]. The 2016 edition of the Future of Open Source survey [18] estimated that 78% of companies use open source software for part or all their management, production, and research and development activities, and that only 3% of the companies use no open-source software. The same study shows that 90% of the respondents consider that open-source improves their efficiency, interoperability, and innovation. The study was carried out by

Black Duck Software and more than 50 collaborators (like Microsoft, Redhat, Ubuntu or the Linux Foundation), and covered 1313 companies worldwide.

One of the main reasons why open-source software has gained a large adoption even outside the IT industry is the diversity and quality of the available solutions. Having the design and inner workings of any software public allows anyone to review and comment on it, much like the peer reviewing system used for academic papers. This public exposure raises the responsibility of the software authors, forcing them to follow best-practice guidelines and avoid cutting corners.

Another important reason why companies opt for open-source solutions is their competitive features and technical capabilities that frequently put them on par with proprietary competitors (e.g. Apache vs Microsoft IIS, VirtualBox vs VMWare, Odoo vs SAP ERP, etc.). Many companies consider open-source a catalyst for software and technology innovation [18], a fact proven by the decision of many software companies to open-source their traditionally proprietary software. (e.g. Chromium and BoringSSL from Google, .Net Framework and ChakraCore from Microsoft, CodeXL from AMD, Java and NetBeans from Sun Microsystems, etc.). Similar moves to open-source have also affected none-software products, like designs of the transportation system Hyperloop from SpaceX, AI personal assistant Mark 2 from MyCroft, ARA smartphone platform from Google, etc.

Finally, the third major reason companies choose open-source is the ability to review, fix and change the behavior of the product according to their needs, in a flexible, cost and time efficient manner.

3 PlagZap - System Architecture

During past decades, various proprietary and open-source textual plagiarism detection systems have been developed, both for universal content (e.g. essays, theses, homework) [19–22], and for special content (e.g. source code files) [23–25].

Textual plagiarism detection systems provide statistics and/or detailed analysis about the similarity of input documents based on their contents and, optionally, their metadata. Figure 1 presents the main components of such a system, which is also used by PlagZap. Implementation details, specific to PlagZap, are presented in the following section.

In general, before doing any comparison, the input documents are preprocessed to remove irrelevant information from them, such as headers and footers, blank pages, small or all images, change mathematical equations to textual form, etc., and, finally, convert the documents to a common format that is suitable as input for the similarity analysis algorithm. Most plagiarism detection systems (e.g. TurnItIn, Viper, QueText) use plain text or HTML as the common format for their analysis algorithm.

The similarity analysis module parses the common format, splitting the text content into text fragments, and searches for copies of each fragment in other documents located in a document library. During the search, the similarity analysis module can use the plain text format of the document or alternative encodings such as n-grams [26] or hashes of the text fragments [27]. The search algorithm may also perform additional

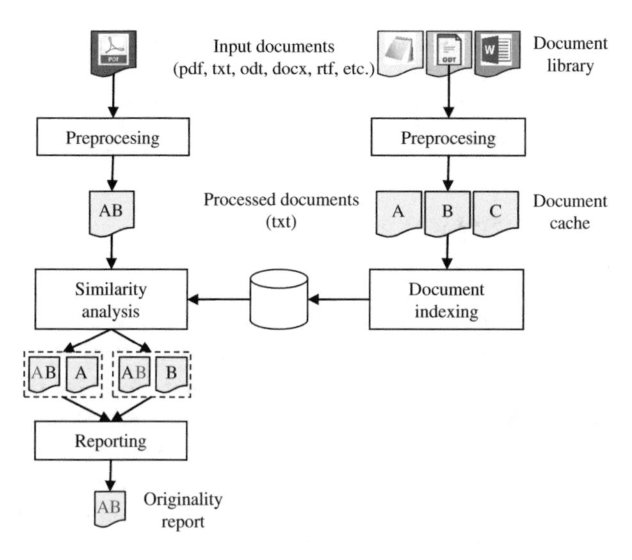

Fig. 1. The architecture of a textual plagiarism detection systems

transformations of the text fragments while searching for similar content, such as phrase extraction, tokenization, synonyms alternatives, case conversion, etc.

After the search for plagiarized text is completed, a reporting module receives the output of the similarity analysis module (i.e. the list of non-original fragments and information about their sources) and summarizes the results, providing statistics about the level of originality, top plagiarized sources and more.

The similarity analysis task is the one that requires the most time and computing power to complete. Searching directly within the original documents available in the document library is very inefficient due to the preprocessing stage required for each document, for each fragment text search. The first step towards improving the search speed is to cache the common format of the preprocessed version of each document. In this way, the preprocessing of documents happens only once per document. If a document is added to, changed or removed from the document library, then the document cache should be updated also. If the preprocessing algorithm or its parameters change, then the document cache has to be completely rebuilt. This is one of the reasons why most commercial plagiarism detection services request in their terms of service a permanent unlimited license on the uploaded documents.

The second step towards improving the search speed is to use a relational or NoSQL database management system that implements full-text indexing and searching for indexing these documents. Full-text indexes use an inverted index data structure to store mappings from content fragments (words, numbers, expressions, ngrams, etc.) to the documents that contain them.

In a general sense, the entire World Wide Web can be viewed as the document library for plagiarism detection. However, indexing all WWW is not practical for plagiarism detection purposes only (or for any purpose for that matter). Instead, some commercial plagiarism detection services rely on paid APIs from Google, Bing or

Yahoo to search the web for documents that match particular text fragments. In fact, some commercial plagiarism detection services (e.g. Plagiarisma, Grammarly) do not offer "search within your private document library" features, because they do not implement their own document libraries and rely solely on 3rd party web search APIs.

4 PlagZap - Implementation Details

The preprocessing module converts the input documents to a common format while also providing some additional filtering. The common format PlagZap uses is plain text. The first implementation of the preprocessing module relied on Universal Office Converter (unoconv), an open-source command line tool that uses the UNO bindings of LibreOffice to convert documents between the different formats it supports. The unoconv integration used temporary files and Linux sub-processes. Additional filtering was implemented using custom code. It functioned correctly, but occasional conversion failures and high CPU usage encouraged us to experiment with other options.

The current implementation for the preprocessing module relies on Apache Tika, a library that can detect and extract metadata and content from over a thousand different file types. Tika offers various integration options: Java library, command line interface (CLI), and RESTful server. We started with the CLI as a drop-in replacement for unoconv, and later moved to RESTful API for performance reasons. The RESTful API also offers better scalability as multiple servers instances can be employed (using containers like Docker, or load-balancers like Varnish). Due to Tika's extensive support for document formats, PlagZap can process virtually any textual documents created by students, in Microsoft Office, Open Document, iWorks, Portable Document, HTML, rich text, or plain text format.

For the document indexing module, we choose to use Apache Solr, a search platform that offers full-text search, index replication, and distributed search features via a RESTful API. Solr has integration with Apache Tika, which would simplify the system design by merging the preprocessing and indexing modules. We decided to not merge them, in order to keep the system's components decoupled, partially because Tika was included later in our project, and partially because decoupled modules allow us to test other options for the indexing module (e.g. ElasticSearch, Sphinx, etc.). Solr uses a configurable pipeline approach to process the input content (i.e. text). For PlagZap, we decided to use the following pipeline elements:

- solr.StandardTokenizerFactory – splits the input (text) into tokens (words)
- solr.ASCIIFoldingFilterFactory – folds accented characters to ASCII equivalents
- solr.StopFilterFactory – ignores common (irrelevant) words
- solr.LowerCaseFilterFactory – lowercase tokens

The Solr project also offers many other filters that can be added to the processing pipeline, to customize it, like multi-word synonym support, stem filters, phonetic filters, and more. We do not use any of them at this moment, but present some future development ideas in the following section.

The similarity analysis module is implemented with custom code in PHP as a standalone library. It splits the input content (text) into variable-length fragments using regular expressions and queries the document library for each fragment using Curl library and Solr's RESTful API. The number of words allowed in each fragment influences the precision of the search. The module can perform single-document or multiple-document source matching. For each fragment, the similarity analysis module gathers statistics about its originality (i.e. number of plagiarized words and characters) and source document(s).

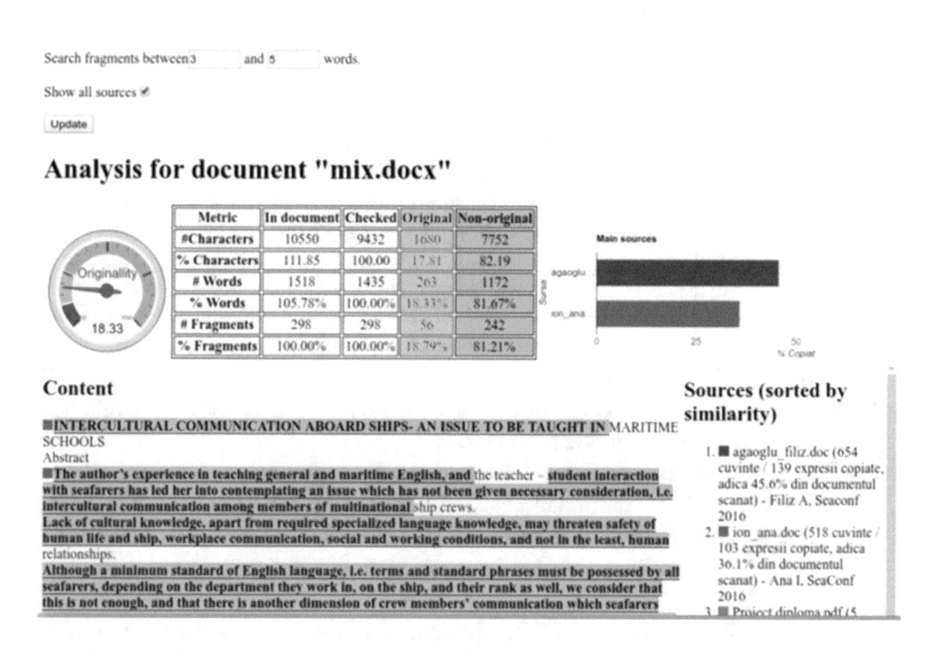

Metric	In document	Checked	Original	Non-original
#Characters	10550	9432	1680	7752
% Characters	111.85	100.00	17.81	82.19
# Words	1518	1435	263	1172
% Words	105.78%	100.00%	18.33%	81.67%
# Fragments	298	298	56	242
% Fragments	100.00%	100.00%	18.79%	81.21%

Fig. 2. The PlagZap report for a document (using text fragments of 3 to 5 words)

The reporting module is the frontend for the entire system, also implemented with custom code in PHP. It integrates the data output from the similarity analysis module into an easy to use, interactive, media-adaptive report. Each report can contain the plagiarism analysis of one or more documents. For each document, the report includes the document name, an originality gauge, a summary table, the top plagiarized sources, the scanned document content, and the list of all matching document libraries. The document content contains interactive markers for plagiarized content that highlight and focus the original source when hovered over. The interactive markers are media-adaptive, having different layouts for screen and printed version of the report. Figure 2 presents an example of an analysis report, which includes the originality metrics (the gauge and metric tables), plagiarism sources (the main sources chart and the list of all sources) and document content, showing plagiarized content in red, with the interactive markers in front of it, linking to the source of the text.

5 Results and Further Work

In order to validate PlagZap, we performed a series of tests using as input documents 64 bachelor theses submitted between 2016 and 2017 by students of "Mircea cel Batran" Naval Academy. The document library we used consisted of about 9000 documents, including study program textbooks and bachelor theses submitted between 2011 and 2014. The total size of original documents from the document library is 9,5 GB, which resulted in a document index of only 58 MB.

For each of the input document, we performed a plagiarism check using Plagiarisma and our system. Plagiarisma can check for plagiarism using Bing or Google API. It uses quite large text fragments (12 words on average), which speeds-up the search process, reduces the usage costs for search-engines APIs, but makes the similarity search biased towards false-negatives. The same method is used by other commercial plagiarism detection services (e.g. StrikePlagiarism, QueText).

We compared the plagiarism reports and made the following observations:

- Because PlagZap document library contains many documents that are not publicly available on the Internet, it provides higher rates plagiarism and more accurate detection than Plagiarisma.
- Plagiarism checking with PlagZap was on average 34 times faster than Plagiarisma, because it queries its own document index.
- PlagZap allows the user to control the size of text fragments, thus balancing speed and precision, with no additional costs; Plagiarisma doesn't have additional costs either, but it doesn't offer any control over the size of text fragments; other service providers (e.g. StrikePlagiarism) offer such options, but at higher costs than a regular check.
- Because PlagZap is limited to its document library (i.e. hermetic plagiarism checking), it did not catch plagiarism from Internet sources unless they were already plagiarized by other documents from the document library; Plagiarisma did catch some of them, but missed others (possible of its large text fragments).

Based on these observations, we conclude that PlagZap can provide a cost-efficient, high-volume, and high-speed plagiarism detection system for students' assignment at small and medium size universities. Large universities might require additional tuning of the system architecture, something we will consider further.

PlagZap is in alpha stage and still undergoing testing and development. We packaged the system as an easy-to-run docker service and published it on GitHub (at https://github.com/abautu/plagzap). Further work will focus on detecting plagiarism obfuscation techniques, like character and image hacks (i.e. replace Latin letters with images or different Unicode characters with similar print), words manipulation and synonyms, back-translation [28] (i.e. disguise the original text by using automatic translation engines), etc.

Bibliography

1. Watson, R.: Beall's list of predatory open access journals: RIP. Nursing Open **4**(2), 60 (2017)
2. Strielkowski, W.: Predatory journals: Beall's List is missed. Nature **544**(7651), 416 (2017)
3. Sattler, S., Wiegel, C., van Veen, F.: The use frequency of 10 different methods for preventing and detecting academic dishonesty and the factors influencing their use. Stud. High. Educ. **42**(6), 1126–1144 (2017)
4. Gasparyan, A.Y., Nurmashev, B., Seksenbayev, B., Trukhachev, V.I., Kostyukova, E.I., Kitas, G.D.: Plagiarism in the context of education and evolving detection strategies. J. Korean Med. Sci. **32**(8), 1220–1227 (2017)
5. Cronan, T.P., Mullins, J.K., Douglas, D.E.: Further understanding factors that explain freshman business students' academic integrity intention and behavior: plagiarism and sharing homework. J. Bus. Ethics **147**(1), 197–220 (2018)
6. Krishan, K., Kanchan, T., Baryah, N., Mukhra, R.: Plagiarism in student research: responsibility of the supervisors and suggestions to ensure plagiarism free research. Sci. Eng. Ethics **23**(4), 1243–1246 (2017)
7. Ehrich, J., Howard, S.J., Mu, C., Bokosmaty, S.: A comparison of Chinese and Australian university students' attitudes towards plagiarism. Stud. High. Educ. **41**(2), 231–246 (2016)
8. Adenle, A.A., Sowe, S.K., Parayil, G., Aginam, O.: Analysis of open source biotechnology in developing countries: an emerging framework for sustainable agriculture. Technol. Soc. **34**(3), 256–269 (2012)
9. Feshbach, E.E.: GroBot: an Open-Source Model for Controlled Environment Agriculture. Massachusetts Institute of Technology, Cambridge (2015)
10. Bonvoisin, J., Boujut, J.-F.: Open design platforms for open source product development: current state and requirements. In: 20th International Conference on Engineering Design (ICED 2015), Milan (2015)
11. Rebollo-Lopez, M.J., Lelievre, J., Alvarez-Gomez, D., et al.: Release of 50 new, drug-like compounds and their computational target predictions for open source anti-tubercular drug discovery. PloS one **10**(12), e0142293 (2015)
12. Giannozzi, P., Baroni, S., Bonini, N., Calandra, M., et al.: QUANTUM ESPRESSO: a modular and open-source software project for quantum simulations of materials. J. Phys. Condens. Matter **21**(39), 19 (2009)
13. Rognes, T., Flouri, T., Nichols, B., Quince, C., Mahe, F.: VSEARCH: a versatile open source tool for metagenomics. Peer J. **4**, e2584 (2016)
14. Ungi, T., Lasso, A., Fichtinger, G.: Open-source platforms for navigated image-guided interventions. Med. Image Anal. **33**, 181–186 (2016)
15. Chin, J.C., Gray, J.S.: Open-source conceptual sizing models for the hyperloop passenger pod. In: 56th AIAA/ASCE/AHS/ASC Structures, Structural Dynamics, and Materials Conference, Kissimmee, Florida (2015)
16. Mede, K., Brandt, T.D.: The Exoplanet Simple Orbit Fitting Toolbox (ExoSOFT): an open-source tool for efficient fitting of astrometric and radial velocity data. Astron. J. **153**(3), 135 (2017)
17. Hoste, J.-J.O., Casseau, V., Fossati, M., Taylor, I.J., Gollan, R.: Numerical modeling and simulation of supersonic flows in propulsion systems by open-source solvers. In: 21st AIAA International Space Planes and Hypersonics Technologies Conference, Xiamen (2017)
18. Black Duck Software: The Tenth Annual Future of Open Source Survey. Black Duck Software, Burlington (2016)

19. Atayero, A.A., Alatishe, A.A., Sanusi, K.O.: Development of iCU: a plagiarism detection software. Int. J. Electr. Comput. Sci. **11**(4), 30–35 (2011)
20. Batane, T.: Turning to Turnitin to fight plagiarism among university students. J. Educ. Technol. Soc. **13**(2), 1 (2010)
21. Choi, S.P., Lam, S.S.: iChecker: an efficient plagiarism detection tool for learning management systems. Int. J. Syst. Serv. Oriented Eng. **6**(3), 16–31 (2016)
22. Velásquez, J.D., Covacevich, Y., Molina, F., Marrese-Taylor, E., et al.: DOCODE 3.0 (DOcument COpy DEtector): a system for plagiarism detection by applying an information fusion process from multiple documental data sources. Inf. Fusion **27**, 64–75 (2016)
23. Liu, C., Chen, C., Han, J., Yu, P.S.: GPLAG: detection of software plagiarism by program dependence graph analysis. In: 12th ACM SIGKDD International Conference on Knowledge Discovery and Data Mining (2006)
24. Yan, L., McKeown, N., Sahami, M., Piech, C.: TMOSS: using intermediate assignment work to understand excessive collaboration in large classes. In: 49th ACM Technical Symposium on Computer Science Education (2018)
25. Ahtiainen, A., Surakka, S., Rahikainen, M.: Plaggie: GNU-licensed source code plagiarism detection engine for Java exercises. In: 6th Baltic Sea Conference on Computing Education Research (2006)
26. Barrón-Cedeño, A., Rosso, P.: On automatic plagiarism detection based on n-grams comparison. In: European Conference on Information Retrieval, Berlin, 2009
27. Bloomfield, L.: WCopyfind. http://plagiarism.bloomfieldmedia.com/wordpress/software/wcopyfind/
28. Jones, M., Sheridan, L.: Back translation: an emerging sophisticated cyber strategy to subvert advances in 'digital age' plagiarism detection and prevention. Assess. Eval. High. Educ. **40**(5), 712–724 (2015)

Smart Contract for Monitoring and Control of Logistics Activities: Pharmaceutical Utilities Case Study

Roberto Casado-Vara[1(✉)], Alfonso González-Briones[1], Javier Prieto[1], and Juan M. Corchado[1,2,3]

[1] BISITE Digital Innovation Hub, University of Salamanca. Edificio Multiusos I+D+i, 37007 Salamanca, Spain
{rober,alfonsogb,javierp,corchado}@usal.es
[2] Osaka Institute of Technology, Osaka, Japan
[3] Universiti Malaysia Kelantan, Kota Bharu, Kelantan, Malaysia
https://bisite.usal.es/es

Abstract. Logistics services involve a wide range of transport operations between distributors and clients. Currently, the large number of intermediaries are a challenge for this sector, as it makes all the processes more complicated. In this paper we propose a system that uses smart contracts to remove intermediaries and speed up logistics activities. In addition, a multi-agent system is used to coordinate entire logistics services, smart contracts and compliance with their terms. Our new model combines smart contracts and a multi-agent system to improve the current logistics system by increasing organization, security and significantly improving distribution times.

Keywords: Blockchain · Smart contract · Multi-agent system Logistical utilities

1 Introduction

Nowadays logistics is a core area for companies which is concerned with transporting products between parties. However, the problem of this sector is that its scale may lead to delays and defaults in the delivery of goods as well as other issues. In addition, large distributors need a large volume of workers to meet all the demands of stores. All this may contribute to big delays in order processing and increases the possibility of losing orders [19]. In an attempt to solve this problem, companies have automated all their processes, contributing to a significant increase in the number of businesses and distributors in the logistics sector. However, an increase in the amount of digitized data and the expansion of Internet companies means that the risk of attacks on their databases is also greater. Hackers may intend to modify, steal or delete data [8,22].

We suggest an alternative way of solving this problem. In our case study (i.e., pharmaceutical utilities), we are going to consider two different scenarios [20].

M. Graña et al. (Eds.): SOCO'18-CISIS'18-ICEUTE'18, AISC 771, pp. 509–517, 2019.
https://doi.org/10.1007/978-3-319-94120-2_49

Firstly, we provide security to the data of the companies involved in the logistics sector with the inclusion of blockchain. Secondly, multi-agent systems will be used for the organization's problem [21]. It has been proven that multi-agent systems provide efficient solutions to a huge variety of problems [37]. These include, but are not limited to, the use of agents for image classification [13,15], decentralized network control [26], real-time problems [6] and Internet of Things applications [14].

In this paper we propose a new model that uses blockchain, smart contract and a multi-agent system to protect the data of the logistics sector [34] and improve logistic activities (i.e., our model allows to remove intermediaries and speed up logistics activities). In addition, multi-agents are capable of coordinating all the logistic services (i.e., our model improves organization, security and distribution times). This model is designed to improve the efficiency of the logistics sector. For this purpose, data is protected and smart contracts are used to remove intermediaries [9,28]. On the other hand, the multi-agent system validates all transactions and the miners' system is responsible for creating the blockchain with the transactions made by the agent system through smart contracts [5]. Although there is a lot of discussion on the use of blockchain in logistical services, there have not been any systems that would implement and evaluate it. Our approach is a functional prototype which has been evaluated empirically. Furthermore, it has been proven that it resists third-party attacks, in the case study context, more efficient than a traditional logistic model. In the methodology section, the model is proposed and in the discussion section, its advantages and disadvantages are evaluated in line with the conventional logistic model.

2 Related Work

A blockchain is a distributed data structure that is replicated and shared among the members of a network [25]. It was introduced with Bitcoin [27] to solve the double-spending problem [11]. As a result of how the nodes in Bitcoin (the so-called miners) mutually validate the agreed transactions, Bitcoin blockchain establishes the owners and states what they own [12]. A blockchain is built using cryptography. Each block is identified by its own cryptographic hash and each block refers to the hash of the previous block. This establishes a link between the blocks, forming a blockchain [2,7]. For this reason, users can interact with a blockchain by using a pair of public and private keys. Miners in a blockchain need to agree on the transactions and the order in which they have occurred. Otherwise the individual copies of this blockchain can diverge producing a fork; miners then have a different view of how the transactions have occurred, and it will not be possible to keep a single blockchain until the fork is not solved [17,30].

To achieve this a distributed consensus mechanism is required in every blockchain network [35]. Blockchain's way to solve the fork problem is that each blockchain node can link next block. Just a correct random number with SHA-256 has to be found [1,8,18] so you have number of zeros that the blockchain expects. Any node that can solve this puzzle has generated the so-called proof-of-work (pow) and gets to shape the chain's next block [27]. Since a one-way cryptographic hash function is involved, any other node can easily verify that the given answer satisfies the requirement. Notice that a fork may still occur in the network, when two competing nodes mine blocks almost simultaneously. Such forks are usually resolved automatically by the next block [4,29].

With the implementation of blockchain, smart contracts are included to make transactions between different users faster and more effective [16]. Nick Szabo introduced this concept in 1994 and defined a smart contract as "a computerized transaction protocol that executes the terms of a contract" [32]. Szabo suggested that the clauses of contracts could be transferred to code, thus reducing the need for intermediaries in transactions between parties. In the blockchain context a smart contract is a script that is stored on a blockchain [33]. Smart contracts have a unique address in a blockchain (i. e., they are in a block with a hash that identifies it). We can trigger a smart contract in a transaction by indicating the address on the blockchain. It is executed independently and automatically in a prescribed manner on every node in the network, according to the data contained in the triggered transaction.

A multi-agent system is a computerized system composed of multiple intelligent agents that interact with each other. Multi-agent systems are used to solve complex problems with very good results. Multi-agent systems are used in a wide range of applications. [31] presents a multi-agent system for the intelligent use of electricity in a Smart home and thus, an increase in its energy efficiency. Another problem that multi-agent systems have solved effectively is the monitoring of sound in a variety of situations [23]. The application of a multi-agent system to logistics is not a new problem, in [24] a multi-agent system is proposed to provide a solution to the logistical problem. In addition, another successful application of multi-agent systems is the problem of distributed computing [3]. Therefore, some of the proposals that we find in the literature combine the advantages of blockchain and multi-agent systems. From a range of systems that integrate blockchain and multi-agent systems the work of [38] is worthy of mention. This work proposes to use both technologies to increase security and privacy in decentralised energy networks. In [36] authors propose a model that employs agents and blockchain for a ride-sharing system. In addition, there are other applications of blockchain and multi-agent systems. In [10] the authors propose an innovative blockchain model for IoT. However, after looking at the state of the art, we believe that the current blockchain and multi-agent system models have some shortcomings. We propose a new model that leverages smart contracts and multi-agent systems, it is aimed at increasing efficiency in the logistics system management. This paper describes a case study which verified the proposed model, it focused specifically on logistics transport in the pharmaceutical sector [8].

3 Methodology

This paper presents a new model which consists of the following elements: a blockchain in which all transactions are stored. Smart contracts that will manage commercial transactions between the different parties. A multi-agent system will enable the executing of all these operations. In this section we describe the how our model works. The members of each business operation have smart devices which monitor the status of each one. The case study was conducted in the pharmaceutical sector, Fig. 1 shows process members. The seller (i. e., pharmacies), the producer (i. e., pharmacists) and the shipping companies. In this case, the sellers have sensors that monitor the number of drugs stored in the pharmacy, the type of drugs sold and the amount of money stored. Regarding the pharmaceuticals, they have sensors in charge of knowing the available stock and amount being produced. Finally, transportation companies have sensors on each of their transport vehicles to monitor the position of the cargo. All these elements make up the Wireless Sensor Network (WSN) that monitors operations in the pharmaceutical sector. Within the WSN there are 2 types of smart devices. Smart devices monitor every single member. Sensors monitor the processes of each member. The smart devices that monitor every single element are part of the miners' network in the blockchain. While smart devices that monitor each of these members' processes are responsible for creating transactions that miners include in a blockchain. Users and miners are in charge of creating the blockchain where all transactions are stored. In addition, the blockchain also includes smart contracts that control all transactions created between members of pharmaceutical sector.

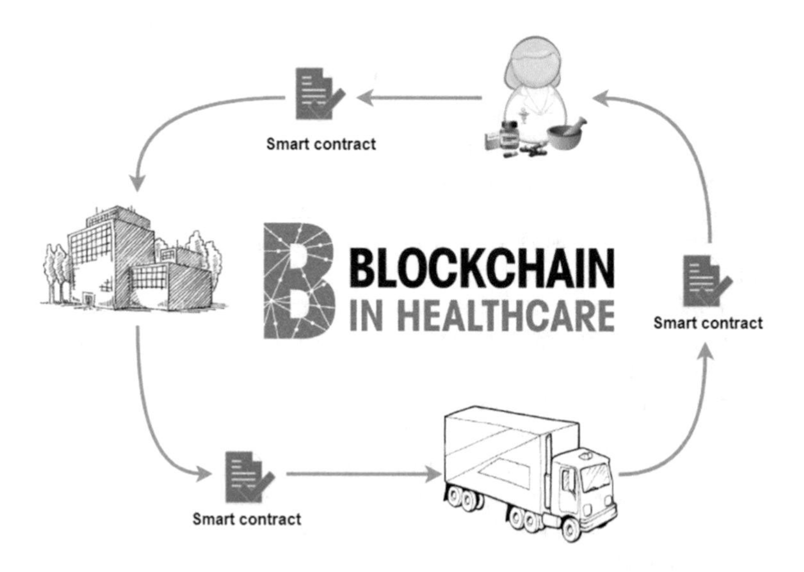

Fig. 1. Graph based on smart contract for pharmaceutical logistics sector.

A multi-agent system controls the whole process. The architecture of the multi-agent system consists of the following layers (see Fig. 2). (1) Client layer: this layer consists of three different types of agents that manage pharmacies.

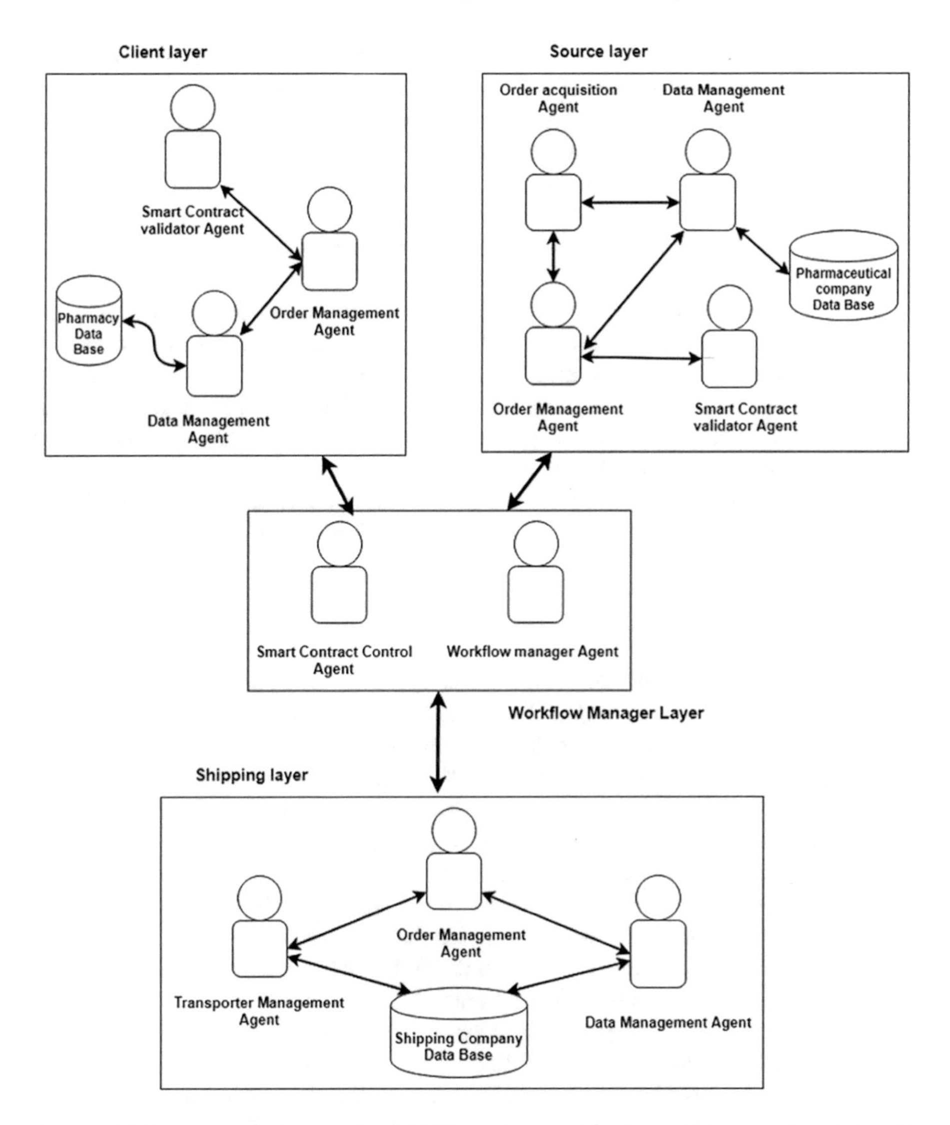

Fig. 2. Multi-agent architecture: (1) Client layer: the pharmacies are found in this layer. (2) Source layer: this layer contains pharmaceutical companies. (3) Shipping layer: this layer manages transport companies. (4) Workflow management layer: this layer contains an agent that controls the entire information flow and an agent which ensures that smart contract conditions are fulfilled.

These include the pharmacy stock control agent, the agent that is responsible for placing orders and that which verifies of smart contracts. (2) The source layer contains the following agents: Two agents receive orders from pharmacists and another places orders with the transport company in order to take the goods to the pharmacies. Another agent's task is the control of stock and production levels. Finally, there is an agent responsible for verifying whether smart contract obligations are fulfilled. (3) Shipping layer has the following agents. An agent that manages the incoming orders, another agent that manages the fleet of vehicles, and finally, an agent that verifies smart contracts. (4) The workflow management layer has a workflow management agent and a smart contract control agent this agent is in charge of creating smart contracts, keeping the money while making transactions and applying penalties in case of non-compliance with the smart contracts.

Thus, one of the agents included in each of the 4 layers verifies that the smart contract terms are abided to. For instance, when a smart contract is initiated between a pharmacy and a pharmaceutical company for the purchase of medicine, both sign a smart contract. The pharmacy pays for the drugs, but money is kept in the blockchain by a control entity, in this case the agent that verifies the smart contract. When the pharmacy receives the drugs it ordered, this agent confirms that the conditions of the smart contract have been fulfilled and automatically pays the pharmacist the agreed sum of money.

4 Discussion

This paper presents a new Smart contract approach to improving logistics services. The novelty of this paper lies in a blockchain to store all transaction information in the logistical process of the proposed case study. In addition, the multi-agent system uses smart contracts to manage the entire logistics process more efficiently, this is because smart contracts remove intermediaries.

Our model can be used to improve any logistics system. The case study conducted in this proposal focuses on the pharmaceutical sector. Our model has improved security and efficiency as it is automated by the agent system. By incorporating blockchain we provide the logistics system with solid security features . Shipments can be tracked, origin and destinations authenticated, and proof of all transactions can be stored and unmanipulated.

Another novelty of this paper are agents who verify that both parties abide to the terms of a smart contract. If the agents detect that either of the parties is not fulfilling the established conditions, a penalty is imposed and the agents keep money in the control entity until the conditions agreed upon are met. This makes our model more efficient than current models. Moreover, it is able to track and authenticate orders. A penalties pattern is introduced for breach of smart contracts.

Future lines of research include improving the multi-agent system by introducing new agents for the monitoring of procedures. In addition, our model could be enhanced by integrating a case-based reasoning system (CBR).

Acknowledgments. This paper has been funded by the European Regional Development Fund (FEDER) within the framework of the Interreg program V-A Spain-Portugal 2014-2020 (PocTep) grant agreement No 0123_IOTEC_3_E (project IOTEC).

References

1. Announcing the Secure Hash Standard, 1 August 2002. http://csrc.nist.gov/publications/fips/fips180-2/fips180-2.pdf
2. Antonopoulos, A.M.: Mastering Bitcoin: Unlocking Digital Cryptocurrencies, 1st edn. O'Reilly Media Inc., Sebastopol (2014)
3. Banerjee S., Hecker J.P.: A Multi-agent system approach to load-balancing and resource allocation for distributed computing. In: Bourgine P., Collet P., Parrend P. (eds) First Complex Systems Digital Campus World E-Conference 2015. Springer Proceedings in Complexity. Springer, Cham (2017)
4. Becerra-Bonache, L., López, M.D.J.: Linguistic models at the crossroads of agents, learning and formal languages. ADCAIJ: Adv. Distrib. Comput. Artif. Intell. J. **3**(4), 67–87 (2014)
5. Durik, B.O.: Organisational metamodel for large-scale multi-agent systems: first steps towards modelling organisation dynamics. ADCAIJ Adv. Distrib. Comput. Artif. Intell. J. **6**(3), 17–27 (2017)
6. Carrascosa, C., Bajo, J., Julián, V., Corchado, J.M., Botti, V.: Hybrid multi-agent architecture as a real-time problem-solving model. Expert Syst. Appl. **34**(1), 2–17 (2008)
7. Cardoso, R.C., Bordini, R.H.: A multi-agent extension of a hierarchical task network planning formalism. ADCAIJ: Adv. Distrib. Comput. Artif. Intell. J. **6**(2), 5–17 (2017)
8. Chamoso, P., Rivas, A., Martín-Limorti, J.J., Rodríguez, S.: A hash based image matching algorithm for social networks. In: Advances in Intelligent Systems and Computing, vol. 619, pp. 183–190. Springer, Cham (2017)
9. Costa, Â., Novais, P., Corchado, J.M., Neves, J.: Increased performance and better patient attendance in an hospital with the use of smart agendas. Logic J. IGPL **20**(4), 689–698 (2012)
10. Daza, V., Di Pietro, R., Klimek, I., Signorini, M.: CONNECT: CONtextual NamE disCovery for blockchain-based services in the IoT. In: 2017 IEEE International Conference on Communications (ICC), pp. 1–6 (2017). ISSN 1938–1883
11. Double-Spending—Bitcoin WiKi. https://en.bitcoin.it/wiki/Double-spending. Accessed 15 Mar 2016
12. Eris Industries Documentation—Blockchains. https://docs.erisindustries.com/explainers/blockchains/. Accessed 15 Mar 2016
13. Coria, J.A.G., Castellanos-Garzón, J.A., Corchado, J.M.: Intelligent business processes composition based on multi-agent systems. Expert Syst. Appl. **41**(4 PART 1), 1189–1205 (2014)
14. Gazafroudi, A.S., Pinto, T., Prieto-Castrillo, F., Prieto, J., Corchado, J.M., Jozi, A., Venayagamoorthy, G.K.: Organization-based multi-agent structure of the smart home electricity system. In: 2017 IEEE Congress on Evolutionary Computation (CEC), pp. 1327–1334. IEEE (2018)
15. González-Briones A., Villarrubia G., De Paz J.F., Corchado J.M.: A multi-agent system for the classification of gender and age from images. Comput. Vis. Image Underst. (2018). https://doi.org/10.1016/j.cviu.2018.01.012, ISSN 1077-3142
16. González, A., Ramos, J., De Paz, J.F., Corchado, J.M.: Obtaining relevant genes by analysis of expression arrays with a multi-agent system. ADCAIJ: Adv. Distrib. Comput. Artif. Intell. J. **3**(3), 35–42 (2014)
17. Greenspan, G.: Avoiding the Pointless Blockchain Project (2015). http://www.multichain.com/blog/2015/11/avoidingpointless-blockchain-project/

18. Hashcash-Bitcoin WiKi. https://en.bitcoin.it/wiki/Hashcash. Accessed 15 Mar 2016
19. Li, T., Sun, S., Bolić, M., Corchado, J.M.: Algorithm design for parallel implementation of the SMC-PHD filter. Signal Process. **119**, 115–127 (2016)
20. Li, T., Sun, S., Corchado, J.M., Siyau, M.F.: Random finite set-based Bayesian filters using magnitude-adaptive target birth intensity. In: FUSION 2014–17th International Conference on Information Fusion (2014). https://www.scopus.com/inward/record.uri?eid=2-s2.0-84910637788&partnerID=40&md5=bd8602d6146b014266cf07dc35a681e0
21. Li, T., Sun, S., Corchado, J.M., Siyau, M.F.: A particle dyeing approach for track continuity for the SMC-PHD filter. In: FUSION 2014–17th International Conference on Information Fusion (2014). https://www.scopus.com/inward/record.uri?eid=2-s2.0-84910637583&partnerID=40&md5=709eb4815eaf544ce01a2c21aa749d8f
22. Lima, A.C.E.S., De Castro, L.N., Corchado, J.M.: A polarity analysis framework for Twitter messages. Appl. Math. Comput. **270**, 756–767 (2015)
23. López Barriuso, A., Villarrubia González, G., Bajo Pérez, J., de Paz Santana, J.F., Corchado Rodríguez, J.M.: Embedded agents to monitor sounds. Simulation of consumers and markets towards real time demand response. In: Proceedings of the First DREAM, GO Workshop, pp. 44–51, 7 April 2016
24. Li, K., Zhou, T., Liu, B.: A multi-agent system for sharing distributed manufacturing resources. Expert Syst. Appl. **99**, 32–43 (2018)
25. Bremer, J., Lehnhoff, S., Li, H.: Decentralized coalition formation with agent-based combinatorial heuristics. ADCAIJ: Adv. Distrib. Comput. Artif. Intell. J. **6**(3), 29–44 (2017)
26. Najafi, S., et al.: Decentralized control of DR using a multi-agent method. In: Sustainable Interdependent Networks, pp. 233–249. Springer, Cham (2018)
27. Nakamoto, S.: Bitcoin: A Peer-to-Peer Electronic Cash System (2008). https://bitcoin.org/bitcoin.pdf
28. Rodríguez, S., De La Prieta, F., Tapia, D.I., Corchado, J.M.: Agents and computer vision for processing stereoscopic images. LNCS (LNAI and LNBI), vol. 6077 (2010)
29. Marin, P.A.R., Duque, N., Ovalle, D.: Multi-agent system for Knowledge-based recommendation of learning objects. ADCAIJ Adv. Distrib. Comput. Artif. Intell. J. **4**(1), 80–89 (2015)
30. Santos, G., Pinto, T., Vale, Z., Praça, I., Morais, H.: Enabling communications in heterogeneous multi-agent systems: electricity markets ontology. ADCAIJ Adv. Distrib. Comput. Artif. Intell. **5**(2), 15–42 (2016)
31. Gazafroudi, A.S., Pinto, T., Prieto Castrillo, F., Corchado Rodríguez, J.M., Jozi, A., Vale, Z., Venayagamoorthy, G.K.: Organization-based multi-agent structure of the smart home electricity system. In: 2017 IEEE Congress on Evolutionary Computation (CEC), pp. 1327–1334. IEEE, 07 July 2017
32. Szabo, N.: Smart Contracts (1994). http://szabo.best.vwh.net/smart.contracts.html
33. Szabo, N.: The Idea of Smart Contracts (1997). http://szabo.best.vwh.net/smart_contracts_idea.html
34. Tapia, D.I., Fraile, J.A., Rodríguez, S., Alonso, R.S., Corchado, J.M.: Integrating hardware agents into an enhanced multi-agent architecture for ambient Intelligence systems. Inf. Sci. **222**, 47–65 (2013)
35. Casado-Vara, R., Corchado, J.M.: Blockchain for democratic voting: how blockchain could cast off voter fraud. Orient. J. Comp. Sci. and Technol. **11**(1) http://www.computerscijournal.org/?p=8042

36. Yuan, Y., Wang F.Y.: Towards blockchain-based intelligent transportation systems. In: 2016 IEEE 19th International Conference on Intelligent Transportation Systems (ITSC), Rio de Janeiro, pp. 2663–2668 (2016)
37. Wooldridge, M., Jennings, N.R.: Intelligent agents: theory and practice. Knowl. Eng. Rev. **10**(2), 115–152 (1995)
38. Aitzhan, N.Z., Svetinovic, D.: Security and privacy in decentralized energy trading through multi-signatures, blockchain and anonymous messaging streams. IEEE Trans. Dependable Secure Comput. **PP**(99), 1 (2016)

ICEUTE 2018

Evaluation as a Continuous Improvement Process in the Learning of Programming Languages

Marcos Gestal, Carlos Fernandez-Lozano(✉), Cristian R. Munteanu, Juan R. Rabuñal, and Julian Dorado

Department of Computation, Faculty of Computer Science, Campus Elviña s/n, 15071 A Coruña, Spain
carlos.fernandez@udc.es

Abstract. Learning a programming language requires a great deal of effort in both the theoretical and practical domains. As far as theory is concerned, a knowledge of the methods, concepts, attributes that are characteristic of the language as well an understanding of the its specific structures and peculiarities is required. On the other hand, mastering the theoretical concepts is not enough as it is necessary to be able to apply them optimally, efficiently and effectively. To adapt the teaching to those aspects that require the most attention, the weaknesses shown by the students must be identified. An exhaustive analysis of their performance – which should go beyond a mere numerical assessment – is required to focus the teaching efforts on those areas where needs are greater. Consequently, to assess the theoretical knowledge a statistical analysis from the results of the theoretical test conducted will be shown (multiple-choice type test) where the analysis is not confined to the number of wrong answers but looks at where they occur and in what percentage. As far as the practical part, a rubric has been designed to exhaustively correct the assignments, which also allows for the introduction of such remarks as are deemed necessary regarding all points of interest.

1 Contextualization

For some years now, the J2EE framework has clearly dominated the development of business applications. In parallel, countless tutorials, handbooks, tools, etc. have emerged which explain why it has been – and most likely will continue to be – the most widely taught programming language at universities. Over the last two years, the emergence of .NET has triggered a strong demand for programmers conversant with this language. However, neither the quantity nor the quality of the documents available was anywhere close to its main alternative (J2EE).

This disadvantage is particularly salient in a subject with such peculiarities as the one addressed in this paper. The course "Development Frameworks" is a compulsory course in the curriculum for the fourth year of the Degree in Computer

M. Graña et al. (Eds.): SOCO'18-CISIS'18-ICEUTE'18, AISC 771, pp. 521–529, 2019.
https://doi.org/10.1007/978-3-319-94120-2_50

Engineering. It is a one-term course comprising 6 ECTS credits. Most crucial for the matter under discussion, this is the first (and last) course where students deal with .NET language after having studied J2EE for the three previous years.

This fact has meant that ever since its introduction, the teaching team has paid particular attention to the selection of contents taught to the students as well as to ensuring that contents are correctly assimilated as students will not be presented with another opportunity to consolidate the contents.

2 Programming in the Curriculum

For a correct learning to occur the different concepts, structures, tools etc. that are required for programming must be introduced gradually and progressively.

Consequently, the current curriculum in force in the School of Computer Sciences (www.fic.udc.es) of the University of A Coruña includes a number of courses that address this issue.

Indeed, in the first year the fundamental notions are introduced in the courses Programming I and Programming II; in the second year, more advance concepts are introduced in the courses Algorithms; Software Design and Programming Paradigms. During the third and four years, a specialty is available that is more specifically focused on developing programming skills where students acquire knowledge about Software Architecture, Advanced Programming, Development Tools or Programming Language Design.

The common denominator of these courses is the prevalence of the Java language in the practical work, which means that the skills of the students with this language progress in parallel with the introduction of new theoretical concepts.

Until the time students study the course discussed in this paper, they have not had any contact whatsoever with the .NET ecosystem. This means that they know that they need to do to develop and application, but they do not know how to do it using this technology. It is for this reason that identifying their background weaknesses is a priority as the curriculum provides no other opportunity to work with this technology.

3 Aims of the Course

The course "Development Frameworks" focuses on the presentation of design and structural patterns for the design and implementation of applications in a Web environment with .NET technologies [Gr02, ZE09].

The course is taught from a very practical approach as the practical domain is where students must show their skills in their professional careers. Consequently, throughout the course practical work consists in implementing a web application of a business nature using .NET, albeit with limited functionality because of time constraints. The correct performance of this practical work will allow students to achieve the following objectives that are considered priority aims of the course:

- Knowing the fundamentals of programming using .NET technologies
- Knowing the architectural principles of business applications
- Knowing the design techniques to develop business applications (particularly as far as Web applications is concerned) using layered architecture.

On the other hand, to attain these objectives the course includes 2 credits devoted to lectures focused on two fundamental issues that students must master to be able to embark on the practical work:

- Design and implementation of the model layer
- Design and implementation of the Web layer.

4 Evaluation

4.1 What It Evaluated?

As we have noted above, the course is eminently practical. Therefore, in keeping with this approach, its assessment is done on the basis of practical work from the outset. To this end, upon the submission of the practical assignment, its correction is done in the defense of the same by the group that submitted the project. This defense includes the verification of the implementation of practical assignment by verifying some critical points. Another element that needs to be verified is that the group has correctly understood the concepts applied for the development of the practical project. The Web application will be assessed on the basis of the seriousness of the mistakes identified, if any, in the Web application on the basis of a marking scale from 0 to 10.

Additionally, a multiple-choice test is administered to ensure that students have correctly assimilated the concepts. This multiple-choice test consists of 20 questions with several choices and only one correct answer. Unanswered questions do not score, and wrong answers score negatively in order to prevent random answers. Questions deal with issues addressed by the student during the development of the practical project.

To pass the course, the following is required: (1) to pass the Web application where the passing mark is 5 out of 10 and (2) a mark of at least 4.5 points (out of 10) in the multiple-choice test. In principle, the final grade of a student that meets these two requirements will be the weighted sum of the practical project mark and the multiple-choice mark. Given the practical nature of the course, the practical project mark is 60% of the final grade while the test accounts for only 40% of the final grade. Although the teaching team would like to see a greater weight given to the practical project, University regulations do not allow it.

4.2 How Is the Evaluation Conducted?

The greater weight of the course in on the practical project. It is for this reason that this is where the focus is when assessing the work conducted by the students.

The development of the practical work for the course in divided on the basis of two iterations or deadlines. In the first of these iterations, to which no mark is associated, the initial part is implemented. The aim of this first submission is to ensure that the student is approaching the project correctly and that the concepts are being applied adequately. This prevents delays at a later stage resulting from core conceptual errors. The lecturer therefore tries to identity these errors and guide the student towards a solution.

At the second iteration the student corrects the errors identified at the first iteration and adds the rest of functionalities. It is during the correction of this second iteration when the student must explain the functionalities implemented; how work has been divided, etc. Based on their work and explanations a mark is given on an individual basis which represents most of the grade for the whole course. Given the width of the aspects to be analyzed, correcting each practical project takes between 1 h and 1:30 h.

To ensure a complete and even-handed correction for all groups, right from the introduction of the course a correction scheme was developed which is based on the correct attainment of a number of points in a checklist. This ensures that a series of points are checked for all groups in the practical project although, of course, slight variations may occur depending on the specificities of the practical project proposed each year. And the deviations in the corrections arising from the fact that they are made on different days, with variations in mood, tiredness or levels of attention are minimized...

5 Using Evaluation as a Source of Continuous Improvement

From the outset, one of the main concerns for the teaching team has been to ensure that the contents taught are correctly acquired, especially as this is the only course in which students work with the .NET platform. More specifically, an attempt has been made to see evaluation as not just a mere numerical mark of the skills shown by the student but as a process aimed at improving the quality and level of learning [Ha04].

Indeed, a variety of strategies have been analyzed and developed throughout time with two clearly defined objectives:

- To determine clearly and in the most balanced manner possible the level of knowledge acquired by the students so that the mark can be assigned in the fairest possible manner.
- To obtain information about which concepts have been least understood to improve their explanation, the making of supporting materials, planning of seminars ... for the next academic years.

These strategies have been implemented for both domains, namely the theoretical and the applied one.

5.1 Evaluating the Theoretical Domain

As pointed out above, the assessment of the theoretical section of the course consists of a multiple-choice test.

This ensures full transparency and even-handedness in the marks as there is no room for interpretation of the results.

It also facilitates the making of a statistical study of the answers provided by the students to measure which concepts have been correctly assimilated and which have not been. To this end, a spreadsheet has been made that facilitates the correction of the tests (see Fig. 1) as well as the extraction of additional information.

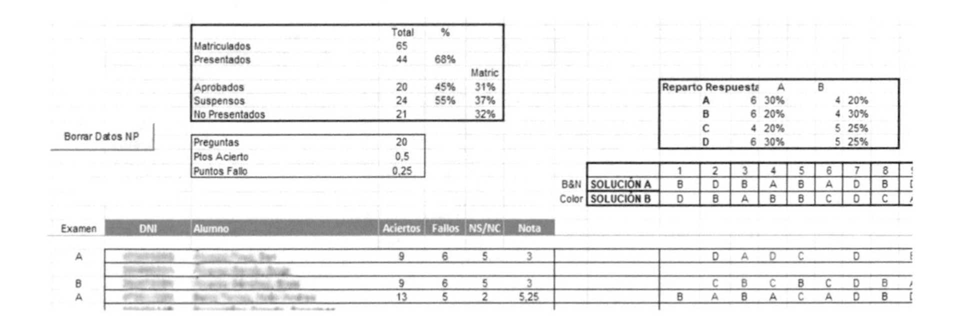

	Total	%	
Matriculados	65		
Presentados	44	68%	
			Matric
Aprobados	20	45%	31%
Suspensos	24	55%	37%
No Presentados	21		32%

Borrar Datos NP

Preguntas	20
Ptos Acierto	0,5
Puntos Fallo	0,25

Reparto Respuesta	A		B	
A	6	30%	4	20%
B	6	20%	4	30%
C	4	20%	5	25%
D	6	30%	5	25%

		1	2	3	4	5	6	7	8
B&N	SOLUCIÓN A	B	D	B	A	B	A	D	B
Color	SOLUCIÓN B	D	B	A	B	B	C	D	C

Examen	DNI	Alumno	Aciertos	Fallos	NS/NC	Nota		1	2	3	4	5	6	7	8
A			9	6	5	3			D	A	D	C		D	
B			9	6	5	3			C	B	C	B	C	D	B
A			13	5	2	5,25		B	A	B	A	C	A	D	B

Fig. 1. Extract from the evaluation spreadsheet

This additional information is based on the answers provided by students and translates into 2 different graphs.

On the one hand, a rate of right answers per questions, which makes it possible to determine the percentage of right answers, wrong answers and the percentage of students who have provided no answer for each question (see Fig. 2). As each question is associated with a specific block of course content (access to BBDD, configuration management, dependencies, etc.) it is possible to count and analyze which concepts have been more poorly grasped so as to underscore them in the future.

Global	1	2	3	4	5
Aciertos	30%	70%	45%	66%	41%
Fallos	36%	18%	48%	11%	36%
NS/NC	34%	11%	7%	23%	23%
Mayor num. Aciertos, fallos o Ns/NC	▼	▲	▼	▲	▲
Aciertos >= (Fallos + Ns/NC)	▼	▲	▼	▲	▼

Fig. 2. Different rates of right, wrong and not answered question in the evaluation.

On the other hand, the percentages with which each of the options is ticked as an answer for each question is analyzed.

An analysis is made to ascertain whether the rate of right answers is lower than the rate of wrong answers and a color scale is assigned to every question to visually identify where wrong, right or blank answers are more frequent or where the level of right answers is very low. This same information is displayed in a graph bar format in Fig. 3.

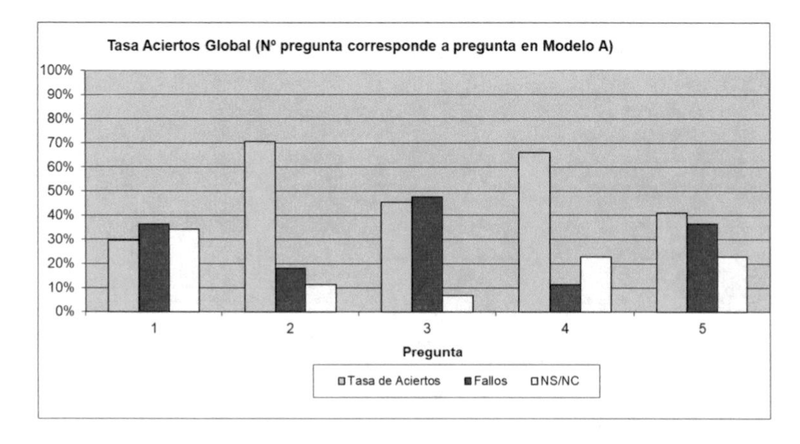

Fig. 3. Different rates (barplot) of right, wrong and not answered question in the evaluation.

Lastly, to analyze in further detail how or why mistakes occur, each of the options chosen by the students in the answers to the test is displayed as shown in Fig. 4.

These graphs, despite their simplicity, make it possible to immediately focus the attention on those questions that are more interesting (e.g. the question with the most wrong answers) and not merely determine for which concepts errors are more frequent but also if the error is massively oriented in one specific direction or whether doubts exist between 2 or more answer options.

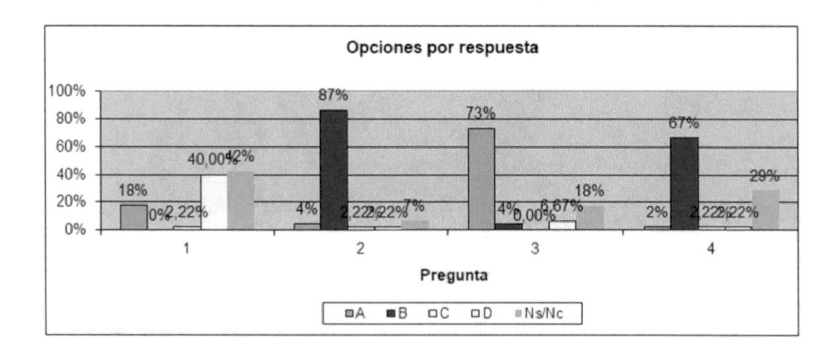

Fig. 4. Options per answers.

5.2 Evaluating the Practical Domain

As far as the practical domain is concerned, its evaluation is supported by the use of checklists [St01] and rubrics [La98, Cu96, BR11].

The checklist (see Fig. 5) ensures that the same points are addressed in all practical projects and that this is done in the proper sequence to facilitate correction.

Marcos de Desarrollo (2ª iteración)

Cód. grupo: mad____		Fecha: ___/01/2018		Se aplican simplificaciones para grupos de 1 persona [ ] Funcionalidad opcional implementada [ ] 1. OAuth [ ] 2. Cacheado búsquedas [ ] 3. Etiquetado de comentarios
	Nombre y Apellidos	Ficha	Nota Examen	
1				
2				

CHECKPOINT	OK	COMENTARIOS
Calidad Memoria:		
1. Impresión general Muy Mala, [ ] Mala, [ ] Normal, [ ] Buena, [ ] Muy Buena [ ]		
2. Diagramas representativos		
3. Diagramas legibles		
Repositorio		
4. Práctica entregada en plazo? (10:00 de Viernes 22/01/2015)		
5. Estructura en repositorio correcta		
6. Repositorio incluye memoria		
Corrección Iteración 1 (anotar las correcciones y soluciones adoptadas)		
7. El modelo de entidades es ahora correcto?		
Aspectos Globales		
8. **[Desactivar cookies]** - La práctica compila y se ejecuta correctamente?		
9. ¿Justificación correcta acerca de las fachadas tiene la aplicación? (User, Group, Comment, Favorite, Recommendation, Tag (Opc. 2), Event)		
10. Desde el controlador únicamente se realiza una llamada al servicio		
11. Los accesos a las prop. Navegación (e.g. Comment.User) se hacen sólo en los DAOs		
Casos de uso		
- Evento		
12. Obtener todos los eventos activos *[búsqueda por keyword del título del evento]*		
12.1. Permite especificar de manera opcional la categoría para restringir la búsqueda		
12.2. Resultado búsqueda incluye		
a) Nombre, categoría y fecha		
b) Enlace "Ver comentarios"		
c) Enlace "Añadir comentario"		
d) Enlace "Recomendar evento"		
- Usuario		
13. Registrar usuario *[deben validarse los campos]*		
14. Modificación de la información de registro		
15. Autenticación (con posibilidad de recordar contraseña) y salida		

Fig. 5. Checklist for the correction of the practical project (extract)

This checklist is of vital importance as far as remarks are concerned as a later analysis of them will make it possible to determine where the greater number of mistakes has occurred in the making of the practical project and what mistakes have been made for each specific point. This will be the information used as feedback.

Once the practical projects have been reviewed and the checklists for each of them have therefore been filled out, the assessment is obtained from a rubric (see Fig. 6) that makes it possible to bring together and weight each and every point noted during the correction.

Plantilla de evaluación unificada de la asignatura

Convocatoria:	MaD	
Grupo:	EXAMEN	Nota
Alumno 1:		
Alumno 2:		
Alumno 3:		

Parte básica: 7 puntos				A1	
				A2	
				A3	
Diseño: 1,5 puntos					
☐ Malo (0) Fallos muy graves en el diseño. No ha corregido ningún fallo de la primera iteración. No se siguen las indicaciones dadas en clase. No se sigue el patrón MVC. Faltan casos de uso.	☐ Regular (0,40) Fallos graves en el diseño. No ha corregido todos los fallos de la primera iteración. No se siguen las indicaciones dadas en clase. No se sigue el patrón MVC. Faltan casos de uso	☐ Normal (0,80) No hay fallos en el diseño o estos son leves. Se siguen las indicaciones dadas en clase. Están la mayor parte de los casos de uso	☐ Bueno (1,2) El diseño es bueno. Se siguen las indicaciones dadas en clase. Se incluyen algunas mejoras de diseño. Están todos los casos de uso.	☐ Muy Bueno (1,5) El diseño es muy bueno. Se siguen las indicaciones dadas en clase. Se incluyen la mayoría de las mejoras de diseño posibles en la práctica. Las decisiones de diseño están fuertemente argumentadas.	
Notas:					

Fig. 6. Assessment rubric (extract)

6 Conclusions

The concept of "continuous assessment" should not only be applied to the students regarding the course they are following but also to the course itself.

Indeed, an analysis of the results achieved by the students will make it possible to determine the suitability of the contents or of the techniques or the support resources used for its teaching.

In the particular case of the course under study, this rationale has over the last few years contributed to prioritizing the making of support tutorials for the correct acquisition of the concepts taught in the course [Ge12].

The first of the tutorials – on the connectivity to Databases through ADO.NET – was implemented after observing a spike in the number of errors made for this block of the course in exams. This tutorial was updated to show the operation of the Entity Framework (the new way of connecting to Databases that was created within .NET a couple of years ago), following the suspicion that fresh doubts will occur regarding this issue as new technology was introduced. Coinciding with this change, an increase in the number of errors was detected regarding the development of the code of access to data and the test questions regarding this point but it did not reach the levels of mistakes made in the first years.

The ASP.NET tutorial was implemented as a response to the low level of knowledge regarding its use, which could be identified thanks to the practical project correction rubrics.

As far as the course itself, and on the basis of the surveys conducted among the students, the course is among the most highly valued of the degree despite the fact (or precisely because of it) that it is one of the courses with the highest workload. This assessment has remained the same in the last few years, which means that, at least, the self-imposed level of quality can be maintained.

Our plan in the near future, after compiling a historical base to be analyzed, is the use of Artificial Intelligence techniques (Machine Learning) in order to optimize students' triangulation.

References

[Cu96] Custer, R.L.: Rubrics: an authentic assessment tool for technology education. Technol. Teach. **55**(4), 27–37 (1996)

[La98] Lazear, D.: The Rubrics Way: Using Multiple Intelligences to Assess Understanding. Zephyr Press, Tucson (1998)

[St01] Stufflebeam, D.L.: Evaluation checklists: practical tools for guiding and judging evaluations. Am. J. Eval. **22**(1), 71–79 (2001)

[Gr02] Grimes, F.: Microsoft .NET for Programmers. Manning Publications (2002)

[Ha04] Hawes, G.: Evaluación de logros de aprendizaje de competencias. Instituto de investigación y desarrollo educacional. Universidad de Talca (2004)

[ZE09] Zeldman, J., Ethan, M.: Designing with Web Standards (2009)

[BR11] Vidales, K.B., Rodríguez, I.R., Jauregui, P.A.: La evaluación de competencias en la educación superior: Las rúbricas como instrumento de evaluación. Alcalá de Guadaira, Madrid (2011)

[Ge12] Gestal, M., et al.: Docencia mediante casos de estudio para el desarrollo de aplicaciones web empresariales. In: IX Foro Internacional sobre la evaluacion de la calidad de la investigacion y de la educacion superior (FECIES), pp. 1375–1380 (2012)

Development of a Workshop About Artificial Intelligence for Schools

J. Estevez(✉), X. Elizetxea, and G. Garate

University of the Basque Country (UPV/EHU), Leioa, Spain
julian.estevez@ehu.eus

Abstract. In this article, a workshop about basic artificial intelligence (AI) concepts for high school students is presented, where basic concepts of these algorithms are explained and practised using simple software tools. The aim of this activity is to proportionate a clear idea about the simple rules of AI and enhance the social trust of science since early ages.

Keywords: Artificial intelligence · Education · Software

1 Introduction

Scientific method is one of the most powerful resources that policy makers can use for making decisions for the welfare of the community [6]. The achievement of a general scientific knowledge by students, however, still remains a challenge for our society. The educational community recognizes that school curricula must move on from traditional expositive classes to more informal contexts. The aim of these new contexts is that the students themselves take an active part in the learning process, instead of the passive function that a formal expositive class leaves to the student. In order to achieve active involvement, the classes should make use of more attractive resources, such as fun, games, social interaction, observation of real problems, novelty, etc.

Surprisingly, there is a quite widespread mistrust of science in the post-truth era we live. This makes it necessary to insist on a bigger effort on divulgation of the benefits of science among younger students [1,2].

The present article presents the design and development of several simple educational exercises to promote understanding and learning of Artificial Intelligence (AI) at schools.

2 What Is Artificial Intelligence?

AI is one of the technologies that will transform our society, economy and jobs in a greater extent along the next decades. Some of the most known examples of AI are driverless cars, chatbots, voice assistants, internet search engines, robot traders, etc. These systems can be embedded in physical machines or in software,

M. Graña et al. (Eds.): SOCO'18-CISIS'18-ICEUTE'18, AISC 771, pp. 530–535, 2019.
https://doi.org/10.1007/978-3-319-94120-2_51

and the promising capacities of both architectures makes it necessary for society and politicians to regulate the functions and limits of these devices [9].

Despite the myth of destructive AI represented in films such as *Terminator* or *I Robot*, the truth is that nowadays most usual smart algorithms consist of a series of simple rules applied to large series of numbers, and the result of that is called *learning* or *artificial intelligence*.

This article will depict three algorithms that can be practised in MS Excel and Scratch [7], in a workshop with students 16–18 years old. After the workshop, the students should be able to understand, play and eventually code these algorithms. In next sections, the techniques used for this workshop will be explained and detailed. Three algorithms are presented:

1. Linear and polynomial regression
2. K-means clustering
3. Neural network.

3 Concepts to Keep in Mind and Understand

3.1 Linear and Polynomial Regression

Linear and polynomial regression are very simple approaches for AI. Though it may seem somewhat dull compared to some of the more modern algorithms, this technique is still a useful and widely used statistical learning method. Depending on whether the relationship between X and Y is linear or non-linear, we can choose among different types of regression lines. Regression is one of the most usual and simple algorithms to make predictions or to automatically classify values, which is a fundamental part of AI. This mathematical tool is very easy to use and teach (Fig. 1).

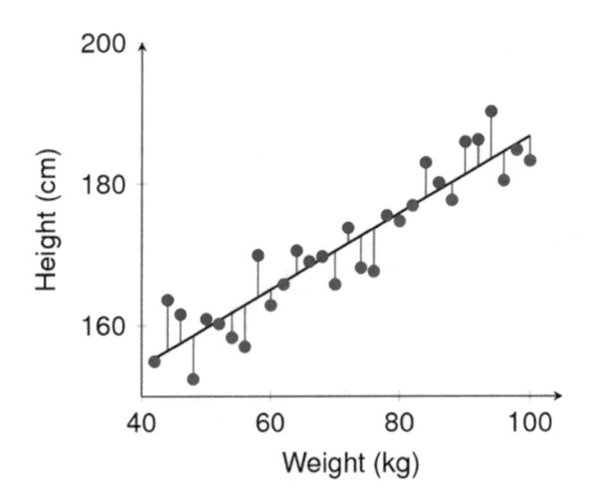

Fig. 1. Linear regression example

Linear regression is made with an assumption that there is a linear relationship between X and Y, while polynomial regression is used for the fitting of values with non-linear relationship.

Root mean square error is used to measure the dispersion of the values around that fitting line or average. To construct this error, first *the residuals* need to be determined. Residuals are the differences between the actual values and the predicted values. The residual is calculated as $\hat{Y}_i - Y_i$, where Y_i is the observed value for the$i - th$ observation and $\hat{Y}_i$ is the predicted value (Eq. 1). Residuals can clearly be seen in Fig. 1.

$$\sqrt{\frac{1}{n}\sum\left(\hat{Y}_i - Y_i,\right)^2} \tag{1}$$

They can be positive or negative as the predicted value under or over estimates the actual value. Then, the stated error (Eq. 1) can be used as a measure of the dispersion of the real values from the predicted values.

3.2 K-Means Clustering

The algorithm K-means, developed in 1967 by MacQueen [3], is one of the simplest unsupervised learning algorithms that solve the well known clustering problem. The procedure follows a simple and easy way to classify a given dataset into a certain number of clusters (assume k clusters) fixed a priori. The algorithm assumes that each feature is given on a numerical scale, and it tries to find classes that minimize the sum-of-squares error when the predicted values for each example are derived from the class to which it belongs.

After the algorithm finishes, it produces these outputs (Fig. 2):

- A group assignation for each data point
- The center for each label.

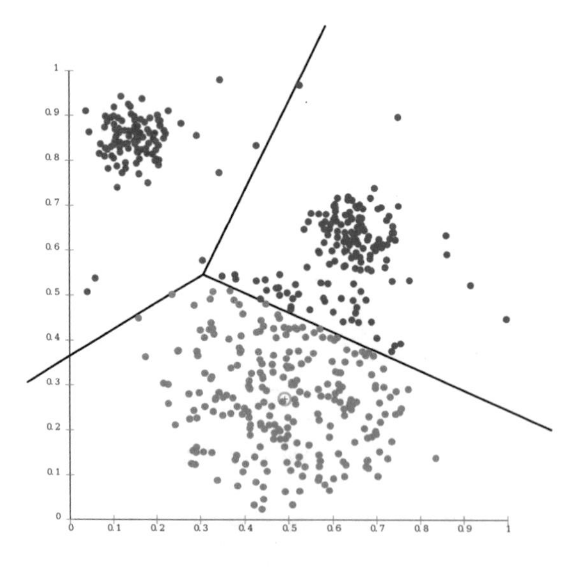

Fig. 2. K-means clustering example

3.3 Neural Network

What is a neural network? The basic idea behind a neural network is to simulate lots of densely interconnected brain cells inside a computer so you can get it to learn things, recognize patterns, and make decisions in a human-like way. The main characteristic of this tool is that a neural network learns all by itself. The programmer just needs to design the physical structure (number of outputs, inputs, hidden layers) and set some very simple rules involving additions, multiplications and derivatives (Fig. 3).

They are based on *perceptrons*, which were developed in the 1950s and 1960s by the scientist Rosenblatt [8], inspired by earlier work by McCulloch and Pitts [4]. It is important to note that neural networks are (generally) software simulations: they are made by programming very ordinary computers. Computer simulations are just collections of algebraic variables and mathematical equations linking them together. One advantage of neural networks is that they are capable of learning in a nonlinear way [5,10].

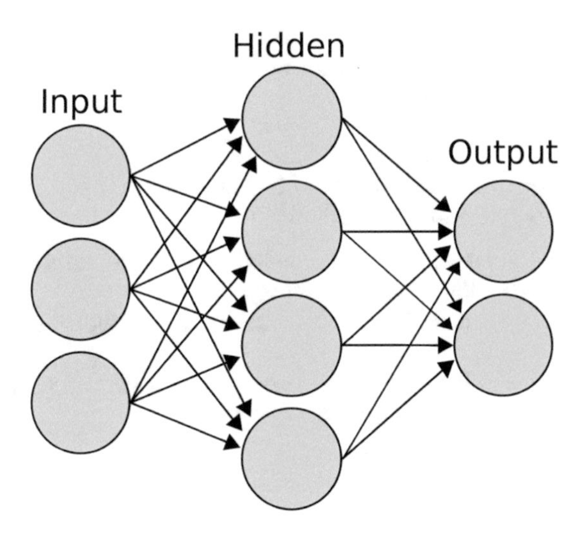

Fig. 3. Neural network example

4 Educational Activity

The educational activity to learn basic artificial intelligence will be developed in May 2018 at a high school in Bilbao, during a two-hours session in a single day. The workshop is prepared for a maximum capacity of 25 students of ages from 16 to 18. The main objective of the activity will be to demystify the AI *blackbox*, and show them some simple mathematical rules that will help them understand what does an “intelligent system” imply. Once the students are sitting individually on the computers, slides will be used to give the explanations needed. Moreover, a short manual explaining each algorithm to be used will be delivered to each student.

The activity will consist in short explanations of each algorithm followed by simple exercises to check the simple rules that make them work. Algorithms will be described individually, and 20 min will be left to the students to practise with the different tasks, with the help of the teachers if necessary. Tools to be used will be MS Excel and Scratch. In next subsections, a description of the taks to perform with each algorithm will be described.

4.1 Linear and Polynomial Regression

MS Excel will be used to learn this algorithm. This software can automatically calculate the straight line or curve that best fits the input data. Teachers will provide a small dataset of $X - Y$ values to the students, and the students will be asked to perform the following tasks:

- adjust the average with a linear or polynomial line and calculate the root mean square error.
- predict next data value.
- demonstrate why the equation that calculates Excel is the best. For that, the teacher calculates the root mean square error, and compares it to the error that creates any other equation that students can guess.

4.2 K-Means

The teachers will provide a dataset of 1,000 values to students. This exercise will consist on the following tasks, in Scratch:

- Students should start applying K-means to first 400 values and choose the number of clusters they wish.
- Next, they must add next 100 values and check how the clusters evolute with extra data.
- Finally, students will have the opportunity to alterate the dataset with values they guess and check how the clusters change.

4.3 Neural Network

This part is the most complex of all the AI algorithms to explain to students, as some mathematical operations (derivatives), are still a big obstacle when they are 16–18 years-old. Teachers will prepare three different datasets. The number of inputs and outputs between them will vary. Using the first dataset, teachers will explain next concepts and propose following tasks using Scratch:

- Explain what a neural network exactly is and what rules does the training process follow.
- Show how varying the number of components in the hidden layer increases or decreases the training speed.
- Demonstrate visually that the neural network weights converge slowly, reducing the error between the output and training value.
- Provide them with new datasets and let the students design using Scratch a self-designed neural network to train and test.

5 Conclusions

The objective of the present activity consists on the presentation of basic AI ideas and algorithms so that high school students can acquire a critic opinion about these machines, their advantages and their social impact in coming years. The activity is intended to emphasize that quite simple engineering and mathematics allow us to create machines that are able to identify patterns and predict data, and to correct the simplistic idea that all AI technology can be reduced to a *blackbox* that we cannot understand. It is essential to highlight that the workshop does not pursue to teach concepts; this is the reason why concepts themselves are not evaluated. The evaluation of the activity is based on the questionnaire about the organization of the event, which focuses on the attendant's perceptions and experiences.

As a future line work, we will include extra algorithms that permit understand more smart applications of everyday life.

References

1. Arimoto, T., Sato, Y.: Rebuilding public trust in science for policy-making. Science **337**(6099), 1176–1177 (2012)
2. Haerlin, B., Parr, D.: How to restore public trust in science. Nature **400**(6744), 499 (1999)
3. MacQueen, J.: Some methods for classification and analysis of multivariate observations. In: Proceedings of the Fifth Berkeley Symposium on Mathematical Statistics and Probability, Volume 1: Statistics, pp. 281–297. University of California Press, Berkeley (1967)
4. McCulloch, W.S., Pitts, W.: A logical calculus of the ideas immanent in nervous activity. Bull. Math. Biophys. **5**(4), 115–133 (1943)
5. Nielsen, M.A.: Neural Networks and Deep Learning. Determination Press (2015)
6. OECD: Scientific advice for policy making: The role and responsibility of expert bodies and individual scientists. Technical Report 21, OECD Science, Technology and Industry Policy Papers (2015)
7. Resnick, M., Maloney, J., Monroy-Hernández, A., Rusk, N., Eastmond, E., Brennan, K., Millner, A., Rosenbaum, E., Silver, J., Silverman, B., et al.: Scratch: programming for all. Commun. ACM **52**(11), 60–67 (2009)
8. Rosenblatt, F.: The perceptron: a probabilistic model for information storage and organization in the brain. Psychol. Rev. **65**(6), 386 (1958)
9. Stone, P., Brooks, R., Brynjolfsson, E., Calo, R., Etzioni, O., Hager, G., Hirschberg, J., Kalyanakrishnan, S., Kamar, E., Kraus, S., et al.: Artificial intelligence and life in 2030. In: One Hundred Year Study on Artificial Intelligence: Report of the 2015–2016 Study Panel (2016)
10. Woodford, C.: How neural networks work-a simple introduction (2016)

Prelude of an Educational Innovation Project: Discussing a Redesign of the Continuous Assessment in Mathematics for Chemistry and Geology Bachelor Degree Students

J. David Núñez-González[1,4](✉), Manuel Graña[2,4], and Jose Manuel Lopez-Guede[3,4]

[1] Department of Mathematics, University of the Basque Country (UPV/EHU), Leioa, Spain
[2] Department of Computer Science and Artificial Intelligence, University of the Basque Country (UPV/EHU), Leioa, Spain
[3] Department of Systems Engineering and Automation, University of the Basque Country (UPV/EHU), Leioa, Spain
[4] Computational Intelligence Group, University of the Basque Country (UPV/EHU), Leioa, Spain
{josedavid.nunez,manuel.grana,jm.lopez}@ehu.eus

Abstract. The jump from Secondary Education to University is noticed by students once they have seen the bad obtained results after the first ordinary exam session they deal with. Many factors such as group integration problems, lack of motivation and/or required study methodology turn the jump a difficult wall to overcome. The aim of this work is start a discussion of some ideas for redesign a continuous assessment in Mathematics for Chemistry and Geology Bachelor degrees so in the near future we have a defined work lines to submit and implement an educational innovation project.

1 Introduction

Studying at University supposses a challenge for anyone who is willing to accept it. The common student profile is one which comes from Secondary School. The jump from Secondary Education to University is noticed by students once they have seen the bad obtained results after the first ordinary exam session they deal with. Many factors suchs as group integration problems, lack of motivation and/or requiered study methodology turn the jump a difficult wall to overcome. It is expected that a student who has chosen to study a Chemistry or Geology Bachelor degree does not feel special motivation for Mathematics and relegate the study of this Subject giving priority to other Subjects more related to their interests. In this sense, it is necessary to draw a work pipeline to get the attention of the student in order to lead their work to a constant and methodical study

M. Graña et al. (Eds.): SOCO'18-CISIS'18-ICEUTE'18, AISC 771, pp. 536–543, 2019.
https://doi.org/10.1007/978-3-319-94120-2_52

throughout the course. In this way, the student will be able to deal with the final exam with more guarantees of success [3–5].

The aim of this work is start a discussion of some ideas for redesign a continuous assessment in Mathematics for Chemistry and Geology Bachelor degrees so in the near future we have some defined work lines in order to submit and implement an Educational Innovation Project. Both Bachelor degrees have two Mathematics Subjects in the first year of the curriculum. One each semester. In this paper we only fix our attention on Mathematics I, related to the first semester [2,6].

The structure of the paper is as follows: Sect. 2 reviews the current situation of the Subject. Section 3 gives the proposal for this Subject. Section 4 discuss the proposal and give expected results and future work.

2 Current Teaching Guide

In this Section we show the main aspects of the current teaching guide[1].

2.1 Contents

The contents of the Subject deepen and extend the mathematical knowledge that is supposed to the student. Here we show the contents table of the Subject. Each content will be consider as a unit here in after.

1. **Numbers and functions.** The complex numbers. Inequalities and inequations. Elementary functions.
2. **Continuity:** real variable functions. Limits and continuity. Fundamental theorems of continuity.
3. **Differential calculus**. Derivation. Derivation rules. Optimization. Representation of functions. Taylor's polynomial.
4. **Integral calculus.** Methods of integration of functions of a real variable. Definite integrals: the integral as an area. Fundamental theorem of the calculation. Applications.
5. **Linear algebra and applications.** Real vectorial spaces. Linear functions Matrices. Matrix calculation. Determinants. Eigenvalues and eigenvectors. Diagonalization of matrices.

2.2 Distribution of Hours

Within the Bologna plan each Subject has a weight in credits (European Credit Transfer System). Thus, 1 ECTS is equivalent to 25 h of work distributed in 10 classroom hours and 15 non-contact activity hours. The Subject we are introducting has a weight of 6 ECTS (60 classroom hours) distributed in 30 Magistral hours, 18 Classroom practice hours, 6 Seminars hours and 6 Computer practices hours (Table 1).

[1] Information available at University website: https://www.ehu.eus/es/web/estudiosdegrado-gradukoikasketak/grado-biologia-asignaturas-por-cursos?p_redirect=consultaAsignatura&p_anyo_acad=20170&p_ciclo=X&p_curso=1&p_cod_asignatura=25141.

Table 1. Distribution of hours

	Classroom hours	Non-contact activity hours
Magistral	30	45
Classroom practice	18	27
Seminars	6	9
Computers practice	6	9

2.3 Evaluation System

At the moment, the 15% of marks are given by a test (done by the middle of the course) and 85% of marks are given by final ordinary exam. No more marks are given by other kind of activities.

3 Proposal

The proposed plan (see Fig. 1) includes three stages: A Welcome Programme, the Readapted Work Plan and the Final Exam. The first stage will not be marked but is compulsory to achieve a successful situation at the end of the course. The second stage covers proposed tasks and test. It will be marked with a weight of 60%. Final exam will take place in the third stage being marked with a weight of 40%.

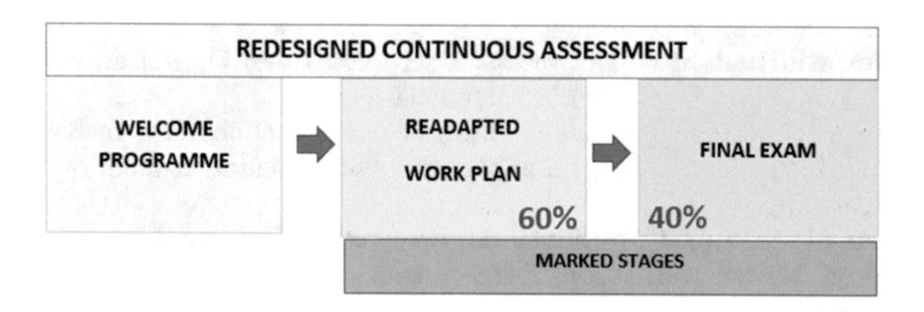

Fig. 1. Proposed general plan

3.1 Welcome Programme

A welcome program is a zero course where the student is given the necessary skills before starting the official course of the subject [1]. Depending on the available resources, the proposal for a welcome program can be launched in face-to-face mode or in a virtual MOOC (Massive Open Online Course) type. A MOOC one can be implemented via Educational Innovation Project. A MOOC course offers the student the freedom to choose when viewing the classes, without being subject to a schedule that could be incompatible according to their personal situation. As we can see in Fig. 2 the Welcome Programme is divided in three modules. Next subsections describe each module. At the end of each module there will be some questionnaires to check the follow-up of the course by the student. The questionnaires are not scoring tests but they are mandatory for students who want to participate in the continuous assessment.

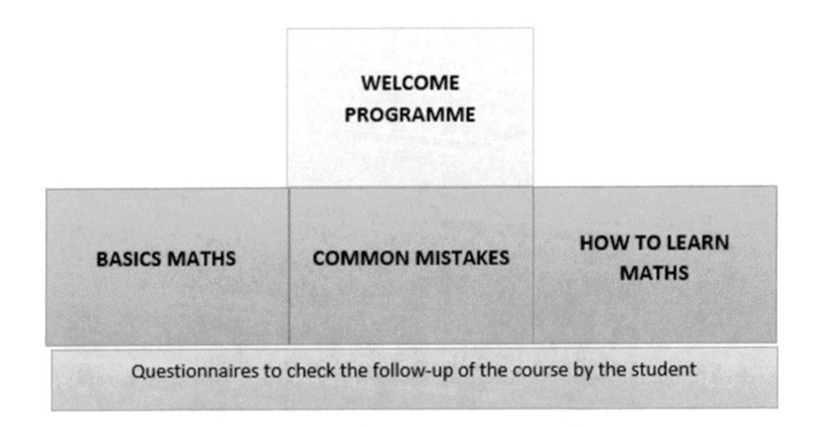

Fig. 2. Welcome programme plan

Basics Maths. This module will be the largest one comparing to other two modules. Three main lines will be teached in order to cover possible knowledge gaps that some students may present. As a task, a theoretical-practical questionnaire will be proposed.

- Mathematical Language. Notation. Logic. Induction.
- Basics on Analysis, Algebra and Geometry to start at University.
- Basics on Statistics and Probability.

Common Mistakes. It will be a session focused as a workshop where the professor will propose diverse and varied examples of typical mistakes made in exams. As a task, a questionnaire will be proposed with examples where the student should detect errors and propose a correction.

How to Learn Maths. It will be a session aimed to explain how the subject should be studied. The professor will show how to:

- Read and proof Theorems, Corolaries, Lemmas, Propositions and Definitions.
- Support definitions with a drawing when possible.
- Going from simple examples to generalizations and vice versa.
- How to support exercises on theory.
- Look for (online) bibliography references at Library (website).

Tasks will be related to proof an easy definitions and look for some bibliography on the Internet.

3.2 Readapted Work Plan

This stage is subdivided in two block (see Fig. 3). Being the 60% the weight of this stage, each block will have a weight of 30%. The aim of this stage is keep students working during the whole semester. Thus, we avoid the typical error of not having a continuous work rhythm and therefore, leaving the entire study process to the end.

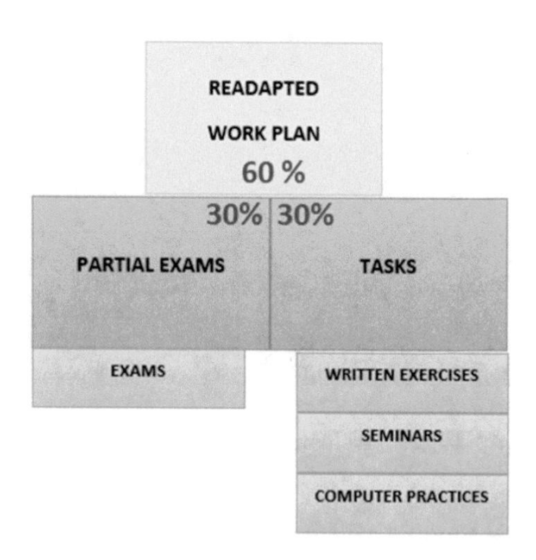

Fig. 3. Readapted work plan

Partial Exams. This block contains 3 partial exams to be done every third of semester (5 weeks). Each partial exam has a weight of 10%. First partial will be related to units 1 and 2 s partial wil be related to units 3 and a half of unit 4. Third partial will be related to the second half of unit 4 and unit 5.

Written Exercises. For each unit a collection of exercises will be proposed to students. Both, theoretical and numerical exercises should be proposed. Starting with most easy ones (e.g. immediate resolution) and increasing the degree of difficulty (e.g. exercises that require more elaboration) to exam exercises (e.g. exercises depending of a parameter) of previous years. Those exercises must be delivered until the corresponding partial exam (3 partial exams suppose 3 collection of exercises). Written exercises have a weight of 10%. (3.333% each collection).

Seminars. There will be 3 seminars of two hours each. Group tasks will be proposed. The first seminar will cover topics 1 and 2. The second seminar will cover topics 3 and 4. The third seminar will cover topic 5. Each seminar session will consist of two tasks. Seminars have a weight of 10%. (3.333% each seminar).

First Task. Students will carry out a group task before the seminar consisting of finding contextualized problems in Chemistry and Geology that have been solved with the mathematical concepts seen in class. Each group will prepare a report on the problems they have encountered and make a presentation in class. A delegate will be appointed in each group. The role of the delegate will be to be in permanent contact with the rest of the delegates to report the problems that will be included in their report. In this way copy cases will be avoided.

Second Task. The professor will give few problems (2–4) to be solved by groups in class. At the end of the class each group will deliver their proposals. The aim of the task is that each group is discussing in a critical and reasoned way the resolution of the proposed problems.

Computer Practices. The computer practices are focused on the use of mathematical software to work on the concepts seen in class. Geogebra could be a good software due to ease to plot functions and use functions without previous knowledge of programming. The proposal consists in carrying out these practices in two blocks of three hours each. The first block will be given at the end of the semester, while the second block will be taught at the end of the semester. These practices have a weight of 10% (5% each block).

3.3 Final Exam

Once the student has been working under the scheduled planning it is time to deal with Final Exam. This stage will take place in period of ordinary call examinations. The exam will include theoretical-practical issues that have been worked on throughout the course. As mentioned before, Final Exam has a weight of 40%.

4 Discussion and Expected Results

4.1 Final Mark

A student will get a $PASS$ mark if and only if the student maintain a constant rhythm of work with satisfactory results. There are 5 items (from B to F) in which it is necessary to reach the minimum qualification (5 marks out of 10 marks) indicated in the formula. Notice that item A is compulsory to remain in continuous assessment. Although each item requires a $PASS$ grade, this system allows any activity to obtain the grade of $FAIL$ as long as the item has an average rating of $PASS$. When the student's performance is not as expected and this leads to the student's consecutive suspension of the different activities, he/she must leave from the continuous assessment and go only to the final exam which will have a weight of 100% in the final mark. In this case, it is desirable that the student continues to participate voluntarily in activities B and E so that he/she maintains a continuous and progressive study of the subject.

$$PASS \Leftrightarrow [A = DONE] \wedge [B = PASS] \wedge [C = PASS] \wedge [D = PASS] \wedge [WeightedAVG[E, F] = PASS] \quad (1)$$

$$MARK = 0.1B + 0.1C + 0.1D + 0.3E + 0.4F \quad (2)$$

where:

- $A = WelcomeProgramme$
- $B = WrittenExercises$
- $C = Seminars$
- $D = ComputerPractices$
- $E = PartialExams$
- $F = FinalExam$

4.2 Discussion, and Expected Results

Motivation. As far as possible, we must contextualize the exercises and activities in matters related to Chemistry and Geology. Providing in this way a more attractive material for the student. A collection of monotonous exercises without any attraction can lead to boredom and the desire to leave the studio. It is important that the student finds the subject he is studying useful. To do this, in this paper we try to answer at all times the question: "and this, what is it for?"

Continuous Work. High motivation is essential to maintain a continuous work rate. Continuous study supposes the student will have more guarantees of success when facing an exam. The student is expected to maintain regular work peaks. In comparison with the current system, the student only made a peak of work in the middle of the semester for the evaluation of the test. Next figure gives the idea of work loads.

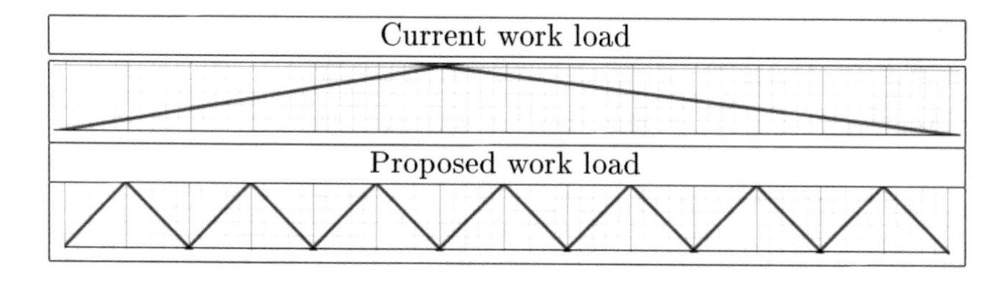

Fig. 4. Work load comparison

Collaborative Working vs Copy. In order to avoid fraudulent deliveries per copy, as far as possible, we propose the use of encoding methods[2] in exercises and problems. Example: We want to ask as a task to solve this exercise:

$$\int \frac{a_1 x}{a_2 x^{a_3}} dx$$

[2] Thanks to Joseba Santisteban Elorriaga (Department of Mathematics, University of the Basque Country - UPV/EHU) for the encoding methods.

$$\int bx^a dx$$

If we propose an exercise with fixed values, this encourages students to copy each other. When coding the exercise, the students will be able to consult the exercises of other classmates but they will not be able to copy them literally since all the exercises will undergo modifications.

Key Examples. We can ask the ID number of a student which format is: $a_8a_7a_6a_5a_4a_3a_2a_1$ to make direct replacements or using defined replacements (e.g. $a = \min\{a_k \neq 0\}, b = \max\{a_k\}$) having $0 \leq a_k \leq 9$.

4.3 Future Work

In the short term we want to end this discussion and leave defined lines of work so that in the medium term this methodology can be implemented through an Educational Innovation Project following the educational model proposed by the University, the Cooperative and Dynamic Learning model.

Acknowledgements. This work has been funded by the European Union's Horizon 2020 research and innovation programme under the Marie Skłodowska-Curie grant agreement No 777720.

References

1. de Larrinoa, E.A-F., Sancho-Saiz, J., Mesanza-Moraza, A., Delgado-Tercero, R., Tazo-Herran, I., Ramos, J., de Eribe-Vazquez, J.I.O., Lopez-Guede, J., Zulueta, E., de Argandona-Gonzalez, J.D.: Welcome program for first year students at the faculty of engineering of vitoria-gasteiz. soft skills. In: Grana, M., Lopez-Guede, J.M., Etxaniz, O., Herrero, A., Quintian, H., Corchado, E. (eds.) International Joint Conference SOCO 2016-CISIS 2016-ICEUTE 2016, vol. 10, pp. 711–717 (2016)
2. Ayalon, M., Hershkowitz, R.: Mathematics teachers' attention to potential classroom situations of argumentation. J. Math. Behav. **49**, 163–173 (2018)
3. Gonzalez-Marcos, A., Olarte-Valentin, R., Sainz-Garcia, E., Mugica-Vidal, R., Castejon-Limas, M.: A virtual learning environment to support project management teaching. In: Pérez García, H., Alfonso-Cendón, J., Sánchez González, L., Quintián, H., Corchado, E. (eds.) International Joint Conference SOCO 2017-CISIS 2017-ICEUTE 2017, pp. 751–760 (2017)
4. Musgrave, S., Carlson, M.P.: Understanding and advancing graduate teaching assistants mathematical knowledge for teaching. J. Math. Behav. **45**, 137–149 (2017)
5. Olarte-Valentin, R., Mugica-Vidal, R., Sainz-Garcia, E., Alba-Elias, F., Fernandez-Robles, L.: Analysis of the online interactions of students in the project management learning process. In Pérez García, H., Alfonso-Cendón, J., Sánchez González, L., Quintián, H., Corchado, E. (eds.) International Joint Conference, SOCO 2017-CISIS 2017-ICEUTE 2017, pp. 743–750 (2017)
6. Wasserman, N.H.: Knowledge of nonlocal mathematics for teaching. J. Math. Behav. **49**, 116–128 (2018)

The Boundary Element Method Applied to the Resolution of Problems in Strength of Materials and Elasticity Courses

J. Vallepuga-Espinosa, Lidia Sánchez-González(✉), Iván Ubero-Martínez, and Virginia Riego-Del Castillo

Universidad de León, León, Spain
lidia.sanchez@unileon.es

Abstract. Strength of materials and elasticity are included within the core subjects in the School of Engineering. Modeling and simulation are fundamental in early stages of engineering design and testing. For this reason, we propose the use of a Java application named BEMAPEC in combination with hand made problems to improve the motivation of the engineering students. BEMAPEC applies the Boundary Element Method (BEM) to solve thermoelastic problems of one 3D solid, among other functionalities. Thus, with this application the students could compare solutions and visualize how the tractions and displacements are distributed on the typical problems solved in Strength of Materials and Elasticity courses. BEMAPEC provides a helpful tool for self-learning and a better understanding of the theoretical concepts.

Keywords: Boundary element method · Strength of materials Elasticity · Simulator · Contact · Thermoelastic

1 Introduction

Strength of Materials and Elasticity courses apply many theoretical concepts difficult to understand for undergraduate students. All the theoretical background is used now to solve actual problems with many considerations to deal with. It is also required 2D or 3D vision in order to provide a full vision of the problem. For that reason, it is widely considered in different universities that those subjects are identified by the students as some of the most difficult ones. There are works to make easier the learning process to the students using simulators [9], learning by teaching techniques [8] or a virtual gaming environment [1]. More recently, the flipped classroom has also been applied to mechanical studies [7].

BEMAPEC (Boundary Element Method Applied to Problems of thermal, elastic and thermoElastic Contact) is a Java application developed by the authors of this paper and their collaborators [5,6]. BEMAPEC is based on the iterative method that uses the BEM proposed by Vallepuga and Foces [4] for solving thermoelastic problems in elasticity applying equilibrium and continuity

M. Graña et al. (Eds.): SOCO'18-CISIS'18-ICEUTE'18, AISC 771, pp. 544–552, 2019.
https://doi.org/10.1007/978-3-319-94120-2_53

and contact problems between 3D solids in which it is possible to introduce a no constant resistance in the contact zone. Linear elastic and potential problems can be also defined.

BEMAPEC operative improvements and updates are friendly implemented since it is based on the use of design patterns. In addition, Java offers a cross-platform application executable in any operating system. Thus, this application makes possible for the students to compare solutions and visualize how the tractions and displacements are distributed on the typical problems solved in Strength of Materials and Elasticity courses.

The method proposed by Vallepuga and Foces [4] to solve thermoelastic contact problems for 3D solids is briefly described in Sect. 2. Section 3 presents how a student would use BEMAPEC to solve a typical elasticity problem [2] with BEMAPEC. Finally, Sect. 4 gathers the final conclusions and future work.

2 The Thermoelastic Contact Problem in 3D

Two elastic bodies A and B are in contact, defined by their boundaries (Γ^A and Γ^B) and both subjected to boundary conditions. Applying the boundary integral equations, the resulting linear equations are coupled by imposing the contact conditions [3].

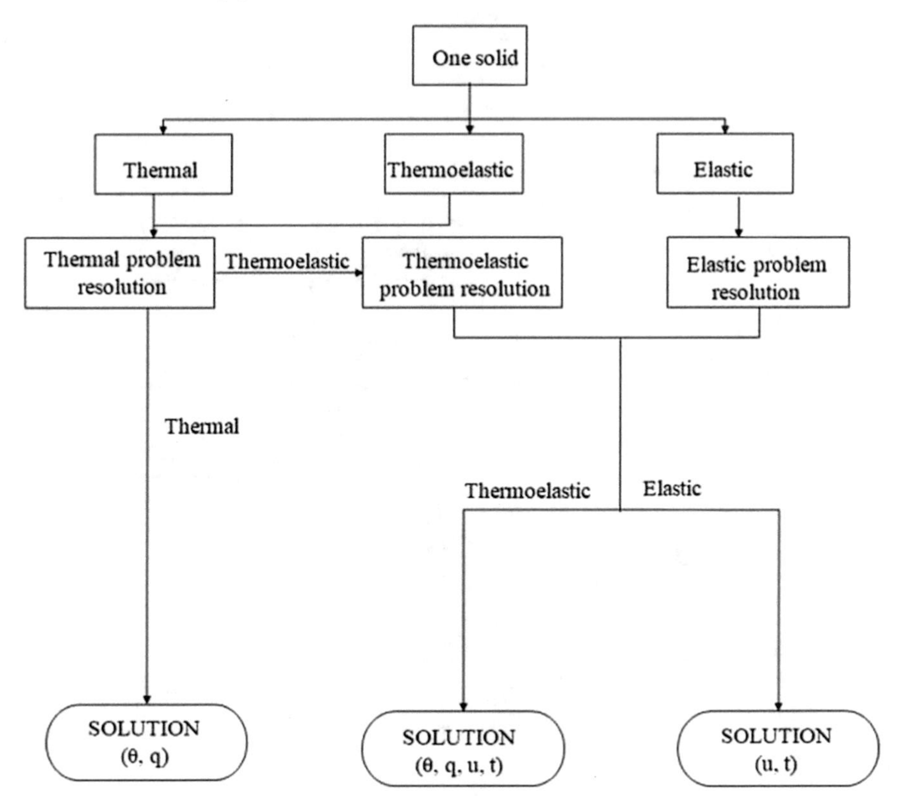

Fig. 1. Schema of the problem for one solid.

For heat conduction problems the equations become:

$$c(\xi)\theta(\xi) + \int_{\Gamma} q^*(\xi,\gamma)\theta(\gamma)\, d\Gamma(\gamma) = \int_{\Gamma} \theta^*(\xi,\gamma)q(\gamma)\, d\Gamma(\gamma) \tag{1}$$

For the three-dimensional thermoelastic problem, considering small deformations due to static loads and stationary conduction thermal, the integral equations are:

$$c_{ij}(\xi)u_j(\xi) - \int_{\Gamma} U^*_{ij}(\xi,\gamma)t_j(\gamma)\, d\Gamma(\gamma) + \int_{\Gamma} T^*_{ij}(\xi,\gamma)u_j(\gamma)\, d\Gamma(\gamma)$$
$$= \int_{\Gamma} P^*_i(\xi,\gamma)\theta(\gamma)\, d\Gamma(\gamma) - \int_{\Gamma} Q^*_i(\xi,\gamma)q(\gamma) d\Gamma(\gamma) \tag{2}$$

where $c_{ij}(\xi)$ is the free term of the elastic problem, $T^*_{ij}(\xi,\gamma)$ and $U^*_{ij}(\xi,\gamma)$ are the singular solution due to Kelvin. More information about the numerical definition of the problem can be found in [4].

Now, after doing the integrations over the elements and imposing the boundary and contact conditions, the Eqs. (1) and (2) become in two linear systems

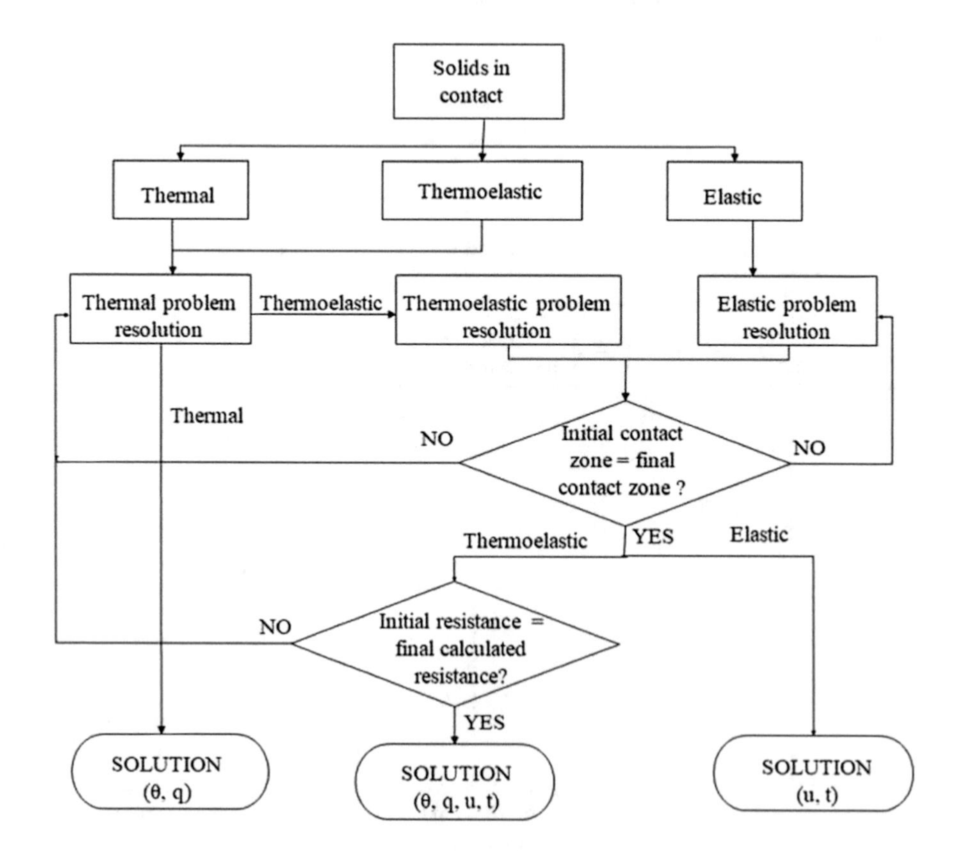

Fig. 2. Schema of the problem for more than one solid.

of equations. Solving the system of equations, geometry of solids, stresses, temperatures and thermal gradients are obtained, including the final contact zone when a contact problem is solved.

The procedure proposed by Vallepuga [4] is summarized in Fig. 1 for 1 solid and in Fig. 2 for more than one.

3 Application example

This section presents how a student would solve a typical problem in Strength of Materials and Elasticity courses by means of BEMAPEC. It is briefly detailed how geometry, properties and mesh parameters are introduced and how the student can analyze the results.

The problem chosen (see Fig. 3) is taken from [2] which is one references book of Strength of Materials and Elasticity courses. This problem corresponds to a thermoelastic contact problem, so both elastic and thermal conditions are included. Geometry of the problem and boundary conditions are shown in Fig. 4.

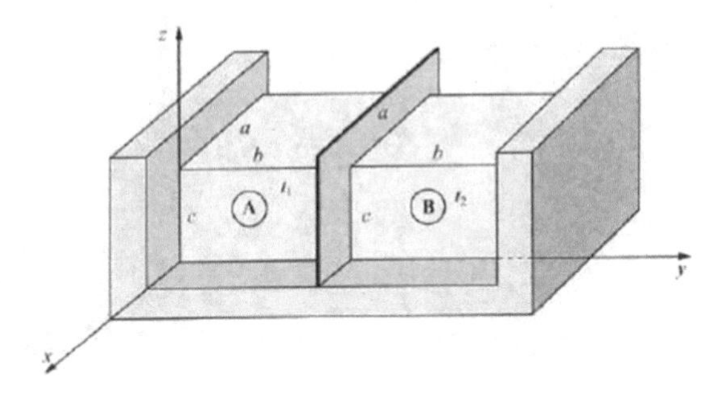

Fig. 3. Problem solved. Image obtained from [2].

In order to solve the problem by means of BEMAPEC, the student has to introduce the geometry of the solids of the problem. The geometry of the solids is defined by means of their surfaces, thus the student can easily give the coordinate points that shape each surface as it is shown in Fig. 5(a). Moreover, the defined geometry is plotted, so it is easy for the student to check if the defined geometry is correct (Fig. 5(b)). Furthermore, the student can use symmetry conditions in order the reduce the geometry and consequently obtain less degrees of freedom of the problem.

Once the geometry has been defined, the student has to introduce the material properties of each solid and define the type of contact if the problem to be solved is a contact problem as Fig. 6 shows. The material properties for this problem has been obtained from [2] and gathered in Table 1.

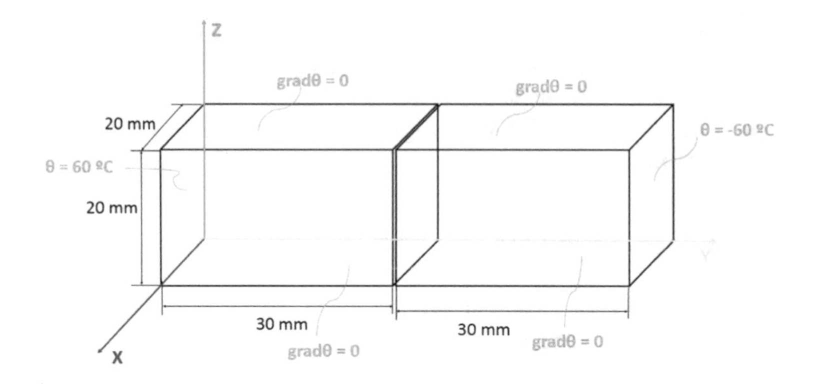

Fig. 4. Geometry and boundary conditions introduced in BEMAPEC.

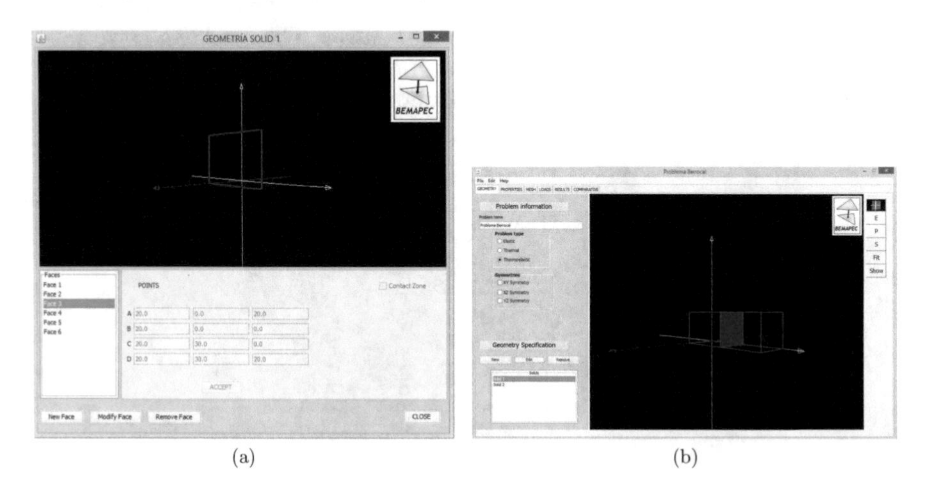

(a) (b)

Fig. 5. (a) The student has to introduce the coordinates of the points of each face that shape the solid. (b) The geometry of the solids of the problem to solve can be visually checked with BEMAPEC.

Table 1. Material Properties for the solved thermoelastic problem

	$E\,[GPa]$	ν	$\alpha\,[^{\circ}C^{-1}]$	$k\,[\frac{W}{^{\circ}Cmm}]$
Solid A	210	0.33	2×10^{-5}	50×10^{-3}
Solid B	210	0.33	1×10^{-5}	35×10^{-3}

The next step is the introduction of the mesh. Meshing is done by planar triangular elements whose node is located at the barycenter (Fig. 7a)). The student can perform the meshing by entering the number of divisions in each axis and its ratio (Fig. 7b)). These parameters can be given or not by the lecturer, since the mesh has great influence on the results so the lecturer can assess the student's ability to relate the quality of the mesh with the obtained results. Normals to

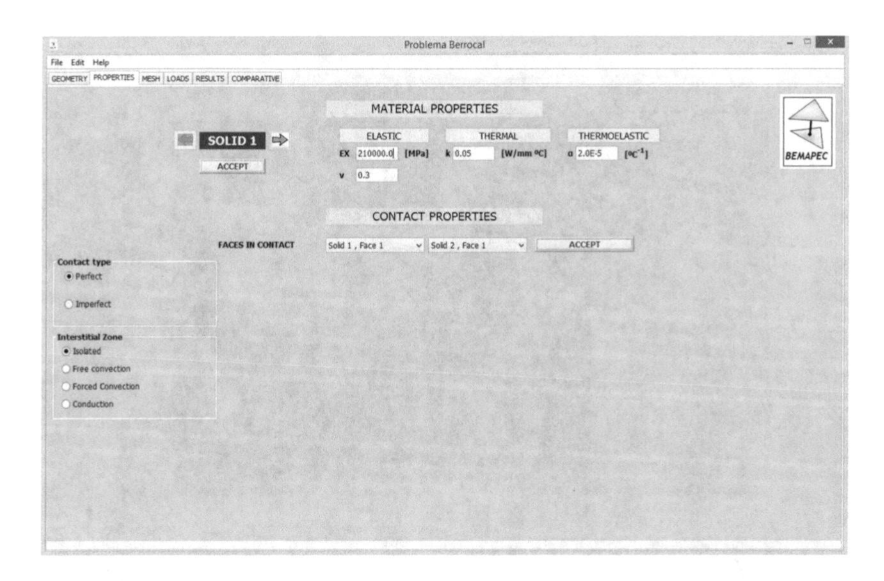

Fig. 6. Type of contact and properties.

the surface can be displayed in BEMAPEC, so the student can also check if the geometry is correctly introduced (Fig. 7c)).

The last step to be performed by the student before solving the problem is to introduce the boundary conditions of the problem. These boundary conditions can be introduced by surfaces or by elements (Fig. 8). There are two different types of boundary conditions: thermal boundary conditions (temperatures, forced convection and temperature gradients) and elastic boundary conditions (displacements and tractions).

Finally the student can solve the problem obtaining thermal and elastic results. The temperatures obtained are shown in Fig. 9(a). It is shown how the temperature is distributed evenly from the solid with the highest temperature to the lowest temperature. Figure 9(b) shows the stresses distribution in the contact zone between both solids. Maximum pressures occur at the edges and are caused by the effect of punching.

BEMAPEC displays all the obtained results using an OpenGL environment (Fig. 9). The student can plot the results using a color scale or a grey scale and if the student wants a more realistic image, the results can be interpolated. Moreover, it is possible to select a surface of the solid and apply zoom to analyze the results in more detail. The user of BEMAPEC can also obtain the results in a tabular output or save them as a csv file which can be open in any spreadsheet program.

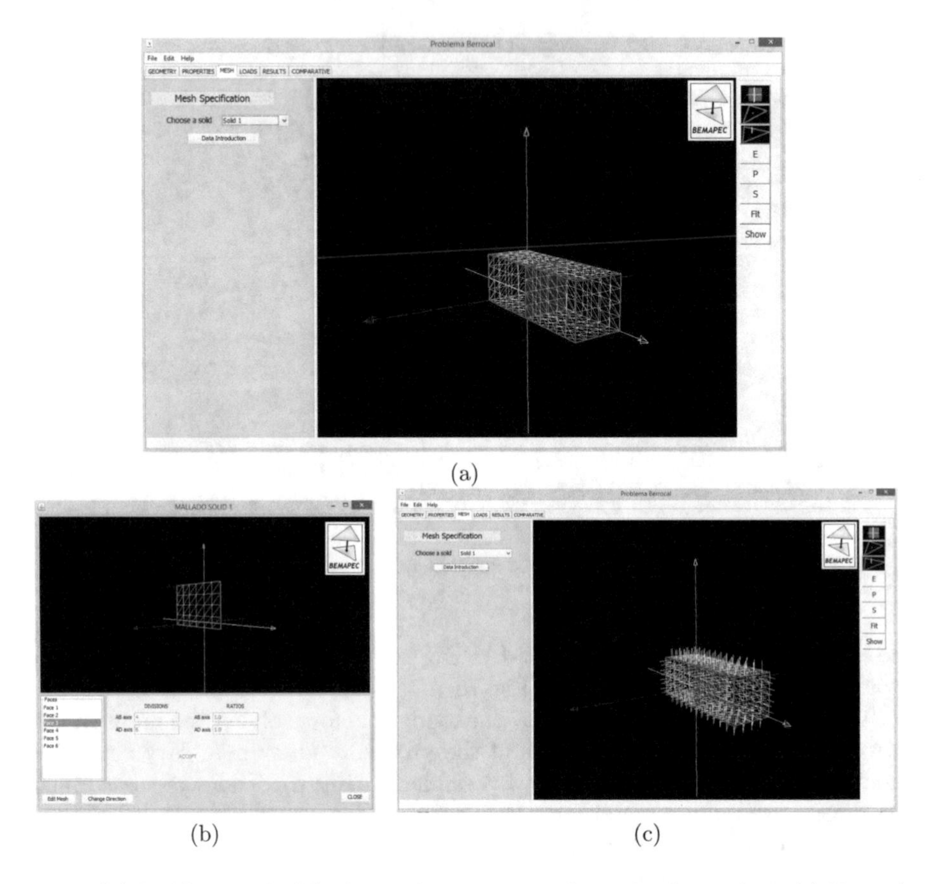

Fig. 7. (a) Problem mesh. (b) The mesh parameters chosen by the student. (c) Normals to the surfaces can be displayed to check if the geometry has been created properly.

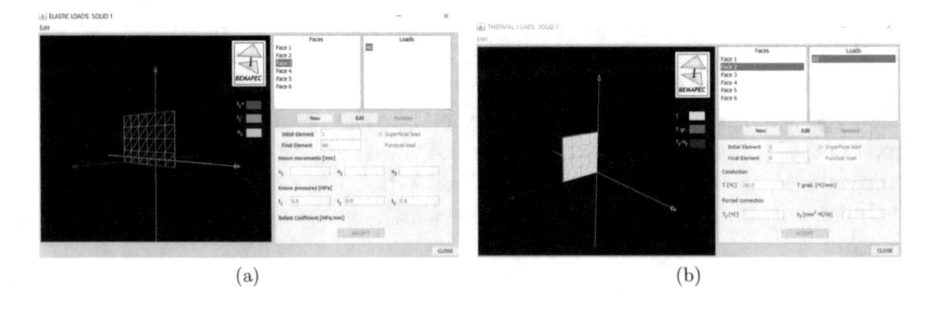

Fig. 8. (a) Elastic boundary conditions. (b) Thermal boundary conditions.

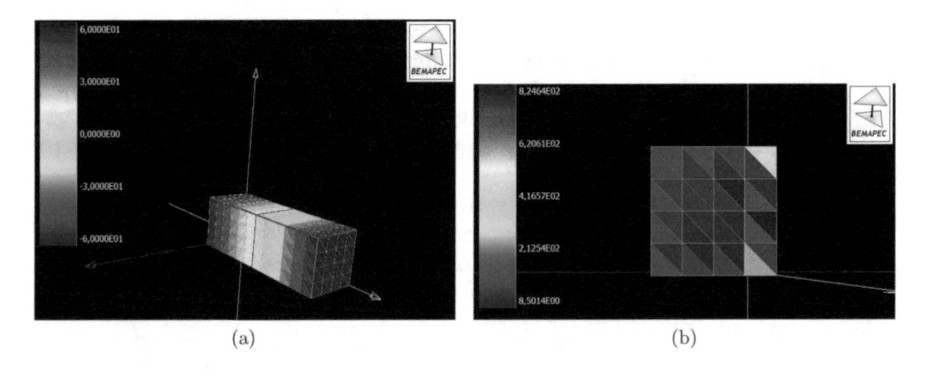

(a) (b)

Fig. 9. Temperatures (a). Normal stresses in contact zone [MPa] (b).

4 Conclusions and future work

In this work it is presented a Java program named BEMAPEC that solves elastic, thermal and thermoelastic 3D problems and contact problems between 3D solids which can be used in any Elasticity and Strength of Materials course. BEMAPEC will give to the student a better understanding of the behaviour of the solved problem.

Moreover, the future work is focused on developing more lab assignments as well as assessing the impact on learning process of our program by comparing the performance of students if they use BEMAPEC or not.

References

1. Barham, W., Preston, J., Werner, J.: Using a virtual gaming environment in strength of materials laboratory. In: Issa, R.R., J.D., Flood, P.I. (eds.) Proceedings of the 2012 ASCE International Conference on Computing in Civil Engineering, p. 105112. American Society of Civil Engineers (2012)
2. Berrocal, L.O.: Elasticidad. Mc Graw Hill, Madrid (1997)
3. Brebbia, C.A., Telles, J.C.F., Wrobel, L.C.: Boundary Element Techniques. Springer, Heidelberg (1984). https://doi.org/10.1007/978-3-642-48860-3
4. Espinosa, J.V., Mediavilla, A.F.: Boundary element method applied to three dimensional thermoelastic contact. Eng. Anal. Bound. Elem. **36**(6), 928–933 (2012)
5. González, R., Sánchez, L., Vallepuga, J.: Desarrollo de un entorno en C++ y OpenGL para la resolucion de problemas de contacto entre solidos 3D. In: Actas del congreso Metodos numericos en Ingenieria (2009)
6. González, R., Sánchez-González, L., Vallepuga, J., Ubero, I.: Parallel performance of the boundary element method in thermoelastic contact problems. In: International Joint Conference SOCO'17-CISIS'17-ICEUTE'17, pp. 524–532. Springer, Cham (2017)
7. Kanelopoulos, J., Zalimidis, P., Papanikolaou, K.A.: The experience of a flipped classroom in a mechanical engineering course on machine design: a pilot study. In: 2017 IEEE Global Engineering Education Conference (EDUCON), pp. 496–501. IEEE (2017)

8. Pollock, M.: Basic mechanics: learning by teaching - an increase in student motivation (a small scale study with technology education students). In: Proceedings 35th Annual Conference on Frontiers in Education, FIE 2005, p. T2F. IEEE (2005)
9. Sinha, R., Paredis, C., Khosla, P.: Behavioral model composition in simulation-based design. In: 35th Annual Simulation Symposium, 2002 Proceedings, pp. 303–315. IEEE (2002)

Impact of Auto-evaluation Tests as Part of the Continuous Evaluation in Programming Courses

C. Rubio-Escudero[1], G. Asencio-Cortés[2], F. Martínez-Álvarez[2(✉)], A. Troncoso[2], and J. C. Riquelme[1]

[1] Department of Computer Science, University of Seville, Seville, Spain
{crubioescudero,riquelme}@us.es
[2] Department of Computer Science, University Pablo de Olavide, 41013 Seville, Spain
{guaasecor,fmaralv,atrolor}@upo.es

Abstract. The continuous evaluation allows for the assessment of the progressive assimilation of concepts and the competences that must be achieved in a course. There are several ways to implement such continuous evaluation system. We propose auto-evaluation tests as a valuable tool for the student to judge his level of knowledge. Furthermore, these tests are also used as a small part of the continuous evaluation process, encouraging students to learn the concepts seen in the course, as they have the feeling that the time dedicated to this study will have an assured reward, binge able to answer correctly the questions in the continuous evaluation exams. New technologies are a great aid to improve the auto-evaluation experience both for the students and the teachers. In this research work we have compared the results obtained in courses where auto-evaluation tests were provided against courses where they were not provided, showing how the tests improve a set of quality metrics in the results of the course.

Keywords: Auto-evaluation · Continuous evaluation
Test questions · New technologies

1 Introduction and Background

In the Education system, the evaluation is the way to know the degree of learning achieved by the student. Traditionally, the evaluation has focused on the final stage of learning and, in general, the student organized his learning according to the type of evaluation followed. As a consequence of the process of convergence towards the European Higher Education Area [4], the evaluation acquires a new dimension by focusing the learning process on the student [5]. In this sense, it must be correctly designed to allow assessing whether the student has reached, as a goal, not only the knowledge but also the competences previously defined by the teacher for a specific course. Therefore, the continuous evaluation appears as an appropriate tool that assesses the progressive assimilation and development

M. Graña et al. (Eds.): SOCO'18-CISIS'18-ICEUTE'18, AISC 771, pp. 553–561, 2019.
https://doi.org/10.1007/978-3-319-94120-2_54

of the contents of the course and of the competences that must be achieved. In this way the teacher can carry out a greater and better follow-up of the progress in the student's learning.

There is a wide range of activities that can be carried out for the continuous evaluation process, such as resolution of practical scenarios, questions to develop theoretical concepts, multiple selection tests, true or false propositions, planning of debates on current issues, critical comment, preparing a topic at home using online resources, oral presentations of topics, preparation of tables and comparative diagrams, fill the gap type exercises, among others. In this work, we focus on the continuous evaluation aided by auto-evaluation test questions, that are provided along with the solutions to the students in the class. There are several publications stating that students show significantly more favorable attitudes towards multiple choice test format compared to essay type formats [11,12].

The continuous evaluation system proposed is based on the use of new technologies, which represent a qualitative step in the learning process for both students and teachers [1,7]. The technology platforms used for the auto-evaluation tests are Blackboard Learn [3], a virtual learning environment and course management system and a blog developed specifically for the programming courses (http://laprogramacionnoesunarte.blogspot.com.es/). Blogs and other Web 2.0 applications have shown to be very successful in aiding the learning process on different areas [2,10].

Our proposal is to provide the students with auto-evaluation test questions throughout the course's development, and to use some of those questions as a small part of the continuous evaluation system itself, thus encouraging students to revise their knowledge on the course concepts based on the results of the auto-evaluation. This methodology has been applied in programming courses of freshmen both in the Computer Engineering and Health Engineering degrees at University of Seville, Spain (US). We compare the results obtained by the students on these courses in terms of different quality metrics to other programming courses of freshmen in the University of Pablo de Olavide of Seville, Spain (UPO), where no auto-evaluation tests have been provided to the students. We will show that using auto-evaluation tests improves the course results related to number of students attending the continuous evaluation and number of students attending the final exam if they did not pass by continuous evaluation.

The rest of the article is organized as follows. Section 2 describes in detail how the auto-evaluation tests have been created and provided to the students. Section 3 shows the comparison of the results obtained by the two groups of freshmen compared: US with auto-evaluation tests vs. UPO without auto-evaluation tests. In Sect. 4 we highlight the main findings of this job.

2 Methodology

In this section we describe the method used to create the auto-evaluation tests, which are taken into account as a small part of the continuous evaluation system. The details of this technique are now presented.

For each of the conceptual blocks of the course, a set of test questions are provided, usually between 20 and 30. The questions include theoretical concepts as well as practical cases. The students gain access to this tests when the block concepts have been explained in class, and they can practice with the questions throughout the rest of the course. The questions follow two formats: fill the gaps (See Fig. 1(a)) or multiple choice (See Fig. 1(b)) [8,9]. The questions can be accessed as many times as the student needs, and the results for each test answered will be provided along with the correct answers for each question.

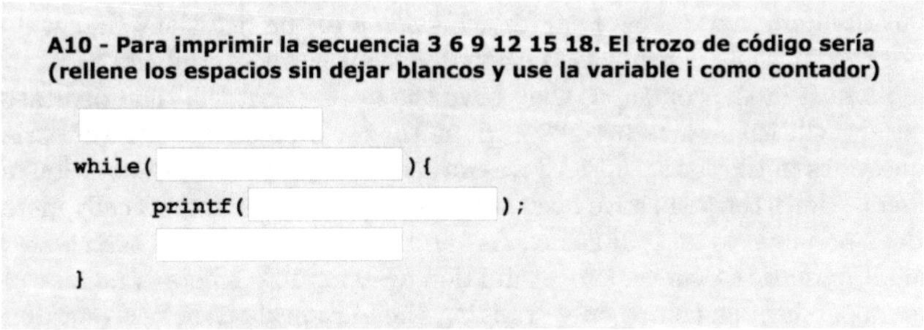

A10 - Para imprimir la secuencia 3 6 9 12 15 18. El trozo de código sería (rellene los espacios sin dejar blancos y use la variable i como contador)

```
[          ]
while( [          ] ){
    printf( [          ] );
    [          ]
}
```

(a) Fill the gaps question

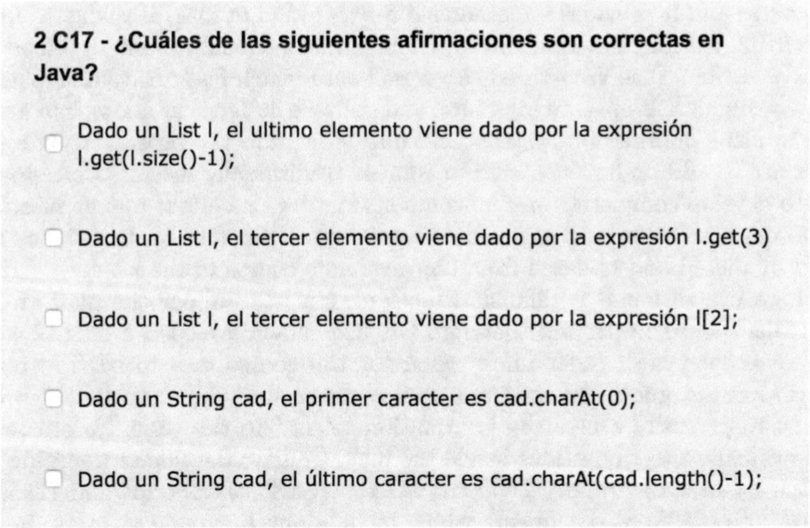

2 C17 - ¿Cuáles de las siguientes afirmaciones son correctas en Java?

- [ ] Dado un List l, el ultimo elemento viene dado por la expresión l.get(l.size()-1);
- [ ] Dado un List l, el tercer elemento viene dado por la expresión l.get(3)
- [ ] Dado un List l, el tercer elemento viene dado por la expresión l[2];
- [ ] Dado un String cad, el primer caracter es cad.charAt(0);
- [ ] Dado un String cad, el último caracter es cad.charAt(cad.length()-1);

(b) Multiple choice question

Fig. 1. Examples of auto-evaluation test questions (screen capture from the blog in the original language).

The continuous evaluation method [6] is made up of several parts. There are theoretical and practical sections, with the theoretical part being 60% of the final mark and the practical being the 40%. The theoretical part consists in 4 exams throughout the course. Each exam has 2 parts: the first one, some test questions very similar to the ones in the auto-evaluation tests, the second one writing some code to solve some problem. The test questions represent 25% of the exam's mark. The practical part consists in an exam taken in the computer lab, where the students have to solve a given problem.

Therefore, a small portion of the final mark is obtained with the test questions provided for each block of the course are asked to the students. In particular, 15% of the final mark. This fact, knowing some of the questions that will take part in the continuous evaluation system, encourages the students to learn the concepts seen in the course, as they have the feeling that the time dedicated to this study will have an assured reward, as they will be able to answer correctly the questions in the tests. In Fig. 2 we can see some graphic representation of the number of times the tests have been accessed. Tests access dramatically increases the days previous to an evaluation. As the test questions in the exams are only an small part of the continuous evaluation system, 15%, there is no risk of the students passing the course only studying the auto-evaluation test questions.

Regarding the platform in which the test questions are developed, we initially used the Blackboard Learn software [3], as it is the virtual platform for our course and it provides a mechanism for this kind of task. Blackboard Learn is a virtual learning environment and course management system developed by Blackboard Inc. It is Web-based server software which features course management, customizable open architecture, and scalable design that allows integration with student information systems and authentication protocols. Its main purposes are to add on-line elements to courses traditionally delivered face-to-face and to develop completely on-line courses with few or no face-to-face meetings. We found it suitable to create a database of test questions for each block that could be reused and updated from one academic course to the next.

However, we found it difficult to extract statistics about the platform use from Blackboard Learn and therefore Dr. Riquelme developed a blog in which the tests along with other information for the course are stored (theoretical descriptions, exercises, exams from past years, media related content), so that we can more easily access the information related to the tests. To create the auto-evaluation test questions inside the blog, we have developed a specific tool written in Java to produce Dynamic HTML (CSS3, JavaScript and HTML5) from Microsoft Word documents which include questions and answers. In first place, all questions for the studied subjects were written in Microsoft Word documents. Then, our tool was executed producing the web pages in HTML. Finally, the page source code was embedded inside the blog.

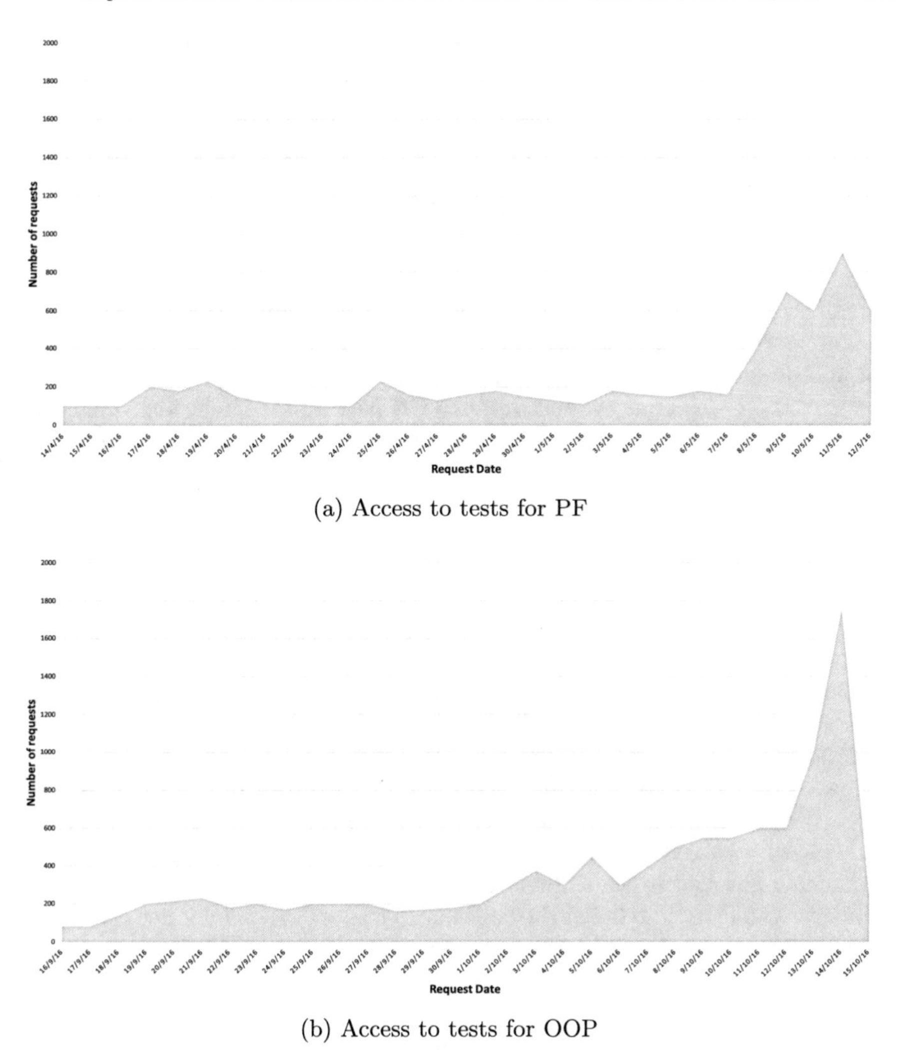

(a) Access to tests for PF

(b) Access to tests for OOP

Fig. 2. Access to tests on the days previous to an exam. Two subjects were analyzed: Programming Fundamentals (PF) and Object Oriented Programming (OOP) (see Sect. 3).

3 Results

Once the auto-evaluation tests were defined and planned properly through the continuous evaluation system of the studied course, a set of quality measures were assessed in order to assess the impact of such auto-evaluation tests into the student learning process.

For such purpose, two subjects were used: Programming Fundamentals (PF) and Object Oriented Programming (OOP). Both belong to the first course of the Computer Engineering degree in the US and UPO Universities. Whereas PF is based on the structured programming paradigm using the C language, OOP is based on the object oriented paradigm using the Java language.

To compare the effect of introducing the auto-evaluation tests, two groups of freshmen from different universities were analyzed. Although the studied courses belong to different universities, their contents are very similar and suitable to be compared. Two academic years were analyzed: 2014/15 and 2015/16.

A set of evaluation metrics were assessed for each course, academic year and university. Specifically, the following five metrics were computed: Enrolled, CE attendance, CE passing students, CE failed & final exam and Only final exam.

The metric Enrolled counts the number of students enrolled in the course. The metric CE attendance refers to the number of students who attended the exams of the continuous evaluation. The CE passing metric means the number of students who passed the course by continuous evaluation. The metric CE failed & final exam counts the students who failed the continuous evaluation and attended final ordinary exam. Finally, the metric named Only final exam is referred to the number of students who did not follow the continuous evaluation and only came to the final ordinary exam.

From the metrics explained above, four indexes were derived: CE Attendance/ Enrolled, CE Passing/Attendance, CE Passing/Enrolled and CE failed†.

The index CE Attendance/Enrolled is the ratio between the number of students who attended the exams of the continuous evaluation and the number of students enrolled in the course. The index CE Passing/Attendance refers to the ratio between the number of students who passed the exams of the continuous evaluation and the students who attended the exams of the continuous evaluation. The CE Passing/Enrolled index means the ratio between the number of students who passed the exams of the continuous evaluation and the number of students enrolled in the course. Finally, the index named CE Failed† is the number of students who failed the continuous evaluation and did not come to the ordinary final exam divided by the number of students who failed the continuous evaluation.

Tables 1 and 2 show the results achieved for each course and course in terms of the metrics and indexes explained before. Specifically, Table 1 shows the values obtained for the course OOP, while Table 2 shows the results achieved for the course PF. Both tables include two subcolumns named *Yes* and *No* that indicate if the autoevaluation tests were applied or not. Results with autoevaluation tests were obtained in the US while those without them were obtained in the UPO. The higher values of indexes were highlighted using text in bold within tables.

As it can be seen from the results shown in Tables 1 and 2, the number of enrolled students were higher in the US than int the UPO. For such reason, the features to be compared in this study were the indexes computed (CE Attendance/Enrolled, CE Passing/Attendance, CE Passing/Enrolled and CE Failed†) rather than the metrics, because the former are relative values.

On the one hand, according to results achieved, we see that the ratio CE Attendance/Enrolled is significantly higher when auto-evaluation tests were applied. Specially, for the subject PF, the attendance to the continuous evaluation in

Table 1. Evaluation metrics and indexes achieved in the course OOP. Columns named *Yes* and *No* indicate whether the auto-evaluation tests were applied.

OOP	2014/15		2015/16	
	Yes	No	Yes	No
Enrolled	131	83	131	81
CE attendance	81	37	71	40
CE passing	25	9	40	17
CE failed & final exam	37	11	16	6
Only final exam	10	1	2	3
CE Attendance/Enrolled	**61.83%**	44.58%	**54.20%**	49.38%
CE Passing/Attendance	**30.86%**	24.32%	**56.34%**	42.50%
CE Passing/Enrolled	**19.08%**	10.84%	**30.53%**	20.99%
CE Failed†	33.93%	**60.71%**	48.39%	**73.91%**

Table 2. Evaluation metrics and indexes achieved in the course PF. Columns named *Yes* and *No* indicate whether the auto-evaluation tests were applied.

PF	2014/15		2015/16	
	Yes	No	Yes	No
Enrolled	112	66	123	73
CE attendance	54	16	95	30
CE passing	23	12	29	11
CE failed & final exam	22	2	28	8
Only final exam	7	5	0	9
CE Attendance/Enrolled	**48.21%**	24.24%	**77.24%**	41.10%
CE Passing/Attendance	42.59%	**75.00%**	30.53%	**36.67%**
CE Passing/Enrolled	**20.54%**	18.18%	**23.58%**	15.07%
CE Failed†	29.03%	**50.00%**	57.58%	**57.89%**

presence of auto-evaluation tests is close to the double of those without them (48.21% vs 24.24% and 77.24% vs 41.10%). Moreover, the ratio CE Passing/Enrolled is also higher when auto-evaluation tests were applied.

On the other hand, the ratio CE Passing/Attendance varies depending on the course. While such ratio was higher with auto-evaluation tests for OOP, it was lower for PF. This can be due to the specific characteristics of the language paradigm. Specifically, auto-evaluation questions in OOP could share more related concepts to the exam questions than PF. Note that the object oriented programming has more language overhead (classes, interfaces, attributes, methods, heritage, polymorphism, among others) than the structured paradigm (variables, functions, among others). As a consequence, OOP has less algorithmic load and the problems solved are simpler than those addressed in PF. Therefore, we conclude that the auto-evaluation questions in the course PF could be improved including more algorithmic load in order to increase the ratio CE Passing/Attendance.

It is desirable that the students who attend the continuous evaluation and fail it attend the final ordinary exam, as a sign of favorable learning during their continuous evaluation. In this sense, the ratio CE Failed† measures the drop-out rate after the continuous evaluation. As it can be seen in the results, the ratio CE Failed† was considerably lower when the auto-evaluation tests were applied (33.93% vs 60.71% and 48.39% vs 73.91% for OOP, 29.03% vs 50.00% for PF).

4 Conclusions

This paper evaluates the impact of using tests as part of the continuous evaluation in Programming courses. Students have been provided with auto-evaluation questions. Some of these tests have taken part of the continuous evaluation system itself. This methodology has been applied in programming courses of freshmen both in the Computer Engineering and Health Engineering degrees at University of Seville, Spain (Java and C language programming). Results obtained by the students on these courses in terms of different quality metrics have been compared to other similar subjects at Pablo de Olavide University of Seville, Spain, where no auto-evaluation tests have been provided to the students. It has been shown that the use of such tests enhance the students performance both in terms of attendance and passing exams.

Acknowledgments. The authors want to thank the financial support given by the Spanish Ministry of Economy and Competitivity project TIN2017-88209-C2-R. Also, this analysis has been conducted under the Innovation Teaching Project helps (2015 and 2016) by the University of Seville and Pablo de Olavide University.

References

1. Acemoglu, D.: Why do new technologies complement skills? Directed technical change and wage inequality. Q. J. Econ. **113**(4), 1055–1089 (1998)
2. Boulos, M.N.K., Maramba, I., Wheeler, S.: Wikis, blogs and podcasts: a new generation of web-based tools for virtual collaborative clinical practice and education. BMC Med. Educ. **6**(1), 41 (2006)
3. Bradford, P., Porciello, M., Balkon, N., Backus, D.: The blackboard learning system: the be all and end all in educational instruction? J. Educ. Technol. Syst. **35**(3), 301–314 (2007)
4. Bologna Declaration: The European higher education area. Joint Declaration of the European Ministers of Education, 19 (1999)
5. Delgado García, A.M., Oliver-Cuello, R.: La evaluación continua en un nuevo escenario docente. RUSC. Univ. Knowl. Soc. J. **3**(1), 1–13 (2006)
6. Deno, S.L., Marston, D., Mirkin, P.: Valid measurement procedures for continuous evaluation. Except. Child. **48**(4), 368–376 (1982)

7. Kellner, D.: New technologies/new literacies: reconstructing education for the new millennium. Teach. Educ. **11**(3), 245–265 (2000)
8. Nicol, D.: E-assessment by design: using multiple-choice tests to good effect. J. Further High. Educ. **31**(1), 53–64 (2007)
9. Prodromou, L.: The Backwash Effect: From Testing to Teaching. Oxford University Press, Oxford (1995)
10. Saeed, N., Yang, Y., Sinnappan, S.: Emerging web technologies in higher education: a case of incorporating blogs, podcasts and social bookmarks in a web programming course based on students' learning styles and technology preferences. Educ. Technol. Soc. **12**(4), 98–109 (2009)
11. Tozoglu, D., Tozoglu, M.D., Gurses, A., Dogar, C.: The students' perceptions: essay versus multiple-choice type exams. J. Baltic Sci. Educ. **6**, 52–59 (2004)
12. Zeidner, M.: Essay versus multiple-choice type classroom exams: the student's perspective. J. Educ. Res. **80**(6), 352–358 (1987)

Social Engagement Interaction Games Between Children with Autism and Humanoid Robot NAO

Chris Lytridis, Eleni Vrochidou(✉), Stamatis Chatzistamatis, and Vassilis Kaburlasos

Human-Machines Interaction (HUMAIN) Laboratory, Department of Computer and Informatics Engineering, Eastern Macedonia and Thrace Institute of Technology (EMaTTech), 65404 Kavala, Greece
{lytridic, evrochid, stami, vgkabs}@teiemt.gr

Abstract. Children with autism are characterized by impairments in communication, social interaction and information processing. This work presents a module to encourage children with autism to improve their social and communication skills, through a specially designed game-based approach. The humanoid robot NAO is utilized to autonomously engage with a child. The proposed module suggests a multiple role for the robot which can act as a teacher, as a toy and as a peer, through a successive set of joint activities. Overall observation encourages the utilization of NAO in the rehabilitation of children with autism.

Keywords: Autism · Human-robot interaction · Special education

1 Introduction

Educational robotics are used worldwide in education as a learning tools [1], but surprisingly rarely in special education. At the moment, the demand for special education remains unsatisfied due to the high cost involved. However, the benefits surpass all costs. This is why recently, Cyber-Physical Systems (CPSs) have been proposed in an educational application domain, particularly in special education [2]. More specifically, social robots that can interact with children, including NAO, Pepper, Jibo, Leka etc., have been suggested as an affordable and efficient solution, due to the dropping prices as well as their increased functionality [3, 4].

Humanoid robots such as NAO, have already demonstrated their effectiveness on treatment of children with Autism Spectrum Disorder (ASD) [4–6]. Conventional emotional learning for children with ASD often uses medical means, and behavioral analysis that could be costlier and less effective. Therapeutic intervention requires significant resources in terms of time as well as money from families. Moreover, results indicate that current robot technology is surprisingly close in achieving autonomous bonding and socialization with children and it would have great potential in special education as an assisting tool for teachers and therapists [7]. In particular, educational applications for robots are promising for students with disabilities in two main ways;

M. Graña et al. (Eds.): SOCO'18-CISIS'18-ICEUTE'18, AISC 771, pp. 562–570, 2019.
https://doi.org/10.1007/978-3-319-94120-2_55

The robots can motivate children to undertake a wide range of tasks that would otherwise deny due to their disability and can lead them to equal participation with peers in robot-based learning activities [8]. Experiments have confirmed that children with autism are more positive to treatment by social robots rather than therapists. The reason is that humans may charge the children emotionally, whereas robots do not. The eminent feature of robots is their high level of repeatability and flexibility, as well as working without getting tired or making complaints. Since robots do not humiliate or belittle people, it is expected that people with autism face less anxiety in interacting with them and are more willing to participate in the learning exercises [5]. Therefore, a robot that could help children with ASD to learn and acquire new developmental skills while having fun, would be appreciated by teachers and parents as well as enriching for the children.

According to the above, a newly designed game-based module for the rehabilitation of children with ASD is presented. The proposed interactive module exploits the benefits of the positive impact between the humanoid robot NAO and children in order to improve the therapeutic effectiveness. This paper is organized as follows. Section 2 reviews some of the basic concepts that are considered important for the understanding of this work, and suggested games from the bibliography for treating ASD children. The proposed gaming module is presented in Sect. 3. Section 4 summarizes the suggested experimental setup. Conclusions are underlined in Sect. 5.

2 Conceptual Background

Game design for children with disabilities is complex, multidimensional process. In order to direct the game designing process towards the rehabilitation of children with special needs, first thing is to understand the learning process for this special category of children and how it works for them. One way to understand the learning process for children and explore the various components of information processing that are impacted by a presence of a disability or exceptionality, is by using the information processing model, illustrated in Fig. 1. Information process explains how students interact with and respond to the world around them and describes the learning process [9]. First the children receive information from their senses input (vision, hearing etc.). Then, they process this information through memory classification and reasoning abilities. Finally, they respond to the input information through output (i.e. speaking, writing, acting). Students are aided in this processing of information by their executive function which is the ability to decide which information to attend to, how to interpret the information and which option to use in response. Information processing talks place within the emotional context that influences every aspect of the system.

Special education is required when a child is unable to process information effectively. The problem of the child may be either in the input of information, in the internal processing of that information or in the output of the information. The executive function is the decision-making aspect of the model that helps the child to attend to the input by choosing what thinking process should call upon and deciding how to react. All of this information processing is done within an emotional context which can help or scramble the other components of the model under conditions of stress, anxiety, or calm and confidence. Every exceptionality, refers to an impacted element of the

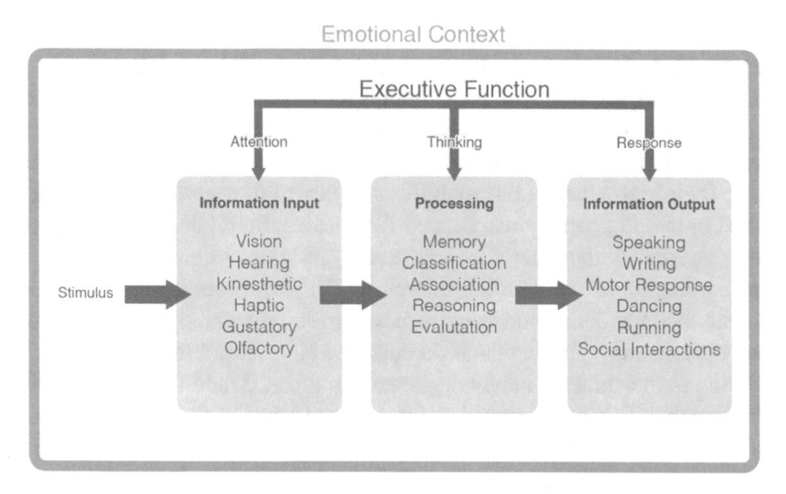

Fig. 1. The information processing model

information processing model. ASD is a developmental disorder characterized by impairments in communication, social interaction and imagination that can occur to different degrees and in a variety of forms. Children with ASD often have accompanying learning disabilities and experience inability to relate to other people, rare eye contact, difficulty in verbal and non-verbal communication and tendencies toward repetitive behavior patterns [24].

Based on the above, various gaming modules for treatment of children with autism have been proposed to the literature. All games are targeting in social and communication skills or at higher order neurocognitive functions. Table 1 summarizes some of the proposed games of the bibliography and their therapeutic objectives.

Table 1. Proposed games and their therapeutic objectives.

Interactive game	Therapeutic objective	Ref.
Imitation	Imitation skills, social-communication skills	[5, 10, 11, 21–23]
Rock-Paper-Scissors	Mindreading, theory of mind, cognitive flexibility	[5, 12, 13]
Turn taking games	Social interaction skills, joint attention	[5, 14, 15, 21]
Simon says	Response inhabitation, imitation	[5, 16, 17]
3 Card Monte	Attention, tracking	[5, 17]
Freeze dance	Response inhibition	[5, 17, 22]
Attention cueing through toys and music	Joint attention	[5, 18–20]
Follow me	Imitation, cognitive flexibility	[22]
Flying objects	Response inhabitation, attention	[5]

3 The Proposed Gaming Module

The proposed approach involves the humanoid robot NAO. NAO is selected due its humanoid, yet cartoon-like appearance, which is rather friendly and fascinating for children. The capability of NAO to focus its sight and comprehend the environment, the change of eye colors and the gestures that can imitate human gestures, make it appealing for children [25]. The main objective of using a robot in therapy sessions is to keep children engaged and focused since they tend to easily loose interest. NAO is programmed to teach and guide the children through the play scenarios and can easily attract the children's attention since it is animated and appears to act autonomously.

The proposed pilot gaming module consists of four sub-modules. All games and activities are selected on the basis of their potential to enhance functioning at a behavioural as well as at a neurocognitive level. The flow chart of the four modules of interaction is illustrated in Fig. 2. Each module aims at improving one or more aspects of the information processing model described in the previous section (Fig. 1).

Interaction begins with NAO introducing himself and waving his hand. Some simple questions are posed the child. The objective is to promote two-way communication and establish a friendly environment for the child during the interaction. After the introductory section, NAO describes the simple rules of the first game. The game is a variation of "Simon says". In this version, NAO instructs the child to touch various body parts and the child is encouraged to touch the robot on its pressure sensors and indicate them. NAO is rewarding the child for every correct touch by performing gestures and exclamations of approval. Studies have shown that positive feedback from the robot on the participants' performance is an effective way to encourage children with ASD to communicate more [20]. When the child touches the wrong body part, the robot encourages the child to try again by stating which body part was touched and which is the correct one. The child is allowed two attempts before the game proceeds with the next command. The objective of this game is to promote kinesthetic skills, touching, associating and identifying the body parts.

Game 2 is a memory game that involves a specially designed card that the NAO robot is revealing to the child. The card presents three illustrations of situations and objects (Fig. 3). The child is encouraged to observe the card for a few seconds, and then NAO hides it, and asks simple questions regarding the card illustrations. The child has two tries to give the correct answer. Depending on whether a correct answer for each question is given or not, the robot responds with sounds and gestures of either approval or consolation. This game is designed to stimulate the memory of children with autism, and improve their capability of identifying emotions, shapes and colours.

Finally, Game 3 is a dance and sing game, that involves imitation of movements suggested by NAO. The chosen song is a well-known Greek children song that involves gestures and is extensively utilized in music-kinetic activities with children. The child is encouraged by NAO to sing and dance together. Songs can help the children to stimulate their brain and identify emotions more effectively [22]. Moreover, this game is selected as the last game of the therapeutic session with the aim of finishing the session on a more pleasant atmosphere and therefore create an expectation for a forthcoming session.

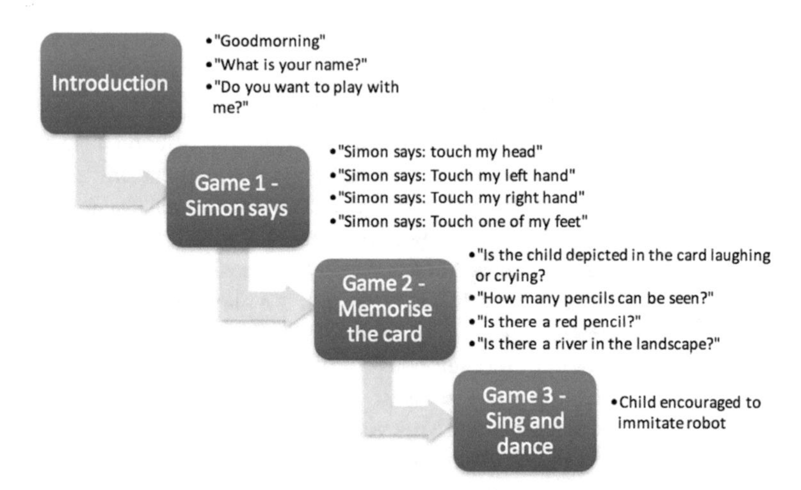

Fig. 2. The flow chart of the proposed game module

Fig. 3. Card example for the proposed Memory Cards Game (Game 2)

Table 2 summarizes the proposed game modules and the corresponding aspects of the Information Processing Model that each module is designed to improve.

4 Proposed Experimental Setup

A fixed distance between child and robot should be tested prior to the experiment to ensure that the child can hear the robot and vice versa. It is important to ensure that the children do not feel threatened or frightened. For this reason, the therapist should always be on the background to help and guide the child during the interaction, if needed. It must be noted that the robot in rehabilitation is not to replace the psychiatrist or any practitioner, but only to use the technology to assist them towards a more

Table 2. Description of the proposed modules.

	Method	Objective	Target skills		
			Attention	Processing	Response
Introduction	Introduction using voice recognition	Two-way communication helps children to interact and creates a friendly environment	Vision Hearing		Speaking Social interaction
Game 1	Variation of "Simon says"	To promote kinesthetic skills, touching, associating and identifying the body parts	Vision Hearing Kinesthetic Haptic	Memory Association	Social interaction Motor response
Game 2	Memory cards	To help identify emotions, shapes, colours, stimulate the memory	Vision Hearing	Memory Classification Evaluation Association Reasoning	Speaking Writing
Game 3	Dance and sing a well-known children song	To observe the imitation ability of children and finish the session pleasantly	Vision Hearing kinesthetic	Association	Dancing Running Motor response

effectively therapy. Additionally, a stable environment for the experiments must be provided, in order to avoid sudden changes for the children with autism.

The robot appears to act autonomously. The programmer is not in the room so as not to put additional pressure to the child. Autonomy of the robot makes the child see it as peer rather than as a machine. NAO is programmed in a way to display all the four gaming modules one after another in a total session time of less than ten minutes, together with the in-between intervals. Short sessions of less than ten minutes promise to keep the child focused in the whole duration of the session, since the shift of the games is quick and there is no time for the child to get bored or distracted.

An experimental study involving subjects, of children with autism and typically developing children, is underway. Figure 4 illustrates a typically developed 4 years old child interacting with NAO, testing the proposed game module. Current game modules presented in this study will be further improved based on inputs and feedback from therapists and medical experts.

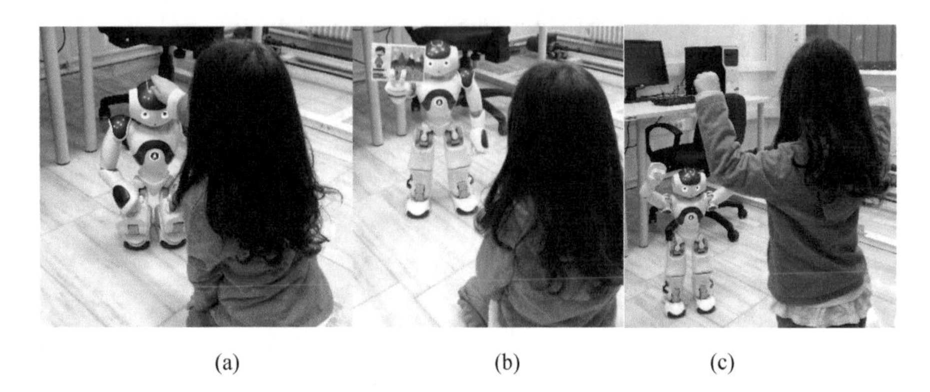

(a) (b) (c)

Fig. 4. Child-robot interaction while playing (a) game 1, (b) game 2 and (c) game 3

5 Conclusions

Research focusing on applying robots in autism treatment points out that robots increase enthusiasm, focus, and attention and cause novel social behaviors such as joint attention and automatic imitation. Research also indicates that children with autism work quite naturally with robotic technologies. This work presents a preliminary pilot study on understanding how humanoid robots can successfully improve social and communication skills among children with ASD through interaction games. The proposed interaction game consists of four sections that its one stimulates a different skill of the child. Future work will consider experimental results involving children with ASD so as to measure the social impact of the interaction on the child, in terms of joint attention, eye contact, facial expressions etc. and, thus, quantify the quality of interaction.

Acknowledgment. This project has received funding from the European Union's Horizon 2020 research and innovation programme under the Marie Skłodowska-Curie grant agreement No. 777720.

References

1. Miller, G., Church, R., Trexler, M.: Teaching Diverse Learners Using Robotics, pp. 165–192. Morgan Kaufmann, San Francisco (2000)
2. Dimitrova, M., Wagatsuma, H.: Designing humanoid robots with novel roles and social abilities. Lovotics **3**(112), 2 (2015). https://doi.org/10.4172/2090-9888.1000112
3. CybSPEED: Cyber-Physical Systems for PEdagogical Rehabilitation in Special EDucation. European Commission approved H2020-MSCARISE-2017 proposal no. 777720
4. Ueyama, Y.: A Bayesian model of the uncanny valley effect for explaining the effects of therapeutic robots in autism spectrum disorder. PloS One **10**(9) (2015). https://doi.org/10.1371/journal.pone.0138642

5. Amanatiadis, A., Kaburlasos, V.G., Dardani, Ch., Chatzichristofis, S.A.: Interactive social robots in special education. In: IEEE 7th International Conference on Consumer Electronics (ICCE), Berlin, pp. 210–213 (2017). https://doi.org/10.1109/icce-berlin.2017.8210609
6. Kaburlasos, V.G., Dardani, Ch., Dimitrova, M., Amanatiadis, A.: Multi-robot engagement in special education: a preliminary study in autism. In: 36th IEEE International Conference on Consumer Electronics (ICCE), Las Vegas, USA, 12–15 January 2018 (2018)
7. Tanaka, F., Cicourel, A., Movellan, J.R.: Socialization between toddlers and robots at an early childhood education center. Proc. Natl. Acad. Sci. **104**(46), 17954–17958 (2007). https://doi.org/10.1073/pnas.0707769104
8. Martyn Cooper, D.K., William Harwin, K.D.: Robots in the classroom-tools for accessible education. In: Assistive Technology on the Threshold of the New Millennium, vol. 6, p. 448 (1999)
9. Kirk, S., Gallagher, J.J., Coleman, M.R., Anastasiow, N.J.: Educating Exceptional Children. Cengage Learning, Belmont (2011)
10. Duquette, A., Michaud, F., Mercier, H.: Exploring the use of a mobile robot as an imitation agent with children with low-functioning autism. Auton. Robots **24**(2), 147–157 (2008). https://doi.org/10.1007/s10514-007-9056-5
11. Ingersoll, B.: The social role of imitation in autism: implications for the treatment of imitation deficits. Infants Young Child. **21**(2), 107–119 (2008). https://doi.org/10.1097/01.IYC.0000314482.24087.14
12. de Weerd, H., Verbrugge, R., Verheij, B.: Higher-order social cognition in rock-paper-scissors: a simulation study. In: 11th International Conference on Autonomous Agents and Multiagent Systems, vol. 3, pp. 1195–1196, June 2012
13. Perry, A., Stein, L., Bentin, S.: Motor and attentional mechanisms involved in social interaction—evidence from mu and alpha EEG suppression. Neuroimage **58**(3), 895–904 (2011). https://doi.org/10.1016/j.neuroimage.2011.06.060
14. Nadel, J.: Early imitation and the emergence of a sense of agency. In: 4th International Workshop on Epigenetic Robotics, vol. 117, pp. 15–16 (2004)
15. Robins, B., Dautenhahn, K., Te Boekhorst, R., Billard, A.: Robotic assistants in therapy and education of children with autism: can a small humanoid robot help encourage social interaction skills? Univ. Access Inf. Soc. **4**(2), 105–120 (2005). https://doi.org/10.1007/s10209-005-0116-3
16. Chao, C., Lee, J., Begum, M., Thomaz, A.L.: Simon plays Simon says: the timing of turn-taking in an imitation game. In: 2011 IEEE RO-MAN, pp. 235–240, July 2011. https://doi.org/10.1109/roman.2011.6005239
17. Halperin, J.M., Marks, D.J., Bedard, A.C.V., Chacko, A., Curchack, J.T., Yoon, C.A., Healey, D.M.: Training executive, attention, and motor skills: a proof-of-concept study in preschool children with ADHD. Attention Disord. **17**(8), 711–721 (2013). https://doi.org/10.1177/1087054711435681
18. Kajopoulos, J., Wong, A.H.Y., Yuen, A.W.C., Dung, T.A., Kee, T.Y., Wykowska, A.: Robot-assisted training of joint attention skills in children diagnosed with autism. In: International Conference on Social Robotics, pp. 296–305, October 2015. https://doi.org/10.1007/978-3-319-25554-5_30
19. Kryzak, L.A., Bauer, S., Jones, E.A., Sturmey, P.: Increasing responding to others' joint attention directives using circumscribed interests. J. Appl. Behav. Anal. **46**(3), 674–679 (2013). https://doi.org/10.1002/jaba.73
20. Vaiouli, P., Grimmet, K., Ruich, L.J.: "Bill is now singing": joint engagement and the emergence of social communication of three young children with autism. Autism **19**(1), 73–83 (2015). https://doi.org/10.1177/1362361313511709

21. Iacono, I., Lehmann, H., Marti, P., Robins, B., Dautenhahn, K.: Robots as social mediators for children with autism-a preliminary analysis comparing two different robotic platforms. In: IEEE International conference on Development and Learning (ICDL), Frankfurt, pp. 1–6, August 2011. https://doi.org/10.1109/devlrn.2011.6037322
22. Ferrari, E., Robins, B., Dautenhahn, K.: Therapeutic and educational objectives in robot assisted play for children with autism. In: 18th IEEE International Symposium on Robot and Human Interactive Communication, pp. 108–114. IEEE, September 2009. https://doi.org/10.1109/roman.2009.5326251
23. Tapus, A., Peca, A., Aly, A., Pop, C., Jisa, L., Pintea, S., David, D.O.: Children with autism social engagement in interaction with Nao, an imitative robot: a series of single case experiments. Interact. Stud. **13**(3), 315–347 (2012). https://doi.org/10.1075/is.13.3.01tap
24. Jordan, R.: Autistic Spectrum Disorders - An Introductory Handbook for Practitioners. David Fulton Publishers, London (2013)
25. Shamsuddin, S., Yussof, H., Ismail, L., Hanapiah, F.A., Mohamed, S., Piah, H.A., Zahari, N.I.: Initial response of autistic children in human-robot interaction therapy with humanoid robot NAO. In: 8th International Colloquium on Signal Processing and its Applications (CSPA), Melaka, pp. 188–193. IEEE, March 2012. https://doi.org/10.1109/cspa.2012.6194716

On Intelligent Systems for Storytelling

Leire Ozaeta and Manuel Graña(✉)

Department of CCIA, Computational Intelligence Group,
University of the Basque Country, Leioa, Spain
mauel.grana@ehu.es

Abstract. We propose storytelling as a central tool in social robotics and its use in educational environments, either for the conventional classroom or for children with special needs. Storytelling is not only a way to convey a message to the audience, but it is also an excellent guide for interaction. Stories provide the context and can be used also to model the child attention and current state of knowledge of the topic, i.e. to achieve user modelling.

1 Introduction

Storytelling has been a privileged way of transmitting ideas and knowledge since the very beginning of human civilizations. It has proven to be an engaging way of teaching complex concepts [16,19,20,29,33], as well as a way to support the development of a wide spectrum of cognitive functions and skills in children, and even a therapeutic tool for a variety of behavioural disorders [15,17,30]. Furthermore, the interactive creation of stories has been a topic of great interest in psychology and games [31], because it is an entertaining activity as well as therapeutic, since it helps the actors to express themselves in a safe imaginary environment where they can face risks and fail without real consequences.

At the same time, social robots have shown benefits as assistant and tutors for children in various studies [7,9,21,22,26]. The interaction with the artificial entities provide a number of benefits in many situations, from lack of emotional load while teaching children with Autism Spectrum Disorders (ASD) [1], to low infection risk while providing emotional support in hospital settings [14]. Robotic storytelling has been approached in the recent years from this perspective. For instance, using a nursing robot for storytelling in kindergarten [7] has shown cognitive benefits, even if storytelling was unidirectional, using recordings and with no hint of interaction.

Conversely, dialogue systems have shown a sudden spread in human-computer interface systems, as seen by the wide use of systems such as Siri, Echo, and others. This spread is linked to the proven effectiveness of recently developed data-driven machine learning methods for natural language processing [24], as well as the use of Reinforcement Learning approaches for the training of the systems [11,27,28], and the reduction of human interaction during said training using different mechanisms such as adversarial approaches [12].

M. Graña et al. (Eds.): SOCO'18-CISIS'18-ICEUTE'18, AISC 771, pp. 571–578, 2019.
https://doi.org/10.1007/978-3-319-94120-2_56

In this context the challenge of interactive storytelling automation is a very interesting one, not only because of its possible application, or the breath of opportunities for innovative research areas, but because of its particularities. Storytelling not only implies the narration of a specific plot but also the adaptation of said narrative to the interactuators that receive the story and a human-like transmission of it. This is our motivation to deepen into the narrative/plot generation models, the Dialogue Systems, and the social robots areas, as interactive storytelling unifies the need of plot structure, the ability to recognize and follow a dialogue, and the natural interaction with another entity.

Intended contribution. The aim of this article is to present the state-of-the-art models in the field of interactive story plot generation, dialogue systems and therapeutic robotics as a sound base towards the build of a model for interactive storytelling.

The paper contents are as follows: Sect. 2 describes the related work; Sect. 3 explains the unified concept of storytelling systems; Sect. 4 gives some conclusions and an idea of future work.

2 Related Work

There is a lack of references on automated/robotic storytelling systems in the scientific community. Seems that the entire field has been thoroughly neglected. However, storytelling systems are deeply linked to a number of different research areas such as Dialogue Systems, Narrative creation and Social robots. In the ensuing paragraphs, we will provide of a series of basic definitions and an overview of the works of each area most related to storytelling.

Dialogue Systems

Dialogue systems, as seen in Fig. 1, can be divided into two categories: goal-driven systems, such as task oriented personal assistants i.e. Siri, Cortana, Alexa, etc., and conversational goal-free systems. The former have some specific goal to be achieved through the dialogue interaction, while the latter have not a specific goal, so that iteration can evolve indefinitely [25]. In the last years conversational systems have integrated the generation power of the current deep learning techniques with great results [23]. Deep learning approaches leave behind hand-crafted features that represented the state and action space and require either a large annotated task-specific corpus or a large number of human subjects willing to interact with the yet unfinished system [25]. However, the neural networks and end-to-end architecture has specially beneficed the goal-free systems, while deep learning lack of assumptions over the domain or dialogue state structure has not provide significant improvements for goal-driven tasks oriented dialogue systems [2].

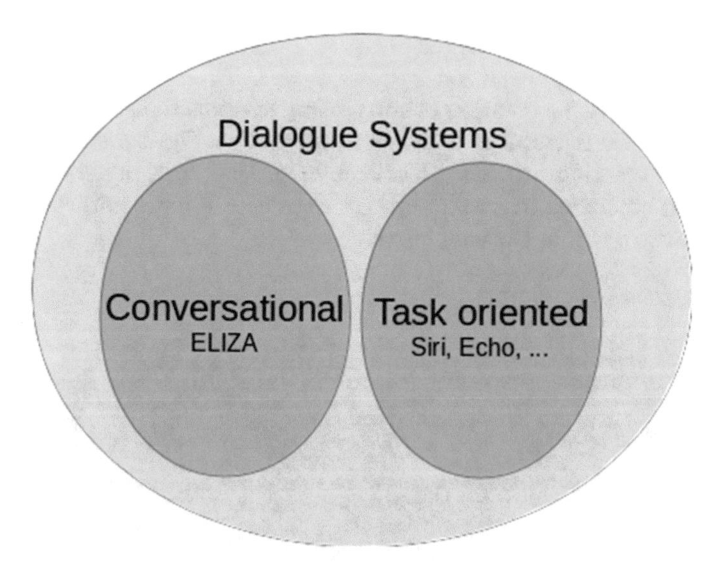

Fig. 1. Dialogue Systems domain

Narrative Creation

The problem of narrative plot creation is a very distinct topic, as it has a close relationship with natural language and dialogue but some closed rules that do not apply to normal speech. The plot is the driver of the most narrative forms in the Western culture [13] and is usually defined by initial situation, conflict and resolution. However, for the plot to be successful, and applauded as such by the audience, it must have coherence and character believability [18]. At this moment, it is not a widely researched topic, although there have been practical works as the one presented in [8], where a repository of existing stories are used to provide with the skeleton of new plots that match a given user query, or even theoretical approaches to the idea of end-to-end system capable of extracting narrative models from text and use those to generate new narratives [31].

Social/Therapeutic Robots

The interest on the use of robots in social and therapeutic areas has increased in the last years [3,4,6]. This is because they provide of a number of proven benefits, specially for children with special needs. They are more readily accepted than a human by ASD children [1] and enhance the learning time and quality of children with potential symptoms of a developmental disability [10]. These complex cognitive technologies are proving to be useful mediators for individuals that need an adaptation to the world [5].

3 Storytelling Systems

What we propose is the creation of storytelling systems that fall in the intersection between various research areas as illustrated in Fig. 2, i.e. the interaction and narrative creation provide a feedback loop, lead by a possible therapeutic objective and embodied in a robot system to provide a more natural experience, with the lowest possible learning curve.

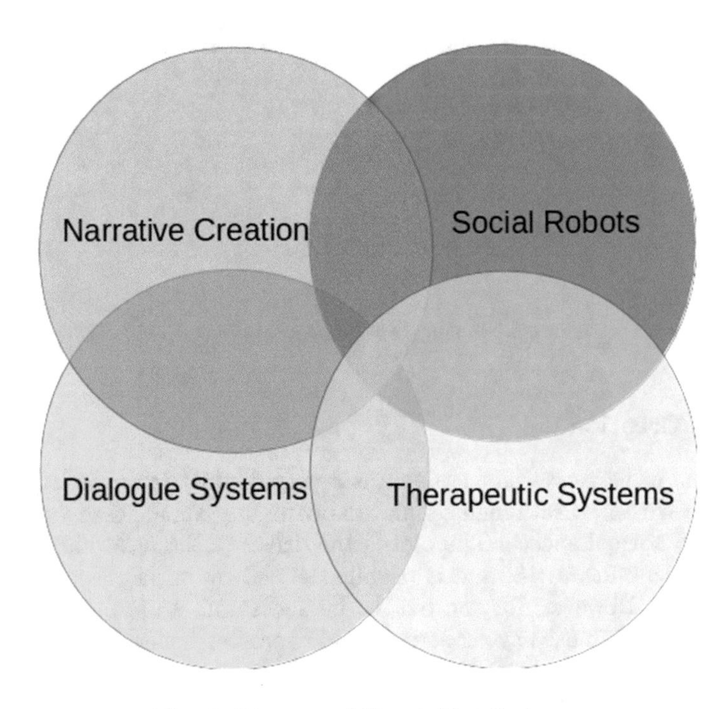

Fig. 2. Diagram of Storytelling Systems

In this context the idea is to use the generative power of the latest Deep Learning techniques, that have already provide good results in vocabulary and speech related tasks [23] for fuelling the creative aspects of the storytelling, i.e. the generation and adaptation of the plot. However, from the Dialogue Systems approach, this implies a more goal-oriented dialogue as an story provides an intrinsic planning problem. In fact, a story can defined by a sequential decision making process and thus, require a consistency in the dialogue over time, which the supervised learning framework does not account of [27].

Furthermore, interactive storytelling systems would be ideally embodied in a robot with a social objective, as this provides that the storytelling can take place in any familiar environment for the children, reducing possible rejection. Also those robots are suggested to provide with a grater enjoyment in the interaction with children [7]. This two are important factors when considering the use of this systems for more complex applications as therapy or special education.

The following are needed components of an interactive storytelling system, which can be embodied in a social anthropomorphic robot:

- Language generation from semantic representations, which can involve also speech synthesis for natural spoken interaction. Commercial robots, such as Nao, already have good speech synthesis resources.
- A semantic representation of the story plot, that may allow to generate alternative paths, and that can be generatively manipulated using some kind of "plot operators". Plot generation may be driven by some kind of popularity or audience engagement measure, so it becomes some kind of optimization problem that can be solved by stochastic search.
- A capability to asses the state of plot knowledge by the audience. An engaged audience may be able to show its engagement by answering simple questions about the plot. Thus, the robotic storyteller may check them and decide which is the best point for story restart, or it may generate alternative plot lines that provide the required knowledge. This is a process alike to student modelling in automated e-learning systems.
- Capability to sense audience engagement. For instance, obtaining the pose of the head it would be possible to decide if the audience is still attentive to the story, or they have lost interest. Audio cues may help additionally. Noisy ambiance is likely the indication of a lost audience. However, the robot must also be able to detect questioning relative to the story, which is a high indicator of attention and engagement.
- Capability to answer questions about the story, and to reschedule/reorganize the story as a consequence.

All these components require specific research efforts, some of them concurrent with research in other areas, such as task oriented dialogue systems, or some tutoring systems for people training.

4 Discussion and Conclusions

Interactive storytelling in a natural interaction is a problem yet to be approached by the scientific community. There are interesting uses of long short term memory (LSTM) architectures regarding dialogue and speech recognition and generation [32], however they lack the intention and message transmission capability of a linear story. Most of those works are focused in the generation or continuity of the dialogue without proper understanding of dialogue semantics or concrete goal.

On the other hand, however, plot generation has been approached with different techniques, but have yet to be combined with the latest generative techniques, such as Deep Learning. Furthermore those research works have been widely directed to video games or set in very computer dependent settings that limits the reach of the system and compromise the natural interaction.

At the same time the use of social robots have revealed itself as an interesting option to reach and connect with children, specially those with special needs. A good acceptance of this robot systems have been observed when taking the role of either peers or tutors, offering a new way for communication, and the possibility to develop new paradigms in various areas such as education, therapy, and entertainment.

In this situation we consider that these research lines can be combined for more flexible and robust systems that can provide a tool with wider applications, such as education, therapy, and/or communication support. Some of our very early work hinted in this direction with the dialogue and storytelling application embodied in a NAO that we presented in [14] with some good initial observations. However, there is a wide room for improvement that is going to be addressed in future work.

Acknowledgments. Leire Ozaeta has been supported by a Predoctoral grant from the Basque Government. This work has been partially supported by the EC through project CybSPEED funded by the MSCA-RISE grant agreement No 777720.

References

1. Amanatiadis, A., Kaburlasos, V.G., Dardani, C., Chatzichristofis, S.A.: Interactive social robots in special education. In: 2017 IEEE 7th International Conference on Consumer Electronics - Berlin (ICCE-Berlin), pp. 126–129, September 2017
2. Bordes, A., Boureau, Y.L., Weston, J.: Learning end-to-end goal-oriented dialog (2016). arXiv preprint. arXiv:1605.07683
3. Boyraz, P., Yiğit, C.B., Biçer, H.O.: Umay1: A modular humanoid platform for education and rehabilitation of children with autism spectrum disorders. In: 2013 9th Asian Control Conference (ASCC), pp. 1–6, June 2013
4. Dickstein-Fischer, L., Fischer, G.S.: Combining psychological and engineering approaches to utilizing social robots with children with autism. In: 2014 36th Annual International Conference of the IEEE Engineering in Medicine and Biology Society, pp. 792–795, August 2014
5. Dimitrova, M.: Gestalt processing in human-robot interaction: a novel account for autism research. BRAIN. Broad Research in Artificial Intelligence and Neuroscience **6**(1–2), 30–42 (2015)
6. Ferrari, E., Robins, B., Dautenhahn, K.: Therapeutic and educational objectives in robot assisted play for children with autism. In: RO-MAN 2009 - The 18th IEEE International Symposium on Robot and Human Interactive Communication, pp. 108–114, September 2009
7. Fridin, M.: Storytelling by a kindergarten social assistive robot: a tool for constructive learning in preschool education. Comput. Educ. **70**, 53–64 (2014)
8. Gervás, P., Díaz-Agudo, B., Peinado, F., Hervás, R.: Story plot generation based on CBR. Knowl.-Based Syst. **18**(4–5), 235–242 (2005)
9. Iacono, I., Lehmann, H., Marti, P., Robins, B., Dautenhahn, K.: Robots as social mediators for children with autism-a preliminary analysis comparing two different robotic platforms. In: 2011 IEEE international conference on Development and Learning (ICDL), vol. 2, pp. 1–6. IEEE (2011)
10. Jimenez, F., Yoshikawa, T., Furuhashi, T., Kanoh, M., Nakamura, T.: Collaborative learning between robots and children with potential symptoms of a developmental disability. In: 2017 IEEE Symposium Series on Computational Intelligence (SSCI), pp. 1–5, November 2017

11. Li, J., Monroe, W., Ritter, A., Galley, M., Gao, J., Jurafsky, D.: Deep reinforcement learning for dialogue generation (2016). arXiv preprint. arXiv:1606.01541
12. Li, J., Monroe, W., Shi, T., Ritter, A., Jurafsky, D. Adversarial learning for neural dialogue generation (2017). arXiv preprint. arXiv:1701.06547
13. Ogata, T.: Analyzing multiple narrative structures of kabuki based on the frameworks of narrative generation systems. In: Proceedings of 2017 International Conference on Artificial Life and Robotics, pp. 629–634 (2017)
14. Ozaeta, L., Graña, M., Dimitrova, M., Krastev, A.: Child oriented storytelling with nao robot in hospital environment: preliminary application results. Prob. Eng. Cybern. Robot. **69**, 21–29 (2018)
15. Parker, T.S., Wampler, K.S.: Changing emotion: the use of therapeutic storytelling. J. Marital Fam. Ther. **32**(2), 155–166 (2006)
16. Parsons, S., Guldberg, K., Porayska-Pomsta, K., Lee, R.: Digital stories as a method for evidence-based practice and knowledge co-creation in technology-enhanced learning for children with autism. Int. J. Res. Method Educ. **38**(3), 247–271 (2015)
17. Reichert, E.: Individual counseling for sexually abused children: a role for animals and storytelling. Child Adolesc. Soc. Work J. **15**(3), 177–185 (1998)
18. Riedl, M.O., Young, R.M.: An intent-driven planner for multi-agent story generation. In: Proceedings of the Third International Joint Conference on Autonomous Agents and Multiagent Systems, vol. 1, pp. 186–193. IEEE Computer Society (2004)
19. Robin, B.: The educational uses of digital storytelling. In: Society for Information Technology & Teacher Education International Conference, pp. 709–716. Association for the Advancement of Computing in Education (AACE) (2006)
20. Robin, B.R.: Digital storytelling: a powerful technology tool for the 21st century classroom. Theor. Pract. **47**(3), 220–228 (2008)
21. Robins, B., Dautenhahn, K., Dickerson, P.: From isolation to communication: a case study evaluation of robot assisted play for children with autism with a minimally expressive humanoid robot. In: Advances in Computer-Human Interactions, 2009. ACHI'09. Second International Conferences on, pp. 205–211. IEEE (2009)
22. Saerbeck, M., Schut, T., Bartneck, C., Janse, M.D.: Expressive robots in education: varying the degree of social supportive behavior of a robotic tutor. In: Proceedings of the SIGCHI Conference on Human Factors in Computing Systems, pp. 1613–1622. ACM (2010)
23. Sainath, T.N., Vinyals, O., Senior, A., Sak, H.: Convolutional, long short-term memory, fully connected deep neural networks. In: IEEE International Conference on Acoustics, Speech and Signal Processing (ICASSP), 2015, pp. 4580–4584. IEEE (2015)
24. Serban, I.V., Lowe, R., Henderson, P., Charlin, L., Pineau, J.: A survey of available corpora for building data-driven dialogue systems (2015). arXiv preprint. arXiv:1512.05742
25. Serban, I.V., Sordoni, A., Bengio, Y., Courville, A.C., Pineau, J.: Building end-to-end dialogue systems using generative hierarchical neural network models. In: AAAI, vol. 16, pp. 3776–3784 (2016)
26. Shamsuddin, S., Yussof, H., Ismail, L., Hanapiah, F.A., Mohamed, S., Piah, H.A., Zahari, N.I.: Initial response of autistic children in human-robot interaction therapy with humanoid robot nao. In: IEEE 8th International Colloquium on Signal Processing and its Applications (CSPA), 2012, pp. 188–193. IEEE (2012)

27. Strub, F., De Vries, H., Mary, J., Piot, B., Courville, B., Pietquin, O.: End-to-end optimization of goal-driven and visually grounded dialogue systems (2017). arXiv preprint arXiv:1703.05423
28. Su, P.H., Vandyke, D., Gasic, M., Mrksic, N., Wen, T.H., Young, S.: Reward shaping with recurrent neural networks for speeding up on-line policy learning in spoken dialogue systems (2015). arXiv preprint. arXiv:1508.03391
29. Tatli, Z., Turan-Güntepe, E., Ozkan, C.G., Kurt, Y.: The use of digital storytelling in nursing education, case of turkey: Web 2.0 practice. Eurasia J. Math. Sci. Technol. Educ. **13**(10), 6807–6822 (2017)
30. Tielman, M.L., Neerincx, M.A., Bidarra, R., Kybartas, B., Brinkman, W.P.: A therapy system for post-traumatic stress disorder using a virtual agent and virtual storytelling to reconstruct traumatic memories. J. Med. Syst. **41**(8), 125 (2017)
31. Valls-Vargas, J., Zhu, J., Ontañón, S.: From computational narrative analysis to generation: a preliminary review. In: Proceedings of the 12th International Conference on the Foundations of Digital Games, pp. 55. ACM (2017)
32. Wen, T.H., Gasic, M., Mrksic, N., Su, P.H., Vandyke, D., Young, S.: Semantically conditioned lstm-based natural language generation for spoken dialogue systems (2015). arXiv preprint arXiv:1508.01745
33. Yuksel, P., Robin, B., McNeil, S.: Educational uses of digital storytelling all around the world. In: Society for Information Technology & Teacher Education International Conference, pp. 1264–1271. Association for the Advancement of Computing in Education (AACE) (2011)

Proposal of Robot-Interaction Based Intervention for Joint-Attention Development

Itsaso Perez[1], Itziar Rekalde[1], Leire Ozaeta[2], and Manuel Graña[2(✉)]

[1] Didactics and School Organization, University of the Basque Country, Leioa, Spain
[2] Computational Intelligence Group, Department CCIA, University of the Basque Country, Leioa, Spain
manuel.grana@ehu.es

Abstract. The Autism Spectrum Disorder (ASD) is a condition that can be highly challenging when facing interaction with others. One of the pivotal social skills that is not fully developed in ASD individuals is the joint-attention, due to high engagement in their own thoughts and emotions. Joint-attention is a base for important social behaviours. At the same time, promising results have arisen from studies on the interaction of social robots with ASD children, arguably because its low emotional evocation. In this paper we discuss an intervention proposal for a robot-interaction aiming to develop join-attention in ASD children. This intervention is based in two activities, each one can be graduated at three different level each, with an increasing level of socialization, in order to help develop this skills in a natural way.

1 Introduction

The Autism Spectrum Disorder (ASD) is a neurobiological disorder that shows up in the kid's early development stages and lasts the whole life cycle. It's most characteristic and conspicuous signs tend to appear between 2 and 3 years of age. In some cases, it can be diagnosed as early as 18 months [3]. Early intervention is of critical importance to provide these children with tools that can ease their interaction with the world around them improving their autonomy and quality of life. Robots, and specially the NAO robot (Aldebaran, Paris, France), have proven to be of high acceptance in this group and to enhance their learning abiities. The hypothesis is that antropomorphic simple robots do not induce emotional stress to these students.

Intended Contribution. In this work we discuss a proposal for an intervention focused on the development of joint attention in ASD children using a NAO robot. ASD students are specially lacking join attention skills. Joint attention is one of the basic areas of development. Thus, we hypothesize that the intervention would be of benefit for people in different cognitive situations, specially for those with special education needs.

M. Graña et al. (Eds.): SOCO'18-CISIS'18-ICEUTE'18, AISC 771, pp. 579–585, 2019.
https://doi.org/10.1007/978-3-319-94120-2_57

This paper is organized as follows: in Sect. 2 we present a small introduction to the ASD. Section 3 provides some insights in the meaning and importance of joint attention. Section 4 presents some related works in robot therapy. Section 5 presents the proposed intervention, and Sect. 6 the linked methods. Finally, Sect. 7 provides some conclusions.

2 Autism Spectrum Disorder

There are different types of autism within the autistic spectrum disorder, the best known are: Asperger syndrome, Childhood disintegrative disorder (CDD), and Pervasive Development Disorder (PDD). Thus we use the term "spectrum" to refer to them all. All ASD variables have in common a range of conditions characterized by challenges with social skill, repetitive behaviours, speech and non-verbal communication.

The Table 1 shows the classification of disorder severity level according to the support level needed [3].

Table 1. The American Psychiatric Association's classification.

Heading level	Social communication	Restricted, repetitive behaviours
Level.1 Requiring support	- Without supports in place, deficits in social communication cause noticeable impairments - Difficulty initiating social interactions, and clear examples of atypical or unsuccessful response to social overtures of others - May appear to have decreased interest in social interactions	- Inflexibility of behaviour causes significant interference with functioning in one or more contexts - Difficulty switching between activities - Problems of organization and planning hamper independence
Level.2 Requiring substantial support	- Marked deficits in verbal and non-verbal social communication skills - Social impairments apparent even with supports in place Limited initiation of social interactions - Reduced or abnormal responses to social overtures from others	- Inflexibility of behaviour - Difficulty coping with change, or other restricted/repetitive behaviours appear frequently enough to be obvious to the casual observer and interfere with functioning in a variety of contexts - Distress and/or difficulty changing focus or action
Level.3 Requiring very substantial support	- Severe deficits in verbal and non-verbal social communication skills cause severe impairments in functioning - Very limited initiation of social interactions, and minimal response to social overtures from others	- Inflexibility of behaviour - Extreme difficulty coping with change or other restricted/repetitive behaviours markedly interfere with functioning in all spheres - Great distress/difficulty changing focus or action

The intervention proposal discussed here is addressed to kids located at the first and second levels. We consider they will get the biggest benefit due to the skills they already have and they are able to improve.

3 Joint Attention

People who grow up and progress in a biologically ordinary way, develop joint attention mostly through simulated play and affective interaction forms with their caregivers. This starts to happen between the ages of 9 and 18 months [2]. On the contrary, this phenomenon does not occur on people suffering from ASD. Unlike typically developing kids, children with ASD are often more engaged by their own thoughts and internal sensations than by other people.

Joint attention is a set of non-verbal behaviours including eye gaze, pointing, and showing, which are used to reference outside objects during a communicative exchange [8]. So we call joint attention to the ability of keeping focused on something, in synchronization with our social partners. This ability allows us to cooperate and socialise with others. Interacting with other kids during childhood, is fundamental to develop the social skills and language skills we need in our adulthood [7].

Furthermore, as seen in Fig. 1, joint attention is a pivotal skill, i.e. a skill that is central to other areas of psychological/personal/cognitive functioning. Developing pivotal skills produces generalized behavioural improvements on the kid [2,11].

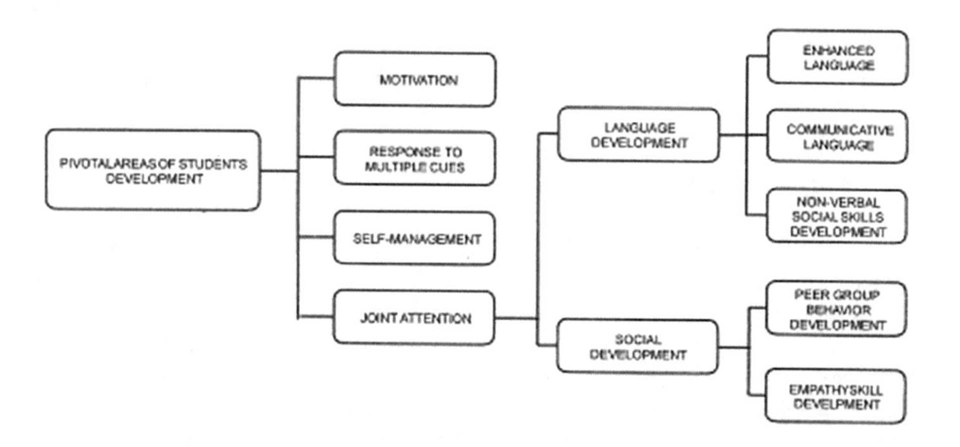

Fig. 1. Structure of pivotal areas.

4 Robotics as Therapy

Understanding and using social skills is the most challenging developmental area for a person with ASD [11]. They feel, live and express affectiveness and emotions in different and personal ways, narrowing the communication channels, which can cause frustration on the participants of those communicative actions.

During the last decades animal assistance for people with special needs has conspicuously grown. The great value of animal-assisted interventions is commonly accepted; the effect of a friendly animal influences positively on the perception of the human in its company and on the stimulation of social behavior. This is also called the "social catalyst effect" when it refers to the facilitation of interpersonal interactions [1]. Dealing with animals makes people with ASD trust others more easily, develops empathy and increases their self-esteem, as they feel capable of doing what is so difficult to them.

Animals that have proven to be more effective are dogs, but we can also found studies about horses, cats, etc, that prove their beneficial effect on people coping with ASD [10].

The biggest difference between using animals and using NAO as social catalyst lies in the control of the situation. We can train pets to be great therapists, but an animal will always have a number of unexpected behaviors facing new situations, that we cannot entirely control. However we can entirely control NAO, and adapt it to respond to situations that are emerging. That gives us flexibility, and a greater ability to reach all users, whatever their circumstances are.

Recent studies show that people with Autism Spectrum Disorder tend to feel comfortable interacting with the NAO robot, due to its low emotional stimuli [6,9]. NAO is a humanoid robot that shows emotions during interaction with people, but less than a human, what makes him accessible to somebody who does not control the wide range oh human emotional reactions.

Interacting with robots can be particularly empowering for children with Autism Spectrum Disorder, because it may overcome various barriers experienced in face-to-face interaction with humans, yet we have a human behind the robot [4,5].

5 Proposed Intervention

The intervention would be composed by the two activities presented below, each one with three levels on socialization, as seen in Fig. 2, starting in a solo scenario, that is thought to help the child get familiar both with the robot and the game, and working towards a peer-group interaction, where more than two children would need to cooperate.

The levelling up would be linked to the children evolution and observed reactions towards other peers, as well as discussed with the teacher and parents, with the possibility of going back if the interaction seems to stress or upset the participants.

Activity 1: Playing with Pictograms

In this scenario there would be different pictograms in the table. The kid knows the meanings of all of the pictograms. The teacher is always in the room, with

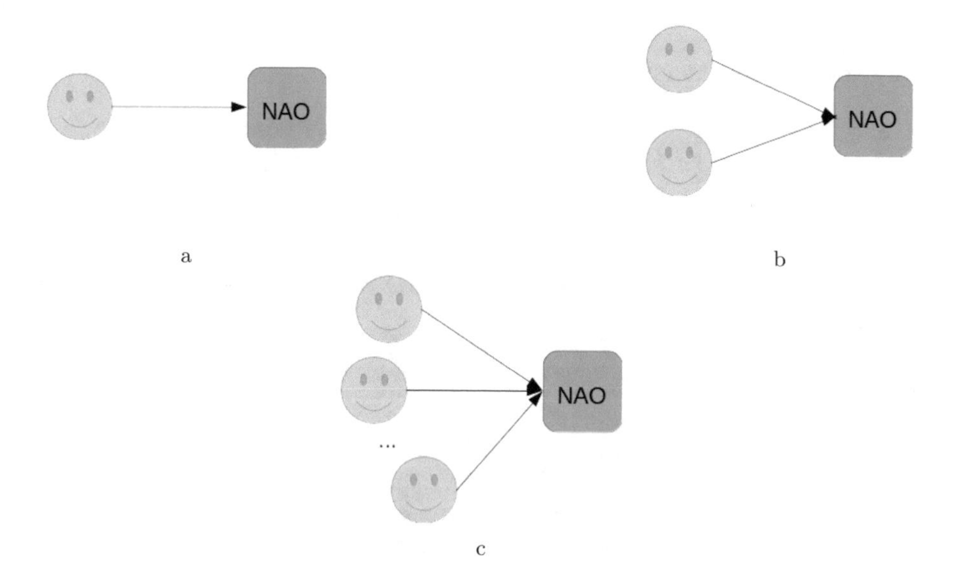

Fig. 2. Structure of the social interaction regarding level 1(a), 2(b), and 3(c) of our proposed intervention.

this familiar presence we want to make the student feel safe and calmed. The robot would be placed in the table in a table in front of the pictograms.

Level 1. We have one student, one NAO, and the teacher in the room. When a pictogram is shown to NAO, it makes the gesture of it, such as; brushing my teeth.

Level 2. We have two students, one NAO, and the teacher in the room. When the two of them show the same pictogram to NAO, it makes the gesture. Otherwise it does not.

Level 3. We have three or more students, one NAO, and the teacher in the room. When all of them show the same pictogram to NAO, it makes the gesture. Otherwise it does not.

Activity 2: Story Telling

In the scenario for this activity we have the students and the teacher in the room with the NAO. The NAO robot is in direct sight of the student, i.e: in the floor or in a table in front of the child. The robot is telling a story.At some point in the story, it will ask the student about the continuation of this story, and the student will have the opportunity to decide the following action from a limited number of options in a choose-your-story type of intereactive game, using the robot's sensors.

Level 1. We have one student and the teacher. The student listens to NAO's story, and decides the following situations.

Level 2. We have two students and the teacher. The students listen to NAO's story, and decides the following situations. NAO will only continue, when the two of them give the same answer.

Level 3. We have three students and the teacher. The students listen to NAO's story, and decide the following situations. NAO will only continue, when all of them give the same answer.

6 Method

We propose our intervention model in the context of running it once a month during a time of 9 months, i.e: the whole school year, in order to be able to observe the possible improvements. We propose to have a treated group and a control group, to be able to compare the natural evolution of the children with the one treated by the proposed robot interaction. For the quantification of the results we propose the measurement of the joint-attention development in this students throughout play time. Recordings on each session of the natural group of the participant during play time would provide of a way of observing behaviour evolution in both control and study group. In this recordings we would focus on behaviours that show joint-attention skill, such as; pointing, shifting gaze between people and objects, drawing another person's attention to something, orienting and attending to a social partner or sharing emotional states. This way there would be a way of observing difference, both between groups as well as between session numbers.

7 Conclusions

The ASD is a fairly common disorder that manifests in early childhood and last during the whole life circle. It is a disorder that can deeply restrict interaction with others an isolate the subject. Although it has been widely studied there is no consensus on preferred interventions to provide the subject of necessary tools to navigate society. However, robot interaction is proving to be of benefit for the education of this and other children with special needs. In this context we provide of an intervention proposal that focuses in the development of joint-attention, as it is a pivotal skill that improves overall the student's situation. Although the presented paper shows an early work we consider our intervention framework of great interest, at it seems transferable to other situation and disorders. In future work we will create the applications needed for the intervention as well as prepare the ethical framework for the first trials.

Acknowledgments. Leire Ozaeta has been supported by a Predoctoral grant from the Basque Government. This work has been partially supported by the EC through project CybSPEED funded by the MSCA-RISE grant agreement No. 777720.

References

1. Beetz, A., Uvnäs-Moberg, K., Julius, H., Kotrschal, K.: Psychosocial and psychophysiological effects of human-animal interactions: the possible role of oxytocin. Front. psychol. **3**, 234 (2012)
2. Charman, T.: Why is joint attention a pivotal skill in autism? Philos. Trans. R. Soc. B: Biol. Sci. **358**(1430), 315–324 (2003)
3. Fourth Edition. Diagnostic and statistical manual of mental disorders. Am Psychiatric Assoc (2013)
4. Hashim, R., Yussof, H.: Humanizing humanoids towards social inclusiveness for children with autism. Procedia Comput. Sci. **105**, 359–364 (2017)
5. Huijnen, C.A.G.J., Lexis, M.A.S., Jansens, R., de Witte, L.P.: How to implement robots in interventions for children with autism? a co-creation study involving people with autism, parents and professionals. J. Autism Dev. Disord. **47**(10), 3079–3096 (2017)
6. Kumazaki, H., Warren, Z., Muramatsu, T., Yoshikawa, Y., Matsumoto, Y., Miyao, M., Nakano, M., Mizushima, S., Wakita, Y., Ishiguro, H., et al.: A pilot study for robot appearance preferences among high-functioning individuals with autism spectrum disorder: implications for therapeutic use. PloS one **12**(10), e0186581 (2017)
7. Matsuda, S., Nunez, E., Hirokawa, M., Yamamoto, J., Suzuki, K.: Facilitating social play for children with PDDS: effects of paired robotic devices. Front. Psychol. **8**, 1029 (2017)
8. Nowell, S.W., Watson, L.R., Faldowski, R.A., Baranek, G.T.: An initial psychometric evaluation of the joint attention protocol. J. Autism Dev. Disord. 1–13 (2018)
9. Salleh, M.H.K., Miskam, M.A., Yussof, H., Omar, A.R.: HRI assessment of asknao intervention framework via typically developed child. Procedia Comput. Sci. **105**, 333–339 (2017)
10. Silva, K., Correia, R., Lima, M., Magalhães, A., de Sousa, L.: Can dogs prime autistic children for therapy? evidence from a single case study. J. Altern. Complement. Med. **17**(7), 655–659 (2011)
11. Weiss, M.J., Harris, S.L.: Teaching social skills to people with autism. Behav. Modif. **25**(5), 785–802 (2001)

Intra and Intergroup Cooperative Learning in Industrial Informatics Area

Jose Manuel Lopez-Guede[1,2(✉)], Jose Antonio Ramos-Hernanz[1], Estibaliz Apiñaniz[1], Amaia Mesanza[1], Ruperta Delgado[1], and Manuel Graña[2]

[1] Faculty of Engineering of Vitoria-Gasteiz, Basque Country University (UPV/EHU), C/Nieves Cano, 12, 01006 Vitoria-Gasteiz, Spain
jm.lopez@ehu.es

[2] Computational Intelligence Group, Basque Country University (UPV/EHU), Leioa, Spain

Abstract. This paper describes an educational experiment carried out by an Educational Innovation Project (EIP), developed in the field of Industrial Informatics during the biennium 2011/2013 at the Faculty of Engineering of Vitoria-Gasteiz (University of the Basque Country, UPV/EHU, Spain). In this paper the general situation regarding the European Higher Education Area (EHEA) at the start of that biennium, as well as the situation and specific problems that occurred in the field of Industrial Informatics at the Faculty are described. It was proposed to rectify the situation by the explicit formulation of ambitious objectives and the use of active learning methods, specifically by intragroup (between members of the same group) and intergroup (between members of different groups) cooperative learning. The paper includes the designing details of the proposed innovation carried out, the designed assessment, and the steps taken for the implementation in each one of the two years of the project. The results have been successful in the academic field since the specific and generic competences have been achieved and even from the point of view of the evaluation of teachers by students.

1 Introduction

This paper describes the work developed during an educational innovation experience carried out within the framework of a call for Educational Innovation Projects (EIP) of the Educational Advisory Service of the Office of the Vicerectorship for Degree and Innovation Studies of the University of the Basque Country (UPV/EHU) during the biennium its call 2011/2013. Such experience was carried out in the Faculty of Engineering of Vitoria-Gasteiz, of the University of the Basque Country (UPV/EHU).

The starting point of the initiative was the verification by one of the authors that the students of the Industrial Informatics subject presented serious deficiencies when it comes to programming tasks with computers. This subject was the core of the third year of the Bachelor's Degree in Industrial Technical Engineering, specializing in Industrial Electronics at the Faculty of Engineering of

M. Graña et al. (Eds.): SOCO'18-CISIS'18-ICEUTE'18, AISC 771, pp. 586–595, 2019.
https://doi.org/10.1007/978-3-319-94120-2_58

Vitoria-Gasteiz (UPV/EHU). That professor had spent the past seven years teaching the subject, and had found that year after year had to invest a relevant part of the time of the subject in reviewing and settling knowledge that in theory had to be entrenched.

That is, it was a circumstance that always occurred, and after careful analysis, it was concluded that the main causes were the following:

- In the curriculum of the aforementioned degree, the only contact that the students have had with questions related to the programming of computers is in the first course of the degree. Taking into account that the area of Industrial Informatics is dealt with in the final year, there is a relatively large temporary distance between when knowledge is acquired and when it is reused.
- The situation is aggravated because on average they usually use one or two years more than those reflected in the study plan to reach the last year, so that four or five calendar years may have passed between both moments.
- There is not a lot of great teaching time in the first course related to computer programming: only the subject of Fundamentals of Computing of 6 ECTS.
- It is typical that all courses have several students who have come from other engineering faculties, even from other universities, so it was usual to work with students with different bases in programming.

To rectify this situation, it was decided to use a methodology based on active and cooperative learning. Active Learning is a broad learning philosophy that groups a number of methods, all based on the responsibility and participation of students in their learning [1,2]. One of them, called cooperative learning, is a paradigm in which learning activities are planned looking for positive interdependence of students [3,4]. More specifically, an implementation was made based on intragroup and intergroup Cooperative Learning, thus giving rise to two distinct areas in which cooperative learning takes place (within a working group and between different working groups).

The remaining of the paper is structured as follows. After the introduction and motivation given in the first section, Sect. 2 poses the objectives of the project, Sect. 3 describes with detail the innovation proposal while Sect. 4 discusses the obtained results. Finally, the main conclusions and promising lines for future work are given in Sect. 5.

2 Objectives of the Innovation

In this section we introduce the objectives of the innovation with regard to the experience carried out, that can be summarized as follows:

- There were a series of inertias acquired by the teaching staff consisting in the conception of a teaching where the teachers themselves are the center of the process. One of the main objectives is to break that dynamic.
- As a result of these dynamics, others are generated in the students, consisting in becoming the students claimants of that type of teaching, where the student

plays a passive role. Therefore, another objective is to present to students another way of conceiving the process of learning and teaching, giving them a much more active role, so that the anomalous or strange is not using active learning methodologies.
- The inclusion of faculty from various departments is also an objective. In this way a break point is obtained in the bachelor's degree where the implantation is carried out, and at the same time the same can be obtained in other bachelor's degrees where the same professors of several departments participate.
- Another objective is to encourage self-learning on the students. The idea to transmit is that in real life the students themselves who will have to take the initiative and worry about self-training.
- The most obvious thing is to associate the idea of active learning methodologies with the new bachelor's degrees and the EHEA. However, the objective was to carry out the implementation in a study plan of Industrial Technical Engineering to be extinguished, given that it was considered opportune that the last Industrial Technical Engineering students knew the active learning methodologies.
- As a consequence, and from a more investigative point of view, the team of professors wanted to investigate an experience of implementation of a project of active methodologies with students that had never had experience with them, so also this aspect was considered as an objective.
- Finally, on the part of the teaching staff there was a desire to carry out an implementation of active learning methodologies in a real subject, beyond the training actions that tend to be on a theoretical level.

3 Innovation Proposal

In this section we describe the characteristics of the innovation proposal that was developed as a result of the EIP implemented in the area of Industrial Informatics in two different subjects, of which one of the members of the teaching team of the project is a teacher at the Faculty of Engineering of Vitoria-Gasteiz (UPV/EHU). This area is contained in the Bachelor's Degree of Industrial Electronics and Automation and in the Bachelor's Degree to be extinguished of Industrial Technical Engineering, specialty of Industrial Electronics.

3.1 Aspects to Address

When the innovation proposal was designed, it was taken into account that it should address the following aspects:

- Project-Based Learning: since small group projects should be the vehicle through which learning should take place.
- Communication, teamwork, entrepreneurship: since students should have to work proactively and with great entrepreneurship, since they should take the step of coming into contact with teachers that they do not know. Therefore,

they should cultivate the communication with other professors, but also with the other groups, since the different projects that will be executed will be interrelated and finally, they will have to fit into another larger one.
- Autonomous student learning: since a server would be configured with the electronic resources (IT) that each group should generate in their project, making them visible to the rest of the groups since all of them will have to be merged when they are finalized and validated.
- Curricular development: given that there should be a special emphasis on the use of tutoring, especially with teachers of the teaching team who are not the usual subjects on which the educational innovation project has been implemented.

3.2 Teaching Team

The teaching team involved was multidisciplinary insofar it belongs to different departments with different areas of technical expertise, all of them necessary for the correct achievement of the projects that should be developed:

- Dept. of Systems Engineering and Automation: it corresponds to the area of Industrial Computing
- Dept. of Computer Languages and Systems: deals with the area of programming and the use of computer tools.
- Dept. of Electronics and Telecommunications: dedicated to the area of digital and analog electronics
- Dept. of Business Organization: specialist in the area of organization of resources and people
- Dept. of Computer Science in Artificial Intelligence: for aspects of the area of artificial intelligence.

3.3 Tasks to Develop

For the design of the innovation proposal, it is assumed that the number of students in the subjects is relatively small (about 20 students) given the enrollments of the last years of these subjects. It is intended that the students have a double interaction: on the one hand, they have never had contact with the professors participating in the departments of Computer Languages and Systems, of Electronics and Telecommunications, of Organization of Companies and of Computer Sciences in Artificial Intelligence, that will require them to interact with unknown people in a technical field, just as it will happen to them in their future professional life. On the other hand, they will have to interact with other students, some of their own group and others from other groups. The basic mechanism for interaction with this faculty would be the tutorship, since these professors will not be present throughout the development of both theoretical and practical classes in their ordinary sessions. To specify the innovation proposal, we can say that the basic idea was to make a division into four groups of all students, once the existing enrollment in the subject is verified. Each one of the groups was commissioned to carry out a project with certain material that was supplied, with totally different statements (all framed in the area of Industrial Informatics). All of them received a basic explanation of the assignment received by all the groups, since finally there are specific dependencies between them and everything has to fit into a final project of more importance. This required them to work on coordination and communication both intragroup (within each group) and intergroup (between groups). A brief explanation of each of the works designed exclusively for each group, all of them focused on working with SR1 autonomous robots (of small size and low cost) is given to all groups. The works are the following:

- The first of the works was related to monitoring the status of the SR1 robot.
- The second work was responsible for designing a control algorithm in a generic way, which can be implemented in any device.
- The third work deals with the realization of an API (Application Program Interface) for the SR1 robot, so that the implementation of the generic control algorithm designed in the second job (or any other control scheme) is trivial.
- The fourth and last work was responsible for the implementation of a communications protocol for the interaction between a central control computer and the SR1 robot by means of a radio frequency modem.

3.4 Tasks Evaluation

Once the tasks design was done, the evaluation was planned to be carried out continuously in time, taking into account three factors:

- The effective interaction that has existed with the professors of the departments of Computer Languages and Systems, of Electronics and Telecommunications, of Organization of Companies and Computer Sciences and Artificial Intelligence (tutorial action)
- Interaction and results obtained at the intragroup level (among members of the same group)
- Interaction and results obtained at the intergroup level (among members of different groups)

3.5 Time Planning

With respect to the implementation and its timing, it was executed as planned. Given the dates of the call and the granting of the EIP by the Educational Advisory Service (Vicerectorship of Degree Studies and Innovation of the UPV/EHU), the initial idea of starting a first approach during the second semester was carried out during the 2011/2012 academic year in the Industrial Informatics course. After having carried out a reflexion and having analyzed the results during the first four months of the 2012/2013 academic year, another implementation was made again in the second semester of the 2012/2013 academic year in the Extended Industrial Information Technology course, thus covering not only a subject, but a branch of knowledge composed of two subjects that is present in two different degrees. Therefore, and being more precise, actually the field work has been carried out throughout the second semesters of the courses 2011/2012 and 2012/2013, although throughout the first semester of the academic year 2012/2013. There has also been work of analysis and synthesis of the experiences carried out in the second semester of the 2011/2012 academic year.

The basic field (classroom) work cycle carried out on both occasions is specified through the following steps:

1. Firstly, the new work methodology was explained to the students, and the work groups were formed within the Industrial Informatics course. This work was carried out by the teacher of the subject.
2. Secondly, an explanation of the four projects (all different) was carried out to facilitate the choice of the project to be developed by each group. After these explanations, the works to be carried out were distributed.
3. Next, and in a continuous way, a persistent monitoring of the work of the students was carried out, collaborative and interdependent within each group and among the groups, since finally all the deliverables should fit into a final product. In the development of this stage, undoubtedly the one that composes the core of the innovation project, counted with the work of the rest of the members of the teaching team participating in the EIP.
4. Next, unitary tests were carried out by the professor in charge of the subject about the technical validity of the solution proposed by each one of the groups on the task entrusted to them. This step was rather formal, since the persistent monitoring ensured in advance the solutions provided.
5. Once all the solutions were validated, two-hour exhibition sessions were held for each working group, in which the sources of information consulted were explained in detail, as well as the technical solution provided to their specific problem.
6. With knowledge about the different works done by all the students, a session was dedicated to explain by the teacher in charge of the subject how it should be the fit of each one of the projects of each group in another of much larger, so that the reasons for the specifications initially provided to each of the groups were clearer.
7. Finally, and also by the professor in charge of the subject, a qualification was established for each one of the works based on the continuous evaluation whose design has been shown previously. The score was modulated for each student thanks to a weighting arising from an intragroup evaluation made by all members of each group, in which each student pondered the work done by all the members of the group.

4 Results

In this section, different types of results achieved and quantified are discussed along the implementation of the teaching innovation experience carried out.

Subsect. 4.1 includes the strictly academic results and those related to transversal competences. Subsect. 4.2 shows the results of the surveys that the students filled out anonymously at the end of each four-month period.

4.1 Academic and Transversal Results

Undoubtedly the first and most important result or product of the innovations carried out is the learning in an autonomous way by the participating students, in relation to the specific competences of the subject as well as to the transversal competences that they have acquired, obviously in different gradations, such as autonomous work, the capacity for self-learning, group work, oral communication and written communication. There is absolute certainty about the fact that all the knowledge and skills have been obtained in an autonomous way by the students, because the teachers only gave instructions to drive the work in specific moments of blockage. Another important result that transcends the limits of this implementation, is the mobilization carried out in the teaching team that was created for the EIP that served as a vehicle for teaching innovation. From our point of view, several teachers have been able to visualize the learning process live through active methodologies. In our view, we have also learned by doing, since one thing is the training courses that can be attended, and another is the real implementations. It has been demonstrated that it was feasible and the positive results achieved, which will undoubtedly encourage future initiatives. Regarding the academic results achieved, in its strictest sense, they have been more than satisfactory in the two courses in which the EIP has been developed, given that for the first time the evaluation has been carried out taking into account the transversal competences. To help the students in this sense, rubrics have been used for the evaluation of written works, oral presentations and work group. The percentage of approved students was 100% in the two courses in which the implementation was made, because the tasks entrusted to each group were carried out successfully by passing unit tests and the results of all of them fitted perfectly into a project of greater scope that included the tasks developed by all the working groups.

4.2 Student Surveys

Another result, but in this case evaluating not explicitly the process or the methodology, but the teachers, are the scores obtained in the surveys that the students fill anonymously at the end of the subject, but before the period of exams.

Figure 1 shows the results obtained in the first cycle of application of the project, while in Fig. 2 those corresponding to the second application cycle. Although the general level of satisfaction remains high in both (4.2 out of 5), in each and every one of the sections of the survey is improved from the first to the second application, except for the first one (Student Self-Assessment), which in reality consists of a self-assessment by the students of the time and dedication given to the subject. That is to say, they became more self-critical with themselves, at the same time they value the action of the teaching staff as better.

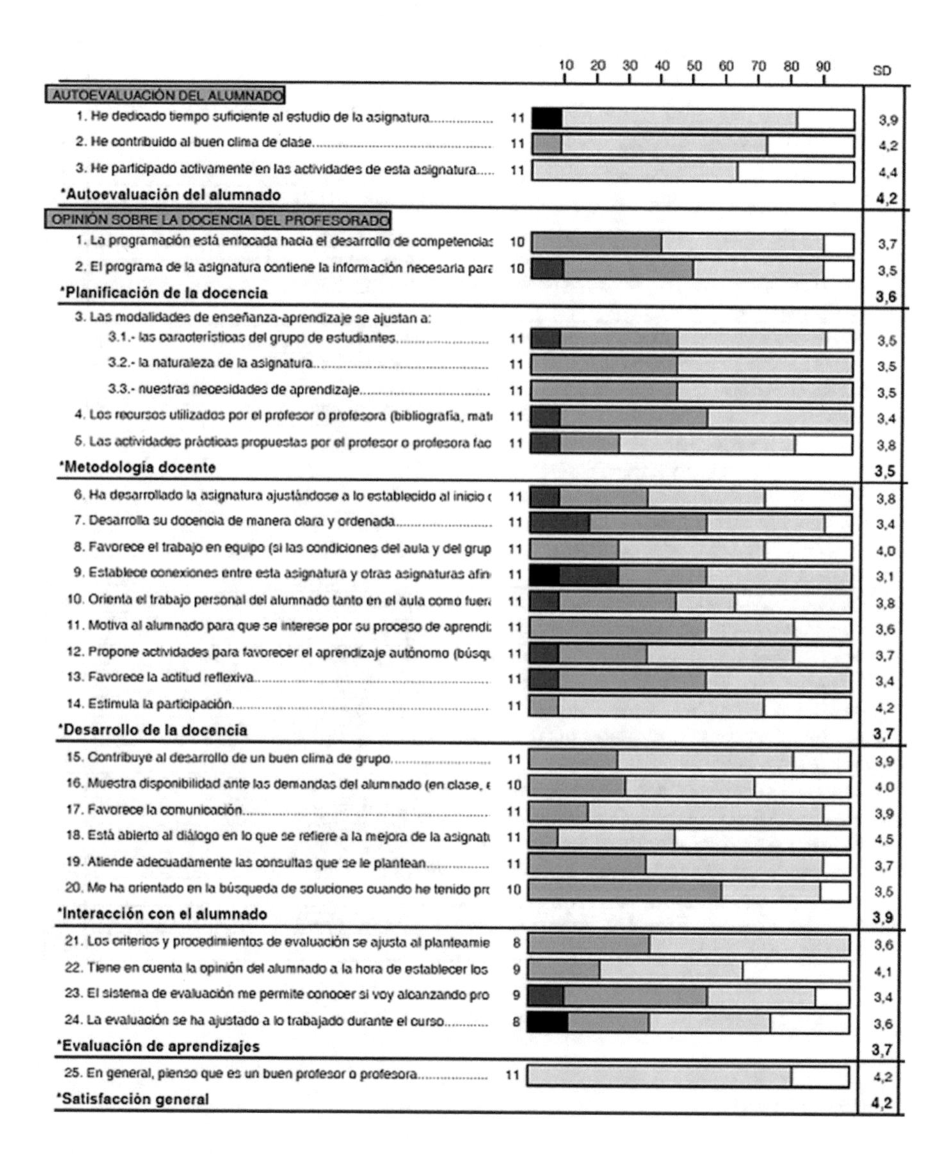

Fig. 1. Result of the students survey for the academic year 2011/2012

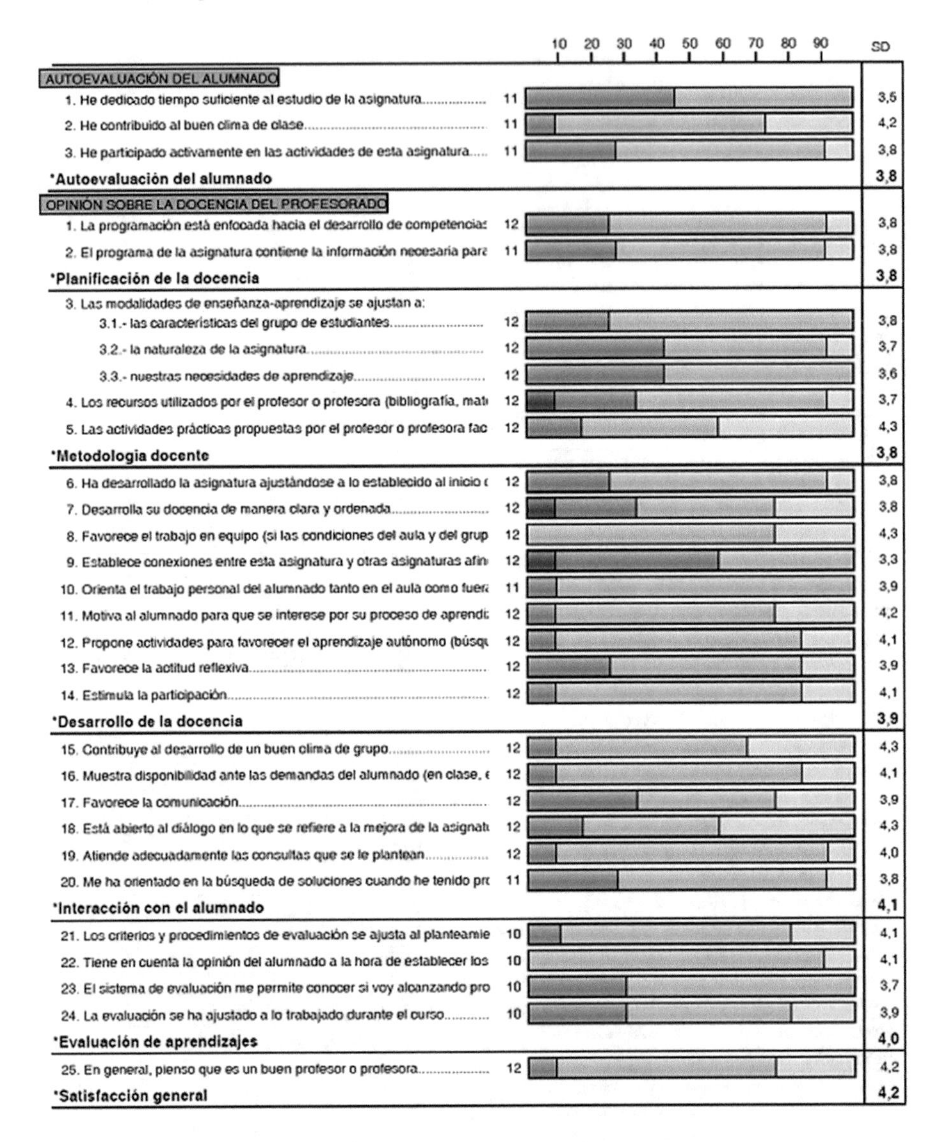

Fig. 2. Result of the students survey for the academic year 2012/2013

5 Conclusions

In this paper the teaching innovation work carried out within the framework of an educational innovation project (EIP) developed following a call for the Educational Advisory Service of the Office of the Vicerectorship for Studies of Degree and Innovation of the University of the Basque Country (UPV/EHU), in its 2011/2013 biennium. In the paper, the motivation of the aforementioned project has been gathered as well as the background of which it starts, the objectives of the innovation, the design and methodology followed. Since the project has already finished at the time of writing this paper, details have also been included about the process followed and the results obtained, understood in their broadest and most generic sense, showing that their implementation has been successful and advantageous for the students, for the teaching staff and for the institution. As a line of progress in the direction that has been drawn up with this project, a new project was proposed for the 2013/2015 call, where similar or even better results than those were achieved, mainly due to these reasons:

- There was a great experience accumulated during these two years.
- The teaching team was made up of more people: there were a greater number of teachers and even two master students.
- There were really valuable and reference members from the point of view of teaching methodologies.
- There was more training behind the back of the teaching team, with participation in the ERAGIN course (Course for Teacher Training of University Teachers, of the University of the Basque Country, UPV/EHU) and in other training activities.

Acknowledgments. The research was supported by the Computational Intelligence Group of the Basque Country University (UPV/EHU) through Grant IT874-13 of Research Groups Call 2013–2018 (Basque Country Government).

References

1. Bonwell, C., Eison, J.: Active learning: creating excitement in the classroom aehe-eric higher education report no. 1 (1991)
2. Felder, R.M., Brent, R.: Active learning: an introduction. ASQ High. Educ. Brief **2**(4), 1–5 (2009)
3. Felder, R.M., Rebecca, B.: Cooperative learning in technical courses: procedures, pitfalls, and payoffs (1994)
4. Felder, R.M., Brent, R.: Effective strategies for cooperative learning. J. Coop. Collab. College Teach. **10**(2), 69–75 (2001)

Dual Learning in the Bachelor's Degree in Automotive Engineering at the Faculty of Engineering of Vitoria-Gasteiz: Quality Requirements

Amaia Mesanza-Moraza(✉), Inmaculada Tazo-Herran, Jose Antonio Ramos-Hernanz, Ruperta Delgado, Javier Sancho-Saiz, Jose Manuel Lopez-Guede, and Estibaliz Apiñaniz

Faculty of Engineering Vitoria-Gasteiz, Basque Country University (UPV/EHU), C/Nieves Cano 12, 01006 Vitoria-Gasteiz, Spain
amaia.mesanza@ehu.es

Abstract. Since 2017–2018 academic year, the Bachelor's Degree in Auto-motive Engineering has been implemented at the Faculty of Engineering of Vitoria-Gasteiz (Basque Country University, UPV/EHU) in dual format, based on the complementarity of the University and company learning environments. With this type of education the aim is to achieve greater motivation of the students and facilitate the work insertion by having greater contact with companies. In this way, companies are fully integrated with the agents who are responsible for the education of university students. The teaching institutions need to accredit Bachelor's degrees or itineraries that are taught using this training format (dual training). In order to get this purpose, the university institutions adapt their studies in dual format to the quality requirements proposed by the state quality agencies of the university systems, based on objective evaluation criteria.

Keywords: Quality · Internship · Company · Dual learning · Accreditation

1 Introduction

During the last years companies are overtaking great changes, in addition, the market is more and more globalized. Therefore, the collaboration between universities and companies is becoming more and more important [1]. It is also very relevant the relationship between students and companies, the earlier the contact the students have with the company the more profitable it will be both for the student and the company. In this way, students will get progressively professional capacities. Dual learning makes the students' adaptation to the companies' working way shorter. Dual training is a teaching modality in which the students alternate between the university and companies, getting a smooth transition between the educative system and the working world. This modality makes the students to adapt their mentality since they start very early to work in companies, not being more time just students since they have both

M. Graña et al. (Eds.): SOCO'18-CISIS'18-ICEUTE'18, AISC 771, pp. 596–602, 2019.
https://doi.org/10.1007/978-3-319-94120-2_59

academic and work responsibilities, they getting both theoretical knowledge and solid working experience [2].

Nowadays, companies change their role from complementary actors to ones actors in the formation of their future workers. In the Autonomous Community of the Basque Country, the president Iñigo Urkullu and the Education counselor Cristina Uriarte, boosted by a document for the Dual Training and created accreditations for Bachelor's Degrees and Master Degrees was created.

In order to obtain the accreditation, universities have to adequate their studies to the quality requirements of Quality Agencies. The Agency for Quality of the Basque University System (UNIBASQ) has a Protocol so that universities can obtain the Accreditation for the Dual Training [3], subject to the following normative:

- Royal Decree 1393/2007, October 29th, offers the possibility to introduce internships with the aim of increasing the working options of the graduates.
- The Statute of the University Student, Royal Decree 1791/2010, December 30th, in the 8th article informs that all Bachelor's Degree students have the right to do internships.
- Royal Decree 592/2014, July 11th, regulates the university students' intern-ships.
- Decree 11/2009, January 20th, regulates the implementation or suppression of university studies for obtaining Bachelor's Degree, Master Degree or Doctorates.

In this work we will first analyze what Dual Learning is for Engineering Degrees and its advantages. Then, we will show the criteria that the Faculty of Engineering Vitoria-Gasteiz presented to UNIBASQ in order to obtain Accreditation for Dual Formation in the Bachelor's Degree in Automotive Engineering. To conclude we present the main conclusions of our work.

2 Dual Training in Bachelor's Degrees in Engineering

Dual training is said to be a gathering of different learning and teaching activities that are implemented to improve the students work qualification. This kind of formation combines the teaching and learning formation both at the university and at the company in order to assure the teaching and learning processes [4].

This type of formation is widely used in Germany or France, but it is not very usual in Spain [5]. In the Autonomous Community of the Basque Country, the Machine Tool Institute-IMH (in Elgoibar) works with this type of formation since 2011, it is also a Faculty of the University of the Basque Country and they teach Bachelor's Degree in Process and Product Innovation. The formation is given by alternating the university and the company work, where the student makes the internship. This type of formation has three main agents [6, 7]:

- The university: It is the connection point between the companies and the students. It is also in charge of looking for the companies so that the student can complete their formation, and it has to assess that the student gains the abilities of the degree.
- The company: It is in charge of teaching the student the real work and the abilities it requires.

- The student: who will gain abilities in both university and company.

After a deep analysis, it is possible to realize that there are a number of advantages of the Dual Training for each agent, more specifically they are the following [6, 7]:

- Advantages for the university:
  - The lecturer that teaches in this type of formation learns a lot due to the continuous contact with the production system.
  - It makes it possible to collaborate in different ways, such as organizing talks of the companies at the university.
- Advances for the company:
  - It will be easier for the company to find a new worker, because the students will have more experience when they finish their studies.
  - The companies will be able to do a better planning when getting new engineers.
- Advantages for the students:
  - The student will have real contact with the working environment; getting working experience. The students will obtain professional abilities and will get a balanced formation as far as theory and practice are concerned. In addition, the student will get money for his/her work.
  - Once the Dual Training is finished, the graduate will have more experience and knowledge as far as the company world is concerned and it will be easier for his/her to get a job.

3 Evaluation Criteria for the Quality Accreditation of the Dual Formation at Universities

The accreditation for the Dual Formation at the university can be obtained for a complete degree or for a dual itinerary (when only some of the students will do the internship at the company). The accreditation agency asks for different characteristics such as:

- Description of the Bachelor's Degree
- Justification of the Study Plan for obtaining the official degree
- Abilities that the graduates must acquire
- Access and admission processes of the students
- Planning of the studies
- Academic and material tools and services

3.1 Description of the Bachelor's Degree

The Bachelor's Degree in Automotive Engineering is a degree that is taught face-to-face and in dual format. The Faculty of Engineering of Vitoria-Gasteiz will guarantee that all students will be able to complete their dual training, having the sufficient number of entities that participate in the shared training. To make this possible, this Bachelor's degree has only 40 seats per year in order to assure that we will have enough companies for all the students. This year has been the first one of this

Bachelor's degree and there have been 40 enrolled students, i.e., it has been 100% covered, showing that there is a big demand by the society for this engineering. Besides, the mark to access to them has been very high, more specifically 9.84.

The syllabus of the Bachelor's Degree in automotive engineering has 240 ECTS: 42 ECTS are completed in companies, in addition, the Final Year Project has 12 ECTS and it is also performed in the same company. That is to say, 54 ECTS are performed in a company (22.5% of the full formation).

In addition, companies will prepare talks for the first and second year of the degree at the university in different subjects. These talks will have 6 ECTS or 60 teaching hours.

3.2 Justification of the Syllabus for Obtaining the Official Degree

Thanks to the Dual Formation students will get different abilities such as initiative, creativity, critical thinking and adaptation to new situations. Graduates will be able to work in multidisciplinary environments and work in different projects linked to Automotive Engineering. One of the aims is to make students feel better as far as the formation and working insertion are concerned.

In order to guarantee the coordination and integration of the activities performed at companies and the ones at the university, a commission will be created with workers from the company and from the university. At the university, this commission will be composed by the Vice-dean of Relations with Companies, the Coordinator of the Bachelor's Degree in Automotive Engineering and a professor who will teach the subject of Project Management. From the company, two supervisors of that participate in the training process of the students.

3.3 Abilities the Graduates Must Acquire

Students will gain the abilities of the degree both in the classroom and in the company. During the class hours both the general abilities and the transversal abilities will be developed by means of active formation such as Project Based Learning, Flipped Classroom, Cooperative Work, etc.

The Bachelor's Degree in Automotive Engineering is divided into the following modules:

- Basic Formation (60 ECTS): these subjects will give a general formation.
- Basic Engineering (60 ECTS): these subjects will give basic knowledge and abilities in industrial engineering.
- Automotive Technologies (42 ECTS): compulsory subjects in automotive technology related to what automotive companies ask the students to know.
- Vehicle Fabrication (24 ECTS): the aim of this module is to teach students enough knowledge to work in an automotive company.
- Elective subjects (18 ECTS): subjects linked to specialized technologies.
- Final Year Project (12 ECTS): this document will show what the student has learn during his 4th year studies.

The basic formation abilities are obtained by means of the different subjects during the first two years. Then, during the last two years by means of dual formation the students will get a more personal and individualized formation.

3.4 Access and Admission of the Students

In the case of degrees called dual training degree, all students enrolled in the degree have right to spend the credits determined by the syllabus in collaboration with a company, so, a procedure that guarantees this will have to be established.

The process of student distribution among the different companies with which there is an alternating training agreement will be shared between the students, the company and the Faculty, to guarantee that all the people enrolled can reach the competences of the syllabus. The main steps are the following:

- The Faculty will make a list with the companies that admit students for each academic year.
- Students can choose 5 different companies in order of preference and they will have to write a brief report that informs about the reasons for making those elections.
- The Faculty will make a preliminary selection taking into account the curriculum of the students.
- Companies will choose the candidate, taking into account a personal interview, his/her curriculum and the abilities that the companies think are necessary de-pending the area that they are working on.

The link between students and companies will be a partial time contract with formation objectives. This contract was created in Law 11/2013, July the 26th- In addition, this contract assures that all the students are paid and that all the students have an insurance to work at the company.

3.5 Planning of the Studies

In order to guaranty the students' formation, students will have to attend lessons during the first two years at the Faculty of Engineering of Vitoria-Gasteiz. Even though from the very first year the companies will collaborate teaching in some of the subjects. The more outstanding periods are the following:

- During the 3rd academic year the formation at the university will alternate with the formation in the company in order to get general abilities and transversal abilities. The first 8 weeks of the 3rd year will take place at the university. During the rest of the weeks both modalities will alternate.
- During the 4th academic year, both classroom formation and company formation will alternate. From the very first week of the academic year students will have to spend 3 days in the company and two at the university. This will last until May, and then the students will start to write their final Year Project.
- During the first period that the student has to spend in the company the university will assign a supervisor, who will develop a Professional Development Plan with the abilities the students has to obtain during that period. As mentioned there are

three periods in which the students have internship. After every period the student has to write a document with the main characteristics of the company, organization and services, together with the work performed by the student. The final mark for each period will be the evaluation for the work performed in the company (the supervisor of the company) and the document (university supervisor).
- The Final Year Project will be performed in the same company where the student has done the internship. The supervisors of the project will be the supervisor in the company and the supervisor at the Faculty.

3.6 Academics

The academic teaching staff that will supervise the training in the company must have an engineering profile training, in addition to a recognized professional experience or perform research in technologies in the automotive sectors, have extensive experience as an internship supervisor or participate in industrial doctorates programs related to the automotive sector, more specifically to vehicle or manufacturing engineering.

The supervisors at the faculty will always be engineers and they will have professional experience or will have performed research in the subject they have to supervise.

The supervisor in the company must have a technical profile and must have an important job in the area in which the student will have the formation.

At the beginning of each academic year the Faculty of Engineering Vitoria-Gasteiz will offer a formation course for Supervisors (both for company supervisors and faculty supervisors). In this course, the main goals of the Dual Formation will be explained and also the roll of the supervisors in the teaching/learning process.

3.7 Material Tools and Services

In order to obtain the Dual Formation Accreditation, the faculty will have to assure that it has enough companies so as to offer Dual Formation to all students that have registered. The faculty will have to show that these companies are prestigious and solvent.

Due to the novelty of the learning methodology in collaboration with the companies, within the Quality Internal Guarantee System of the Faculty of Engineering Vitoria-Gasteiz, the procedure: "2.2. PES-D07 Dual Formation" will be available, included within the subprocess "2.2 of Higher Education Development", in the process "2. Higher Education Process of the Quality Assurance Management System".

4 Conclusions

In the near future the universities will have to take into account the needs and demands of the society and companies. The work precariousness, the unemployment of the young people and the transformation of the companies to the industry 4.0 will induce changes in the requirements of the university students.

This is the reason why the dual teaching at the university will be an alternative that can help to make these changes, it will make the insertion to work market easier and it will help the student to develop capacities that companies require. In this way, the students social integration will be better and also the economy efficiency will in-crease. That is to say, the university studies will be adapted to the new social and economy realities.

Taking the latter into account during 2017–2018 academic year we started a Bachelor's Degree in Automotive Engineering. This Degree has been created together with companies such as Michelin, Mercedes-Benz, RPK or Antolín Group. It is a dual teaching Degree in which 22.5% of the ECTS of the Degree are obtained by doing compulsory internship in companies, this ECTS are divided in the three different periods.

In November 2017, UNIBASQ evaluation agency published the "Accreditation protocol for dual training for university degrees". In this work we have shown how the Faculty of Engineering Vitoria-Gasteiz has taken the previous Protocol into account in order to design its new degree in Automotive Engineering.

References

1. Vilalta, J.M.: ¿Es posible (y deseable) la formación universitaria dual? Blog de Studia XXI (2017)
2. ACUP: Promoción y desarrollo de la formación dual en el sistema universitario catalán. Asociación Catalana de Universidades Públicas Universitat de Lleida (2015)
3. Unibasq: Protocolo para la obtención del reconocimiento de formación dual para títulos universitarios oficiales de grado y máster (2017)
4. Fundación Bertelsmann: Guía básica sobre la formación profesional dual (2014)
5. Rego Agraso, L., Barreira Cerqueiras, E.M., Rial Sánchez, A.F.: Formación profesional dual: comparativa entre el sistema alemán y el incipiente modelo español. Revista Española de Educación Comparada **25**, 149–166 (2015)
6. Dual Training: Guía metodológica alumnos formación profesional dual (2015)
7. Durán López, P., Santos Primo, J.R., Gil Pérez, R.: Guía de la formación dual. Cámaras de Comercio, Ministerio de Educación, Cultura y Deporte y Fondo Social Europeo (2017)

Educational Project for the Inclusion of the Scientific Culture in the Bachelor's Degrees

Jose Manuel Lopez-Guede[1,5(✉)], Unai Fernandez-Gamiz[2], Ana Boyano[3], Ekaitz Zulueta[1], and Inmaculada Tazo[4]

[1] Department of Automatic Control and System Engineering, Faculty of Engineering of Vitoria-Gasteiz, University of the Basque Country (UPV/EHU), C/Nieves Cano 12, 01006 Vitoria-Gasteiz, Spain
jm.lopez@ehu.es

[2] Department of Nuclear Engineering and Fluid Mechanics, Faculty of Engineering of Vitoria-Gasteiz, University of the Basque Country (UPV/EHU), C/Nieves Cano 12, 01006 Vitoria-Gasteiz, Spain

[3] Department of Mechanical Engineering, Faculty of Engineering of Vitoria-Gasteiz, University of the Basque Country (UPV/EHU), C/Nieves Cano 12, 01006 Vitoria-Gasteiz, Spain

[4] Department of Thermal Engineering, Faculty of Engineering of Vitoria-Gasteiz, University of the Basque Country (UPV/EHU), C/Nieves Cano 12, 01006 Vitoria-Gasteiz, Spain

[5] Computational Intelligence Group, Basque Country University (UPV/EHU), Donostia-San Sebastian, Spain

Abstract. In this paper authors introduce an ongoing Educational Innovation Project (EIP) that is being carried out at the Faculty of Engineering of Vitoria-Gasteiz (Basque Country University, UPV/EHU, Spain). Since it is interesting from the point of view of universities, research institutes and even traditional companies, the aim of the project is to introduce the scientific culture in the Bachelor's Degrees of the Faculty. The project has been granted in the call for Educational Innovation Projects 2018–2019 of the Educational Advisory Service, Vicerectorship for Teaching Quality and Innovation. In the paper, the key points of the granted project are explained in order to encourage similar actions, but since it has just started neither practical issues nor experiences can be still reported.

1 Introduction

In this paper we introduce an educational innovation project (EIP) that is being carried out in the scope of Engineering, at the Faculty of Engineering of Vitoria-Gasteiz (Basque Country University, UPV/EHU, Spain).

This project starts from the real need to move towards a knowledge society. The university should to provide that service to society by training young people with the capacity to meet that need. Nowadays engineering is not only based on

M. Graña et al. (Eds.): SOCO'18-CISIS'18-ICEUTE'18, AISC 771, pp. 603–610, 2019.
https://doi.org/10.1007/978-3-319-94120-2_60

specific disciplines of the engineering field, but it is also associated with other professions, as stated in [1]. Science is not based on mere theoretical conceptions, but should be applied in everyday life. Engineering is a professional practice that uses science for its work, applying it in all kinds of creations that we use daily. On the other hand, one of the quality indicators of the universities is the number of scientific publications generated. Other indicators highly valued within the university are the number of doctors and number of six-year terms. In addition, scientific culture is increasingly in demand even by traditional companies. In places where the industry is highly developed, such as Basque Country (Spain), the importance of scientific research for the maintenance of industrial efficiency is well known. It has been widely demonstrated that industry can not be completely successful if it does not continuously improve its manufacturing procedures in order to reduce production costs, improve the quality of its products and develop new ones that meet new needs. One of the complementary merits of teaching valued by National Agency for Quality Assessment and Accreditation of Spain (ANECA) consists in the participation in competitive EIP calls with verifiable results. The fact that the graduates have internalized the foundations of scientific research will give them an added value, as if they want to continue their training and pursue a postgraduate studies (master or doctorate), as well as for their employment in technology centers or companies with high technological level. Up to date, the team of this project is not aware of the existence of a similar EIP neither in the UPV/EHU nor in other universities. Another important aspect to mention supporting it, is the plan of actions to improve the different grades of the Faculty of Engineering of Vitoria-Gasteiz. Some of the actions included in this plan that are consistent with the objectives of this project are the following:

- Advance in the development of transversal competences
- Advance in the implementation of new teaching methodologies
- Schedule research days and sessions to exchange experiences and knowledge
- Promote the participation of the Faculty of Engineering of Vitoria-Gasteiz in scientific and technical dissemination activities.

Therefore, this project can strengthen the philosophy of the Faculty of Engineering of Vitoria-Gasteiz focused on individualized training. This project is part of the lines LP1 (active methodologies) and LP3 (active and autonomous learning) of the competitive call for Educational Innovation Projects 2018–2019 of the Educational Advisory Service, Vicerectorship for Teaching Quality and Innovation, University of the Basque Country (UPV/EHU). So, an EIP that includes the foundations of research and scientific writing in undergraduate studies is considered as needed.

As a tool that intrinsically fit in a perfect way to the aim of the project, it was decided to use a methodology based on active and cooperative learning. Active Learning is a broad learning philosophy that groups together several methods, all based on the responsibility and participation of students in their learning [2,3]. One of them, called cooperative learning, is a paradigm in which learning activities are planned looking for positive interdependence of students [4,5].

The paper is structured as follows. Section 2 provides a global view about the institutional context in which the EIP will be carried out. Section 3 poses formally the objectives of the innovation proposal, while Sect. 4 gives a global description of the project that is detailed by tasks in Sect. 5. The expected improvements in learning outcomes are enumerated in Sect. 6 and finally, our main conclusions are given in Sect. 7.

2 Context for Developing IEPs in UPV/EHU

In order to solve the problems described above, we have applied for a EIP and we have obtained the grant, so now it must be developed in a context and under some specific conditions. This framework will be explained in this section, specifying in aspects of the University as institution in Subsect. 2.1 of the Faculty of Engineering of Vitoria-Gasteiz, where it going to be implemented, Subsect. 2.2.

2.1 Institutional Framework: The University of the Basque Country

The project is being carried out at the University of the Basque Country (UPV/EHU, Spain), financially supported by a grant obtained in the competitive call for Educational Innovation Projects 2018–2019 of the Education Advisory Service, Vicerectorship for Teaching Quality and Innovation. This Vicerectorship is structured in four main areas:

- Quality Cathedra: Contributes to the knowledge, implementation and improvement of quality management in all areas of the organization of the university, helping to achieve the highest standards of excellence.
- Institutional Evaluation Service: A service of the university which aims to guide and promote the process of evaluation, verification and accreditation as well as those related to improving the quality of higher education. It also works with reference quality agencies in developing their programs in the university.
- Faculty Evaluation Service: A service dedicated to promote, design, develop, advise, facilitate and train faculty evaluation process with the desire to contribute to the improvement of teaching quality.
- Educational Advisory Service: It is a service which manages courses and training according to the needs of the faculty. It puts in place processes to gather information about which are the formation necessities.

All these areas converge towards a methodology named IKD-Ikasketa Kooperatibo eta Dinamikoa in Basque (Dynamic and Cooperative Teaching-Learning in English), characterized by the following principles:

- Active Education: IKD invites students to become the architects of their own learning and an active element in the governance of the university. To get this, it encourages learning through active methodologies, ensures continuous and formative evaluation, articulates the acknowledgment of its previous experience (academic, professional, vital and cultural), and promotes mobility programs (Erasmus and SENECA) and cooperation.

- Territorial and social development: The IKD model development requires an ongoing process through which the university is committed to its social environment and community, with public vocation and economic and social sustainability criteria, promoting values of equality and inclusion. It also takes into consideration peculiar characteristics of each of the three provinces where sits the university, to contribute to their empowerment and to extract from them their formative potential. A curricula development responsible with the social environment is done through internships, collaboration with social initiatives, social networks, the relationship with companies and mobility programs that promote international experience and cooperation of our students.
- Institutional Development: IKD curricula development drives institutional policies that promote cooperation between the agents involved in teaching, in an environment of confidence and dynamism. It promotes programs that encourage institutional structuring through the figures of the course or module coordinator, quality commissions and promoting teaching teams, which are key elements in this new teaching culture. Other institutional actions such as offering different types of education (part-time attendance, semi-face, non-face), significant and sustainable use of information and communication technologies (ICTs), institutional regulations concerning assessment, infrastructure design of educational institutions and public spaces (IKDguneak-IKDplaces), the extension of hours of use of space, should be considered from a perspective that encourages IKD culture.
- Professional development: First, the continuous training of the people involved in teaching activities (faculty and support staff to teaching), in order to promote adequate professional development. Training programs (ERAGIN, BEHATU, FOPU), EIPs and assessment tools for teaching (DOCENTIAZ), among others, are actions that support the construction of IKD.

2.2 Faculty Framework: Faculty of Engineering of Vitoria-Gasteiz

This experience is being developed at the Faculty of Engineering of Vitoria-Gasteiz. This Faculty has a long teaching career of over 60 years of existence, so it has known various curricula. The Faculty has a wide range of studies that currently have finished as the Master degree in Industrial Engineering Organization and the following Bachelor's degrees: Technical Industrial Engineering, specialization in Industrial Electronics; Technical Industrial Engineering, specialization in Electricity; Technical Industrial Engineering, specialization in Mechanics; Technical Industrial Engineering, specialization in Industrial Chemistry; Technical Engineering in Topography and Technical Engineering in Computer Management.

At this time, after the adaptation to the European Higher Education Area (EHEA) the Faculty offers the Master in Production Organization and Industrial Management and the following Bachelor's Degrees: Degree in Industrial Electronics and Automatics; Degree in Automotive Engineering, Degree in Mechanical Engineering, Degree in Chemical Engineering and Degree in Computer Science and Information Systems.

3 Objectives of the Innovation Proposal

The innovation proposal of this project is oriented to incorporate the fundamentals of research and scientific writing in form of articles in the different grades as previously indicated. It is intended to train students in tasks of information search, literature review and scientific writing, so it is mandatory to incorporate activities in the undergraduate subjects in order to work on such tasks. It is intended that students are able to write their final degree work (FDW) in the form of a scientific article oriented to a congress or a publication, depending on the theme and scope of the FDW developed. In this sense, the work developed in this EIP can be related to the work developed in another of the EIPs granted by another team of the Faculty of Engineering of Vitoria-Gasteiz in which the gender perspective will be included in teaching. The possibility of relationship of both projects would be through the realization of a FDW in the Faculty including the gender perspective. In addition, it could even participate in the Francisca de Aculodi Awards for the inclusion of the gender perspective in the final projects of the UPV/EHU.

4 Description of the EIP

Once that the objectives of the project have been clearly stated, its the reaching team has structured it taking into account that it should include activities that promote scientific culture in the grade subjects of Table 1 and whose global outline is given in this section.

Table 1. Subjects of the scope of the project

Bachelor's degree	Subject	Number of students
Mechanical Engineering	Installations and Hydraulic Machines	60
	Pneumatic and Hydraulic Systems	15
Automotive Engineering	Fluid Mechanics	40
Industrial Electronics and Automatics Engineering	Automatics and Control	30
Computer Engineering Management and Information Systems	Computer Architecture	60

These tasks in the first phase of the project will basically consist on training students in the following terms:

- On the part of administration staff and library services, in the form of courses about searching for information in scientific databases and bibliographic managers.
- From the faculty, in terms of research topics related to the subject in question.

Once this training is done, the students will have to obtain a state of the art for a topic related to the subject, and in this way apply the concepts learned and work on the competencies of information search, analytical and synthesis skills, critical thinking and scientific writing in a second language.

The second part of the project aims to present the work and results of the FDW in the form of a scientific article. For this, depending on the subject and the scope of the work and with the help of the teaching staff, the publication objective will be chosen: a national conference, international conference or an international journal with or without impact. In addition to the written article, the oral presentation will also be worked on both from the teaching point of view and from the research point of view, taking into account, for example, the requirements of the conference that has been chosen.

5 Tasks of the EIP

Once that the general guidelines of the project have been summarized in the previous section, in this section we give a much more detailed and specific list of the tasks to be completed, that can be enumerated as follows:

- The students involved in the present project will go to the library service to receive information on the most effective methods of searching for scientific information. They will be formed with the aim of searching high quality scientific information in the following databases among others: Web of Knowledge, ScienceDirect, Scopus, Proquest, etc. They will also receive training from different bibliographic managers such as Refworks or End-Note.
- The students, arranged by groups, will make a state of the art of the chosen topic freely related to the subject. Therefore they should use the tools learned in the library service.
- An electronic magazine/web page dedicated to publish the scientific works derived from this project will be designed so that students are aware of the importance of disseminating results.
- The students will make several works with scientific article format and send them to a target dissemination forum (conference or journal). The process of sending and publishing the works will follow the usual process of a scientific conference or journal, that is, the students must send their works online to the conference or journal and the editor will send the works to the teachers of this EIP so that perform the corresponding revision of the manuscripts. These revisions will be sent to the students who will have to correct their works following the instructions of the revisions and send them to the conference or journal again.

- Finally, the editor or publisher of the conference or journal (which will be one of the components of the EIP team) will decide which works are scientifically of high quality and will be published in its online version in the journal.

6 Expected Improvements in Learning Outcomes

In this section we discuss the improvements in learning outcomes that are expected to be reached after carrying out the tasks described in the previous section.

The academic preparation of the students participating in this project will improve considerably, since it will be extended and they will not only have knowledge of subjects, but they will be also able to incorporate the scientific perspective. This will make the students better prepared for the completion of a postgraduate or master's degree or to face work in universities, research institutes or in the research, development and innovation (R + D + I) department of any company. The fact of working the skills of analysis, synthesis and the ability to communicate and write in a second language, is something that is highly valued at the time of the labor insertion of students. A second language is encouraged because most of the relevant scientific documentation is in English. Therefore, through this project, the mentioned transversal competences will be reinforced and the quality of the FDW will be increased. From the EIP team, we understand that this project will enhance the emergence of the research vocation among students.

More specifically, the improvements that the teaching team of the EIP expect to reach are the following:

- Introduce the scientific culture in the engineering degrees of the Faculty
- Complete the training of graduates in engineering and improve their preparation in order to transfer scientific culture to companies in the environment
- Arousal investigative vocations among students
- Two FDWs with the possibility of sending their papers to two international conferences
- One FDW with the possibility of being promoted to be sent to a journal.

7 Conclusions

In this paper we introduced an ongoing Educational Innovation Project (EIP) that is being carried out at the Faculty of Engineering of Vitoria-Gasteiz (Basque Country University, UPV/EHU, Spain). It has the main objective of introducing the scientific culture in the Bachelor's Degrees of that Faculty. We have exposed the context of the project, the main objectives, the tasks to be carried out in order to reach those objectives and finally the expected improvements in the learning outcomes. At the moment, as the project has just started authors cannot report any practical aspect of the implementation, remaining as future work to analyze and discuss the development of the project and its effectiveness.

Acknowledgments. This research project on educational innovation is being financially supported by the grant 10 of the competitive call for Educational Innovation Projects 2018–2019 of the Educational Advisory Service, Vicerectorship for Teaching Quality and Innovation, University of the Basque Country (UPV/EHU), support that is greatly acknowledged.

References

1. Cross, H.: Engineers and Ivory Towers. Books for Libraries (1969)
2. Bonwell, C., Eison, J.: Active learning: Creating excitement in the classroom AEHE-ERIC higher education report no. 1 (1991)
3. Felder, R.M., Brent, R.: Active learning: an introduction. ASQ High. Educ. Brief **2**(4), 1–5 (2009)
4. Felder, R.M., Brent, R.: Cooperative learning in technical courses: procedures, pitfalls, and payoffs (1994)
5. Felder, R.M., Brent, R.: Effective strategies for cooperative learning. J. Coop. Collab. Coll. Teach. **10**(2), 69–75 (2001)

Author Index

M. Graña et al. (Eds.): SOCO'18-CISIS'18-ICEUTE'18, AISC 771, pp. 611–613, 2019.
https://doi.org/10.1007/978-3-319-94120-2

Printed in the United States
By Bookmasters